BASIC LIE THEORY

Hossein Abbaspour

Max Planck Institut für Mathematik, Germany

Martin Moskowitz

The City University of New York, USA

BASIC LIE THEORY

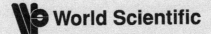 **World Scientific**

NEW JERSEY · LONDON · SINGAPORE · BEIJING · SHANGHAI · HONG KONG · TAIPEI · CHENNAI

Published by

World Scientific Publishing Co. Pte. Ltd.

5 Toh Tuck Link, Singapore 596224

USA office: 27 Warren Street, Suite 401-402, Hackensack, NJ 07601

UK office: 57 Shelton Street, Covent Garden, London WC2H 9HE

Library of Congress Cataloging-in-Publication Data
Abbaspour, Hossein.
    Basic lie theory / by Hossein Abbaspour & Martin Moskowitz.
      p. cm.
    Includes bibliographical references and index.
    ISBN-13: 978-981-270-698-0 (hardcover : alk. paper)
    ISBN-10: 981-270-698-4 (hardcover : alk. paper)
    ISBN-13: 978-981-270-699-7 (pbk. : alk. paper)
    ISBN-10: 981-270-699-2 (pbk. : alk. paper)
  1. Lie groups. I. Moskowitz, Martin A. II. Title.
  QA387.A32 2007
  512'.482--dc22

                        2007023103

British Library Cataloguing-in-Publication Data
A catalogue record for this book is available from the British Library.

Printed in Singapore.

To Gerhard Hochschild

Contents

Contents

Preface and Acknowledgments

In the view of the authors, and as we hope to convince the reader, Lie theory, broadly understood, lies at the center of modern mathematics. It is linked to algebra, analysis, algebraic and differential geometry, topology and even number theory, and applications of some of these other subjects are crucial to many of the arguments we shall present here. This also holds true in the opposite direction: Lie theory can be used to clarify or derive results in these other areas. In a philosophical sense Lie groups are pervasive within much of mathematics, for whenever one has some system, the "automorphisms" of it will frequently be a Lie group. This even occurs in the oldest deductive system in mathematics, namely Euclidean geometry. Here the key issue is congruent figures, particularly triangles. Two such planar triangles are congruent if and only if they differ by an element of the Lie group $E(2) = O(2, \mathbb{R}) \ltimes \mathbb{R}^2$, the group of rigid motions of the Euclidean plane. The reader will find in these many interrelations a vast panorama well worth studying.

This book is the result of courses taught by one of the authors over many years on various aspects of Lie theory at the City University of New York Graduate Center. The primary reader to which it is addressed is a graduate student in mathematics, or perhaps physics, or a researcher in one of these subjects who wants a comprehensive reference work in Lie theory. However, by a judicious selection of topics, some of this material could also be used to give an introduction to the subject to well-grounded advanced undergraduate mathematics majors.

For example, Chapters 3 and most of 7 could form a semester's course in Lie algebras. Similarly, Chapters 0, 2 and 5 (respectively 0, 2 and 8) could be a semester's course in integration in topological groups and their homogeneous spaces (respectively lattices and their applications). For the reader's convenience we have included a diagram of the interdependence of the chapters. We have also tried to make the text as self-contained as possible even at the cost of increased length. We shall assume the reader has some knowledge of basic group theory, topology, and linear algebra, and a general acquaintance with the grammar of mathematics. While reading this book one may wish to consult some of the other books on the subject for clarification, or to see another viewpoint or treatment; especially useful books are listed in the bibliography. We have not attempted to detail the historical development of our subject, nor to systematically give credit to the individual researchers who discovered these results.

The book is organized as follows:

Chapter 0 introduces the players; topological and Lie groups, coverings, group actions, homogeneous spaces, and Lie algebras.

Chapter 1 deals with the correspondence between Lie groups and their Lie algebras, subalgebras and ideals, the functorial relationship determined by the exponential map, the topology of the classical groups, the Iwasawa decomposition in certain key cases, and the Baker Campbell Hausdorff theorem, and local Lie groups.

Chapter 2 concerns Haar measure both on a group and on cocompact and finite volume homogeneous spaces together with a number of applications.

Chapter 3 gives the elements of Lie algebra theory in some considerable detail (except for the detailed structure of complex semisimple Lie algebras, which we defer until Chapter 7).

Chapter 4 deals with the structure of a compact connected Lie group in terms of a maximal torus and the Weyl group.

Chapter 5 contains the representation theory of compact groups.

Chapter 6 concerns symmetric spaces of non-compact type.

Chapter 7 presents the detailed structure of complex semisimple Lie algebras.

Chapter 8 gives an introduction to lattices in Lie groups.

Chapter 9 presents a "density theorem" for cofinite volume subgroups of certain Lie groups.

Although we have included a rather detailed and extensive index, it might be helpful to inform the reader of what we are not doing here. We do not deal extensively with the theory of algebraic groups, nor with transformation groups, although each of these makes some appearance. Similarly, we do not prove that any connected Lie group is, as a manifold, the direct product of a Euclidean space and a maximal compact subgroup, K; nor that any two such Ks are conjugate. But we do this in important special cases. We do not deal, except by example, with the theory of faithful representations. We have also omitted the Weyl character formula, the universal enveloping algebra, the classification of complex simple algebras and, with the exception of one example, branching theorems.

We would like to thank Frederick Greenleaf, Adam Korányi, Keivan Mallahi and Grigory Margulis for reading various chapters of our book and making a number of valuable suggestions. Of course any errors or misstatements are the sole responsibility of the authors.

We would like to thank Richard Mosak for his help in the final preparation of this manuscript. We also thank Isabelle and Anita for their extraordinary patience during the several years that it took to bring this project to completion.

Hossein Abbaspour
Max-Planck Institut für Mathematik
Bonn

Martin Moskowitz
The Graduate School
The City University of New York
New York

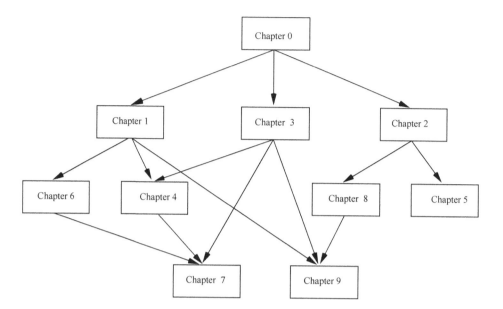

Notations

For a complex number z, $\Re z$ is the real part and $\Im z$ is the imaginary part.

The derivative of a differentiable map $f : M \to N$ at a point $x \in M$ will be denoted $d_x f : T_x M \to T_{f(x)} N$ and takes a tangent vector $v \in T_x M$ to $d_x f(v) \in T_{f(x)} N$.

We use capital letters $G, H, K \dots$ for Lie groups and the corresponding German letters $\mathfrak{g}, \mathfrak{h}, \mathfrak{k} \dots$ for Lie algebras. Lowercase letters such as $a, b, c, \dots, g, h, k \dots$ are used for group elements and uppercase letters $X, Y, Z, \dots$ are reserved for vectors of Lie algebras. For a Lie group G, G_0 is the connected component containing the identity element.

For a Lie group G with Lie algebra $\mathfrak{g}$, $\mathrm{Ad} : G \to \mathrm{GL}(\mathfrak{g})$ is the adjoint representation taking $x \in G$ to $\mathrm{Ad}\, x \in \mathrm{GL}(\mathfrak{g})$, and its image, the adjoint group, is denoted $\mathrm{Ad}\, G$. If $H \subset G$ then $\mathrm{Ad}_G(H)$ is the image of H under Ad and where is no risk of confusion we will simply write $\mathrm{Ad}(H)$. The Lie algebra representation $\mathrm{ad} : \mathfrak{g} \to \mathfrak{gl}(\mathfrak{g})$ takes a vector X to $\mathrm{ad}\, X \in \mathfrak{gl}(\mathfrak{g})$. This is the map defined by the rule $Y \mapsto \mathrm{ad}\, X(Y) = [X, Y]$ and its image is $\mathrm{ad}\, \mathfrak{g}$. To avoid any ambiguity we may write $\mathrm{ad}_{\mathfrak{g}}$ to indicate that where the action is taking place. For a subalgebra $\mathfrak{h} \subset \mathfrak{g}$, the restriction of ad to $\mathfrak{h}$ is denoted $\mathrm{ad}\,|_{\mathfrak{h}}$ which is different from $\mathrm{ad}_{\mathfrak{h}}$.

For a real entry matrix A, A^t is the transpose of A and if A has complex entries then A^* is the transpose conjugate of A.

For a vector space V over a field k and $S \subseteq V$ then $l.s._k(S)$ is the k linear span of S. Finally, $M_n(k)$ is the set of $n \times n$ matrices with entries in the field k. For $T \in \mathrm{End}_k(V)$, define $\mathrm{Spec}(T)$ to be the set of all eigenvalues of T in the algebraic closure of k.

Additional notations are introduced through the index at the end of the book.

Chapter 0

Lie Groups and Lie Algebras; Introduction

0.1 Topological Groups

Before dealing with the generalities concerning topological groups and Lie groups which occupy the next sections of this chapter, we provide some key definitions and examples.

Definition 0.1.1. Let G be a group and at the same time a Hausdorff topological space. Suppose in addition that the group operations

(1) $(g, h) \mapsto gh$

(2) $g \mapsto g^{-1}$

are continuous, where in (1) we take the product topology on $G \times G$. We then call G a *topological group*.

Exercise 0.1.2. Prove that continuity of (1) and (2) is equivalent to that of $(g, h) \mapsto gh^{-1}$.

Exercise 0.1.3. Define the direct product of a finite number of topological groups equipped with the product topology and show that it is a topological group.

We shall almost always be interested in *locally compact groups*. Note that a closed subgroup of a locally compact topological group is again a locally compact topological group.

Examples of topological groups abound. Any abstract group is a topological group having discrete topology. So the theory of topological groups includes abstract groups. The additive group of real numbers $\mathbb{R}$ is also a topological group. The only point that needs to be checked is the continuity of 1.

Exercise 0.1.4. Prove that the continuity of $(x, y) \mapsto x + y$ follows from the triangle inequality, $|x + y| \leq |x| + |y|$.

As indicated above further examples can be gotten by taking direct products so $\mathbb{R}^n$ is a topological group. Of course, this also follows from the triangle inequality $|x + y| \leq |x| + |y|$, where $|\cdot|$ denotes the norm. The multiplicative group of real numbers, $\mathbb{R}^{\times} = \mathbb{R} \setminus \{0\}$, as well as the multiplicative group of complex numbers $\mathbb{C}^{\times} = \mathbb{C} \setminus \{0\}$ (both with the relative topology) are also topological groups. These things follow from the triangle inequality together with $|xy| = |x||y|$.

Exercise 0.1.5. Prove that $\mathbb{R}^{\times}$ and $\mathbb{C}^{\times}$ with the usual multiplication form topological groups. Show that a similar argument works for the multiplicative group of quaternions, $\mathbb{H}^{\times} = \mathbb{H} \setminus \{0\}$.

Notice that $\mathbb{R}^{\times}$ is disconnected while $\mathbb{C}^{\times}$ is connected, but is not simply connected while $\mathbb{H}^{\times}$ is connected and simply connected. This last example is our first noncommutative group. Notice that all these groups are locally compact by the Heine-Borel theorem (see [74]). The (closed) subgroup $\mathbb{R}_{+}^{\times}$ consisting of positive reals however is connected. It has index 2 in $\mathbb{R}^{\times}$. The subgroup of $\mathbb{T}$ of $\mathbb{C}^{\times}$ consisting of elements of norm 1 is a closed and therefore also a locally compact topological group. It is actually compact (and homeomorphic to the circle S^1) as is the finite direct product, $\mathbb{T}^n$. Similarly, the subgroup of $\mathbb{H}^{\times}$ consisting of elements of norm 1 is a compact topological group which is homeomorphic to the 3-sphere S^3.

Exercise 0.1.6. Prove that:

(1) $\mathbb{R}^\times$ is the direct product of ± 1 with $\mathbb{R}_+^\times$.

(2) $\mathbb{C}^\times$ is the direct product of $\mathbb{T}$ with $\mathbb{R}_+^\times$.

(3) $\mathbb{H}^\times$ is the direct product of S^3 with $\mathbb{R}_+^\times$.

Exercise 0.1.7. Show that any closed subgroup H of $G = \mathbb{R}^n$ is isomorphic to a subgroup of the form $\mathbb{Z}^k \times \mathbb{R}^j$, where $k + j \leq n$. In particular, if H is connected it must be $\mathbb{R}^j$, and if H is discrete it must be $\mathbb{Z}^k$. As a result G/H is compact if and only if $k + j = n$ in which case $G/H = \mathbb{T}^k$, and the only closed subgroups of $\mathbb{R}$ are either 0, $\mathbb{R}$ itself or $\{na : n \in \mathbb{Z}\}$, where $a \neq 0$.

Exercise 0.1.8. What are the closed subgroups of $\mathbb{T}$?

Turning to an important source of noncommutative topological groups we consider the real *general linear group*, $\mathrm{GL}(n, \mathbb{R})$. This is the group of invertible $n \times n$ real matrices with the relative topology from $M_n(\mathbb{R})$ identified with $\mathbb{R}^{n^2}$ equipped with its product topology. This is a locally compact topological group because matrix multiplication is given by polynomial functions in the coordinates, inversion by rational functions with nonvanishing denominators and because an open set in a Euclidean space is locally compact by Heine-Borel. The same applies to the complex general linear group, $\mathrm{GL}(n, \mathbb{C}) \subset M_n(\mathbb{C})$. Thus any closed subgroup of either of these groups is again a locally compact group. Of course when $n = 1$, these are nothing more than $\mathbb{R}^\times$ and $\mathbb{C}^\times$ and again $\mathrm{GL}(n, \mathbb{R})$ is disconnected while $\mathrm{GL}(n, \mathbb{C})$ is connected.

Other important examples are $\mathrm{SL}(n, \mathbb{R})$ and $\mathrm{SL}(n, \mathbb{C})$. These are the groups of $n \times n$ real or complex matrices of determinant 1. Additional examples are $\mathrm{O}(n, \mathbb{R})$ and $\mathrm{O}(n, \mathbb{C})$, the real and complex *orthogonal groups* which preserve the bilinear form

$$\langle x, y \rangle = \sum_{i=1}^{n} x_i y_i$$

for $x = (x_1, \ldots, x_n)$ and $y = (y_1, \ldots, y_n) \in \mathbb{R}^n$ or $\mathbb{C}^n$, respectively.

Then $\mathrm{SO}(n, \mathbb{R}) = \mathrm{O}(n, \mathbb{R}) \cap \mathrm{SL}(n, \mathbb{R})$ and $\mathrm{SO}(n, \mathbb{C}) = \mathrm{O}(n, \mathbb{C}) \cap \mathrm{SL}(n, \mathbb{C})$. Further examples are the *unitary group* $\mathrm{U}(n, \mathbb{C})$, the subgroup of $\mathrm{GL}(n, \mathbb{C})$ preserving the Hermitian form

$$\langle x, y \rangle = \sum_{i=1}^{n} x_i \bar{y}_i,$$

for $x = (x_1, \ldots, x_n)$ and $y = (y_1, \ldots, y_n) \in \mathbb{C}^n$, and $\mathrm{SU}(n, \mathbb{C}) = \mathrm{U}(n, \mathbb{C}) \cap \mathrm{SL}(n, \mathbb{C})$. Finally we have the *symplectic groups* $\mathrm{Sp}(n, \mathbb{R})$ and $\mathrm{Sp}(n, \mathbb{C})$. These are respectively the subgroups of $\mathrm{GL}(2n, \mathbb{R})$ and $\mathrm{GL}(2n, \mathbb{C})$ which preserve the *symplectic form*

$$\langle x, y \rangle = \sum_{i=1}^{n} (x_i y_{n+i} + x_{n+i} y_i).$$

A final example is $\mathrm{SO}(p, q)$. This is the subgroup of $\mathrm{GL}(p + q, \mathbb{R})$ which preserves the bilinear form

$$\langle x, y \rangle = \sum_{i=1}^{p} x_i y_i - \sum_{i=p+1}^{p+q} x_i y_i.$$

This is also a nondegenerate bilinear form, making $\mathrm{SO}(p, q)$ a topological group.

Exercise 0.1.9. Prove that each of these is a locally compact topological group.

This will give the reader some idea of what we have in mind when we refer to a topological group. Now as with any category one must specify the morphisms as well. Let G and H be topological groups. We shall call $f : G \to H$ *topological group homomorphism* if it is a group homomorphism and continuous. So for example the exponential map, $\exp : \mathbb{R} \to \mathbb{R}_+^\times$, or $\exp : \mathbb{C} \to \mathbb{C}^\times$ are topological group homomorphisms. The first of these is bijective and has a continuous inverse, namely log. Such a map is called a *topological group isomorphism*, while the second, although surjective, is not one-to-one since $\{2\pi n i \mid n \in \mathbb{Z}\}$ maps to 1. Evidently, isomorphic topological groups share the same properties as topological groups.

Exercise 0.1.10. Show that the kernel of this map is exactly $\{2\pi n i | n \in \mathbb{Z}\}$.

Another example of a topological group homomorphism to keep in mind is given by the identity map from the discrete group of additive reals to $\mathbb{R}$.

If $f : G \to H$ is a topological group homomorphism, then $\mathrm{Ker}\, f$ is a closed normal subgroup of G and $f(G)$ is a subgroup of H. If G is a topological group and N is a closed normal subgroup, we can form the quotient group G/N equipped with the quotient topology.

Exercise 0.1.11. If G is locally compact, show that G/N is a locally compact topological group. Also demonstrate why the projection $\pi :\ G \to G/N$ is a continuous, open, surjective homomorphism.

So, for example, taking $t \mapsto \exp(2\pi i t)$ shows that $\mathbb{R}/\mathbb{Z}$ is a topological group that is isomorphic to $\mathbb{T}$. More generally, if $f : G \to H$ is a topological group homomorphism then this induces an injective topological group homomorphism $f^* : G/\mathrm{Ker}\, f \to H$ making a commutative diagram,

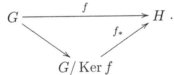

This is called the *first isomorphism theorem* for topological groups. An important special case of this is when the induced map f^* is actually an isomorphism (as was exp just above) which is treated in Corollary 0.4.7.

Exercise 0.1.12. Prove that $\mathbb{T}$ is isomorphic with $\mathbb{R}/\mathbb{Z}$. The subgroup $\mathbb{Q}/\mathbb{Z}$ of $\mathbb{R}/\mathbb{Z}$ is the torsion subgroup. The other elements generate dense cyclic subgroups.

Proposition 0.1.13. *If G is a connected topological group and U is any symmetric neighborhood of 1, then $G = \bigcup_{n=1}^{\infty} U^n$. Of course since $U \supseteq U \cap U^{-1}$, which is symmetric, the result actually holds for any neighborhood U of 1 in G.*

Proof. Note that $\bigcup_{n=1}^{\infty} U^n$ is an open subgroup of G. Therefore it is closed. Because of the connectedness of G, it must be G (see Exercise 0.2.5). □

0.2   Lie Groups

Analogously to the definition of a topological group we define a *real Lie group* as a group G which is also a finite dimensional real differentiable manifold whose operations are smooth, *i.e.* C^{∞}. That is, the map $G \times G \to G$ given by $(g, h) \mapsto gh^{-1}$ is C^{∞} (product manifold structure). We call the dimension n of the manifold the dimension of G. For Lie groups G and H, a *Lie homomorphism* $f : G \to H$ is a smooth group homomorphism. If in addition f is bijective and its inverse is smooth we say that f is an isomorphism. A *Lie subgroup H* of G is a submanifold which is a subgroup as well.

Although the formal parts of this theory emulate those of topological groups there are some not so obvious aspects. For example in the latter closed subgroups are taken as a convenience to insure local compactness while in the case of Lie groups it is an important theorem of Élie Cartan that a closed subgroup of a Lie group has the structure of a Lie group. We mention some variants of the definition. One could consider real analytic manifolds and real analytic maps instead of just C^{∞} ones. This is done in Hochschild [33]. In fact, it does not matter which one does as the theory and the category of the real analytic and C^{∞} Lie groups coincide (see Appendix D).

Another variant which does get one somewhere is to consider the notion of a *complex Lie group*. Here one simply takes complex manifolds and holomorphic maps. The result is called a *complex Lie group*.

Clearly a complex Lie group is a real Lie group. Another variant would be to not limit the manifolds to be finite dimensional. This approach has had only limited success and will not be pursued here.

Exercise 0.2.1. Prove that $(g, h) \mapsto gh^{-1}$ is C^{∞} if and only if multiplication $(g, h) \mapsto gh$ and inversion $g \mapsto g^{-1}$ are C^{∞}.

Given a Lie group G, the left translations $L_g : G \to G$ defined by

$L_g(h) = gh$ are global diffeomorphisms on G. Since they can take any point a to any other point b by taking $g = ba^{-1}$ we see that all local topological properties valid at a single point such as the identity are valid at all other points. The same applies to right translations.

Of course a Lie group is a (very special kind of) locally compact topological group. The converse question is Hilbert's fifth problem which was solved in 1953 by Gleason-Montgomery-Zippin and Yamabe. For the details of this see [26]. As all manifolds, Lie groups are also locally connected. In particular the *identity component*[1] G_0 of G is open.

Lie groups arise in various ways. For example the isometry group of a Riemannian manifold is always a Lie group (see [67, 32]). Similarly, the automorphism group of a Lie group is also a Lie group (see [33]).

As we shall see in Chapter 5, one way of studying Lie groups is through their *representations*. A representation of a Lie group G is a smooth homomorphism $\rho : G \to \mathrm{GL}(V)$. We call $V = V_\rho$ the *representation space* of ρ and $d_\rho = \dim V_\rho$ its degree and $\rho_g : V \to V$ is the map

$$\rho_g(v) = \rho(g)v.$$

A representation ρ is said to be *faithful* if it is injective.

It is appropriate now to give a few simple examples. As usual, when one has an open subset of Euclidean space the manifolds structure consists of a one chart atlas. The various assertions concerning these examples should also be regarded as exercises.

Example 0.2.2.

(1) $\mathbb{R}$, or more generally $\mathbb{R}^n$, is a Lie group with the usual manifold structure.

(2) $\mathbb{T}$, or more generally $\mathbb{T}^n$, is a Lie group with the usual manifold structure and in fact the natural map $\pi : \mathbb{R}^n \to \mathbb{T}^n$ given by projection in each coordinate is a smooth group homomorphism. It is the universal covering of $\mathbb{T}^n$ by $\mathbb{R}^n$. Both $\mathbb{R}^n$ and $\mathbb{T}^n$ are connected and $\mathbb{R}^n$ simply connected. As we saw above, $\mathbb{R}/\mathbb{Z}$ is isomorphic as a topological group to the multiplicative group S^1

---

[1]The connected component containing the identity element.

of all complex numbers of modulus 1; the isomorphism is given by $t \mapsto e^{2\pi i t}$, $t \in \mathbb{R}$ when regarded as a map $\mathbb{R}/\mathbb{Z} \to S^1$. Since this map and its inverse are smooth these are isomorphic as Lie groups.

(3) Any discrete group G is a Lie group of dimension zero. In particular $\mathbb{Z}$ or more generally $\mathbb{Z}^n$, is a Lie group. It is a closed subgroup of $\mathbb{R}^n$ which is the kernel of π above.

(4) The multiplicative group $\mathbb{R}^\times$ is a Lie group. It is not connected, but has two components. Its identity component $\mathbb{R}^\times_+$ of positive real numbers is also a Lie group. This group is isomorphic with $\mathbb{R}$ via the usual exponential map. Similarly, $\mathbb{C}^\times$ is a (2 dimensional) Lie group which is connected. It is actually a complex Lie group. As we saw here however the exponential map, $\exp : \mathbb{C} \to \mathbb{C}^\times$ is not an isomorphism. However, it is a smooth (actually holomorphic) homomorphism, surjective and a local diffeomorphism at each point. The reader should prove this. Since $\mathbb{C}$ is simply connected we have just constructed the universal covering group of $\mathbb{C}^\times$ (see Section 0.3 on covering spaces). As we saw previously, regarding $\mathbb{C}^\times$ as a real Lie group, it is isomorphic via the polar decomposition to $\mathbb{R}^\times_+ \times \mathbb{T}$ (direct product). Later in Chapter 6 we shall see this can be generalized considerably.

(5) More generally, $\mathrm{GL}(n, \mathbb{R})$, the group of invertible $n \times n$ real matrices is a (dense) open subset of Euclidean space $M_n(\mathbb{R})$ and thus acquires a manifold structure in which multiplication is a polynomial function of the coordinates. Moreover, inversion is a rational function of the coordinates with a nowhere vanishing denominator. As an exercise the reader should verify all of these facts including that $\mathrm{GL}(n, \mathbb{R})$ is open and dense. Hence $\mathrm{GL}(n, \mathbb{R})$ is a real Lie group of dimension n^2. When $n = 1$ we get $\mathbb{R}^\times$. We shall see that it has 2 components because $\mathrm{GL}^+(n, \mathbb{R})$, the ones with positive determinant, is connected. To see that $\mathrm{GL}(n, \mathbb{R})$ is not connected we just observe that its image under the smooth (check!) map $A \mapsto \det(A)$ has two components.

(6) Similarly, $\mathrm{GL}(n, \mathbb{C})$, the group of invertible $n \times n$ complex matrices

is a (dense) open subset of Euclidean space of $n \times n$ complex matrices $M_n(\mathbb{C})$ and thus acquires a complex manifold structure in which multiplication is a polynomial function of the coordinates. Moreover, inversion is a rational function of the coordinates with a nowhere vanishing denominator. Hence $\mathrm{GL}(n, \mathbb{C})$ is a complex Lie group of complex dimension n^2. When $n = 1$ we get $\mathbb{C}^\times$. Later we shall see that, just as with $n = 1$, it is connected. Thus all the examples given in Section 0.1 above are also Lie groups.

(7) Further examples are provided by the group of real $T_n(\mathbb{R})$ or complex $T_n(\mathbb{C})$ triangular matrices, or the group of strictly real $N_n(\mathbb{R})$ or complex triangular matrices $N_n(\mathbb{C})$. In all these cases we have an atlas with one coordinate patch. The latter three are connected while $T_n(\mathbb{R})$ has 2^n components.

(8) One can form *semidirect products* to get additional examples. For instance $G = \mathrm{GL}(n, \mathbb{R}) \times \mathbb{R}^n$. This is a manifold with the usual product manifold structure and with group operation $(g, v)(g', v') = (gg', gv' + v)$. It is called the *affine group* of $\mathbb{R}^n$ and is a Lie group of dimension $n^2 + n$. Similarly one can construct the complex Lie group semidirect product $\mathrm{GL}(n, \mathbb{C}) \ltimes \mathbb{C}^n$.

(9) Let G and H be Lie groups and $\eta : G \to \mathrm{Aut}(H)$ be a smooth homomorphism where $\mathrm{Aut}(H)$ takes on a natural Lie group structure. Or more directly, we can give $G \times H$ the product manifold structure and just assume $(g, h) \mapsto \eta(g) \cdot h$ is a smooth map $G \times H \to H$. Then define $(g, h)(g', h') = (gg', \eta(g)h' \cdot h)$. This is a Lie group and is called the *semidirect product* $G \ltimes H$ of G and H and contains a closed subgroup isomorphic to G and a closed normal subgroup isomorphic to H. Notice that in general G is not normal.

(10) Of course when the action η is trivial we get the direct product.

Exercise 0.2.3. Show that a semidirect product $G \ltimes H$ is direct if and only if G is normal.

Exercise 0.2.4. What are the smooth homomorphisms $f : \mathbb{R} \to \mathbb{T}$, and $f : \mathbb{T} \to \mathbb{T}$? Suggestion: use Exercise 0.1.7 above to find $\mathrm{Ker}\, f$. In the first case we have for $t \in \mathbb{R}$, $f_x(t) = e^{2\pi i x t}$, $x \in \mathbb{R}$. In the second we have for $t \in \mathbb{R}$, $f_x(\bar{t}) = e^{2\pi i x t}$, $x \in \mathbb{Z}$, where $\bar{t}$ is the image of t under the covering.

Exercise 0.2.5. Show that:

(1) An open subgroup of a topological group is closed.

(2) If H is a closed subgroup then H is open if and only if G/H is discrete.

(3) Let G be a Lie group and G_0 denote the identity component of 1. Then G_0 is an open normal subgroup of G. In particular, G_0 is a connected Lie group of the same dimension as G.

We close our remarks on Lie groups with an example of a locally compact (in fact compact and commutative) group which is not a Lie group. Let p be a prime number and consider $(\mathbb{Z}, |\cdot|_p)$, where $|\cdot|_p$ is the p-adic norm on $\mathbb{Z}$ which is defined as follows. We take $|0|_p = 0$ and if $z \neq 0 \in \mathbb{Z}$ the prime factorization of $z = p^n s$, where s is relatively prime to p. Then $|z|_p = p^{-n}$. It is easy to see that this gives a norm on $\mathbb{Z}$ with all the usual properties, but instead of the triangle inequality one has the stronger $|x + y|_p \leq \max(|x|_p, |y|_p)$. Then the *$p$-adic integers* $\mathbb{Z}_{(p)}$ is the completion of $\mathbb{Z}$ with respect to the metric $d_p(x, y) = |x - y|_p$. This group has a neighborhood basis at 0 of nested subgroups (p^m), where $m \in \mathbb{Z}_+$. But as we shall see, a Lie group must have a sufficiently small neighborhood of 1 which contains no nontrivial subgroup. This is impossible for $\mathbb{Z}_{(p)}$.

0.3   Covering Maps and Groups

We begin this section by reviewing the notions of *covering spaces* and *covering maps*. Then we will study the covering maps for Lie groups and their relation to the group structure. We refer the reader to [40] for a detailed account of the covering theory. In this section all the topological spaces are path connected and locally path connected.

Suppose that X and Y are two topological spaces and $e : Y \to X$ a continuous map. We shall say e is a covering map if every point $x \in X$ has a neighborhood set U such that e is a homeomorphism on each

connected component of $e^{-1}(U)$ to U. Such an open set U is called an
admissible neighborhood. So by definition a covering map is surjective.
We say that Y is a covering space for X. If $e : Y \to X$ and $e' : Y' \to X$
are two covering maps,

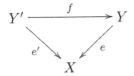

then a continuous map $f : Y' \to Y$ with $e' = e \circ f$ is said to be *fiber
preserving*. Two covering spaces $e : Y \to X$ and $e' : Y' \to X$ are said
to be *equivalent* if there is a fiber preserving map $f : Y' \to Y$ which is
a homeomorphism.

Lifting property: The most fundamental property of the covering is
the lifting property as follows. Suppose that $e : (Y, y_0) \to (X, x_0)$ is a
covering map for the based spaces X and Y that is $x_0 \in X_0$, $y_0 \in Y$
and $e(y_0) = x_0$. Let $g : (P, p_0) \to (X, x_0)$ be a continuous map of based
spaces such that $f_*(\pi_1(P, p_0)) \subset e_*(\pi_1(Y, y_0))$, where f_* and e_* are the
induced maps on the fundamental groups.

Then there exists a unique continuous map $f : P \to Y$ such that
$g = e \circ f$.

A *universal cover* of a topological space X is a covering space $e :
Y \to X$, where Y is a simply connected manifold.

Exercise 0.3.1. Prove that any two universal covers of a topological
space are equivalent.

Given a covering map $e : Y \to X$ with smooth base X, one can make
Y into a smooth manifold such that the covering map is a smooth map.
Let (U, ϕ) be a chart on X so that U is an admissible neighborhood.

Then the connected components of $e^{-1}(U)$ with the map $\phi \circ e$ form an atlas for Y and by construction e is a smooth map.

Now we go one step further and consider the covering maps whose base spaces are Lie groups.

Proposition 0.3.2. *Suppose that $e : \tilde{G} \to G$ is a universal cover of a Lie group G. Then there is a unique Lie group structure on $\tilde{G}$ which makes e a group homomorphism.*

Proof. Earlier we described the unique smooth structure on $\tilde{G}$ which makes e smooth. We still need to introduce the group structure and verify that structural maps are smooth. We choose $\tilde{1} \in e^{-1}(1)$ where 1 is the identity element of G. In what follows 1 is the base point of G and $\tilde{1}$ is the base point of $\tilde{G}$.

Suppose that $(g, h) \mapsto g.h$ denotes the multiplication of G, then we define the smooth map $\psi : \tilde{G} \times \tilde{G} \to G$ to be

$$\psi(x, y) = e(x).e(y) \tag{1}$$

which sends the base point $(\tilde{1}, \tilde{1})$ to the base point 1.

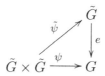

Because $\tilde{G}$ is simply connected, by the lifting property of covering maps ψ lifts to a map $\tilde{\psi} : \tilde{G} \times \tilde{G} \to \tilde{G}$ such that

$$\psi = e \circ \tilde{\psi}. \tag{2}$$

Since $\tilde{\psi}$ is the lift of a smooth map, therefore it is a smooth map with respect to the natural smooth structure of $\tilde{G}$. Similarly, one can lift the smooth map $\tilde{i} : \tilde{G} \to G$ given by $x \mapsto (e(x))^{-1}$ to $\tilde{G}$

to get the smooth map $\tilde{i} : \tilde{G} \to \tilde{G}$ for which

$$e(x)^{-1} = e(\tilde{i}(x)) \tag{3}$$

holds.

We claim that $\tilde{G}$, with $\tilde{1}$ as the identity element, $\tilde{\psi}$ as the multiplication, and $\tilde{i}$ as the inverse element map is a group.

1) *Associativity*: We must show that $\tilde{\psi}(\tilde{\psi}(x,y),z) = \tilde{\psi}(x,\tilde{\psi}(y,z))$. Note that by the associativity of the multiplication for G, (2) and (1),

$$e(\tilde{\psi}(\tilde{\psi}(x,y),z)) = e(\tilde{\psi}(x,y)).e(z) = (e(x).e(y))e(z) = e(x).e(y).e(z)$$

and

$$e(\tilde{\psi}(x,\tilde{\psi}(y,z))) = e(x).e(\tilde{\psi}(y,z)) = e(x).(e(y).e(z)) = e(x).e(y).e(z).$$

Therefore, $\tilde{\psi}(\tilde{\psi}(x,y),z)$ and $\tilde{\psi}(x,\tilde{\psi}(y,z))$ are both liftings of the map $(x,y,z) \mapsto e(x).e(y).e(z)$. Thus by the uniqueness of the lifting they are equal.

2) *Inverse*: We prove that $\tilde{\psi}(x,\tilde{i}(x)) = \tilde{1}$ and the proof is similar to the one above. By (3)

$$e(\tilde{\psi}(x,\tilde{i}(x))) = e(x).e(\tilde{i}(x)) = e(x).e(x)^{-1} = 1$$

which means that $\tilde{\psi}(x,\tilde{i}(x))$ is a lift of the constant map $x \mapsto 1$ just like the map $x \mapsto \tilde{1}$. Therefore $\tilde{\psi}(x,\tilde{i}(x)) = \tilde{1}$.

3) *Identity element*: It is a direct check that $x \mapsto \tilde{\psi}(x,\tilde{1})$ and the identity maps are both liftings of $x \mapsto e(x)$. Therefore they are the same, proving that $\tilde{1}$ is the identity element for the multiplication $\tilde{\psi}$. $\qquad\square$

Proposition 0.3.3. *If $e : \tilde{G} \to G$ is a universal covering homomorphism then any group homomorphism $f : H \to G$ can be lifted to a group homomorphism $\tilde{f} : H \to \tilde{G}$.*

Proof. By the lifting property there is a unique lifting $\tilde{f} : H \to \tilde{G}$ for f. Consider the maps $\psi_1(x, y) = \tilde{f}(x).\tilde{f}(y)$ and $\psi_2(x, y) = \tilde{f}(xy)$, both are a lifting of $\phi : H \times H \to G$ given by

$$\phi(x, y) = f(x)f(y).$$

Therefore, by uniqueness, we have $\psi_1 = \psi_2$ which implies that $\tilde{f}$ is a group homomorphism. $\qquad\square$

Proposition 0.3.4. *Let Γ be a discrete subgroup of a connected Lie group G. Then the natural projection map $p : G \to G/\Gamma$ is a covering map.*

Proof. Let U be an open connected neighborhood of the identity with $U \cap \Gamma = \{1\}$. This is possible as Γ is discrete. Because of the continuity of the multiplication we can choose an open neighborhood V of 1 such that $V^{-1}V \subset U$. Therefore $V^{-1}V \cap \Gamma = \{1\}$. Consider the set of the form $Vh \subset G$ for $h \in \Gamma$. These sets are disjoint because if $v_1 h_1 = v_2 h_2$ then $v_2^{-1} v_1 \in V^{-1}V \cap \Gamma = \{1\}$ which says that $v_1 = v_2$ and $h_1 = h_2$. Moreover, $p|_{Vh}$ is injective for the same reason as above. To get the admissible opens we use the left translation. More precisely for a coset $g\Gamma$ the open set of cosets $gV\Gamma = \{gv\Gamma | v \in V\}$ is an admissible open and $p^{-1}(gV\Gamma) = \cup_{h \in \Gamma}(gVh)$ which is a disjoint union and the restriction of p to each gVh is a diffeomorphism. $\qquad\square$

Remark 0.3.5. In the previous proposition, p is not necessarily a group homomorphism as Γ is not a normal subgroup. The next result addresses this case, and proves something stronger namely that Γ has to be central.

Lemma 0.3.6. *Let G be a connected topological group and Γ be a discrete normal subgroup. Then Γ is central in G.*

Proof. To see that Γ is central, consider the continuous map $G \to \Gamma$ given by $g \mapsto g\gamma g^{-1}$ for a fixed $\gamma \in \Gamma$. This is well-defined as Γ is normal, and constant as Γ is discrete and G connected, so γ is in the center. $\qquad\square$

Corollary 0.3.7. *Let G be a connected Lie group and $f : G \to H$ be a local isomorphism. Then f is a covering map and $\Gamma = \mathrm{Ker}\, f$ is a discrete central subgroup of G. If G is a simply connected group, then $\pi_1(H) = \Gamma$.*

Proof. Since f is a local diffeomorphism then there is an open neighborhood U of the identity such that $f|_U$ is injective. Therefore, $\Gamma \cap U = \{1\}$ and Γ is discrete. We have the isomorphism $\tilde{f} : G/\Gamma \to H$ and by Proposition 0.3.4, $p : G \to G/\Gamma$ is a covering map, and thus f is a covering map. Γ is central by the previous lemma.   □

0.4   Group Actions and Homogeneous Spaces

Here we discuss a notion which is central to many areas of mathematics, namely, that of a *group action* (either of a locally compact group on a space, or of a Lie group on a manifold).

Let G be a locally compact group, X be a (usually locally compact) space and $\phi : G \times X \to X$ be a *jointly continuous* mapping, called the *action* of G on X. Writing $\phi(g, x)$ as $g \cdot x$ or even gx we shall assume that an action, (G, X), satisfies the following:

(1) For all $x \in X$, $1 \cdot x = x$

(2) For all $g, h \in G$ and $x \in X$, $(gh) \cdot x = g \cdot (h \cdot x)$

We shall call X a G-space or equivalently, say that G acts as a *transformation group* on X. An action is *transitive* if for all $x, y \in X$ there exists $g \in G$ such that $g.x = y$. If such g is unique, then it is called a *simply transitive* action.

If G is a Lie group, X a smooth manifold and the action is jointly smooth, then we shall simply call this a *smooth* action.

Exercise 0.4.1. Show that each $\phi(g, .)$ is a homeomorphism of X. Thus G operates on X by homeomorphisms.

In applications X could, for example, be some geometric space and G a group of transformations of X which preserves some geometric property such as length, angle, or area, etc. Such transformations always form a group and it is precisely the properties of this group which is the key to understanding length, angle, or area, respectively. This is the essential idea of *Klein's Erlanger Program*, named after the late 19th century mathematician, Felix Klein. One can also turn this on itself and use it as a tool to study group theory. The first example of a group action is provided by a group G acting on itself by *left translation*. Since G is a topological group this is evidently an action and one sees that the conditions above are designed exactly to reflect this. More generally we can consider a locally compact group G, a closed subgroup H and the space of left cosets $X = G/H$ and let G act on G/H by left translation, $g_1 \cdot gH = g_1gH$, where $g_1 \in G$ and $gH \in G/H$. It is an easy check to show that this is also an action. Another example is provided by taking a locally compact group G and letting $\mathrm{Aut}(G)$ the automorphism group (suitably topologized), or some subgroup of $\mathrm{Aut}(G)$ such as $I(G)$, the inner automorphisms, act in the natural way on G. Or one might take a real or complex (usually finite dimensional) representation $\rho : G \to \mathrm{GL}(V)$ of a locally compact group. Then we evidently have an action of G on V. Such actions are called *linear actions*. In particular, this would be the case if $G \subseteq \mathrm{GL}(V)$, *i.e.* the identity representation. Finally, let G, X be a G-space and $\mathcal{F}(X)$ be some real or complex vector space of functions defined on X. We can define an action of G on $\mathcal{F}(X)$ by $g \cdot f = f_g$, the left translate of the function f, where $f_g(x) = f(g^{-1}x)$. All that is required here is that $\mathcal{F}(X)$ be G-stable. An easy check shows that this is a linear action called the induced action on functions. We shall give other examples of actions in the sequel.

Exercise 0.4.2. Let G be a locally compact group, H a closed subgroup and G/H the space of left cosets having quotient topology. Then the projection $\pi : G \to G/H$ is a continuous and open map and G/H is a locally compact Hausdorff space. Show that G/H is discrete if and only if H is open in G. In particular, this means that every open subgroup is closed. Also show that at the other extreme G/H is not discrete if and only if H is dense in G.

We remark that there is a theorem of Pontrjagin [70] strengthening this considerably. Namely, that for a closed subgroup H of a topological group G the quotient space, G/H, is actually $T_{3\frac{1}{2}}$. In particular, since taking H to be trivial gives us the usual action of a topological group on itself by left translation, we see that a locally compact group itself is $T_{3\frac{1}{2}}$.

If G acts on X a subset Y of X is called G-*invariant* if $G \cdot Y \subseteq Y$. When we have an invariant set we get a new action of G on Y. Similarly, if (G, X) is a group action and H is a closed subgroup of G, then (H, X) is also a group action. Let (G, X) be a group action. For $x \in X$ we define the G-*orbit* of x, written $\mathcal{O}_G(x)$, as $\{g \cdot x : g \in G\}$. When the action is transitive there is only one orbit. Choosing an $x_0 \in X$ gives rise to a map $G \to \mathcal{O}_G(x) \subseteq X$ given by $g \mapsto g(x_0)$ and called the *orbit map*. Evidently each $x \in X$ is in its $\mathcal{O}_G(x)$, so that each orbit is non-empty. Also if $\mathcal{O}_G(x) \cap \mathcal{O}_G(y) \neq \emptyset$, then $\mathcal{O}_G(x) = \mathcal{O}_G(y)$. For suppose $g_1 \cdot x = g_2 \cdot y$. Then $g_2^{-1} g_1 \cdot x = y$ and therefore $g_3 g_2^{-1} g_1 \cdot x = g_3 \cdot y$, proving that $\mathcal{O}_G(y) \subset \mathcal{O}_G(x)$. Similarly, $\mathcal{O}_G(x) \subset \mathcal{O}_G(y)$ so $\mathcal{O}_G(x) = \mathcal{O}_G(y)$. Thus X is the disjoint union of the orbits. (In the case that G and X are finite this gives a useful counting principle which plays a role in proving the Sylow theorems.) For example, for the standard action of $\mathrm{SO}(n, \mathbb{R})$ on $\mathbb{R}^n$ the orbits are 0 and the various spheres are centered at 0. Whereas if $\mathrm{GL}(n, \mathbb{R})$ acts on $\mathbb{R}^n$, there are only two orbits $\{0\}$ and $\mathbb{R}^n \setminus \{0\}$. We observe that the orbits of an action need not be closed in X and can even be dense.

Exercise 0.4.3. Give an example of an action whose orbits are not closed.

Just as for groups we must now also decide when two group actions (G, X) and (G, Y) are essentially the same. We shall say that (G, X) and (G, Y) are G-*equivalent* or *equivariantly equivalent* if there is a bijective bi-continuous map $\pi : X \to Y$ which for all $g \in G$ and $x \in X$ satisfies

$$\pi(g \cdot x) = g \cdot \pi(x). \tag{4}$$

We call such a π a G-*equivariant isomorphism* or *equivalence*. When π is merely a continuous map $X \to Y$ satisfying (4), we say that it is a morphism of G-spaces. Evidently each orbit map is a continuous morphism.

Let G be a group, H be a closed subgroup and G acts on G/H by left translation, as above. A moment's reflection tells us that this is a transitive action. As we shall see it is essentially the only one. When G acts transitively on X, since it acts by homeomorphisms, the local topological properties of X are the same at every point. So, for example, if local compactness or local connectedness holds at one point, then it holds at all points. In particular, this is so for a topological group itself.

Proposition 0.4.4. *Let (G, X) be an action and $x \in X$ be fixed. Then the orbit, $\mathcal{O}_G(x)$ is a G-invariant set; in fact it is the smallest G-invariant set containing x. Hence this gives a transitive action $(G, \mathcal{O}_G(x))$. If (G, X) is a transitive action and $x \in X$, then the stabilizer,*

$$\mathrm{Stab}_G(x) = \{ g \in G : g \cdot x = x \}$$

is a closed subgroup of G. If y is another element of X, then by transitivity we can choose $g \in G$ so that $g \cdot x = y$. Then $g \, \mathrm{Stab}_G(x) g^{-1} = \mathrm{Stab}_G(y)$.

Proof. The fact that $\mathcal{O}_G(x)$ is G-invariant and G acts transitively on $\mathcal{O}_G(x)$ is immediate. The same may be said as far as the stabilizer being a subgroup of G. If g_n is a net in G converging to g with $g_n(x) = x$, then $g(x) = x$, by continuity of the action. Finally, if $g'(x) = x$, then $g g' g^{-1}(y) = g g'(x) = g(x) = y$. This proves $g \, \mathrm{Stab}_G(x) g^{-1} \subseteq \mathrm{Stab}_G(y)$. Hence $\mathrm{Stab}_G(x) \subseteq g^{-1} \mathrm{Stab}_G(y) g$ which by the same reasoning is contained in $\mathrm{Stab}_G(x)$. Thus $g \, \mathrm{Stab}_G(x) g^{-1} = \mathrm{Stab}_G(y)$. $\qquad \square$

We now come to a useful result which will show that when there is a compact, or even locally compact σ-compact group[2] G operating transitively and continuously on a locally compact space X then the topology of X is determined by that of G. It is the quotient topology on $G/\mathrm{Stab}(x)$ transferred to X. If G happens to be a Lie group then as a closed subgroup H is also a Lie group (see Theorem 1.3.5) and X gets a manifold structure as G/H where G is a Lie group and H is a Lie subgroup.

---

[2]A group is said to be σ-compact if it is a countable union of compact subsets.

Theorem 0.4.5. *Let G be a locally compact group, X a locally compact Hausdorff space, (G, X) a transitive G-space and $x \in X$ fixed. If G is compact, or even σ-compact, then (G, X) is equivariantly equivalent to the action of G on $G/\operatorname{Stab}_G(x)$ by left translation.*

Proof. We first deal with the formal part. To see that we have an equivariant equivalence of actions let $\pi : G \to X$ be the orbit map $g \mapsto g \cdot x$. By transitivity, π is onto. Moreover, $\pi(g) = \pi(g')$ if and only if $g^{-1}g' \in \operatorname{Stab}_G(x)$. Hence π induces a bijection $\tilde{\pi} : G/\operatorname{Stab}_G(x) \to X$. To check the commutativity of the diagram we must see that $g \cdot \tilde{\pi}(g' \operatorname{Stab}_G(x)) = \tilde{\pi}(gg' \operatorname{Stab}_G(x))$ for all g and $g' \in G$. But the former is just $g \cdot \pi(g')$ while the latter is $\pi(gg')$. Since π is a G-map, $\tilde{\pi}$ is G-equivariant. Because π is continuous so is $\tilde{\pi}$. Thus $\tilde{\pi}$ is a continuous, bijective G-equivariant map. All that remains is to see that is open. Of course, if G is compact, or even if $G/\operatorname{Stab}_G(x)$ is compact, then this is so because $\tilde{\pi}$ is continuous and bijective and X is Hausdorff. This already suffices for many, but not all applications. For this reason we also make the very mild assumption that G is σ-compact.

Exercise 0.4.6. Show that this σ-compact result is rather general. For instance it shows that it applies whenever G is locally compact and second countable.

To deal with the σ-compact case we recall a version of the Baire Category theorem [32], p. 110.

Baire's Category Theorem: Let X be a σ-compact space. That is $X = \cup C_n$, where each C_n is compact. If X is locally compact, then one of the C_n must have a non-void interior.

Continuing the proof in the σ-compact case, we have the following: Since the orbit map π is G-equivariant and the action of G on itself as well as the action of G on X are both transitive, to show openness we may consider a neighborhood of 1 in G. Let U be such a neighborhood. We show that $U \cdot x$ contains a neighborhood of x in X. Choose a smaller neighborhood U_1 of 1 in G which is compact and $U_1^{-1}U_1 \subseteq U$. Because G is σ-compact and U_1 is a neighborhood there is a countable number of g_n such that $G = \bigcup g_n U_1$. Since $\tilde{\pi}$ is continuous and surjective the $\tilde{\pi}(g_n U_1)$ are compact and cover X. By Baire's theorem there is some

n for which $\tilde{\pi}(g_n U_1) \supseteq V$, a nontrivial open set in X. If $u_1 \in U_1$ and $\tilde{\pi}(g_n u_1) \in V$, then $(g_n u_1)^{-1} V$ is a neighborhood of x in X and

$$(g_n u_1)^{-1} V = u_1^{-1} g_n^{-1} V \subseteq u_1^{-1} U_1 x \subseteq U x.$$

Thus $\tilde{\pi}$, and of course also π, is open.   □

We now apply this to derive the *Open mapping theorem* for group homomorphisms.

Corollary 0.4.7. *Suppose G and H are locally compact groups with G σ-compact. Let $f : G \to H$ be a continuous surjective group homomorphism. Then f is open.*

Proof. Let G act on H by left multiplication through f. Thus if $h = f(g')$, $g \cdot h = f(g)h$. This is clearly a transitive action whose stability group is $\operatorname{Ker} f$. Moreover, taking $h = f(1) = 1$ the orbit map for this action given in the theorem above is f. Thus f is open.   □

We can now apply these methods to some important examples of compact homogeneous G-spaces, where G is a compact Lie group. These are spheres, projective spaces, the Stiefel and Grassmann manifolds and the Flag manifolds.

Spheres: We consider $\mathbb{R}^n$ and $\mathbb{C}^n$ to be inner product and Hermitian inner product spaces, respectively. Evidently, $O(n, \mathbb{R})$ acts transitively on S^{n-1}, the unit sphere in $\mathbb{R}^n$ and $U(n, \mathbb{C})$ acts transitively on S^{2n-1}, the unit sphere in $\mathbb{C}^n$, since these groups can take any orthonormal basis $\{v_1, \ldots, v_n\}$ to another one $\{w_1, \ldots, w_n\}$. If it happens that $\det g \neq 1$, where $g \in O(n, \mathbb{R})$ or $U(n, \mathbb{C})$, which we call G, replace $\{v_1, \ldots, v_n\}$ by $\{v_1, \ldots, \lambda v_n\}$, where $|\lambda| = 1$. Then this is again an orthonormal basis and there is an $h \in G$ such that $h(v_i) = w_i$, for $i < n$ and $h(\lambda v_n) = w_n$. Then $h(v_n) = \frac{1}{\lambda} w_n$ and $\det h = \frac{1}{\lambda} \det g$. Choosing $\lambda = \det g$ which has absolute value 1 since $g \in G$, makes $\det h = 1$.

Thus $SO(n, \mathbb{R})$ and $SU(n, \mathbb{C})$ act transitively on S^{n-1} and S^{2n-1}, respectively. What is the isotropy group of such an action? Since it is transitive we can choose any unit vector as the base point. Since G fixes v_1 then it must stabilize its orthocomplement, the subspace W spanned by $\{v_2, \ldots, v_n\}$ and the restriction of such a g to W is in $SO(n-1, \mathbb{R})$ or $SU(n-1, \mathbb{C})$, respectively.

Applying Theorem 0.4.5 to the various compact groups then yields the following:

(1) $O(n, \mathbb{R})/O(n-1, \mathbb{R}) = S^{n-1}$
(2) $SO(n, \mathbb{R})/SO(n-1, \mathbb{R}) = S^{n-1}$
(3) $U(n, \mathbb{C})/U(n-1, \mathbb{C}) = S^{2n-1}$
(4) $SU(n, \mathbb{C})/SU(n-1, \mathbb{C}) = S^{2n-1}$

In a similar way one sees that $Sp(n)$ acts transitively on S^{4n-1}, $n \geq 1$. Since $Sp(1) = SU(2, \mathbb{C}) = S^3$ it follows that all $Sp(n)$ are compact by Proposition 2.4.5. Also for similar reasons all $Sp(n)$ are connected and since all the spheres from $n \geq 3$ are simply connected it also follows from the results of the next section that the $Sp(n)$ are also all simply connected.

Projective space: This line of argument gives real $\mathbb{R}P^{n-1}$ and complex $\mathbb{C}P^{n-1}$ projective spaces as homogeneous spaces as follows. $SU(n, \mathbb{C})$ acts transitively on $\mathbb{C}P^{n-1}$. Let $p_0 = \mathbb{C}v_n$, where $v_1, \ldots, v_n$ is an orthonormal basis of $\mathbb{C}^n$, and let $p = \mathbb{C}v$, where v has norm 1. Then, as we showed above, by transitivity $g(v_n) = v$ for some g. Therefore for the induced action on $\mathbb{C}P^{n-1}$ we see that $g(p_0) = p$. Similarly, $SO(n, \mathbb{R})$ acts transitively on $\mathbb{R}P^{n-1}$.

What is the isotropy group of p_0? In the complex case it is

$$\{g \in SU(n, \mathbb{C}) : g(v_n) = \lambda v_n\}.$$

Clearly such a g is block diagonal and the upper block g' is unitary. The only condition is that $\lambda = \frac{1}{\det g'}$. Since this is no condition on g', we see that the isotropy group is $U(n-1, \mathbb{C})$ so $SU(n, \mathbb{C})/U(n-1, \mathbb{C}) = \mathbb{C}P^{n-1}$. Similarly, $SO(n, \mathbb{R})/O(n-1, \mathbb{R}) = \mathbb{R}P^{n-1}$.

Corollary 0.4.8. *(1)* $SU(n, \mathbb{C})/U(n-1, \mathbb{C}) = \mathbb{C}P^{n-1}$
(2) $SO(n, \mathbb{R})/O(n-1, \mathbb{R}) = \mathbb{R}P^{n-1}$

The next two corollaries depend on some results in the following section.

Corollary 0.4.9. $\mathbb{C}P^{n-1}$ *is a compact, connected and simply connected complex manifold. The simple connectivity follows from the fact that* $\mathrm{SU}(n,\mathbb{C})$ *is simply connected and* $\mathrm{U}(n-1,\mathbb{C})$ *is connected.* $\mathbb{R}P^{n-1}$ *is merely a compact, connected real manifold.*

Corollary 0.4.10. *As real manifolds,* $S^3/S^1 = S^2$ *(Hopf fibration).*

Taking $n = 2$ in $\mathrm{SU}(n,\mathbb{C})/\mathrm{U}(n-1,\mathbb{C}) = \mathbb{C}P^{n-1}$ yields $S^3/S^1 = \mathbb{C}P^1$. Since, as above, $\mathbb{C}P^1$ is a compact, connected and simply connected manifold of real dimension 2 it is clearly the 2 sphere, S^2.

Grassmann Space: Let V be a real or complex vector space of dimension n. For each integer $1 \le r \le n$ consider the Grassmann space $\mathcal{G}(r,n)$, the set of all subspaces of V of dimension r. Of course, when $r = 1$ we have a real or complex projective space. How can we topologize this space in a convenient way and study it? Evidently $\mathrm{GL}(V)$ acts transitively and continuously on $\mathcal{G}(r,n)$ by natural action. Hence $\mathcal{G}(r,n)$ is a homogeneous space of this group. What is the isotropy group? If $g \in \mathrm{GL}(V)$ fixes an r-dimensional subspace W, *i.e.* a point in the Grassmann space, it must stabilize W. Hence

$$g = \begin{pmatrix} A & B \\ 0 & C \end{pmatrix}$$

and obviously any such g will do. This gives a manifold structure on Grassmann space defined by the quotient structure from $\mathrm{GL}(V)$. Notice this shows that in the complex case, $\mathcal{G}(r,n)$ is actually a complex manifold. Also it will follow that $G/\mathrm{Stab}_G(W)$ is actually compact even though G is not. However, we will prove the compactness of $\mathcal{G}(r,n)$ in another way in a moment. But we can already see that $\dim(\mathcal{G}(r,n)) = r(n-r)$ (over $\mathbb{R}$ respectively $\mathbb{C}$) and that $\mathcal{G}(r,n)$ is connected. The latter follows from the fact that $\mathrm{GL}(n,\mathbb{C})$ is connected and $\mathrm{GL}(n,\mathbb{R})_0$, the identity component of $\mathrm{GL}(n,\mathbb{R})$, also acts transitively in the real case. To see that $\mathcal{G}(r,n)$ is compact we need only show that a compact group acts transitively. We consider a positive definite real symmetric (respectively Hermitian symmetric) form on V. Choose an orthonormal basis of W and by Gram-Schmidt extend this to an orthonormal basis of V. If W_1 is another r-dimensional subspace of V and we make a similar construction then there exists an orthogonal operator (respectively unitary operator) taking one orthonormal basis on V to the other and taking W to W_1. Since $O(n, \mathbb{R})$ (respectively $U(n, \mathbb{C})$) is compact it follows that $\mathcal{G}(r, n)$ is compact.

Exercise 0.4.11. Calculate the isotropy group when $O(n, \mathbb{R})$ (respectively $U(n, \mathbb{C})$) acts.

Flag manifolds: Again let V be a finite dimensional real or complex vector space of dim $= n$ and consider $V_0 = \{0\} < V_1 \ldots < V_n = V$, where dim $V_i = i$. Such a thing is called a *flag*. Let $\mathcal{F}(V)$ be the set of all flags on V. By choosing a basis for V_1, extending this to a basis of V_2, and eventually to V, we see that $GL(V)$ operates transitively and continuously on $\mathcal{F}(V)$. The isotropy subgroup is easily seen to be the group of upper triangular matrices B in $GL(V)$. Arguing, as in the Grassmann manifold we see that $\mathcal{F}(V)$ is connected and its dimension is $\frac{n(n-1)}{2}$. By using the Gram-Schmidt method to get an orthonormal basis compatible with a flag we see that $O(n, \mathbb{R})$ (respectively $U(n, \mathbb{C})$) acts transitively. Hence, as before, $\mathcal{F}(V)$ is compact. Notice that this cuts both ways. We have shown that $GL(V)/B$ is compact without actually having looked at it. Just as in the case of the Grassmann manifold this quotient gives a manifold structure on the flag manifold.

Exercise 0.4.12. Generalize both these constructions by considering a partition $n_1, \ldots, n_s$ of n. That is, $n = n_1 + \ldots + n_s$, where all $n_i > 0$ and consider generalized flags made up of s subspaces of n_i dimensions. Formulate and prove a result that is analogous to what we have just done above.

Stiefel manifolds: This is defined as follows. Let V be a finite dimensional real Euclidean or complex Hermitian space and r be an integer $1 \leq r \leq n = \dim V$. We consider the set $\mathcal{S}_r^{\dim V}$ of r-frames, by which we mean the set of all orthonormal r-tuples of vectors of V. Clearly, $O(n, \mathbb{R})$ (respectively $U(n, \mathbb{C})$) acts transitively on $\mathcal{S}_r$. Since a $g \in G$ which fixes an r-frame (it fixes each of the vectors which make up the r-frame) must also be fixed and hence stabilize the space which they span. Therefore it also stabilizes the ortho-complement. Its restriction to the orthocomplement is something in $O(n-r, \mathbb{R})$ (respectively $U(n-r, \mathbb{C})$). Clearly anything of this type can occur. Thus $\mathrm{Stab}_G(x_0) = O(n-r, \mathbb{R})$ (respectively $U(n-r, \mathbb{C})$), so that since G is compact $G/\mathrm{Stab}_G(x_0) = O(n, \mathbb{R})/O(n-r, \mathbb{R})$ (respectively $U(n, \mathbb{C})/U(n-r, \mathbb{C})$). The dimension of $\mathcal{S}_r^{\dim V}(\mathbb{R})$ as a real manifold is $n(n-1)/2 - (n-r)(n-r-1)/2 = r(2\dim V - r - 1)/2$, while that of the real manifold $\mathcal{S}_r^{\dim V}(\mathbb{C})$ is $n^2 - (n-r)^2 = r(2\dim V - r)$.

We close this section with examples of a non-compact homogeneous space associated with the group $G = \mathrm{SL}(2, \mathbb{R})$. Let $H^+ = \{z = x + yi : x \in \mathbb{R}, y > 0\}$ denote the Poincaré upper half plane and

$$g = \begin{pmatrix} a & b \\ c & d \end{pmatrix},$$

with $\det g = 1$. Then, as we show in Chapter 6, G acts transitively and continuously on H^+ via fractional linear transformations,

$$g \cdot z = \frac{az + b}{cz + d},$$

with $\mathrm{Stab}(i) = \mathrm{SO}(2)$. Since $\mathrm{SL}(2, \mathbb{R})$ is separable, $\mathrm{SL}(2, \mathbb{R})/\mathrm{SO}(2) = H^+$. As we shall see in Section 1.5 since $\mathrm{SO}(2)$ and H^+ are both connected we have:

Corollary 0.4.13. $\mathrm{SL}(2, \mathbb{R})$ *is connected.*

In the following example we will consider the $\mathbb{R}$ matrices, the complex case is identical. Evidently $\mathrm{GL}(n, \mathbb{R})$ acts transitively on $\mathbb{R}^n \setminus \{0\}$, by the natural action. However, certain subgroups also act transitively on $\mathbb{R}^n \setminus \{0\}$.

Example 0.4.14. For $n \geq 2$, $\mathrm{SL}(n, \mathbb{R})$ acts transitively on $\mathbb{R}^n \setminus \{0\}$ and for $n \geq 1$, $\mathrm{Sp}(n, \mathbb{R})$ acts transitively on $\mathbb{R}^{2n} \setminus \{0\}$. Suppose $v \neq 0 \in \mathbb{R}^n$. There are two possibilities. Either $\{v, e_1\}$ are linearly independent, or they are not. If they are, then $v = \lambda e_1$, where $\lambda \neq 0$. In this case let

$$g = \begin{pmatrix} \lambda & 0 & 0 \\ 0 & \frac{1}{\lambda} & 0 \\ 0 & 0 & I \end{pmatrix}.$$

Hence $g(e_1)$ = v.   Otherwise, enlarge this to a basis,
$\{e_1, v, v_3, \ldots, v_n\}$, and let

$$
g = \begin{pmatrix} 0 & -1 & 0 \\ 1 & 0 & 0 \\ 0 & 0 & I \end{pmatrix},
$$

where I is the identity matrix of order $n - 2$. Then again, $g(e_1) = v$
and in either case $\det g = 1$.

Now suppose $G = \mathrm{Sp}(n, \mathbb{R})$ and $v \in \mathbb{R}^{2n} \setminus \{0\}$. Then making the
same choices, but this time with I being the identity matrix of order
$2n - 2$, we see that in the first instance the diagonal matrix g preserves
the symplectic form. In the second instance g is also in the symplectic
group. In fact, g lies in the compact subgroup $\mathrm{U}(n, \mathbb{C})$, of $\mathrm{Sp}(n, \mathbb{R})$ and
in either case $g(v) = e_1$.

Notice, however, that if $G = \mathrm{SO}(1, 1)$, a 1-parameter group of hyperbolic rotations, then already for dimension reasons G cannot act
transitively on the 2-dimensional manifold, $\mathbb{R}^2 \setminus \{0\}$. Moreover, in general, under the natural action $G = \mathrm{SO}(p, q)$, does not act transitively
on $\mathbb{R}^{p+q} \setminus \{0\}$. This is because G preserves a (p, q)-form for all $c \in \mathbb{R}$,
and therefore it must preserve all the varieties

$$
V_c = \{x \in \mathbb{R}^{p+q} \setminus \{0\} : x_1^2 + \ldots + x_p^2 - x_{p+1}^2 \cdots - x_{p+q}^2 = c\},
$$

and these invariant subvarieties of $\mathbb{R}^{p+q} \setminus \{0\}$ are disjoint for different
c's.

0.5   Lie Algebras

Here we define Lie algebras and give a few examples of them. The basic
theory of Lie algebras will be presented in Chapter 3. Let $\mathfrak{g}$ be a vector
space over a field k having a zero characteristic.

Definition 0.5.1. We will say that $\mathfrak{g}$ is a *Lie algebra* if it possesses an
anti-symmetric bilinear product $[\cdot, \cdot] : \mathfrak{g} \times \mathfrak{g} \to \mathfrak{g}$, called the *Lie bracket*,
which satisfies the Jacobi identity,

$$
[[X, Y], Z] + [[Y, Z], X] + [[Z, X], Y] = 0
$$

for all x, y and z in $\mathfrak{g}$.

This is the moral equivalent of the associative law in the case of associative algebras. We shall call $\mathfrak{h}$ a subalgebra of $\mathfrak{g}$ if it is a subspace and is closed under the bracket. Obviously any subalgebra of a Lie algebra is itself a Lie algebra. A Lie algebra with a trivial Lie bracket is called an *abelian* Lie algebra.

Example 0.5.2. Let V be a finite dimensional vector space over a field k as above and let $\mathfrak{gl}(V)$ denote the space of all k-endomorphisms of V. If $n = \dim V$ we usually denote $\mathfrak{gl}(V)$ as $\mathfrak{gl}(n, k)$.

Let $[T, S] = T \circ S - S \circ T$, where T and S are in $\mathfrak{gl}(V)$ and $\circ$ is the composition of maps. Then $\mathfrak{gl}(V)$ is a Lie algebra by virtue of the fact that $\mathfrak{gl}(V)$ is an associative algebra under $\circ$. Thus $\mathfrak{gl}(V)$ and any of its subalgebras provide a wide class of examples of Lie algebras. These are called *linear Lie algebras*.

Exercise 0.5.3. Any associative algebra can be made into a Lie algebra by a similar construction as above. We leave it to the reader to verify this.

Let $\mathfrak{g}$ be a Lie algebra and $X_1, X_2, \ldots, X_n$ be a basis of $\mathfrak{g}$ and consider the expansion of $[X_i, X_j]$ in terms of the basis,

$$[X_i, X_j] = \sum_{k=1}^{n} c_{ij}^k X_k,$$

and c_{ij}^k's are called *structure constants* of $\mathfrak{g}$ with respect to the basis $X_1, X_2, \ldots, X_n$.

Exercise 0.5.4. Let $\mathfrak{g}$ be a vector space of dimension n and $X_1, X_2, \ldots, X_n$ be a basis. Define $[X_i, X_j] = \sum_{k=1}^{n} c_{ij}^k X_k$, where c_{ij}^k's are in the field. Extend $[\cdot, \cdot]$ bilinearly to $\mathfrak{g} \times \mathfrak{g}$. What are the requirements on the c_{ij}^k's in order that $\mathfrak{g}$ be a Lie algebra?

Example 0.5.5. Let $\mathfrak{g} = \mathfrak{gl}(V)$ and consider the basis consisting of the matrices e_{ij}. These are the matrices which are 1 in the (i, j) spot and zero elsewhere. A direct calculation shows that

$$[e_{ij}, e_{kl}] = \delta_{jk} e_{il} - \delta_{il} e_{jk}.$$

This shows that structure constants with respect to this basis are all 0 or ± 1. In particular notice that

$$[e_{ij}, e_{ji}] = e_{ii} - e_{jj}. \tag{5}$$

Example 0.5.6. A Lie algebra of dimension 1 is evidently abelian. As for Lie algebras $\mathfrak{g}$ of a dimension 2, we first show that $[\mathfrak{g}, \mathfrak{g}]$ has dimension less than 1. If $\dim[\mathfrak{g}, \mathfrak{g}] = 2$ and $\{X, Y\}$ is a basis for $\mathfrak{g}$, one sees directly that $[\mathfrak{g}, \mathfrak{g}] = \mathfrak{g} = l.s.[X, Y]$ and this is a contradiction as $l.s.[X, Y]$, has dimension 1. Now if $\mathfrak{g}$ is a Lie algebra of dimension 2 then either $\mathfrak{g}$ is abelian or $[\mathfrak{g}, \mathfrak{g}]$ is of dimension 1 generated by a vector U. In this case we can take U as the first element of a basis $\{U, V\}$. Then $[V, U] = \lambda U$ where $\lambda \neq 0$. So $[1/\lambda V, U] = U$. Thus there is a basis X, Y with $[X, Y] = Y$. This Lie algebra is called the $ax + b$-*Lie algebra*. As we shall see it is solvable.

Example 0.5.7. Let V be a finite dimensional vector space over a field k of characteristic 0 which is equipped with a nondegenerate symmetric bilinear form $(\cdot, \cdot)$. For $A \in \mathrm{End}(V) = \mathfrak{gl}(V)$, one can consider the endomorphism A^t such that for all u and v in V,

$$(Au, v) = (u, A^t).$$

Since the bilinear is nondegenerate A^t is well-defined. A^t is called the *adjoint* or *transpose* of A. The following properties can be easily verified:

(1) $(xA + yB)^t = xA^t + yB^t$ or in other words taking adjoint is a linear operator.

(2) $(A^t)^t = A$.

(3) $(AB)^t = B^t A^t$.

(4) $[A, B]^t = -[A^t, B^t]$.

$A \in \mathrm{End}(V)$ is said to be a *symmetric operator* if $A = A^t$. Similarly it is said to be a *skew symmetric operator* if $A = -A^t$. We denote the set of skew symmetric operators by $\mathfrak{k}$ and that of symmetric operators by $\mathfrak{p}$. It follows from property (1) that $\mathfrak{k}$ and $\mathfrak{p}$ are linear subspaces of $\mathfrak{gl}(V)$.

We have $\mathfrak{k} \cap \mathfrak{p} = 0$, since $A = A^t$ and $A = -A^t$ implies that $A = 0$. On the other hand for any $A \in \mathfrak{gl}(V)$ one can write,

$$A = \frac{A - A^t}{2} + \frac{A + A^t}{2}.$$

It follows from (1) and (2) that $\frac{A-A^t}{2} \in \mathfrak{k}$ and $\frac{A+A^t}{2} \in \mathfrak{p}$. Hence, $\mathfrak{gl}(V) = \mathfrak{k} \oplus \mathfrak{p}$. From (4) it follows easily that

(1) $[\mathfrak{k}, \mathfrak{k}] \subseteq \mathfrak{k}$

(2) $[\mathfrak{k}, \mathfrak{p}] \subseteq \mathfrak{p}$

(3) $[\mathfrak{p}, \mathfrak{p}] \subseteq \mathfrak{k}$

These relations are called the Cartan relations and will play an important role in symmetric spaces. In particular $\mathfrak{k}$ is a subalgebra of $\mathfrak{gl}(V)$. When V has dimension n one also writes $\mathfrak{o}(n, k)$ for $\mathfrak{k}$. Since in $\mathfrak{o}(n, k)$, the trace of every element of $\mathfrak{o}(n, k)$ is zero, we sometimes also write $\mathfrak{so}(n, k)$.

Similar arguments apply in the case of a vector space $V = \mathbb{C}^n$ over the complex field and a Hermitian form $\langle \cdot, \cdot \rangle$. Then we get the real Lie algebra of skew hermitian operators on $V = (\mathbb{C}^n, \langle \cdot, \cdot \rangle)$ which we denote by $\mathfrak{u}(n)$. We leave it to the reader to check that this is not a complex Lie algebra.

Exercise 0.5.8. For positive integers n, p and q where $p + q = n$, consider the matrix $A = \text{diag}(\underbrace{1, \ldots, 1}_{p\text{-times}}, \underbrace{-1, \ldots, -1}_{q\text{-times}})$. Prove that $\mathfrak{o}(p, q) \subset \mathfrak{gl}(n, \mathbb{R})$ and consists of matrices X satisfying

$$XA + AX^t = 0.$$

These form a subalgebra of $\mathfrak{gl}(n, \mathbb{R})$. Prove that $\mathfrak{so}(p, q) \subset \mathfrak{o}(p, q)$, matrices of trace zero, form an ideal of $\mathfrak{o}(p, q)$.

Definition 0.5.9. Let $\mathfrak{g}$ and $\mathfrak{h}$ be Lie algebras over the same field and f be a k-linear map from $\mathfrak{g}$ to $\mathfrak{h}$. We shall say that f is a *Lie homomorphism* if $f([X, Y]) = [f(X), f(Y)]$ for all X and Y in $\mathfrak{g}$. If the Lie homomorphism is bijective we shall call it an isomorphism and say that $\mathfrak{g}$ and $\mathfrak{h}$ are *isomorphic* Lie algebras. An isomorphism $f : \mathfrak{g} \to \mathfrak{g}$ is called *automorphism* and $\mathrm{Aut}(\mathfrak{g})$ denotes the set of automorphisms of $\mathfrak{g}$. Clearly under composition $\mathrm{Aut}(\mathfrak{g})$ forms a subgroup of $\mathrm{GL}(\mathfrak{g})$.

Example 0.5.10. It follows from the Jacobi identity that for $X \in \mathfrak{g}$, $\mathrm{ad}\, X : \mathfrak{g} \to \mathfrak{g}$, defined by

$$\mathrm{ad}\, X(Y) = [X, Y],$$

is a Lie homomorphism.

Evidently isomorphic Lie algebras share all Lie algebra theoretic properties. If $\mathfrak{h}$ is a subspace of $\mathfrak{g}$, then the inclusion map from $\mathfrak{h}$ to $\mathfrak{g}$ is a Lie homomorphism if and only if $\mathfrak{h}$ is a subalgebra.

Definition 0.5.11. A Lie homomorphism $\rho : \mathfrak{g} \to \mathfrak{gl}(V)$ is a called *Lie algebra representation*. The dimension of the representation is the dimension of V and V itself is called the *representation space* of ρ. The Lie algebra representation $\rho : \mathfrak{g} \to \mathfrak{gl}(V)$ is said to be *faithful* if it is injective. For every $X \in \mathfrak{g}$ we denote the map $\rho(X) : V \to V$ by ρ_X.

Obviously the inclusion map of a linear subalgebra of $\mathfrak{gl}(V)$ is a Lie representation.

Example 0.5.12. Let $\mathfrak{g}$ be a Lie algebra. The map $\mathrm{ad} : \mathfrak{g} \to \mathfrak{gl}(\mathfrak{g})$ defined as follows:

$$\mathrm{ad}\, X : \mathfrak{g} \to \mathfrak{g}$$

$$\mathrm{ad}\, X(Y) = [X, Y]$$

is easily seen to be a Lie algebra representation, called the *adjoint representation*. The image of ad, denoted $\mathrm{ad}\,\mathfrak{g}$, is called *adjoint algebra* and is clearly a linear Lie algebra. The kernel of ad is $\mathfrak{z}(\mathfrak{g})$, the center of $\mathfrak{g}$.

Exercise 0.5.13. Let $\mathfrak{g} \subset \mathfrak{gl}(V)$ be a linear Lie algebra and $X \in \mathfrak{g}$. Then the eigenvalues of $\mathrm{ad}\, X$ are all of the form $\lambda_i - \lambda_j$ where λ_i and λ_j are eigenvalues of X. Hint: One may assume that the field is algebraically closed. Then use the Jordan form.

Chapter 1

Lie Groups

In this chapter we define a Lie group and study its basic properties.

1.1 Elementary Properties of a Lie Group

We first give the definition of a smooth invariant vector field on a Lie group. As in Appendix A, a smooth vector field X can be regarded as a first-order differential operator, X_g, each operating on the space of smooth functions on G. As in the page on Notation, T_gG denotes the tangent space to G at g and d_gf is the derivative of a map f at $g \in G$.

Definition 1.1.1. Let G is a Lie group and X a smooth vector field on G. We say that X is *left invariant* if $d(L_h)X_g = X_{hg}$ for all h and $g \in G$. Here L_g denotes left translation by $g \in G$ and $d(L_g)$ its derivative acting on tangent spaces.

We want to describe all such vector fields. Since the L_g act simply transitively on G it follows that a left invariant vector field is completely determined by $X_1 \in T_1G$ and, conversely, that any $v \in T_1G$ determines a unique left invariant vector field defined by $X_g = d(L_g)_1 v \in T_gG$. This vector field is evidently smooth. Also, since $L_{hg} = L_hL_g$, we have left invariance, $d_gL_hX_g = (d_gL_h)d_1(L_g)v = d_1(L_{hg})v = X_{hg}$.

Thus the linear map $X \mapsto X_1$ is a vector space isomorphism from the left invariant vector fields onto T_1G which we call the *Lie algebra*

$\mathfrak{g}$ of G, and hence $\mathfrak{g}$ is a finite dimensional subspace of the space of all vector fields of dimension $= \dim G$. Since G_0, the identity component of G, is open in G, it follows that $T_1 G = T_1 G_0$. Therefore, because of this vector space isomorphism, the Lie algebras of G_0 and G coincide. Here $\mathfrak{g}$ is an invariant of G; that is, it depends intrinsically on G.

Now the left invariant vector fields form a subalgebra of the Lie algebra of all vector fields (see Appendix A). For let X and Y be left invariant vector fields on G.

$$d(L_h)[X_g, Y_g] = d(L_h)(X_g Y_g - Y_g X_g) = d(L_h)(X_g Y_g) - d(L_h)(Y_g X_g)$$
$$= X_{hg} Y_{hg} - Y_{hg} X_{hg} = [X_{hg}, Y_{hg}].$$

Hence $[X, Y]$ is again left invariant and a vector field. We now come to a concept that is of fundamental importance to our subject, namely, the exponential map and 1-parameter subgroups.

Definition 1.1.2. Let G be a Lie group. We say that $\phi : \mathbb{R} \to G$ is a 1-parameter subgroup of G if ϕ is a smooth homomorphism.

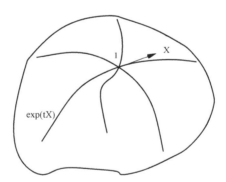

Figure 1.1: 1-parameter subgroups

For example if $G = \mathbb{R}^n$ and $\phi : \mathbb{R} \to \mathbb{R}^n$ is a smooth homomorphism, then ϕ takes 0 to 0. Hence its derivative $d_0\phi$ is a linear map of the linear spaces $d_0\phi : \mathbb{R} \to \mathbb{R}^n$. Therefore, there is a vector $X = (x_1, \ldots, x_n)$ so that $d_0\phi(t) = tX$. Identifying the Lie groups $\mathbb{R}$ and $\mathbb{R}^n$ with their respective tangent spaces at 0 we get $\phi(t) = tX$ for all $t \in \mathbb{R}$ (see Figure 1.1).

We can handle $\mathbb{T}^n$ similarly. If $\phi : \mathbb{R} \to \mathbb{T}^n$ is a smooth homomorphism and $\pi : \mathbb{R}^n \to \mathbb{T}^n$ is the universal covering (see Section 0.3), then ϕ lifts to a smooth homomorphism $\tilde{\phi} : \mathbb{R} \to \mathbb{R}^n$ so that $\pi\tilde{\phi} = \phi$ (see Proposition 0.3.3). Since, as above, $\tilde{\phi}(t) = tX$, it follows that $\phi(t) = \pi(tX)$ for all $t \in \mathbb{R}$. This example shows that in order to find the 1-parameter subgroups of a Lie group G it is sufficient to do it for the universal covering group and apply the projection (see Section 0.3).

Finally, we consider 1-parameter subgroups of the non-abelian Lie groups, $G = \mathrm{GL}(n, \mathbb{R})$ or $\mathrm{GL}(n, \mathbb{C})$. As we shall see, this is more complicated than the examples given above. Let X be a tangent vector (*i.e.* an $n \times n$ real (respectively complex) matrices) to $I = I_{n \times n}$ of G. This is because G is open in the space of matrices. Then $\phi(t) = \mathrm{Exp}\, tX$ is a 1-parameter group. Here Exp is the power series applied to matrices $A \in M_n(\mathbb{C})$.

$$\mathrm{Exp}\, A = \sum_{k=0}^{\infty} \frac{A^k}{k!}.$$

Now this series is absolutely convergent and uniformly on compacta, therefore it defines an entire holomorphic function $M_n(\mathbb{C}) \to M_n(\mathbb{C})$. To see this consider the finite partial sums $\sum_{k=0}^{m} \frac{A^k}{k!}$. Calculating the norm we get

$$\left\| \sum_{k=n}^{m} \frac{A^k}{k!} \right\| \le \sum_{k=n}^{m} \frac{\|A\|^k}{k!}.$$

Since the series $\sum_{k=0}^{\infty} \frac{\|A\|^k}{k!}$ converges, the sequence of partial sums is a Cauchy sequence and by the completeness of $M_n(\mathbb{C})$, we see that $\mathrm{Exp}\, A$ converges absolutely and $\|\mathrm{Exp}\, A\| \le e^{\|A\|}$, for $A \in M_n(\mathbb{C})$. Moreover, the same argument shows that the series for Exp converges uniformly on compacta.

If A and B commute then $\mathrm{Exp}\, A \cdot \mathrm{Exp}\, B = \mathrm{Exp}(A + B)$. Therefore, since A and $-A$ commute $\mathrm{Exp}\, A\, \mathrm{Exp}(-A) = \mathrm{Exp}\, 0 = I$ so that $\mathrm{Exp}\, A$ is always invertible and its inverse is $\mathrm{Exp}(-A)$. Thus $\mathrm{Exp} : M_n(\mathbb{C}) \to$

$\mathrm{GL}(n, \mathbb{C})$. Since, tX and sX commute for all t and s. It follows that $\phi(t) = \operatorname{Exp} tX$ is a 1-parameter group.

The proof of this is quite similar to that of the functional equation for the ordinary numerical exp. Because the convergence is absolute we may perform rearrangements by the Weierstrass theorem,

$$\operatorname{Exp} A \operatorname{Exp} B = (\sum_{k=0}^{\infty} \frac{A^k}{k!})(\sum_{l=0}^{\infty} \frac{B^l}{l!}) = \sum_{k,l=0}^{\infty} \frac{A^k B^l}{k!l!}.$$

On the other hand,

$$\operatorname{Exp}(A + B) = \sum_{p=0}^{\infty} \frac{(A + B)^p}{p!}$$

and since A and B commute it follows from the binomial theorem that

$$\frac{(A + B)^p}{p!} = \sum_{j=0}^{p} \frac{A^j}{j!} \frac{B^{p-j}}{(p - j)!}.$$

Hence $\operatorname{Exp}(A + B) = \operatorname{Exp} A \operatorname{Exp} B$.

Conversely, suppose $\phi(t)$ is a 1-parameter group. Then for all s and $t \in \mathbb{R}$, $\phi(s + t) = \phi(s)\phi(t)$. Differentiating with respect to s at $s = 0$ gives $\phi'(t) = \phi'(0)\phi(t)$, for all real t. Also $\phi(0) = I$. This is a first-order linear matrix differential equation with constant coefficients (or a system of such numerical equations) and hence has the unique *global* solution $\phi(t) = \operatorname{Exp} tX$ where X is the tangent vector, $\phi'(0)$. For clearly by absolute and uniform convergence we can differentiate $\phi(t) = \operatorname{Exp} tX$ term by term and get $\phi'(t) = X\phi(t)$. Thus $\operatorname{Exp} tX$ satisfies the differential equation on all of $\mathbb{R}$ and $\phi(0) = I$. If another function ψ did this then $\frac{d}{dt}(\operatorname{Exp}(-tX)\psi(t)) = \operatorname{Exp}(-tX)\psi'(t) + -X(\operatorname{Exp}(-tX))\psi(t) = \operatorname{Exp}(-tX)(X + -X)\psi(t)$, since $\operatorname{Exp} -tX$ and X commute and ψ satisfies the differential equation. Since this is zero, $\operatorname{Exp} -tX\psi(t)$ is a constant. Evaluating at $t = 0$ shows the constant is I. Hence $\psi(t) = \operatorname{Exp}(tX)$. Thus, once again, the 1-parameter groups are uniquely determined by a tangent vector at the origin, but this time by an ordinary differential equation rather than a linear algebraic one. As we shall see, when properly understood, these examples are typical.

Exercise 1.1.3. Prove that for an $n \times n$ matrix X and for m a positive integer, one has $\mathrm{Exp}\, X = \lim_{m \to \infty} (I + \frac{X}{m})^m$.

Finally, notice also that the derivative of Exp at zero is the identity, this is because the power series expression for $\mathrm{Exp}(X)$ shows that $\mathrm{Exp}(X) = \mathrm{Exp}(0) + (X - 0)(I) + (X - 0)(\frac{X}{2!} + \frac{X^2}{3!} + \ldots)$. By the linear approximation theorem if $\frac{X}{2!} + \frac{X^2}{3!} + \ldots$ tends to 0 as $X \to 0$, we conclude $d_0 \mathrm{Exp} = I$. But $\|\frac{X}{2!} + \frac{X^2}{3!} + \ldots\| \le \frac{\|X\|}{2!} + \frac{\|X\|^2}{3!} + \ldots$, which definitely tends to zero since it is the tail of e^t, $t \in \mathbb{R}$, which is differentiable at 0 with derivative 1 so that $\lim_{t \to 0}(\frac{t}{2!} + \frac{t^2}{3!} + \ldots) = 0$.

We include two other useful facts about Exp and linear Lie groups here. Namely, for any complex $n \times n$ matrix A and $P \in \mathrm{GL}(n, \mathbb{C})$, $P \mathrm{Exp}(A) P^{-1} = \mathrm{Exp}(PAP^{-1})$. That is, Exp commutes with conjugation. This is because conjugation by P is an automorphism of the associative algebra $M_n(\mathbb{C})$. Hence for any positive integer j, and any constant $c \in \mathbb{C}$ we have $PcA^j P^{-1} = c(PAP^{-1})^j$. Hence for any polynomial $p(A) = \sum c_j A^j$ we get $Pp(A)P^{-1} = p(PAP^{-1})$. Taking limits gives the result. Actually, we see that this holds for any absolutely convergent power series. That is, if $f(z)$ is an entire function then $f(A)$ commutes with conjugation.

Secondly, for any complex $n \times n$ matrix A, $\det(\mathrm{Exp}\, A) = e^{\mathrm{tr}(A)}$. When A is triangular with eigenvalues $\lambda_1, \ldots, \lambda_n$, then a direct calculation shows that $\mathrm{Exp}\, A$ is also triangular with eigenvalues $e^{\lambda_1}, \ldots, e^{\lambda_n}$. Hence $\det(\mathrm{Exp}\, A) = e^{\lambda_1} \ldots e^{\lambda_n} = e^{\lambda_1 + \ldots + \lambda_n} = e^{\mathrm{tr}(A)}$. In general, we can apply the 3rd Jordan canonical form to get PAP^{-1} triangular. Then $P \mathrm{Exp}(A) P^{-1} = \mathrm{Exp}(PAP^{-1})$. The determinant and trace of this triangular matrix is the same as that of $\mathrm{Exp}\, A$. Hence the result.

Let $\mathbb{R}$ be the additive group of real numbers, considered to be parameterized by t and let its Lie algebra be generated by the tangent vector field $D_\tau = \frac{d}{dt}|_{t=\tau}$. In this way we can identify $\mathbb{R}$ with its Lie algebra. Let G be a Lie group, $\mathfrak{g}$ be its Lie algebra, and suppose $t \mapsto f(t)$ is a smooth curve in G defined everywhere on $\mathbb{R}$. From now on we shall write $f'(\tau) = d_\tau f(D_\tau)$. Thus by evaluating at $\tau = 0$ we get a vector $f'(0) \in T_1 G$.

For $X \in T_1 G$ consider the associated invariant vector field X_g on G.

Let $\phi_X(t)$ be an integral curve for this vector field, which passes through 1 at $t = 0$. We check that $t \mapsto \phi(t)$ is a group homomorphism, that is, $\phi_X(t+s) = \phi_X(t)\phi_X(s)$. Because of the left invariance of X the curves $s \mapsto \phi_X(t+s)$ and $s \mapsto \phi_X(t)\phi_X(s)$ are both integral curves passing through $\phi_X(t)$ at $s = 0$. Therefore, by the uniqueness of solution of ODEs, they are equal in a neighborhood of $s = 0$. Then Proposition 1.1.4 below shows that they are equal everywhere. Conversely if f were any smooth homomorphism $\mathbb{R} \to G$ and we take the derivative, then $f'(0) = X \in \mathfrak{g}$. Since $f(0) = 1 = \phi_X(0)$ and $\dot{\phi}_X(0) = X = f'(0)$. By uniqueness of local solutions to ODEs, this means that $f = \phi_X$ in some neighborhood of zero. By Proposition 1.1.4 below, smooth homomorphisms which agree in a neighborhood of zero must coincide on all of $\mathbb{R}$.

In fact more generally one has,

Proposition 1.1.4. *Let G be a connected Lie group, H be a Lie group and f and g be globally defined smooth homomorphisms $G \to H$ which coincide in a neighborhood U of 1 in G. Then $f \equiv g$. (See a previous exercise on how U generates H.)*

Proof. Because G is connected, the symmetric neighborhood $V = U \cap U^{-1}$ generates G. Then by Proposition 0.1.13 $G = \bigcup_{n=1}^{\infty} V^n$. Since f and g are homomorphisms which agree on V, they agree on V^n for every n. $\square$

Thus we get,

Proposition 1.1.5. *There is a bijection $X \mapsto \phi_X$ from $\mathfrak{g}$ to the set of all smooth 1-parameter subgroups of G, subject to the requirement that $\phi'_X(D) = X$ for $X \in \mathfrak{g}$.*

Notice, however, that if $s \in \mathbb{R}$ is fixed and $t \mapsto \phi_X(t)$ is a 1-parameter group, then $t \mapsto \phi_X(st) = f(t)$. Since $f' = sX$, we see by the injectivity of the correspondence that $\phi_{sX}(t) = \phi_X(st)$, $s, t \in G$, $X \in \mathfrak{g}$. We can now define the *exponential map* $\exp : \mathfrak{g} \to G$ of a Lie group G.

Definition 1.1.6. For $X \in \mathfrak{g}$ we define $\exp(X) = \phi_X(1)$.

Since for all real t, $\exp(tX) = \phi_X(t)$, by Proposition 1.1.5 above we can identify all the 1-parameter subgroups of G. Also, $\exp(0) = 1$. By connectedness all 1-parameter subgroups lie in G_0 so the range of exp is in G_0. Therefore exp does not really help much in non-connected Lie groups and for this reason one often simply assumes one is working with a connected Lie group.

Corollary 1.1.7. *The 1-parameter subgroups of G are precisely the maps $t \mapsto \exp(tX)$, for $X \in \mathfrak{g}$.*

This, together with Definition 1.1.6, enable us to determine the exponential map in the case of the various examples discussed above:

The exponential map for $\mathbb{R}^n$ is the identity, for $\mathbb{T}^n$ it is π and for $GL(n, \mathbb{C})$ or $GL(n, \mathbb{R})$ it is Exp. This is the reason the exponential map has its name.

Corollary 1.1.8. *The exponential map is smooth and its derivative at 0 is the identity, i.e. $d_0 \exp = I$. Since exp is smooth, the inverse function theorem tells us that it is a local diffeomorphism of a ball about 0 in $\mathfrak{g}$ with a neighborhood of 1 in G. We shall call its local inverse log.*

Proof. Let TG denote the tangent bundle of G. The map, $(g, X) \mapsto d_1 L_g(X)$, going from $G \times \mathfrak{g} \to TG$ is smooth. Now $\phi_X(t) = \exp tX$ is the integral curve of the vector field $X_g = d_1 L_g(X)$ with the initial data $\phi_X(0) = 1$ and $\exp(X) = \phi_X(1)$. So it follows from the smooth dependence on initial data of solutions of ODEs that exp is itself smooth (see Appendix A). The directional derivative in the direction $X \in \mathfrak{g}$ is $\frac{d}{dt}(\exp(tX))|_{t=0} = X$. Therefore $d_0 \exp = id_{\mathfrak{g}}$. $\qquad\square$

Notice that all of our discussion works equally well for complex Lie groups; just substitute complex Lie algebras for $\mathfrak{g}$, complex 1-parameter groups (namely, $z \mapsto \exp(zX)$, $z \in \mathbb{C}$) for ϕ_X, and the simply connected group $\mathbb{C}$ for $\mathbb{R}$. Here again G_0 is also open in G so these groups have the same Lie algebra.

Exercise 1.1.9. If the Lie algebra of a real Lie group G is a complex Lie algebra then G is a complex Lie group.

Proposition 1.1.10. *Let G be a complex connected Lie group and ρ a holomorphic representation of G on the complex vector space V. If $\rho(G)$ is bounded, then it is trivial.*

In particular, a compact, complex connected Lie group must be abelian. This follows by taking for ρ the adjoint representation. Then $\operatorname{Ad} G$ is trivial and hence $G = Z(G)$ is abelian. (In fact it is a torus of even dimension.)

Proof. To prove this we may replace G by the complex connected subgroup of $\operatorname{GL}(V)$ and show that G is trivial. Consider the 1-parameter subgroup $\operatorname{Exp}(zX)$, where $X \in \mathfrak{g}$. This is a bounded entire function of $z \in \mathbb{C}$ taking values in the finite dimensional vector space $\operatorname{End}_{\mathbb{C}}(V)$. Applying Liouville's theorem to each of the finitely many numerical coordinate functions, we conclude that $\operatorname{Exp}(zX)$ is constant (see [55]). Evaluating at $z = 0$ tells us this constant is I. Taking the derivative $\frac{d}{dz}$ at $z = 0$ shows $X = 0$. Since X was arbitrary $\mathfrak{g} = \{0\}$ and since G is connected $G = \{I\}$. $\square$

Exercise 1.1.11. Prove that if a homomorphism on a connected Lie group is smooth in a neighborhood of 1 then it is smooth everywhere.

Proposition 1.1.12. *Let $X_1, \ldots, X_n$ be a basis for $\mathfrak{g}$. Then for suitably small t_i the map $p : (t_1, \ldots, t_n) \mapsto \exp_G(t_1 X_1) \cdots \exp_G(t_n X_n)$ is a diffeomorphism onto an open neighborhood of $1 \in G$.*

Proof. This follows from Corollary 1.1.8 and the fact that the derivative of p at $(0, \ldots, 0)$ is a block diagonal matrix of the derivatives of $\exp(t_i X_i)$'s, therefore it is the identity map and the conclusion follows from the inverse function theorem. $\square$

A similar argument proves

Corollary 1.1.13. *Suppose G is a Lie group and its Lie algebra, $\mathfrak{g}$, is the direct sum of subspaces, $\mathfrak{a}_1 \oplus \ldots \oplus \mathfrak{a}_j$. Then we can find small balls $U_{\mathfrak{a}_1} \ldots, U_{\mathfrak{a}_j}$ about 0 in $\mathfrak{a}_1 \ldots, \mathfrak{a}_j$ such that $(a_1, \ldots a_j) \mapsto \exp a_1 \ldots \exp a_j$ is a diffeomorphism.*

Corollary 1.1.14. *Let G be a connected Lie group and H any Lie group. Then a continuous homomorphism $f : G \to H$ is smooth.*

Proof. We note that this is true for a 1-parameter subgroup $\phi : \mathbb{R} \to H$. Since $\phi_X(t) = \exp_H(tX)$, for $X \in \mathfrak{h}$ and $\exp_H$ is smooth, this is true. Now let $\{X_1, \ldots, X_n\}$ be a basis of $\mathfrak{g}$. Then for each i, $t_i \mapsto f(\exp_G(t_i X_i))$ is a smooth 1-parameter subgroup of H. Hence for each i, $f(\exp_G(t_i X_i)) = \exp_H(t_i Y_i)$, where $Y_i \in \mathfrak{h}$. Since f is a homomorphism $f(\exp_G(t_1 X_1)) \cdots \exp_G(t_n X_n)) = \exp_H(t_1 Y_1) \cdots \exp_H(t_n Y_n)$. By Corollary 1.1.12, for small t_i, $p : (t_1, \ldots, t_n) \mapsto \exp_G(t_1 X_1) \cdots \exp_G(t_n X_n)$ is a diffeomorphism onto a small neighborhood U of 1 in G and we have $f \circ p(t_1, \ldots, t_n) = \exp_H(t_1 Y_1) \cdots \exp_H(t_n Y_n)$ which is smooth as $\exp_H$ is smooth. Therefore f is smooth in a neighborhood of 1 and by Exercise 1.1.11 is smooth everywhere. $\qquad\square$

1.2   Taylor's Theorem and the Coefficients of expX expY

We first deal with Taylor's theorem on a Lie group. Throughout this section $\tilde{X}, \tilde{Y}, \ldots$ denotes the left invariant vector fields associated to $X, Y \ldots \in \mathfrak{g}$. Consequently $\tilde{X}, \tilde{Y} \ldots$ act on $C^\infty(G)$ as first-order differential operators,

$$(\tilde{X}.f)(x) = d_x f(\tilde{X}(x))).$$

Proposition 1.2.1. *Let G be a Lie group with Lie algebra $\mathfrak{g}$. Suppose $X \in \mathfrak{g}$ and f is a smooth function on G. Then for every positive integer m and $g \in G$,*

$$\tilde{X}^m f(g \exp(tX)) = (\tilde{X} \cdots \tilde{X} f)(g \exp(tX)) = \frac{d^m}{dt^m}(f(g \exp(tX))).$$

Moreover, for each positive integer m,

$$f(\exp(X)) = \sum_{k=0}^{m} \frac{1}{k!} \tilde{X}^k f(1) + R_m(X),$$

where $\|R_m(X)\| \leq c_m \|X\|^{n+1}$ and c_m is a positive constant depending only on m.

In particular, taking $m = 1$ and $t = 0$ gives $\tilde{X}f(g) = \frac{d}{dt}(f(g\exp(tX)))|_{t=0}$, for each $g \in G$.

Proof. To prove the first equation we may assume $g = 1$ by replacing the C^∞ function f by f_g and using left invariance of $\tilde{X}$. So for the first equation it remains to show that $\tilde{X}^m f(\exp(tX)) = \frac{d^n}{dt^n}(f(\exp(tX)))$. This is obvious from the definition of $\tilde{X}f$.

Turning to the second equation, we consider the mth order Taylor expansion of $f(\exp(tX))$ about $t = 0$ with the integral remainder and evaluate at $t = 1$.

$$f(\exp(X)) = \sum_{k=0}^{m} \frac{1}{k!}\tilde{X}^m f(1) + R_m(X),$$

where $R_m(X) = \frac{1}{(m+1)!}\int_0^1 (1-s)^m \frac{d^{m+1}}{ds^{m+1}}(f(\exp(sX)))ds$. By the first equation $R_m(X) = \frac{1}{(m+1)!}\int_0^1 (1-s)^m \tilde{X}^{m+1} f(\exp(sX))ds$. Let $\{X_1, \ldots, X_n\}$ be a basis of $\mathfrak{g}$ and write $X = \sum_{i=1}^{n} x_i X_i$. Then $\tilde{X}^{m+1} = (\sum_{i=1}^{n} x_i \tilde{X}_i)^{m+1}$ which is a finite sum of products of $\tilde{X}_i$ of order $m+1$ indexed by the various partitions of $m+1$ into n parts with coefficients, the product of the corresponding x_i's. Now each of these coefficients is $\leq ||X||^{n+1}$. Since $(1-s)^m \geq 0$, letting $d_{n,m}$ be the number of partitions and using the Banach algebra properties of $|| \cdot ||$ and the fact that $f(\exp(sX))$ is bounded, say, by c, on the interval $[0.1]$ we get $||R_m(X)|| \leq \frac{||X||^{m+1}}{(m+1)!}cd_{n,m}\int_0^1 (1-s)^m ds$. $\qquad\square$

In what follows $O(t^k)$ indicates any smooth function of t in a symmetric interval about 0 with the property that $\frac{O(t^k)}{t^k}$ remains bounded at $t \to 0$. We now come to a key lemma which can be considered as the second order approximation to the *Baker-Campbell-Hausdorff* (BCH) formula which reads

$$\exp(X)\exp(Y) = \exp(A + B + C_2(A, B) + C_3(A, B) + \cdots) \qquad (1.1)$$

where each $C_n(A, B)$ is a finite linear combination of the expressions $[X_1, [X_2[\cdots[X_{n-1}, X_n]\cdots]]] = (\operatorname{ad} X_1)(\operatorname{ad} X_2)\cdots(\operatorname{ad} X_{n-1})X_n,$

for $X_i = A$ or B and when A and B are sufficiently close to the identity. The remarkable fact is that C_n does not depend on G, A or B and its coefficients are rational. For instance

$$C_2(A, B) = \tfrac{1}{2}[A, B]$$

$$C_3(A, B) = \tfrac{1}{12}[A, [A, B]] + \tfrac{1}{12}[B, [B, A]] \tag{1.2}$$

$$C_4(A, B) = -\tfrac{1}{24}[A, [B, [A, B]]].$$

Since every Lie group is locally isomorphic to a subgroup of some $GL(k, \mathbb{R})$ and the exponential map of a subgroup is the restriction of the exponential group of the group, one therefore only has to verify this formula for $GL(k, \mathbb{R})$; it turns out that the formula for $\exp(X) \exp(Y)$ in $GL(k, \mathbb{R})$ is independent of k.

Lemma 1.2.2. *For X and $Y \in \mathfrak{g}$ and $t \in \mathbb{R}$ we have*

(1) $\exp(tX) \exp(tY) = \exp(t(X + Y) + \tfrac{1}{2}t^2[X, Y] + O(t^3))$
(2) $\exp(tX) \exp(tY) \exp(-tX) = \exp(tY + t^2[X, Y] + O(t^3))$
(3) $\exp(tX) \exp(tY) \exp(-tX) \exp(-tY) = \exp(t^2[X, Y] + O(t^3))$.

We remark that the third relation gives a geometric interpretation of the bracket: $[X, Y]$ is the tangent vector at 1 to the curve $t \mapsto \exp(\sqrt{t}X) \exp(\sqrt{t}Y) \exp(-\sqrt{t}X) \exp(-\sqrt{t}Y)$, $t \geq 0$. It also shows that if, for all small t, $\exp(tX)$ and $\exp(tY)$ commute in G then $[X, Y] = 0$. In particular, if for all small t, $\exp(tX)$ and $\exp(tY)$ commute then by (1), we get,

$$\exp(tX) \exp(tY) = \exp(t(X + Y)). \tag{1.3}$$

As a result $\exp(tX)$ and $\exp(-tX)$ commute and are mutual inverses of one another.

Another remark to be made is that the first relation implies $\exp(tX) \exp(tY) = \exp(t(X + Y) + O(t^2))$. This means that the tangent vector at 0 to the curve $t \mapsto \exp(tX) \exp(tY)$ is $X + Y$.

Proof. Let f be a smooth function defined in a neighborhood of 1 and $X \in \mathfrak{g}$. By Proposition 1.2.1 for $n \geq 0$, $(\tilde{X}^n f)(g \exp tX) =$

$\frac{d^n}{dt^n} f(g \exp tX)$. Hence $(\tilde{X}^n \tilde{Y}^m f)(1) = \frac{d^n}{dt^n} \frac{d^n}{ds^n} f(\exp tX \exp sY)_{t=0,s=0}$.
Therefore,

$$f(\exp(tX) \exp(sY)) = \sum_{n \geq 0, m \geq 0} \frac{t^n}{n!} \frac{s^m}{m!} (\tilde{X}^n \tilde{Y}^m) f(1).$$

On the other hand since exp is smooth and invertible in a neighborhood of 0 and group multiplication is smooth we see that for $|t|$ sufficiently small $\exp(tX) \exp(tY) = \exp(Z(t))$ where $Z(t)$ is a smooth function $Z : U \to \mathfrak{g}$, and U is a symmetric interval about 0. Evidently $Z(0) = 0$. Taking the Taylor expansion of the second order of $Z(t)$ about $t = 0$ gives $Z(t) = tZ_1 + t^2 Z_2 + O(t^3)$, where Z_1 and Z_2 are constants in $\mathfrak{g}$. Let $\{X_1, \ldots, X_n\}$ be a basis of $\mathfrak{g}$ and f be any of the coordinate functions $\exp(x_1 X_1 + \cdots + x_n X_n) \mapsto x_i$. Then $f(\exp Z(t)) = f(\exp(tZ_1 + t^2 Z_2)) + O(t^3)$. But $f(\exp(tZ_1 + t^2 Z_2)) = \sum_{n=0}^{\infty} \frac{1}{n!} (t\tilde{Z}_1 + t^2 \tilde{Z}_2) f(1)$. Hence as above,

$$f(\exp Z(t)) = \sum_{n=0}^{\infty} \frac{1}{n!} (t\tilde{Z}_1 + t^2 \tilde{Z}_2) f(1) + O(t^3). \tag{1.4}$$

Whereas,

$$\exp(tX) \exp(sY) = \sum_{n \geq 0, m \geq 0} \frac{t^n}{n!} \frac{s^m}{m!} (\tilde{X}^n \tilde{Y}^m) f(1). \tag{1.5}$$

Letting $s = t$ in (1.5) and comparing coefficients with (1.4) yields $Z_1 = X + Y$ and $Z_2 + \frac{1}{2} Z_1^2 = \frac{1}{2}(X^2 + 2XY + Y^2)$. Therefore $2Z_2 + (X + Y)^2 = X^2 + 2XY + Y^2$ and since $(X + Y)^2 = X^2 + XY + YX + Y^2$ we see that $Z_2 = \frac{1}{2}[X, Y]$. This proves (1).

Part (3) follows by applying (1) twice. To prove (2) observe that since $\exp(tX) \exp(tY) \exp(-tX) \exp(-tY) = \exp(t^2 [X, Y] + O(t^3))$ we know that $\exp(tX) \exp(tY) \exp(-tX) = \exp(t^2 [X, Y] + O(t^3)) \exp(tY)$. Reasoning as before this is $\exp(tY + t^2([X, Y] + \frac{Y^2}{2}) + O(t^3))$. But also as before,

$$\exp(tX) \exp(tY) \exp(-tX) = \exp Z(t),$$

where $Z(t)$ is again a smooth function with $Z(0) = 0$ and $Z(t) = tZ_1 +$

$t^2 Z_2 + O(t^3)$. Then as above,

$$\exp(tZ_1 + t^2 Z_2)) + O(t^3) = \exp(tY + t^2[X,Y] + \frac{Y^2}{2} + O(t^3))$$

and comparing coefficients we get $Z_1 = Y$ and $Z_2 + \frac{Y^2}{2} = [X,Y] + \frac{Y^2}{2}$ so $Z_2 = [X,Y]$. $\qquad\square$

From (1) and (3) and the continuity of exp we conclude

Corollary 1.2.3. *Let G be a Lie group. Then for X and $Y \in \mathfrak{g}$, and sufficiently small $t \in \mathbb{R}$ and $n \in \mathbb{Z}$ we have*

(1) $\exp(t(X + Y)) = \lim_{n \to \infty}(\exp(\frac{1}{n}tX)\exp(\frac{1}{n}tY))^n$

(2) $\exp(t[X,Y]) = \lim_{n \to \infty}[\exp(\frac{1}{n}tX), \exp(\frac{1}{n}tY)]^{n^2}$.

Corollary 1.2.4. *Let G be a Lie group with Lie algebra $\mathfrak{g}$ and H be a Lie subgroup. Then the Lie algebra of H is $\{X \in \mathfrak{g} : \exp(tX) \in H \text{ for all } t\}$.*

Proof. Calling this set S we see immediately that $\mathfrak{h} \subseteq S$. On the other hand if $X \in \mathfrak{g}$ has the property that the whole curve is in H, then its tangent vector at $t = 0$ lies in $T_1(H)$. Thus $\mathfrak{h} \subseteq S \subseteq T_1(H)$. Hence the result. $\qquad\square$

Exercise 1.2.5. Show that $\exp(tX) \in H$ for all small t then for $\exp(tX) \in H$ for all $t \in \mathbb{R}$.

We now give an example of a general type of group which will have a certain importance.

Definition 1.2.6. One calls a subgroup $G \subseteq \mathrm{GL}(n, \mathbb{C})$ an *algebraic group* if it is the simultaneous zero set within $\mathfrak{gl}(n, \mathbb{C})$ of a family of polynomials with complex coefficients in the x_{ij} coordinates of the matrices in $\mathfrak{gl}(n, \mathbb{C})$. Clearly, such a group is a closed subgroup of $\mathrm{GL}(n, \mathbb{C})$ in the usual Euclidean topology and hence is a Lie group by a theorem of E. Cartan, Theorem 1.3.5. Furthermore, we shall call $G_{\mathbb{R}} = G \cap \mathrm{GL}(n, \mathbb{R})$ the $\mathbb{R}$-*points* or the *real points* of G. Similarly, $G_{\mathbb{R}}$ of an algebraic group G is also a Lie group. If the family of polynomials defining G happens to have all its coefficients lying in some subfield F of $\mathbb{C}$, we then say that G is defined over F. A group G is said to be *essentially algebraic* if it is either an algebraic group or it has finite index in the real points of an algebraic group.

Typical examples of algebraic groups are $\mathrm{GL}(n, \mathbb{C})$ itself (the empty set of polynomials) and $\mathrm{SL}(n, \mathbb{C})$ itself (the single polynomial $\det -1 = 0$). These groups are defined over $\mathbb{Q}$. The respective real points are $\mathrm{GL}(n, \mathbb{R})$ and $\mathrm{SL}(n, \mathbb{R})$. If V is a finite dimensional vector space over $\mathbb{C}$ and we have a *nondegenerate* bilinear form $\beta : V \times V \to k$. Let $G_\beta = \{g \in \mathrm{GL}(V) : \beta(gv, gw) = \beta(v, w) \text{ for all } v, w \in V\}$. It is obvious that G_β is a subgroup of $\mathrm{GL}(V)$ which is algebraic. These evidently include $\mathrm{O}(n, \mathbb{C})$ and $\mathrm{Sp}(n, \mathbb{C})$ with real points $\mathrm{O}(n, \mathbb{R})$ and $\mathrm{Sp}(n, \mathbb{R})$ (see [15] for the geometric significance of these groups).

Exercise 1.2.7. Prove that in fact G_β is an algebraic group defined over $\mathbb{Q}$.

By Cartan's theorem, Theorem 1.3.5, G_β is a Lie subgroup. We compute its Lie algebra, $\mathfrak{g}_\beta$. By our criterion $\mathfrak{g}_\beta$ is $\{X \in \mathrm{End}(V) : \beta(\mathrm{Exp}(tX)v, \mathrm{Exp}(tX)w) = \beta(v, w)\}$. Calculating the derivative at $t = 0$ tells us that $\beta(v, Xw) + \beta(Xv, w) = 0$. Because Exp is faithful on a neighborhood of the identity and $\mathrm{Exp}\, X^t = (\mathrm{Exp}\, X)^t$, the converse is also true. We leave the details to the reader.

When β is the symplectic form, a $2n \times 2n$ matrix g preserves β if and only if $g^t J g = J$, where J is the $2n \times 2n$ matrix consisting of the following $n \times n$ blocks:

$$J = \begin{pmatrix} 0 & I \\ -I & 0 \end{pmatrix}.$$

This description makes it more convenient to calculate things. For example, it shows easily that $\mathrm{Sp}(1, \mathbb{C}) = \mathrm{SL}(2, \mathbb{C})$, but for higher n, $\mathrm{Sp}(n, \mathbb{C}) \neq \mathrm{SL}(2n, \mathbb{C})$. The Lie algebra, $\mathfrak{sp}(n, \mathbb{C})$, of $\mathrm{Sp}(n, \mathbb{C})$ is the subalgebra of the $2n \times 2n$ matrices, $M_{2n}(\mathbb{C})$, consisting of

$$X = \begin{pmatrix} X_1 & X_2 \\ X_3 & X_4 \end{pmatrix},$$

and satisfying $X^t J + J X = 0$. This means X_2 and X_3 are symmetric while $X_1 = -X_4^t$. Thus X_4 is arbitrary, X_1 is determined by X_4 and

X_2 and X_3 have $\frac{n(n+1)}{2}$ free parameters each. Hence the Lie algebra $\mathfrak{sp}(n, \mathbb{C})$ of $\mathrm{Sp}(n, \mathbb{C})$ has complex dimension $2n^2 + n$. Similarly $\mathfrak{sp}(n, \mathbb{R})$, the Lie algebra of $\mathrm{Sp}(n, \mathbb{R})$, has real dimension $2n^2 + n$.

1.3 Correspondence between Lie Subgroups and Subalgebras

In this section we characterize the Lie subalgebras of the Lie algebra of a Lie group. In fact we show that there is a one-to-one correspondence between Lie subalgebras of $\mathfrak{g} = \mathrm{Lie}\, G$ and connected Lie subgroups of G.

A k-dimensional *distribution* $\mathcal{D}$ on a smooth manifold M is a choice of k-dimensional subspace $\mathcal{D}(m)$ of $T_m M$ for each $m \in M$. We shall say $\mathcal{D}$ is *smooth* if each $m \in M$ has a neighborhood U with vector fields $X_1, \ldots, X_k$ defined on U which span $\mathcal{D}$ at all points $m \in U$. A vector field X is in $\mathcal{D}$ if $X(m) \in \mathcal{D}(m)$ for all $m \in M$.

A natural class of smooth distributions is given by foliations, more precisely if $\mathcal{F}$ is a foliation of M with leaves $\mathcal{F}_i$ then let $\mathcal{D}(m) = T_x \mathcal{F}_i$, the tangent space at the leaf containing m. These types of distributions are called *integrable i.e.* for every $m \in M$ there is a submanifold N which passes through m and $T_m N = T_m M|_N = \mathcal{D}(m)$; the submanifold N is said to be an *integral manifold* for $\mathcal{D}$.

One would like to characterize such distributions and Frobenius' theorem addresses this matter.

Definition 1.3.1. A smooth distribution $\mathcal{D}$ is called *involutive* if the set of vector fields in $\mathcal{D}$ is closed under Lie bracket.

Obviously integrable distributions are involutive as the vector fields on a submanifold (or any manifold) form a Lie algebra.

Theorem 1.3.2. *An involutive distribution on a smooth manifold M is integrable. Moreover, for every $m \in M$ there is a unique connected maximal integral manifold containing m and the set of maximal integral manifolds of $\mathcal{D}$ gives rise to a foliation of M.*

This is known as Frobenius' theorem [80].

Here is one of the main results in this section

Theorem 1.3.3. *Let G be a Lie group with Lie algebra $\mathfrak{g}$. Then there is bijection between the connected subgroups of G and the subalgebras of $\mathfrak{g}$.*

Proof. Suppose that H is a connected Lie subgroup of G and $\tilde{X}, \tilde{Y}$ are two left invariant vector fields on H which correspond to the vectors $X, Y \in \mathfrak{h} = T_1 H \subset \mathfrak{g} = T_1 G$. By the definition of the Lie bracket in $\mathfrak{g}$ we have

$$[\tilde{X}, \tilde{Y}](g) = dL_g[X, Y] \tag{1.6}$$

for all $g \in G$ which uniquely determines $[X, Y] \in \mathfrak{g}$. By restricting (1.6) to H we observe that $[X, Y]$ has to be the Lie bracket of X and Y in $\mathfrak{h}$, in particular $[X, Y] \in \mathfrak{h}$.

Conversely, a subalgebra $\mathfrak{h}$ of $\mathfrak{g}$ is indeed the Lie algebra of a unique connected subgroup of G. Consider the smooth distribution $\mathcal{D}$ on G which consists of all the left invariant vector fields generated by the vectors in $\mathfrak{h}$. Then $\mathcal{D}$ is involutive by (1.6) and due to the assumption that $\mathfrak{h}$ is a subalgebra. Hence by Frobenius' theorem there is a maximal connected integral manifold H which contains $1 \in G$. We claim that H is subgroup and to prove that it suffices to show that $xH = H$ for all $x \in H$. Note that xH is also an integral manifold since the distribution $\mathcal{D}$ is left invariant. On the other hand, since $1 \in H$ therefore $x \in xH$ and by the uniqueness part of the Frobenius' theorem we have $H = xH$.

Now we must show that H is a Lie group, or equivalently that $\tau : H \times H \to H$, the map given by $\tau(x, y) = x^{-1} y$ is smooth. Consider the diagram

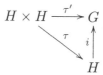

Obviously τ is continuous. Note that $\tau' : H \times H \to G$ is smooth as the inclusion $H \times H \hookrightarrow G \times G$ is smooth. Now we are going to introduce a chart for H which makes τ smooth. Suppose that G has dimension

$k+n$ and k is the dimension of $\mathfrak{h}$. By Frobenius theorem, for each $x \in H$ there is a chart $\phi : U \to \mathbb{R}^{k+m}$

$$U \cap H = \phi^{-1}(\{(x_1, \ldots, x_{k+m}) | x_{k+1} = \cdots = x_{k+m} = 0\}).$$

Consider $V = U \cap H$ and $\pi : \mathbb{R}^{k+m} \to \mathbb{R}^k$ the projection on the first k coordinates, then $(V, \psi = \pi \circ \phi \circ i)$ is the desired chart. We have

$$\psi \circ \tau = \pi \circ \phi \circ i \circ \tau = \pi \circ \phi \circ \tau'$$

which is smooth as τ' and π are.

For uniqueness, let K be another such connected Lie group. As K is an integral manifold then there is smooth inclusion $K \subset H$. Since $T_1 H = T_1 K = \mathfrak{h}$ the inclusion is a local isomorphism thus surjective which proves that $K = H$. □

We now come to Cartan's theorem mentioned earlier. For that we need the following lemma.

Corollary 1.3.4. *Let H be a closed subgroup of a Lie group G and $\mathfrak{h} = \{X \in \mathfrak{g} : \exp(tX) \in H \text{ for all } t \in \mathbb{R}\}$. Then $\mathfrak{h}$ is a Lie subalgebra of $\mathfrak{g}$.*

Proof. Clearly $\mathfrak{h}$ is closed under scalar multiplication. Since H is a closed subgroup of G, Corollary 1.2.3 part 1, shows $\mathfrak{h}$ is closed under addition, while part 2 shows $\mathfrak{h}$ is closed under bracketing. □

Theorem 1.3.5. *A closed subgroup H of a Lie group G is a Lie group with relative topology.*

Proof. It is enough to prove the theorem when H is a connected subgroup. Consider the Lie subalgebra $\mathfrak{h}$ as it is defined in Corollary 1.3.4. Then by Theorem 1.3.3 there exists a Lie subgroup $H' \subset G$ such that $T_1 H' = \mathfrak{h}$. Let $\mathfrak{s}$ be a complementary subspace for $\mathfrak{h}$ in $\mathfrak{g}$ such that $\mathfrak{g} = \mathfrak{h} \oplus \mathfrak{s}$. Let U and V be two sufficiently small neighborhoods of 0 in $\mathfrak{h}$ and $\mathfrak{s}$ such that the restriction of $\exp_G$ to $U \times V$ is a diffeomorphism onto its image (see Corollary 1.1.13). We prove that $H \cap \exp_G(U \times V) = \exp U$. If $\exp(X + Y) \in H$, $X \in U$ and $Y \in V$ then by Corollary 1.2.3, $\exp Y = \lim_{n\to\infty}(\exp(\frac{1}{n}(X+Y))\exp(-\frac{1}{n}X))^n$ is in H as H is a closed subgroup, hence $Y \in \mathfrak{h} \cap \mathfrak{s} = \{0\}$. Consequently, $\exp U$ is an open set in H. On the other hand $\exp U$ is an open set in H' as $\exp_G|_{\mathfrak{h}} = \exp_{H'}$. Therefore, $H = \bigcup_{n=1}^{\infty} U^n = H'$ as both are connected. $\square$

Corollary 1.3.6. *A Lie group has no small subgroups.*

Proof. Let U be a ball about 0 in $\mathfrak{g}$ on which exp is a global diffeomorphism and suppose that H were a subgroup of G contained in $\exp(U)$. Let $h \in H$, $h = \exp X$, then for any positive integer n, $h^n = \exp nX$. Since everything takes place in a neighborhood where exp is a global diffeomorphism, and $h^n \in H \subset \exp(U)$ therefore $\exp nX \in \exp(U)$ so $nX \in U$ for all n. This is impossible for a ball. $\square$

1.4   The Functorial Relationship

We next turn to functoriality questions. We will deal with the real case, but this works just as well in the complex case. For a Lie homomorphism $f : G \to H$ we denote the derivative of f at the identity element 1 by

$$f' = d_1 f : \mathfrak{g} \to \mathfrak{h}$$

which is a linear map. Since $d_0 \exp = id$, one can write

$$f'(X) = \frac{d}{dt}|_{t=0} f(\exp_G(tX)).$$

Theorem 1.4.1. *Let $f : G \to H$ be a smooth homomorphism between Lie groups and $f' : \mathfrak{g} \to \mathfrak{h}$ be its derivative at 0. Then*

(1) f' is a Lie algebra homomorphism.

(2) If $\exp_G$ and $\exp_H$ denote the respective exponential maps then $f \circ \exp_G = \exp_H \circ f'$.

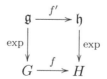

(3) f' is uniquely determined by (1) and (2).

(4) If e is any other smooth homomorphism $e : H \to L$, then ef is a smooth homomorphism $G \to L$ and $(ef)' = e'f'$.

Proof. Let $X \in \mathfrak{g}$ and consider the corresponding 1-parameter group $t \mapsto \exp_G(tX)$. As we saw earlier, because f is a smooth homomorphism $t \mapsto f(\exp_G(tX))$ is a 1-parameter subgroup of H and hence its infinitesimal generator $\frac{d}{dt}|_{t=0} f(\exp_G(tX)) = f'(X) \in \mathfrak{h}$. Thus for all t,

$$\exp_H(tf'(X)) = f(\exp_G(tX)). \tag{1.7}$$

Taking $t = 1$ in (1.7) gives the commutativity of the diagram above. We now show that f' is a Lie algebra homomorphism. Using (1.7) but replacing X by cX gives $\exp_H(tcf'(X)) = \exp_H(tf'(cX))$. Differentiating at $t = 0$ shows $cf'(X) = f'(cX)$. Let X and $Y \in \mathfrak{g}$. Now recall that, for any Lie group, the tangent vector at 0 to the curve $t \mapsto \exp(tX)\exp(tY)$ is $X+Y$. Applying f and using (1.7) we get $f'(X+Y) = f'(X)+f'(Y)$. In a similar manner for $t \geq 0$ applying

$$\exp(\sqrt{t}X)\exp(\sqrt{t}Y)\exp(-\sqrt{t}X)\exp(-\sqrt{t}Y) = \exp(t[X,Y] + O(t^{\frac{3}{2}}))$$

we conclude that $f'([X,Y]) = [f'(X), f'(Y)]$. We leave the verification of this last statement to the reader. The Chain rule also proves (4). To prove (3), we suppose $\phi : \mathfrak{g} \to \mathfrak{h}$ is any other Lie algebra homomorphism which commutes the diagram. Let B be a ball about 0 in $\mathfrak{g}$ on which $\exp_G$ is a diffeomorphism with its image $\exp_G(B) = U$. Then by the commutativity of the diagram on $\log(U) = B$, $\phi = f'$. Since these maps are linear and B contains a basis of $\mathfrak{g}$ they coincide. $\square$

Corollary 1.4.2. *(1) $f(G)$ is a Lie group whose Lie algebra is isomorphic to $f'(\mathfrak{g})$.*

(2) If H is connected, then f is surjective if and only if f' is surjective.

(3) Ker f is Lie group whose Lie algebra is Ker f'.

(4) f is locally one-to-one if and only if f' injective.

Proof. Proof of (1). In Theorem 1.4.1 let $H = f(G)$ which is a closed subgroup. Therefore it is a Lie group by Cartan's theorem, 1.3.5. Let $\mathfrak{h}$

be its Lie algebra. Then $\mathfrak{h} \supseteq f'(\mathfrak{g})$. On the other hand by Proposition 1.4.8,

$$\dim f(G) = \dim G - \dim \operatorname{Ker} f = \dim \mathfrak{g} - \dim \mathfrak{h} = \dim \mathfrak{g}/\mathfrak{h} = \dim f'(\mathfrak{g}).$$

Thus $\mathfrak{h}$ and $f'(\mathfrak{g})$ have the same dimension and hence they are equal. In particular, f' is surjective if and only if $\mathfrak{h} = f'(\mathfrak{g})$. Thus $f(G)$ has the same dimension as H. Therefore $f(G)$ is open in H which proves (2).

   Proof of (3). By our criterion for a Lie subalgebra, the Lie algebra of Ker is $\{X \in \mathfrak{g} : \exp_G(tX) \in \operatorname{Ker} f\}$, *i.e.* $f(\exp_G(tX)) \equiv 1$. That is, $\exp_H tf'(X) \equiv 1$. Differentiating gives $f'(X) = 0$. Conversely if $f'(X) = 0$, then $f(\exp_G(tX)) \equiv 1$. Therefore the Lie algebra of $\operatorname{Ker} f$ is $\operatorname{Ker} f'$. Finally, (4) follows from the inverse function theorem. $\qquad\square$

   The proof of the following corollary uses Theorem 1.4.1 and Proposition 0.1.13 and is left to the reader.

Corollary 1.4.3. *Let G be a connected Lie group and e and f be two smooth homomorphisms $G \to H$, with $e' = f'$. Then $e \equiv f$.*

   As the next result, we show that for simply connected Lie groups, there is a one-to-one correspondence between Lie homomorphisms and Lie algebra homomorphisms.

Theorem 1.4.4. *For Lie groups G and H with G simply connected, every Lie algebra homomorphism $\phi : \mathfrak{g} \to \mathfrak{h}$ is the derivative of a Lie homomorphism $f : G \to H$, i.e. $\phi = f'$.*

   This theorem follows from an important result called the *monodromy principle*. Here is a formulation of it as it appears in [15] p. 46.

Theorem 1.4.5. *(Monodromy Principle) Let X be a simply connected space. Suppose that we are given a collection of sets M_p, $p \in X$, parameterized by the elements of X. Assume that $D \subset X \times X$ is a connected subset containing the diagonal such that for each $(p, q) \in D$ there is a map $\phi_{pq} : M_p \to M_q$ satisfying the following conditions:*

   (1) ϕ_{pq} is a one-to-one map and $\phi_{pp} = id$.

   (2) If ϕ_{pq}, ϕ_{qr} and ϕ_{pr} are all defined, then $\phi_{pr} = \phi_{qr} \circ \phi_{pq}$.

Then there is a map ψ which assigns to each $p \in X$ an element $\psi(p) \in M_p$ in such a way that

$$\psi(q) = \phi_{pq}(\psi(p)),$$

whenever ϕ_{pq} is defined. If we required that $\psi(p_0) = m_0$ for some fixed elements $p_0 \in X$ and $m_0 \in M_{p_0}$, then ψ is unique.

Here is a consequence of the monodromy principle.

Lemma 1.4.6. *Let G be a simply connected topological group and f a local homomorphism from G to a topological group H. Then f can be extended to all of G. Local homomorphism means that f is defined on an open neighborhood of the identity element and $f(ab) = f(a)f(b)$ on the neighborhood.*

Proof. Let U be the open neighborhood where f is defined. We may assume that U is symmetric. If not, we can replace U by $U \cap U^{-1}$. Let $D \subset G \times G$ be the subset consisting of (p, q) such that $qp^{-1} \in U$. Evidently D contains the diagonal and is connected. To every $p \in G$, we associate the map $\phi_{pq} : x \mapsto f(qp^{-1})x$ on H. One checks directly that D, $M_p = H$ and ϕ_{pq} satisfy the conditions of the previous theorem. Therefore, there is a unique map $\psi : G \to H$ such that $\psi(1) = 1$ and

$$\psi(p) = f(qp^{-1})\psi(q), \qquad (1.8)$$

whenever $qp^{-1} \in U$. We shall prove that ψ is a group homomorphism which extends f. By taking $q = 1$ and $p \in U$ in (1.8) we obtain $f(p) = \psi(p)$, thus ψ extends f. Let $r = qp^{-1}$, we get

$$\psi(rq) = \psi(r)\psi(q)$$

for $r \in U$. In particular we have,

$$\psi(r_1 r_2 \cdots r_n) = \psi(r_1) \cdots \psi(r_n)$$

if $r_i \in U$. Since $G = \bigcup_{n=1}^{\infty} U^n$, every element $g \in G$ can be written as $g = r_1 r_2 \cdots r_n$ for some n. Therefore,

$$\psi(g) = \psi(r_1 r_2 \cdots r_n) = \psi(r_1) \cdots \psi(r_n)$$

which implies that ψ is a group homomorphism. $\qquad \square$

Proof of Theorem 1.4.4: Given $\phi : \mathfrak{g} \to \mathfrak{h}$, and using the exponential map one can construct a local homomorphism f from G to H. More explicitly, let $U \subset \mathfrak{g}$ and $V \subset \mathfrak{h}$ be two neighborhoods of the origin where the corresponding exponential maps are diffeomorphisms and $\phi(U) \subset V$. Then for $g = \exp X$, $X \in U$, we define,

$$f(g) = \exp(\phi(X)).$$

It follows from the properties of the exponential map that f is a local homomorphism and $f' = \psi$. By the previous lemma, f can be extended to all of G which is unique by Corollary 1.4.3.

The next result follows immediately.

Corollary 1.4.7. *Let G be a simply connected Lie group. Then there is a one-to-one correspondence between the automorphims of G and the automorphisms of $\mathfrak{g}$, the Lie algebra of G.*

Here we give the Lie theoretic analogue of the *first isomorphism theorem* whose proof is left to the reader.

Proposition 1.4.8. *If $f : G \to H$ is a smooth surjective homomorphism between connected Lie groups, then it induces a bijective Lie homomorphism $f^* : G/\mathrm{Ker} \to H$. In particular, $\dim G = \dim \mathrm{Ker}\, f + \dim H$.*

Proof. Let $K = \mathrm{Ker}\, f$ which is a closed normal subgroup. Therefore a Lie group with Lie algebra $\mathfrak{k}$. Then we have the isomorphism $f^* : G/K \to H$ as groups which is a smooth map by definition of the smooth structure of the quotient. To prove that it is an isomorphism we must show that it is a local isomorphism at the identity. The derivative of f^* at the identity is basically the induced map on the Lie quotient $f' : \mathfrak{g}/\mathfrak{k} \to \mathfrak{h}$. The latter is an isomorphism by the first isomorphism theorem for Lie algebras, Theorem 3.1.6.   □

The other isomorphism theorems are formulated and proven similarly.

Corollary 1.4.9. *(The second isomorphism theorem) If G is connected Lie group with connected subgroups K and H such that HK is closed and H normalizes K in G, then*

$$H/H \cap K \simeq HK/K.$$

Proof. Apply Corollary 1.4.8 to the natural projection map $H \to HK/H \cap K$. $\qquad\square$

Corollary 1.4.10. *(The third isomorphism theorem) If G is a connected Lie group with connected normal subgroups $K \subset H$ then $(G/K)/(H/K) \simeq G/H$.*

Proof. Apply the previous result to then natural map $G/K \to G/H$ induced by the inclusion $K \to H$. $\qquad\square$

We now specialize Theorem 1.4.1 to finite dimensional representations.

Proposition 1.4.11. *Let $\rho : G \to \mathrm{GL}(V)$ be a smooth representation of the Lie group G on V over k, where $k = \mathbb{R}$ or $\mathbb{C}$, and let $\rho' : \mathfrak{g} \to \mathfrak{gl}(V)$, be its derivative. If G is connected, then a subspace W of V is ρ-invariant if and only if it is ρ'-invariant.*

Proof. Here we use the following commutative diagram

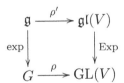

Now W is ρ'-invariant if and only if it is stable under $\rho'(B)$, where B is a ball about 0 in $\mathfrak{g}$. This is because if $X \in \mathfrak{g}$, then $X = cY$, where $Y \in B$ and ρ' is linear. Choosing B small enough $B = \log U$, where U is a canonical neighborhood of 1 in G we see, by the commutativity of the diagram, that this is equivalent to W being stable under $\rho(U)$. Since U is symmetric, $G = \bigcup_{n=1}^{\infty} U^n$. Hence because ρ is a homomorphism, this condition is, in turn, equivalent to the invariance of W under $\rho(G)$. $\qquad\square$

Corollary 1.4.12. *If G is connected, then ρ is irreducible (respectively completely reducible) if and only if ρ' is irreducible (respectively completely reducible).*

Definition 1.4.13. If ρ and σ are representations of a Lie group on V_ρ and V_σ, respectively, then $C_{\rho,\sigma}$, the space of intertwining operators, consists of all linear maps $T : V_\rho \to V_\sigma$, such that $T\rho_g = \sigma_g T$, for all $g \in G$.

If $\rho : G \to \mathrm{GL}(V)$ is a representation of G then $\rho' : \mathfrak{g} \to \mathfrak{gl}(V)$ denotes the derivative of ρ at the identity element. Note that ρ' is a Lie algebra representation. Similar reasoning, together with the fact that Exp commutes with conjugation, yields the following result whose proof is left to the reader.

Corollary 1.4.14. *If G is connected and ρ and σ are finite dimensional representations of G, then ρ and σ are equivalent if and only if ρ' and σ' are equivalent. More generally, if ρ and σ are representations then $C_{\rho,\sigma} = C_{\rho',\sigma'}$.*

The following corollary stems directly from Theorem 1.4.4.

Corollary 1.4.15. *For a simply connected Lie group G, there is a one-to-one correspondence between the representations of G and its Lie algebra $\mathfrak{g}$.*

We now need the following lemma.

Lemma 1.4.16. *Let $X \in \mathrm{End}_k(V)$, where $k = \mathbb{R}$ or $\mathbb{C}$ and W be a k-subspace of V. Then $X(W) \subseteq W$ if and only if $\mathrm{Exp}\, tX(W) = W$ for all $t \in k$.*

Proof. Suppose $\mathrm{Exp}(tX)(W) = W$ for all $t \in k$. Let $w \in W$ and consider the smooth curve $t \mapsto \mathrm{Exp}(tX)(w)$ in W. Then $\frac{d}{dt}|_{t=0}$ of this curve is $X(w) \in W$ because the tangent space of an open set in Euclidean space is the Euclidean space. Conversely, if $X(W) \subseteq W$, then $X^n(W) \subseteq W$ and since W is a subspace, any polynomial $p(X)(W) \subseteq W$. Because W is closed, for any entire function f, $f(X)(W) \in W$. Applying this to tX as the X gives the conclusion. $\qquad\square$

Definition 1.4.17. Let ρ be a representation of G on V, ρ' a representation of a Lie algebra, $\mathfrak{g}$, and W be a subspace of V. We write $\mathrm{Stab}_G(W)$ for the set of all g's that stabilize W and similarly for $\mathrm{Stab}_{\mathfrak{g}}(W)$. We also shall write $\mathrm{Fix}_G(v)$ and $\mathrm{Fix}_{\mathfrak{g}}(v)$ for the things that fix $v \in V$, in each case.

Proposition 1.4.18. *Let ρ be a representation of G on V, ρ' be its derivative and W a subspace of V. Then*

(1) $\mathrm{Stab}_G(W) = \mathrm{Stab}_{\mathfrak{g}}(W)$

(2) $\mathrm{Fix}_G(v) = \mathrm{Fix}_{\mathfrak{g}}(v)$, $v \in V$

The proof of the first statement follows from Lemma 1.4.16, and the second can be seen directly. We leave the details to the reader.

Now we turn to the most important representation of a Lie group, namely the adjoint representation. If G is a Lie group (not necessarily connected) and α is a smooth automorphism of G then α' is an automorphism of $\mathfrak{g}$ and as usual we get a commutative diagram. Now take $\alpha = \alpha_g$, the inner automorphism gotten throught conjugation by $g \in G$. We call the differential $\mathrm{Ad}\, g$. Since $\alpha_g \alpha_h = \alpha_{gh}$ we see that $\mathrm{Ad}(gh) = \mathrm{Ad}\, g\, \mathrm{Ad}\, h$. Thus we get a linear representation $\mathrm{Ad} : G \to \mathrm{GL}(\mathfrak{g})$ called the *adjoint representation* and the image of G under Ad is denoted $\mathrm{Ad}\, G$. For g near 1, $\mathrm{Ad}\, g = \log \cdot \alpha_g \cdot \exp$, a composition of smooth functions so Ad is smooth in a neighborhood of 1 in G. Hence by an exercise above Ad is a smooth representation of G in $\mathfrak{g}$. This works equally well over $\mathbb{R}$ or $\mathbb{C}$.

Corollary 1.4.19. *Let G be a connected Lie group. Then $Z(G) = \mathrm{Ker}\, \mathrm{Ad}$ and its Lie algebra is $\mathfrak{z}(\mathfrak{g}) = \mathrm{Ker}\, \mathrm{ad}$.*

Proof. Now because of connectedness, $g \in Z(G)$ if and only if for all $t \in k$ and $X \in \mathfrak{g}$ we have

$$\exp(tX) = g \exp(tX) g^{-1} = \exp(t\, \mathrm{Ad}\, g)(X).$$

Differentiating at $t = 0$, we see this is equivalent to $\mathrm{Ad}\, g(X) = X$ for all X. That is, $g \in \mathrm{Ker}\, \mathrm{Ad}$.

Also, the Lie algebra of $Z(G)$ is $\{X \in \mathfrak{g} : \mathrm{Ad}(\exp(tX)) \equiv$ for all $t\}$. So, again, differentiating at $t = 0$ gives $\mathrm{ad}\, X = 0$. Since all steps are reversible this proves the second statement. $\qquad\square$

Corollary 1.4.20. *If G is a connected abelian Lie group, then $\mathfrak{g}$ is abelian and $\exp$ is a surjective homomorphism.*

Proof. It follows from Theorem 1.4.25 below, that the Lie algebra of $[G, G]$ is $[\mathfrak{g}, \mathfrak{g}]$. Therefore since G is abelian, so is $\mathfrak{g}$ and $\exp(X + Y) = \exp(X)\exp(Y)+$ higher order terms all involving commutators. Since $\mathfrak{g}$ is abelian, these all vanish. $\qquad\square$

Corollary 1.4.21. *Any connected abelian Lie group is of the form $\mathbb{R}^j \times \mathbb{T}^k$.*

Proof. Since $\mathfrak{g}$ is abelian we can regard it as $\mathbb{R}^n$. Now the homomorphism, $\exp$ is locally one-to-one since it is one-to-one in a neighborhood of 0. Therefore, its kernel is discrete. Hence, its image, G, is $\mathbb{R}^j \times \mathbb{T}^k$, where $j + k = n$. $\qquad\square$

It is sometimes convenient to have an explicit description of the adjoint representation which we get in the next corollary.

Corollary 1.4.22. *If G is a linear group, then $\mathrm{Ad}\, g(X) = gXg^{-1}$, for $X \in \mathfrak{g}$.*

Proof.

$$\exp t(\mathrm{Ad}\, g(X)) = \exp \mathrm{Ad}\, g(tX) = \alpha_g(\exp tX)$$
$$= g(\exp tX)g^{-1} = \exp(gtXg^{-1}) = \exp t(gXg^{-1}).$$

Calculating the derivative at $t = 0$ gives the conclusion. $\qquad\square$

We now come to the question of what is Ad'?

Corollary 1.4.23. *In a connected Lie group $\mathrm{Ad}\exp X = \mathrm{Exp}(\mathrm{ad}\, X)$, $X \in \mathfrak{g}$.*

Proof. Since Ad is a smooth representation of G on $\mathfrak{g}$ its derivative Ad' is a Lie algebra representation of $\mathfrak{g}$ on $\mathfrak{g}$ making the appropriate diagram commutative. For $t \in k$ and $X \in \mathfrak{g}$ we know $\mathrm{Exp}[t\,\mathrm{Ad}'(X)] = \mathrm{Exp\,Ad}'(tX) = \mathrm{Ad\,exp}(tX)$. Let $Y \in \mathfrak{g}$. Then since $\mathrm{Ad}\,G$ is a linear Lie group $\mathrm{Exp}[t\,\mathrm{Ad}'(X)](Y) = \mathrm{Ad\,exp}(tX)(Y) = \exp(tX)(Y)\exp(-tX)$. By a fact proved earlier this last term is $(1 + tX + O(t^2))(Y)(1 - tX + O(t^2))$. But this is just $Y + t[X,Y] + O(t^2)$. On the other hand, $\mathrm{Exp}\,t\,\mathrm{Ad}'(X) = (1 + t\,\mathrm{Ad}'\,X + O(t^2))(Y)$. Taking derivatives at $t = 0$ yields $\mathrm{Ad}'(X)(Y) = [X,Y]$ for all Y. Thus $\mathrm{Ad}' = \mathrm{ad}$ and we have the commutativity relation $\mathrm{Exp}(\mathrm{ad}\,X) = \mathrm{Ad\,exp}(X)$, $X \in \mathfrak{g}$.   □

Corollary 1.4.24. *Let G be a connected Lie group and H be a connected Lie subgroup with Lie algebras $\mathfrak{g}$ and $\mathfrak{h}$. Then the following conditions are equivalent.*

(1) H is normal in G.

(2) $\mathfrak{h}$ is an ideal in $\mathfrak{g}$.

(3) $\mathfrak{h}$ is an Ad-invariant subspace of $\mathfrak{g}$.

Proof. Since $\mathfrak{h}$ is a subspace of $\mathfrak{g}$, we know from the above that $\mathfrak{h}$ is Ad-invariant if and only if it is ad-invariant, proving that (2) and (3) are equivalent. Suppose H is normal in G, and U is a canonical neighborhood of 1 in H and $B = \log U$ is a ball about 0 in $\mathfrak{h}$. For $g \in G$, $\exp\mathrm{Ad}\,g(B) = \alpha_g(U) \subseteq H$. Then $\exp\mathrm{Ad}\,g(tX) = \exp t\,\mathrm{Ad}\,g(X) \in H$ for all $X \in B$ and $|t| \leq 1$. Hence $\mathrm{Ad}\,g(B) \subseteq \mathfrak{h}$ and, by linearity of $\mathrm{Ad}\,g$ and the fact that $\mathfrak{h}$ is a vector space, $\mathrm{Ad}\,g(\mathfrak{h}) \subseteq \mathfrak{h}$. Conversely, if each $\mathrm{Ad}\,g$ preserves $\mathfrak{h}$ then reversing all these steps tells us that $\alpha_g(U) \subseteq H$ and since U generates H, $\alpha_g(H) \subseteq H$ so H is normal. This proves parts (1) and (2) are equivalent and completes the proof.   □

We now identify the Lie algebras of some other commonly encountered Lie subgroups. The following result shows, in particular, that a connected Lie group is solvable (respectively nilpotent) if and only if Lie algebra being solvable (respectively nilpotent). In the case of a semisimple (or reductive) Lie group we merely take as the definition the corresponding property of the Lie algebra.

However, in contrast to all the other theorems in this book, Theorem 1.4.25 requires the full statement of the BCH formula (1.9), not merely its second order approximation.

Theorem 1.4.25. *Let G be a connected Lie group and N a connected normal Lie subgroup, with respective Lie algebras $\mathfrak{g}$ and $\mathfrak{n}$. Then $[G, N]$ is a normal connected Lie subgroup whose Lie algebra is the ideal $[\mathfrak{g}, \mathfrak{n}]$.*

Proof. That $[G, N]$ is normal and $[\mathfrak{g}, \mathfrak{n}]$ is an ideal follow directly from the fact that N is normal and $\mathfrak{n}$ is an ideal, Corollary 1.4.24. We leave the verification of this to the reader. By Theorem 1.3.3, there is a connected Lie subgroup, H of G whose Lie algebra is $[\mathfrak{g}, \mathfrak{n}]$ which by Corollary 1.4.24 is normal. We next show $[G, N] \subseteq H$. Since G and H are both connected in order to show $[G, N] \subseteq H$ it is sufficient to prove that for a small ball B about 0 in $\mathfrak{n}$ that

$$[\exp X, \exp Y] \in H, \text{ for } X \in B \text{ and } Y \in B \cap \mathfrak{n}.$$

Choose the ball small enough so that the BCH formula (1.9) is valid in it. Then $[\exp X, \exp Y] = \exp([X, Y] + \eta(X, Y))$, where η is a convergent power series consisting of commutators of various orders in X and Y (such as $[X, [X, Y]]$ or $[X, [Y, [X, Y]]]$) with rational coefficients. In particular, since $\mathfrak{n}$ is an ideal, $[\mathfrak{g}, \mathfrak{n}]$ is closed, $\eta(X, Y)$ and $[X, Y] \in [\mathfrak{g}, \mathfrak{n}]$. Hence $[\exp X, \exp Y] \in H$. This means $[G, N] \subseteq H$.

Now choose $X_1, \ldots, X_p$ from $\mathfrak{g}$ and $Y_1, \ldots, Y_p$ from $\mathfrak{n}$ so the $[X_i, Y_i]$ are a basis for $[\mathfrak{g}, \mathfrak{n}]$. Modify the X_i by scalar multiplication to insure that they also lie in B. Let $\phi_i(t) = [\exp X_i, \exp tY_i]$, for $i = 1, \ldots, p$, $\phi_i : \mathbb{R} \rightarrow [G, N]$. Let $\phi(t_1, \ldots, t_p) = \phi_1(t_1) \ldots \phi_p(t_p)$ then ϕ is a smooth $\mathbb{R}^p \rightarrow [G, N]$. We can identify $\mathbb{R}^p$ with $[\mathfrak{g}, \mathfrak{n}]$ via the basis $[X_1, Y_1], \ldots, [X_p, Y_p]$. Choose $t_1, \ldots, t_p$ small enough so that the BCH formula applies. Now $\log(\phi) : B_1 \rightarrow \mathbb{R}^p$, where B_1 is a small ball about 0 in $[\mathfrak{g}, \mathfrak{n}]$ on which exp is invertible. Then, as above, for each i, $\log[\exp X_i, \exp tY_i] = [X_i, tY_i] + \eta(X_i, tY_i)$ has a nonzero derivative at $t = 0$. Hence $d(\log(\phi))(0)$ is invertible. By the inverse function theorem there is a ball B_2 about 0 in $\mathbb{R}^p$ so that $\log(\phi)(V)$ contains a neighborhood of 0 in $[\mathfrak{g}, \mathfrak{n}]$. Hence $\phi(B)$ contains a neighborhood of 1 in H. This neighborhood is in $[G, N]$ since ϕ takes values there. Because H is connected and therefore the neighborhood generates H we get $H \subseteq [G, N]$. Therefore $H = [G, N]$ so the Lie algebra of $[G, N]$ is $[\mathfrak{g}, \mathfrak{n}]$. □

Proposition 1.4.26. *Over $\mathbb{C}$, $\mathrm{Aut}(\mathfrak{g})$ is a complex algebraic group, while over $\mathbb{R}$, $\mathrm{Aut}(\mathfrak{g})$ is the real points of an algebraic group over $\mathbb{R}$. In any case, $\mathrm{Aut}(\mathfrak{g})$ is closed in $\mathrm{GL}(\mathfrak{g})$ and hence is either a complex Lie group, or a real Lie group, respectively.*

Proof. Let $X_1, \ldots, X_n$ be a basis for $\mathfrak{g}$ and $[X_i, X_j] = \sum_k c_{ij}^k X_k$ be structure constants for this basis. For an automorphism, α, $\alpha(X_k) = \sum_p \alpha_{kp} X_p$, where α_{kp} are the matrix coefficients of α relative to this basis. Thus

$$\alpha([X_i, X_j]) = \sum_k c_{ij}^k \alpha(X_k) = \sum_k c_{ij}^k \sum_p \alpha_{kp} X_p = \sum_p (\sum_k c_{ij}^k \alpha_{kp}) X_p.$$

These relations are determinative. On the other hand, $\alpha([X_i, X_j]) = [\alpha(X_i), \alpha(X_j)]$. Applying the same reasoning to $\alpha(X_i)$ and $\alpha(X_j)$ and then using the linearity of the bracket and equating coefficients of the basis gives a finite number of equations which on one side are (quadratic) polynomials in the matrix coefficients of α and on the other are linear. Thus $\alpha \in \mathrm{Aut}(\mathfrak{g})$ if and only if its coefficients satisfy this (finite) set of polynomial equations. □

Definition 1.4.27. A map $D : \mathfrak{g} \to \mathfrak{g}$ is called a *derivation* of $\mathfrak{g}$ if D is linear and for all $X, Y \in \mathfrak{g}$, $D[X, Y] = [D(X), Y] + [X, D(Y)]$. We denote by $\mathrm{Der}(\mathfrak{g})$ the derivations of $\mathfrak{g}$. Evidently $\mathrm{Der}(\mathfrak{g})$ is a subspace of the vector space $\mathrm{End}(\mathfrak{g})$. Under taking the commutator of operators, $\mathrm{Der}(\mathfrak{g})$ is actually a Lie subalgebra of $\mathrm{End}(\mathfrak{g})$.

For example $\mathrm{ad}\, X$ is a derivation for each $X \in \mathfrak{g}$. Such a derivation is called an *inner derivation*. It is not difficult to see that $\mathrm{Der}(\mathfrak{g})$ is a subalgebra of $\mathfrak{gl}(\mathfrak{g})$ and that $\mathrm{ad}\, \mathfrak{g}$ is a subalgebra of it, called the algebra of *inner derivations*. In fact $\mathrm{ad}\, \mathfrak{g} \subset \mathrm{Der}(\mathfrak{g})$ is an ideal: if $D \in \mathrm{Der}(\mathfrak{g})$ and $X \in \mathfrak{g}$ we get

$$[D, \mathrm{ad}\, X] = \mathrm{ad}\, D(X)$$

If $\mathfrak{g}$ is an abelian Lie algebra then any endomorphism of $\mathfrak{g}$ is a derivation.

Exercise 1.4.28. Let D be a derivation $\mathfrak{g}$ and n a positive integer. Then,

$$D^n[X,Y] = \sum_{i=0}^{n} \binom{n}{i} [D^i X, D^{n-i} Y].$$

Corollary 1.2.4 enables us to see the relationship of automorphisms and derivations.

Theorem 1.4.29. *Let $\mathfrak{g}$ be any real or complex Lie algebra. Then the Lie algebra of* $\mathrm{Aut}(\mathfrak{g})$ *is* $\mathrm{Der}(\mathfrak{g})$.

Proof. Let $D \in \mathrm{Der}(\mathfrak{g})$. As we saw $\mathrm{Exp}\, D \in \mathrm{GL}(\mathfrak{g})$. If $X, Y \in \mathfrak{g}$, then $D[X,Y] = [D(X),Y] + [X,D(Y)]$. It follows from the binomial theorem $D^n[X,Y] = \sum_{i=0}^{n} \frac{n!}{i!(n-i)!}[D^i(X), D^{n-i}Y]$, for all n.
Hence

$$\mathrm{Exp}\, D[X,Y] = \sum_{n=0}^{\infty} \frac{1}{n!} \sum_{i=0}^{n} \frac{n!}{i!(n-i)!}[D^i(X), D^{n-i}Y].$$

Arguing exactly as in the proof of $\mathrm{Exp}(A+B) = \mathrm{Exp}\,A\,\mathrm{Exp}\,B$ when A and B commute we see that $\mathrm{Exp}\, D[X,Y] = [\mathrm{Exp}\, D(X), \mathrm{Exp}\, D(Y)]$ so $\mathrm{Exp}\, D \in \mathrm{Aut}(\mathfrak{g})$.
   Now suppose $T \in \mathrm{End}(\mathfrak{g})$ and $\mathrm{Exp}\, tT \in \mathrm{Aut}(\mathfrak{g})$ for all t. Then we show $T \in \mathrm{Der}(\mathfrak{g})$. Since $\mathrm{Exp}\, tT[X,Y] = [\mathrm{Exp}\, tT(X), \mathrm{Exp}\, tT(Y)]$ for all t we can differentiate both sides of the equation with respect to t at $t = 0$ and get $T[X,Y] = [T(X),Y] + [X,T(Y)]$. Thus T is a derivation. $\square$

Remark 1.4.30. We remark that the Lie algebra of Haar measure preserving automorphisms of a group consists of the derivations of the Lie algebra of trace 0 (see [47]).

1.5   The Topology of Compact Classical Groups

Here we will determine a number of global topological properties of important *compact* Lie groups. In Chapter 6 we shall see how these global topological properties propagate to non-compact groups.

Let G be a connected Lie group and H a closed subgroup. We shall show that the smooth map $\pi : G \to G/H$ always has a smooth local cross-section which implies that $\pi : G \to G/H$ is a *fibration* [76].

Let $\mathfrak{g}$ be the Lie algebra of G. Since H is a closed subgroup it is also a Lie group. Let $\mathfrak{h}$ be its Lie algebra and choose a vector space complement W to $\mathfrak{h}$ in $\mathfrak{g}$. Then there is a *local* diffeomorphism $\phi : W \to G/H$ such that $\pi \exp = \phi d\pi$, where exp is the exponential map of G and $d\pi$ is the differential of π at the identity. Also by the choice of W, $d\pi$ has a global cross section on W, namely i, the injection of W into $\mathfrak{g}$.

Let $\sigma = (\exp i)\phi^{-1}$. Then σ is a smooth map locally defined in a neighborhood of the coset H in G/H and $\pi\sigma = \pi(\exp i)\phi^{-1}$. But since $\pi \exp = \phi d\pi$, we have $\pi(\exp i) = \phi$ so that $\pi\sigma = \pi(\exp i)\phi^{-1} = I$.

A fibration gives rise to a *long exact* sequence of homotopy groups (see [76])

$$\ldots \pi_2(H) \to \pi_2(G) \to \pi_2(G/H) \to \pi_1(H) \to \pi_1(G) \to$$
$$\to \pi_1(G/H) \to \pi_0(H) \to \pi_0(G) \to \pi_0(G/H) \to 0$$

From this we can draw a number of important conclusions. For example, by looking at the exactness of $\pi_0(H) \to \pi_0(G) \to \pi_0(G/H)$ we see that if H and G/H are both connected then so is G.

We will prove this directly so that the reader will be confident that this is true. Let U and V be an open partition of G by non-empty sets. Then since π is open and surjective, $\pi(U)$ and $\pi(V)$ are open non-empty sets whose union is G/H. But since G/H is connected these must intersect at least in some coset gH. Thus there is a $u \in U$ and a $v \in V$ which are each congruent to $g \bmod H$. But then $U \cap gH$ and $V \cap gH$ are non-empty and so give an open partition of gH. Since gH is homeomorphic to H, it is also connected so this is impossible.

Corollary 1.5.1. *For all $n \geq 1$, $\mathrm{SO}(n,\mathbb{R})$, $\mathrm{U}(n,\mathbb{C})$ and $\mathrm{SU}(n,\mathbb{C})$ are connected. $\mathrm{O}(n,\mathbb{R})$ has 2 components.*

This follows by induction from the fact that the spheres are connected and $\mathrm{SO}(1,\mathbb{R})$, $\mathrm{U}(1,\mathbb{C})$ and $\mathrm{SU}(1,\mathbb{C})$ are connected. Since $\mathrm{SO}(n,\mathbb{R})$ is connected and det $: \mathrm{O}(n,\mathbb{R}) \to \pm 1$ is a surjective and a continuous map, $\mathrm{O}(n,\mathbb{R})$ has 2 components.

Now let us look at other parts of the *long exact sequence*. Consider $\pi_1(H) \to \pi_1(G) \to \pi_1(G/H)$. This tells us that if H and G/H are both simply connected then so is G.

Corollary 1.5.2. *For all $n \geq 1$, $\mathrm{SU}(n, \mathbb{C})$ is simply connected.*

To see this we must know that S^n is simply connected for $n \geq 3$. In fact this is so for $n \geq 2$ and follows from the Van Kampen theorem (see [40]).

The exactness of $\pi_1(G) \to \pi_1(G/H) \to \pi_0(H) \to \pi_0(G)$ tells us that if G is connected and simply connected then $\pi_1(G/H) = H/H_0$. In particular, if H is also connected then G/H is simply connected. If Γ is a discrete subgroup, then $\pi_1(G/\Gamma) = \Gamma$. For example, this latter fact tells us that if G is a simply connected Lie group and Γ is a discrete central subgroup then the fundamental group of the quotient is Γ. Thus if $(\tilde{G}, \pi)$ is the universal covering group of G then $\pi_1(G) = \mathrm{Ker}\,\pi$.

The exactness of $\pi_2(G/H) \to \pi_1(H) \to \pi_1(G) \to \pi_1(G/H)$ tells us that if $\pi_2(G/H) = \{1\}$ and $\pi_1(G/H) = \{1\}$ then $\pi_1(H)$ and $\pi_1(G)$ are isomorphic.

Corollary 1.5.3. *For $n \geq 1$, $\pi_1(\mathrm{U}(n, \mathbb{C})) = \mathbb{Z}$ and for $n \geq 3$, $\pi_1(\mathrm{SO}(n, \mathbb{R})) = \mathbb{Z}_2$, $\pi_1(\mathrm{SO}(2, \mathbb{R})) = \mathbb{Z}$ (since $\mathrm{U}(1, \mathbb{C}) = \mathrm{SO}(2, \mathbb{R})$).*

To prove this we need only show in addition to what we know already that $\pi_1(\mathrm{U}(1, \mathbb{C})) = \mathbb{Z}$, $\pi_1(\mathrm{SO}(3, \mathbb{R})) = \mathbb{Z}_2$ and $\pi_2(S^n) = \{1\}$ for $n \geq 3$. For this last fact see [76]. Concerning the first, since the universal covering $\mathbb{R} \to \mathrm{U}(1, \mathbb{C})$ is $t \mapsto e^{2\pi i t}$ and $\mathbb{R}$ is simply connected then as above $\pi_1(\mathrm{U}(1, \mathbb{C})) = \mathbb{Z}$. The second fact will follow in a similar way by constructing the universal covering group of $\mathrm{SO}(3, \mathbb{R})$ below.

We shall use this last principle to calculate both $\pi_1(\mathrm{SO}(3, \mathbb{R}))$ and $\pi_1(\mathrm{SO}(4, \mathbb{R}))$. Consider the quaternions $\mathbb{H}$ and the nonzero quaternions, $\mathbb{H}^\times$. Given a quaternion

$$q = a_0 + a_1 i + a_2 j + a_3 k,$$

we define its conjugate,

$$\bar{q} = a_0 - (a_1 i + a_2 j + a_3 k)$$

and its norm,

$$N(q) = a_0^2 + a_1^2 + a_2^2 + a_3^2.$$

Then clearly $N(q) = q\bar{q}$ and the real number $N(q) \geq 0$ and $= 0$ if and only if $q = 0$. From this we see immediately that each nonzero quaternion q is invertible with $q^{-1} = \dfrac{\bar{q}}{N(q)}$. Thus $\mathbb{H}^\times$ is a group. It is easy to see from the formulas for multiplication and inversion that $\mathbb{H}^\times$ is in fact a Lie group. Also $^-$ is an anti-automorphism of this group; $\bar{q}r = \bar{r}\bar{q}$ and from this it follows that N is a homomorphism of this group to the multiplicative group $\mathbb{R}^\times$; $N(qr) = N(q)N(r)$. Let G denote the elements of unit norm in $\mathbb{H}$. Then G is a subgroup of $\mathbb{H}^\times$ which topologically is the 3-sphere S^3. In particular, G is a compact connected and simply connected Lie group which is noncommutative. Incidentally, like the 1-sphere this also shows that the 3-sphere carries a Lie group structure. This is not so, for example, of the 2-sphere.

Now we define the left, right and two-sided regular representations respectively of the $\mathbb{R}$-algebra $\mathbb{H}$ as follows: $L_q(x) = qx$, $R_q(x) = xq^{-1}$ and $T_{(q,r)}(x) = qxr^{-1}$. It is easy to see that each of these is an $\mathbb{R}$-linear representation $\mathbb{H}^\times$ or $\mathbb{H}^\times \times \mathbb{H}^\times$ on $\mathbb{H}$. Restricting the two-sided regular representation to $G \times G$ we get a smooth homomorphism of the latter to $O(4, \mathbb{R})$ and by connectedness to $SO(4, \mathbb{R})$. Since $G \times G$ and $SO(4, \mathbb{R})$ are both connected and have dimension 6 and the kernel of this map is $\{\pm(1,1)\}$ which is finite we see that this map is onto and a covering. Since $G \times G$ is simply connected (see [40]) it follows that $\pi_1(SO(4, \mathbb{R})) = \mathbb{Z}_2$. Now further restrict this representation to the diagonal subgroup of $G \times G$. This gives a representation of G on $\mathbb{H}$ (by conjugation) which leaves the center fixed. Since G is compact it must leave the orthocomplement stable and preserve the norm. Thus we have a smooth homomorphism $G \to O(3, \mathbb{R})$. As above it takes values in $SO(3, \mathbb{R})$ and its kernel is $\{\pm(1,1)\}$. Since G and $SO(3, \mathbb{R})$ are both connected and have dimension 3 and the kernel of this map is finite the map is onto and a covering. Because G is simply connected it follows as above that $\pi_1(SO(3, \mathbb{R}))$ is also $\mathbb{Z}_2$. Thus we have constructed the universal covering group $G \to SO(3, \mathbb{R})$ and $G \times G \to SO(4, \mathbb{R})$. Finally, taking the differential of the first of these shows that the Lie algebra

$\mathfrak{su}(2, \mathbb{C})$ of G is isomorphic with $\mathfrak{so}(3, \mathbb{R})$. Taking the differential of the second yields an important decomposition of the Lie algebra of $\mathfrak{so}(4, \mathbb{R})$ as the direct sum of ideals isomorphic with $\mathfrak{so}(3, \mathbb{R}) \oplus \mathfrak{so}(3, \mathbb{R})$.

We note that the fundamental group of these compact semisimple groups seem to be finite, while that of the non-semisimple ones are finitely generated abelian, but infinite. We shall see later that this is not an accident.

We now turn to the identification of the group G. First observe that $\mathbb{H}$ contains the field $\mathbb{C}$ in the form $\{a_0 + a_1\sqrt{-1}\}$ and so is a vector space over $\mathbb{C}$ and since $\mathbb{H}$ is a 4-dimensional algebra over $\mathbb{R}$ it has dimension 2 over $\mathbb{C}$. To be specific we shall take the scalar multiplication by $\mathbb{C}$ on the right. The associative and distributive laws of $\mathbb{H}$ tell us that, in this manner, $\mathbb{H}$ is a vector space over $\mathbb{C}$. Here any $q = a_0 + a_1 i + a_2 j + a_3 k \in \mathbb{H}$ can be written as follows:

$$q = 1(a_0 + a_1\sqrt{-1}) + j(a_2 - a_3\sqrt{-1}).$$

In this way $\{1, j\}$ are a basis for $\mathbb{H}$ over $\mathbb{C}$. Now consider the left regular representation L of the $\mathbb{R}$-algebra $\mathbb{H}$. This representation is faithful. For if $L_q = 0$, then $qr = 0$ for every r. But, if $q \neq 0$ taking $r = q^{-1}$ would give a contradiction. Hence $q = 0$ and so L is faithful. It is actually a $\mathbb{C}$-representation since $L_q(xc) = L_q(x)c$, $c \in \mathbb{C}$ by associativity. Thus each T_q can be represented by a 2×2 complex matrix whose coefficients are determined by $T_q(1)$ and $T_q(j)$ as follows:

$$T_q(1) = 1z_{11}(q) + jz_{21}(q)$$

and

$$T_q(j) = 1z_{12}(q) + jz_{22}(q).$$

Now since $q = 1(a_0 + a_1 i) + j(a_2 - a_3 i)$, $T_q(1) = q1 = q = 1z_{11}(q) + jz_{21}(q)$. So

$$z_{11}(q) = a_0 + a_1 i \text{ and } z_{21} = a_2 - a_3 i,$$

while

$$T_q(j) = qj = (a_0 + a_1 i)j + j(a_2 - a_3 i)j = a_0 j + a_1 k - a_2 - a_3 i.$$

Since $T_q(j) = -a_2 - a_3 i + a_0 j + a_1 k$ we see that $z_{12}(q) = -a_2 - a_3 i$ and $z_{22}(q) = a_0 - a_1 i$. Denoting $a_0 + a_1 i$ by a and $a_2 - a_3 i$ by b, the matrix of L_q with respect to this basis is

$$L_q = \begin{pmatrix} a & -\bar{b} \\ b & \bar{a} \end{pmatrix}$$

and we have a faithful matrix representation of H. In particular, $\det L_q = N(q)$. This gives an independent proof of the fact that things with nonzero norm are invertible and $N(qr) = N(q)N(r)$. Also $q \in G$ if and only if $|a|^2 + |b|^2 = 1$. So G is isomorphic to this group of 2×2 complex matrices.

Finally, we identify the latter. The set G of these matrices

$$g = \begin{pmatrix} a & -\bar{b} \\ b & \bar{a} \end{pmatrix}$$

clearly forms a subgroup of $SU(2, \mathbb{C})$ which is homeomorphic to the 3 sphere and in particular is connected and compact and hence closed. Thus G is a connected Lie subgroup of $SU(2, \mathbb{C})$. Since the latter is also connected and both these groups have dimension 3 they coincide and $G = SU(2, \mathbb{C})$.

Turning to $Sp(n)$, similar arguments show that $Sp(1) = SU(2, \mathbb{C})$ and $Sp(n)/Sp(n-1) = S^{4n-1}$, $n \geq 2$. Hence $Sp(n)$ is a compact connected and simply connected Lie group.

We conclude this section with the calculation of some covering groups of non-compact groups. Consider $SO(1,2)_0$ and $SO(1,3)_0$, the connected components of the group of isometries of hyperbolic 2 and 3 space, H^2 and H^3. That is, we consider the forms $q_{12}(X) = x^2 - y^2 - z^2$ on $\mathbb{R}^3$ and $q_{13}(X) = x^2 - y^2 - z^2 - t^2$ on $\mathbb{R}^4$ and the corresponding groups of isometries.

We let $SL(2, \mathbb{R})$ act on the space of 2×2 real symmetric matrices S by $\rho_g(S) = gSg^t$. Then S is a real vector space of dimension 3 and ρ is a continuous real linear representation of $SL(2, \mathbb{R})$. Since $\det g = 1$, $\det(\rho_g(S)) = \det(gSg^t) = \det S$. Now if

$$S = \begin{pmatrix} a & z \\ z & b \end{pmatrix},$$

then $\det S = ab - z^2$. Consider the change of variables with $x = \frac{a+b}{2}$ and $y = \frac{a-b}{2}$. This is an $\mathbb{R}$ *linear change of variables* and $a = x+y$ and $b = x-y$. Therefore, $\det S = x^2 - y^2 - z^2$. Thus ρ preserves a $(1,2)$ form on $\mathbb{R}^3$. Since $SL(2,\mathbb{R})$ is connected so is its image $\rho(G)$ in $GL(3,\mathbb{R})$. A direct calculation shows $\mathrm{Ker}\,\rho = \pm I$. Since this is finite and therefore discrete $\rho(G)$ has dimension 3 just as $SL(2,\mathbb{R})$. But $\rho(G) \subseteq SO(1,2)_0$. Since this connected group also has dimension 3 ρ is onto and therefore a covering. It induces an isomorphism between $SO(1,2)_0 = SO(1,2)$ and $SL(2,\mathbb{R})/\{\pm I\} = PSL(2,\mathbb{R})$.

Similarly, let $SL(2,\mathbb{C})$ act on the space of 2×2 complex Hermitian matrices $\mathcal{H}$ by $\rho_g(H) = gHg^*$. Then $\mathcal{H}$ is a real vector space of dimension 4 and ρ is also a continuous real linear representation of $SL(2,\mathbb{C})$. Since $\det g = 1$, $\det(\rho_g(H)) = \det(gHg^*) = \det H$. Now if

$$H = \begin{pmatrix} a & z \\ \bar{z} & b \end{pmatrix},$$

then $\det H = ab - |z|^2$. Consider the change of variables $x = \frac{a+b}{2}$ and $y = \frac{a-b}{2}$. Then this is an $\mathbb{R}$ linear change of variables, $a = x+y$ and $b = x - y$. Therefore, $\det H = x^2 - y^2 - |z|^2 = x^2 - y^2 - u^2 - v^2$. Thus ρ preserves a $(1,3)$ form on $\mathbb{R}^4$. Since $SL(2,\mathbb{C})$ is connected so is its image $\rho(G)$ in $GL(4,\mathbb{R})$. A direct calculation shows $\mathrm{Ker}\,\rho = \pm I$. Since this is finite and therefore discrete $\rho(G)$ has dimension 6 just as $SL(2,\mathbb{C})$ does. But $\rho(G) \subseteq SO(1,3)_0$. Since this connected group also has dimension 6 ρ is onto and therefore a covering. It induces an isomorphism between $SO(1,3)_0$ and $SL(2,\mathbb{C})/\{\pm I\} = PSL(2,\mathbb{C})$. Since as we shall see $SL(2,\mathbb{C})$ is simply connected (see Corollary 6.3.7), in fact it is the universal cover of $SO(1,3)_0$.

Finally we apply the same method to $SU(2,\mathbb{C})$ to get another way of calculating the universal cover of $SO(3,\mathbb{R})$.

Let $SU(2,\mathbb{C})$ act on $\mathcal{H}$, the space of 2×2 complex skew-Hermitian matrices of trace 0 by $\rho_g(H) = gHg^* = gHg^{-1}$. Then $\mathcal{H}$ is a real vector space of dimension 3 and ρ is also a continuous real linear representation of $SU(2,\mathbb{C})$. Then $\det(\rho_g(H)) = \det(gHg^{-1}) = \det H$. Now if

$$H = \begin{pmatrix} ia & z \\ -\bar{z} & -ia \end{pmatrix},$$

We see that $\det H = -a^2 - |z|^2$. Thus ρ preserves a negative definite form and therefore also a positive definite form on $\mathbb{R}^3$. Thus $\rho(G) \subseteq$ O$(3, \mathbb{R})$. Since SU$(2, \mathbb{C})$ is connected so is its image $\rho(G)$. A direct calculation shows $\operatorname{Ker} \rho = \pm I$. Since this is finite and therefore discrete $\rho(G)$ has dimension 3 just as SU$(2, \mathbb{C})$ does. But $\rho(G) \subseteq$ SO$(3, \mathbb{R})$. Since this connected group also has dimension 3, ρ is onto and therefore a covering. It induces an isomorphism between SO(3) and SU$(2, \mathbb{C})/\{\pm I\} = Ad(\text{SU}(2, \mathbb{C}))$. Since SU$(2, \mathbb{C})$ is simply connected it is the universal cover of SO$(3, \mathbb{R})$.

Exercise 1.5.4. Prove that the group SO$(1, 1)_0$ is $\{g : g = \begin{pmatrix} \cosh t & \sinh t \\ \sinh t & \cosh t \end{pmatrix}\}$, where $t \in \mathbb{R}$, and therefore SO$(1, 1)_0$ is isomorphic with $\mathbb{R}$ and is simply connected.

1.6   The Iwasawa Decompositions for GL$(n, \mathbb{R})$ and GL$(n, \mathbb{C})$

Here we shall prove that the manifold, GL$(n, \mathbb{R})$, is the direct product of three submanifolds K, A_0 and N, where each of these is actually a closed subgroup. $K =$ O$(n, \mathbb{R})$, the orthogonal group, A_0 is the diagonal matrix with positive entries, and N is the subgroup of unitriangular matrices. Note that A_0 is the identity component of A, the group of diagonal matrices with nonzero entries on the diagonal. Moreover, the diffeomorphism of $K \times A_0 \times N$ with GL$(n, \mathbb{R})$ is given by multiplication.

Similarly, for GL$(n, \mathbb{C})$, we get a diffeomorphism of $K \times A_0 \times N$ with GL$(n, \mathbb{C})$ given by multiplication. Here K is the unitary group, U$_n(\mathbb{C})$, A_0 is as before, and N is the unitriangular matrices on GL$(n, \mathbb{C})$.

Exercise 1.6.1. Show that:

(1) Over $\mathbb{R}$, A has 2^n components, while over $\mathbb{C}$, A is connected.

(2) O$_n(\mathbb{R})$ has two components while U$_n(\mathbb{C})$ is connected.

(3) In both cases, over $\mathbb{R}$ and $\mathbb{C}$, N is connected.

(4) In either case $A_0 N$ is diffeomorphic to Euclidean space and determine the dimension.

Proposition 1.6.2. $G = KA_0N$ *where the diffeomorphism is given by multiplication.*

Before proving these facts we mention that they can be used to tell much about the topology of the non-compact groups $\mathrm{GL}(n, \mathbb{R})$ and $\mathrm{GL}(n, \mathbb{C})$, if one knows something about the compact group, K, because in both cases K is a deformation retract of G. In Section 1.5 we dealt with the topology of compact Lie groups.

Another consequence is that in either case A_0N is a subgroup (since A_0 normalizes N). It consists of the triangular matrices with positive diagonal entries. Since $G = KA_0N$ it is also true that $G = KB$ where B is the group of all triangular matrices in G. Since $G = KB$ the second isomorphism theorem tells us that $G/B = K/K \cap B$ and in particular that G/B is compact.

Proof. Our proof shall deal with both the real and complex cases simultaneously. We let V stand for either $\mathbb{R}^n$ or $\mathbb{C}^n$. Let $\{e_i : i = 1, \ldots, n\}$ be the standard basis of V and $g \in G$. Then $v_i = g^{-1}e_i$, $i = 1, \ldots, n$ is also a basis of V. We apply the Gram-Schmidt orthogonalization process to $\{v_i : i = 1, \ldots, n\}$. Let $u_1 = v_1/\|v_1\|$ and for $i = 2, \ldots, n$,

$$u_i = \frac{v_i - \sum_{j=2}^{i-1}(v_i, u_j)u_j}{\|v_i - \sum_{j=2}^{i-1}(v_i, u_j)u_j\|}.$$

Then the u_i's form an orthonormal basis of V, depending smoothly on $g \in G$, and by the formulas above, one can write

$$u_i(g) = \sum_{j \leq i} a_{ji}(g)v_i(g),$$

where $a_{ii} > 0$. Let $a(g)$ be the diagonal matrix with entries a_{ii} and $n(g) = a(g)^{-1}(a_{ji}(g))$. Then a and n depend smoothly on g, $a(g) \in A_0$ and $n(g) \in N$, all $g \in G$. Also, for all $g \in G$ and $i = 1, \ldots, n$, $a(g)n(g)v_i = (a_{ji}(g))(v_i(g)) = u_i(g)$. Now since $\{e_i : i = 1, \ldots, n\}$ and $\{u_i : i = 1, \ldots, n\}$ are both orthonormal basis there exists a *unique* $k(g) \in K$ so that $k(g)(u_i(g)) = e_i$, for all i. k also depends smoothly on g. Moreover, $k(g)a(g)n(g)(v_i) = k(g)u_i(g) = e_i$ for all i. But also $g(v_i) = e_i$. Since v_i's form a basis we get $g = k(g)a(g)n(g)$. $\quad\square$

We remark that similar reasoning applies to $\mathrm{SL}(n, \mathbb{R})$. Just replace $\mathrm{O}_n(\mathbb{R})$ by $\mathrm{SO}_n(\mathbb{R})$ and A_0 by $\{a \in A_0 : \det a = 1\}$. Similarly, for $\mathrm{SL}(n, \mathbb{C})$. Just replace $\mathrm{U}_n(\mathbb{C})$ by $\mathrm{SU}_n(\mathbb{C})$ and A_0 by $\{a \in A_0 : \det a = 1\}$.

We also get corresponding decompositions of the respective Lie algebras, $\mathfrak{g} = \mathfrak{gl}(n, \mathbb{R})$ or $\mathfrak{gl}(n, \mathbb{C})$. We let $\mathfrak{a}$ denote the diagonal elements of either one, $\mathfrak{n}$ denote the strictly triangular elements of either one and $\mathfrak{k}$ denote the skew symmetric elements in the case of $\mathbb{R}$ and the skew Hermitian symmetric elements in the case of $\mathbb{C}$.

As the reader can easily check these are always *real* subspaces of $\mathfrak{g}$. Note, however, that in the complex case, $\mathfrak{k}$ is not a complex subspace of $\mathfrak{g}$. It follows immediately from the previous result that,

Corollary 1.6.3. *For $\mathfrak{g} = \mathfrak{gl}(n, \mathbb{R})$ or $\mathfrak{gl}(n, \mathbb{C})$ we have $\mathfrak{g} = \mathfrak{k} \oplus \mathfrak{a} \oplus \mathfrak{n}$.*

1.7 The Baker-Campbell-Hausdorff Formula

In order to prove the Baker-Campbell-Hausdorff formula, as well as for other purposes, we first calculate the derivative of the exponential map.

Let G be a connected Lie group, which for convenience we shall assume to be *linear* and $\mathfrak{g} = T_1(G)$ its Lie algebra. We now calculate the derivative, $d_X \exp$, of the exponential map at a point $X \in \mathfrak{g}$. Since $d_0 \exp = I$ and in particular is nonsingular it follows by continuity that for small X, $d_X \exp$ is also invertible. We can do much better than this with an explicit formula for $d_X \exp$. This will tell us how near zero we have to be for $d_X \exp$ to be invertible and will be important in its own right. Under this identification, if $f : G \to H$ is a smooth homomorphism, $X \in \mathfrak{g}$, and $f(\exp_G(tX)) = \exp_H(tf'(X))$ is the corresponding 1-parameter subgroup of H, then

$$d_1 f(X) = \frac{d}{dt} \exp_H(tf'(X))|_{t=0} = f'(X).$$

So $d_1 f = f'$. We shall also make the further convention that we identify the tangent space $T_g(G)$ of a point g of G with $\mathfrak{g}$ by applying the left translation by g^{-1} to $T_g(G)$. This will enable us to normalize the situation and for every $X \in \mathfrak{g}$ view $d_X \exp$ as an operator on $\mathfrak{g}$.

Before beginning the calculation of $d_X \exp$ a word must be said about functional calculus. Let V be a finite dimensional complex vector space and $\mathfrak{gl}(V)$ be its endomorphism algebra. For each complex analytic function $f(z) = \sum_{k=0}^{\infty} a_k z^k$ on some disk D about 0 in $\mathbb{C}$ and linear operator L on V. We may assume $\sum_{k=0}^{\infty} a_k z^k$ absolutely convergent by taking the radius of convergence smaller. We may form, for any L with $\|L\| <$ radius of D, $f(L) = \sum_{k=0}^{\infty} a_k L^k$. Each such $f(L)$ is a linear operator on V and the series $\sum_{k=0}^{\infty} a_k L^k$ is absolutely convergent. The resulting map $(f, L) \mapsto f(L)$ is called the operational calculus. By looking at the Jordan triangular form of such an L and taking into account the fact that $Pf(L)P^{-1} = f(PLP^{-1})$, we see easily that

$$\operatorname{Spec} f(L) = \{f(\lambda) : \lambda \in \operatorname{Spec} L\}.$$

Now, for fixed L, the map $f \mapsto f(L)$ is clearly an algebra homomorphism from the holomorphic functions about 0 to $\mathfrak{gl}(V)$. In particular, if f is holomorphic and $f(0) \neq 0$ then $1/f$ is also holomorphic and

$$(1/f)(L) = f(L)^{-1}.$$

For example, if f were exp or log then we have already applied this functional calculus to study Exp and Log. Now let $\phi(z) = (e^z - 1)/z = 1 + z/2! + z^2/3! + \dots$. Then ϕ is an entire function with a removable singularity at $z = 0$, $\phi(0) = 1$.

Theorem 1.7.1. *For each $X \in \mathfrak{g}$, $d_X \exp = \phi(-\operatorname{ad} X)$.*

This formula will be important in proving the Baker-Campbell-Hausdorff formula. We shall also use it in studying the geometry of the symmetric spaces associated with certain Lie groups. As a corollary we have

Corollary 1.7.2. *$d_X \exp$ is nonsingular if and only if $\operatorname{ad} X$ has no eigenvalues of the form $2\pi i n$ for some nonzero integer n.*

Proof. $d_X \exp$ is nonsingular if and only if $\phi(\lambda) \neq 0$ for all $\lambda \in \operatorname{Spec}(-\operatorname{ad} X) = -\operatorname{Spec}(\operatorname{ad} X)$. Since $\phi(0) = 1$, $\phi(z) = 0$ if and only if $e^z = 1$, $z \neq 0$ if and only if $z = 2\pi i n$ for some nonzero integer n. $\square$

We need the following lemma.

Lemma 1.7.3. *For X and $Y \in \mathfrak{g} = \mathfrak{gl}(V)$ and $t \in \mathbb{R}$,*

$$\frac{d}{dt}\exp(X + tY)|_{t=0} = Y + 1/2!(XY + YX)+$$
$$1/3!(X^2Y + XYX + YX^2) + \dots$$

The terms of this convergent series are in $\mathfrak{gl}(V)$ in our case, or in the universal enveloping algebra in general.

Proof. Write $\exp(X + tY) = c + t(\dots) + O(t^2)$ where $(\dots) = Y + 1/2!(XY+YX)+1/3!(X^2Y+XYX+YX^2)+\dots$ where c is a cosntant. Differentiating and evaluating at $t = 0$ gives the result.   □

Lemma 1.7.4. *Let X and Y be fixed in $\mathfrak{g}$. Then*

$$\exp(-X) \circ \frac{d}{dt}(\exp(X + tY))|_{t=0} = \phi(-\operatorname{ad}X)(Y).$$

Proof. In particular, from the previous lemma

$$\frac{\partial}{\partial t}\exp s(X + tY)_{t=0} = sY + s^2/2!(XY + YX)+$$
$$s^3/3!(X^2Y + XYX + YX^2) + \dots.$$

For $s \in \mathbb{R}$, let $\Phi(s) = \exp(-sX) \circ \frac{\partial}{\partial t}(\exp s(X + tY))|_{t=0}$. Then Φ is analytic and $\Phi(0) = 0$. We shall show that Φ satisfies the global linear nonhomogeneous differential equation with constant coefficients $\frac{d\Phi}{ds} = -[X, \Phi(s)] + Y$, for $s \in \mathbb{R}$. Now

$$\Phi'(s) = \exp(-sX)(Y + s(XY + YX)$$
$$+ s^2/2!(X^2Y + XYX + YX^2) + \dots)$$
$$+ \exp(-sX)(-X)(sY + s^2/2!(XY + YX)$$
$$+ s^3/3!(X^2Y + XYX + YX^2) + \dots)$$
$$= \exp(-sX)(Y + sYX + s^2/2!(YX^2) + \dots).$$

On the other hand,

$$-[X, \Phi(s)] + Y = Y + \exp(-sX)(sY + s^2/2!(XY + YX) +$$
$$s^3/3!(X^2Y + XYX + YX^2) + \ldots)X$$
$$- X \exp(-sX)(sY + s^2/2!(XY + YX) +$$
$$s^3/3!(X^2Y + XYX + YX^2) + \ldots)$$
$$= Y + \exp(-sX)(s(YX - XY)$$
$$+ s^2/2!(YX^2 - X^2Y) + \ldots).$$

Hence, $\exp(sX)\Phi'(s) = Y + sYX + s^2/2!(YX^2) + \ldots$. Whereas,

$$\exp(sX)(-[X, \Phi(s)] + Y) = \exp(sX)Y + s(YX - XY) +$$
$$s^2/2!(YX^2 - X^2Y) + \ldots$$
$$= Y + sYX + s^2/2!(YX^2) + \ldots.$$

Since $\exp(sX) \circ \Phi'(s) = \exp(sX)(-[X, \Phi(s)] + Y)$ we see that $\Phi'(s) = -[X, \Phi(s)] + Y$ for all $s \in \mathbb{R}$. Now the lemma follows from the next proposition. $\qquad \square$

Proposition 1.7.5. *Let X and Y be fixed in $\mathfrak{g}$ and $\Phi : \mathbb{R} \to \mathfrak{g}$ be a map which satisfies the differential equation $\Phi'(s) = -[X, \Phi(s)] + Y$ and initial condition that $\Phi(0) = 0$. Then for all $s \in \mathbb{R}$,*

$$\Phi(s) = \phi(-s \operatorname{ad} X)(sY)$$

and, conversely this Φ does satisfy the differential equation with this initial condition. In particular, $\Phi(1) = \phi(-\operatorname{ad} X)(Y)$. (Notice that $\phi(-s \operatorname{ad} X)(sY) \in \mathfrak{g}$ for all s.)

Proof. This is a system of nonhomogeneous equations with *constant* coefficients so it has a global analytic solution where $X_i \in \mathfrak{g}$ and $s \in \mathbb{R}$. (See [69] for further detail.)

$$\Phi(s) = X_0 + sX_1 + s^2X_2 + s^3X_3 + \ldots$$

Since $\Phi(0) = 0$, we know $X_0 = 0$. Now $\Phi'(s) = X_1 + 2sX_2 + 3s^2X_3 + \ldots$ and $-[X, \Phi(s)] + Y = Y - s[X, X_1] - s^2[X, X_2] - \ldots$. Therefore, $X_1 = Y$,

$X_2 = -\frac{1}{2}[X, X_1] = -\frac{1}{2}\operatorname{ad}X(Y)$, $X_3 = -\frac{1}{3}[X, X_2] = \frac{1}{6}\operatorname{ad}^2 Y$ and in general

$$X_n = (-1)^{n-1}\frac{1}{n!}\operatorname{ad}X^{n-1}(Y).$$

So $\Phi(s) = sY - s^2/2!\operatorname{ad}X(Y) + s^3/3!\operatorname{ad}X^2(Y) + \dots$. Applying $-\operatorname{ad}X$ we get

$$-\operatorname{ad}X \circ \Phi(s) = -s\operatorname{ad}X(Y) + s^2/2!\operatorname{ad}X^2(Y) - s^3/3!\operatorname{ad}X^3(Y) + \dots$$
$$= \exp(-s\operatorname{ad}X - I)Y.$$

Hence,

$$-s\operatorname{ad}X \circ \Phi(s) = (\exp(-s\operatorname{ad}X) - I)(sY).$$

So $\Phi(s) = \phi(-s\operatorname{ad}X)(sY)$. Conversely, if $\Phi(s) = \phi(-s\operatorname{ad}X)(sY)$, then $\Phi(0) = 0$ and $\Phi'(s) = -[X, \Phi(s)] + Y$. $\square$

We now turn to the proof of Theorem 1.7.1.

Proof. Since for $t \in \mathbb{R}$, $X + tY$ is a smooth curve passing through $X \in \mathfrak{g}$, $\exp(X + tY)$ is a smooth curve in G. Hence the directional derivative of $\exp(X)$ in the direction Y is given by

$$\frac{d}{dt}\exp|_{t=0} \in T_{\exp X}(G).$$

This means that according to our conventions

$$\exp(-X) \circ \frac{d}{dt}\exp(X + tY)|_{t=0} \in \exp(-X)T_{\exp X}(G) = T_1(G) = \mathfrak{g}.$$

From the proposition above we see that for all $s \in \mathbb{R}$

$$\exp(-sX) \circ \frac{d}{dt}\exp(s(X + tY))|_{t=0} = \phi(-s\operatorname{ad}X)(sY),$$

and so taking $s = 1$, we see that for each $Y \in \mathfrak{g}$

$$d_X\exp(Y) = \exp(-X) \circ \frac{d}{dt}\exp(X + tY)|_{t=0} = \phi(-\operatorname{ad}X)(Y).$$

$\square$

Finally, we deal with the Baker-Campbell-Hausdorff formula itself. Let X and Y be fixed and $|t|$ be small. As every Lie group is analytic (see Appendix D), $\exp(tX)\exp(tY)$ is an analytic function of t which tends to 1 as $t \to 0$. So by injectivity of exp near 0 we see that

$$\exp(tX)\exp(tY) = \exp F(t,X,Y),$$

where $F(t,X,Y) = F(t)$ is an analytic function of t. Expanding F in a power series for small $|t|$ we get

$$F(t,X,Y) = \sum_{n \geq 0} t^n C_n(X,Y),$$

where $C_n(X,Y) = \frac{1}{n!}\frac{d^n}{dt^n}F(t,X,Y)$ at $t = 0$. Since we have already seen that $F(t,X,Y) = t(X+Y) + \frac{1}{2}t^2[X,Y] + O(t^3)$, it follows that $C_0(X,Y) = 0$, $C_1(X,Y) = X+Y$, and $C_2(X,Y) = \frac{1}{2}[X,Y]$. Our objective now is to calculate the higher $C_n(X,Y)$. We shall see that for each n, $C_n(X,Y)$ is a fixed homogeneous polynomial in the coordinates of X and Y with rational coefficients consisting of brackets of degree n. They are universal for all linear Lie groups because it is really a statement about $GL(n,\mathbb{C})$ alone. We would thus get

$$\exp(tX)\exp(tY) = \exp(t(X+Y) + \frac{1}{2}t^2[X,Y] + \ldots)$$

valid for small t. Replacing tX and tY by a new X and Y yields

$$\exp(X)\exp(Y) = \exp(X+Y+\frac{1}{2}[X,Y]+\ldots), \qquad (1.9)$$

an absolutely convergent series, involving higher-order brackets of X and Y, valid for *small* X and Y. Or since log inverts exp locally at the origin and X and Y are small, it follows that an absolutely convergent series of these brackets is also small. Hence

$$\log(\exp(X)\exp(Y)) = X + Y + \frac{1}{2}[X,Y] + \ldots$$

We shall write this last expression as $X \circ Y$. The Baker-Campbell-Hausdorff formula refers to any of these equations. Wherever valid it generalizes a fact which we have already proven, namely, if X and Y commute in $\mathfrak{g}$ then

$$\exp(X)\exp(Y) = \exp(X + Y).$$

To see this observe that since X and Y commute so do tX and tY for real t. Taking t small enough so that tX and tY lie in an appropriate neighborhood and applying the BCH formula we get $\exp(tX)\exp(tY) = \exp(t(X + Y))$ for small t because all the other terms will be zero. Now since we have real analytic functions this must hold for all t by the identity theorem (see [55]). Taking $t = 1$ yields the result. It might be worthwhile to mention here that the BCH formula shows in certain cases that the converse is also true (see [16]). Notice that the converse statement does not follow from $\exp(tX)\exp(tY) = \exp(t(X + Y) + \frac{1}{2}t^2[X, Y] + O(t^3))$, since if $\exp(X)\exp(Y) = \exp(X + Y)$ we cannot conclude that $\exp(tX)\exp(tY) = \exp(t(X+Y))$ for all small t and then merely take d^2/dt^2 at $t = 0$.

In fact, once we know the BCH formula is valid in the case of a linear group we will see that it is valid for an arbitrary Lie group. This is because everything in this chapter is local and it is a theorem of Sophus Lie that every Lie group is locally isomorphic to a linear Lie group. This is due to the fact that the Lie algebra of this group has a faithful representation by Ado's theorem. It is for this reason we write exp rather than Exp. The same remarks of course apply to the formula $d_X \exp = \phi(-\operatorname{ad} X)$.

It is also worth noting that if G is a connected nilpotent Lie group, then $\mathfrak{g}$ is a nilpotent Lie algebra and so, for sufficiently high n, all commutators of order n equal zero. This means that the BCH formula becomes a polynomial and hence converges everywhere on $\mathfrak{g}$, not just near 0. It is also well known, that when G is a connected and simply connected nilpotent Lie group, the exponential map is an analytic diffeomorphism. Together with the BCH formula this gives an alternative way of defining a simply connected nilpotent Lie group, namely, one that is modeled on Euclidean space with polynomial multiplication.

We now begin the proof of the BCH formula. Just as before it will be desirable to blow things up by adding a new variable. We write

$$\exp(uX)\exp(vY) = \exp Z(u,v,X,Y) = \exp Z(u,v).$$

Then Z is an analytic function of (u,v) for small u and v. Letting $u = t = v$ we get $F(t) = Z(t,t)$. We also let $g(z) = (1 - e^{-z})/z = \phi(-z)$. Since $\phi(0) = 1$, $g(0) = 1$ and so g is an entire function. Also $d_X \exp = g(\operatorname{ad} X)$. The zeros of g are $z = 2\pi in$ where n is a nonzero integer. Hence $g \neq 0$ in a neighborhood of 0 so $h = 1/g = z/(1 - e^{-z})$ is analytic there and $h(0) = 1$. Finally, let $f(z) = z/(1 - e^{-z}) - 1/2z$. Then f is also analytic there and $f(0) = 1$.

Lemma 1.7.6. *The function f is even and all its Taylor coefficients are rational numbers.*

Proof. To see that $f(z) = f(-z)$ we show that

$$\frac{z}{(1 - e^{-z})} - \frac{1}{2}z = -\frac{z}{(1 - e^z)} + \frac{1}{2}z.$$

That is, $z(\frac{1}{1-e^{-z}} + \frac{1}{1-e^z}) = z$. Or, $\frac{1}{1-e^{-z}} + \frac{1}{1-e^z} = 1$. This last equation is clearly an identity. It follows that all odd Taylor coefficients equal zero. Hence $f(z) = 1 + \sum_{p \geq 1} k_{2p} z^{2p}$. Now all the Taylor coefficients of g are rational. We show the same is true of h. Since $f = h +$ a polynomial with rational coefficients this will prove the lemma.

Now $h(z)g(z) = 1$. Differentiating n times we see that

$$h^n(z)g(z) + \sum_{0 \leq i \leq n-1} \frac{n!}{i!(n-i)!} h^i(z) g^{n-i}(z) = 0$$

so that

$$\frac{h^n(0)}{n!} = -\frac{1}{g(0)} \sum_{0 \leq i \leq n-1} \frac{h^i(0)}{i!} \frac{g^{n-i}(0)}{(n-i)!}.$$

By induction $\dfrac{h^n(0)}{n!} \in \mathbb{Q}$. $\square$

Proposition 1.7.7. *For X and $Y \in \mathfrak{g}$ and $|t|$ small, F satisfies the (nonlinear) differential equation*

$$\frac{dF}{dt} = f(\operatorname{ad} F)(X + Y) + \frac{1}{2}[X - Y, F]$$

with initial condition $F(0) = 0$ and analytic coefficients.

Proof. We have already seen that $F(0) = 0$. Since for z near zero,

$$f(z) + \frac{1}{2}z = \frac{1}{g(z)},$$

we see that for small $Z \in \mathfrak{g}$

$$f(\operatorname{ad} Z) + \frac{1}{2}\operatorname{ad} Z = g(\operatorname{ad} Z)^{-1}$$

and also

$$f(\operatorname{ad} Z) = 1 + \sum_{p \geq 1} k_{2p}(\operatorname{ad} Z)^{2p}.$$

Now

$$\frac{\partial}{\partial v}(\exp(uX)\exp(vY)) = \frac{\partial}{\partial v}(\exp Z(u, v))$$

so that

$$\exp(uX)\exp(vY)Y = d_{Z(u,v)}\exp \frac{\partial Z}{\partial v}.$$

Identifying $T_{\exp(uX)\exp(vY)}(G)$ with $\mathfrak{g}$ by left translation by $(\exp(uX)\exp(vY))^{-1}$ we get

$$Y = d_{Z(u,v)}\exp \frac{\partial Z}{\partial v}.$$

Now for small u and v, $g(\operatorname{ad} Z(u, v))^{-1}$ exists and equals

$$h(\operatorname{ad} Z(u, v)) = f(\operatorname{ad} Z(u, v)) + \frac{1}{2}\operatorname{ad} Z(u, v).$$

Therefore,

$$f(\operatorname{ad} Z)(Y) + \frac{1}{2}\operatorname{ad} Z(Y) = \frac{\partial Z}{\partial v}.$$

On the other hand, $\exp(-vY)\exp(-uX) = \exp(-Z(u,v))$. Hence taking $\frac{\partial}{\partial u}$ on both sides gives

$$\exp(-vY)\exp(-uX)(-X) = d_{-Z(u,v)}\exp(-\frac{\partial Z}{\partial u}).$$

Since $d_{-Z(u,v)}\exp = g(-\operatorname{ad} Z)$, after identification this gives

$$-X = g(-\operatorname{ad} Z)(-\frac{\partial Z}{\partial u}).$$

So

$$X = g(-\operatorname{ad} Z)(\frac{\partial Z}{\partial u}).$$

Inverting, as before, we get

$$f(-\operatorname{ad} Z)(X) - \frac{1}{2}\operatorname{ad} Z(X) = \frac{\partial Z}{\partial u}.$$

Now let $u = t = v$. Then $F(t) = Z(u,v)$ so

$$F'(t) = \frac{\partial Z}{\partial u}u'(t) + \frac{\partial Z}{\partial v}v'(t) = \frac{\partial Z}{\partial u} + \frac{\partial Z}{\partial v}.$$

This means that

$$\begin{aligned}
\frac{dF}{dt} &= f(-\operatorname{ad} F)(X) - \frac{1}{2}[F,X] + f(\operatorname{ad} F)(Y) + \frac{1}{2}[F,Y] \\
&= f(\operatorname{ad} F)(X+Y) + \frac{1}{2}[X-Y,F].
\end{aligned}$$

$$(1.10)$$

$\square$

We now complete the proof of the BCH formula. Let $F(t) = \sum_{n \geq 0} t^n C_n(X,Y)$ where $F : (-\epsilon, \epsilon) \to \mathfrak{g}$ is the local analytic solution to the differential equation $\frac{dF}{dt} = f(\operatorname{ad} F)(X+Y) + \frac{1}{2}[X-Y,F]$ with initial condition $F(0) = 0$. Then $C_0(X,Y) = 0$, and $C_{n+1}(X,Y)$ satisfies the following formula where $S(n) = \{p \in \mathbb{Z}^+ : 2p \leq n\}$ and $T(n) = \{(a(1),\ldots,a(2p)) : a(i) \in \mathbb{Z}^+, \sum a(i) = n\}$.

$$(n+1)C_{n+1}(X,Y) = \frac{1}{2}[X-Y,C_n(X,Y)]+ \qquad (1.11)$$

$$\sum_{S(n)} k_{2p} \sum_{T(n)} [C_{a(1)}(X,Y),[\ldots[C_{a(2p)}(X,Y),X+Y]]\ldots].$$

These recursive relations clearly determine the $C_n(X,Y)$ (and hence also F) uniquely, since $C_1(X,Y) = X+Y$.

For example, if $n = 1$ then $S = \phi = T$ and hence

$$2C_2(X,Y) = \frac{1}{2}[X-Y,X+Y] = [X,Y]$$

so $C_2(X,Y) = \frac{1}{2}[X,Y]$. If $n = 2$ then $S = 1$ i.e. $p = 1$ and $T = (1,1)$. Hence $3C_3(X,Y) = \frac{1}{2}[X-Y,\frac{1}{2}[X,Y]] + k_2[X+Y,X+Y]$. Since k_2 does not actually enter this formula we get

$$C_3(X,Y) = \frac{1}{12}[X,[X,Y]] - \frac{1}{12}[Y,[X,Y]].$$

Proof of (1.1). Fix an integer n. Then

$$\frac{dF}{dt}(t) = C_1 + 2tC_2 + \ldots (n+1)t^n C_{n+1} + O(t^{n+1}). \qquad (1.12)$$

Since ad is linear and continuous

$$\mathrm{ad}\, F(t) = t\,\mathrm{ad}\, C_1 + t^2\,\mathrm{ad}\, C_2 + t^n\,\mathrm{ad}\, C_n + O(t^{n+1}).$$

Hence for any positive integer p with $2p \le n$

$$(\mathrm{ad}\, F(t))^{2p} = \sum_{S(n)} t^s \sum_{T(s)} \mathrm{ad}\, C_{a(1)} \ldots \mathrm{ad}\, C_{a(2p)} + O(t^{n+1}).$$

But also $\mathrm{ad}\, F(t) = O(t)$ so applying the power series definition of f we find that

$$f(\mathrm{ad}\, F(t)) = I + \sum_{S(n)} k_{2p}(\mathrm{ad}\, F(t))^{2p} + O(t^{n+1}).$$

Hence,

$$f(\mathrm{ad}\, F(t)) = I + \sum_{1 \le s \le n} t^s \sum_{S(s)} k_{2p} \sum_{T(s)} \mathrm{ad}\, C_{a(1)} \ldots \mathrm{ad}\, C_{a(2p)} + O(t^{n+1}).$$

$$(1.13)$$

Substituting (1.12) and (1.13) into the differential equation for F and equating the coefficients of t^n on both sides yields (1.11).

We have already observed that for small X and Y, that $X \circ Y$ is an analytic function of X and Y. By induction from (1.11) it follows that.

Corollary 1.7.8. *For each* n, $C_n(X,Y)$ *is a degree* n *homogeneous polynomial in the coordinates of* X *and* Y *with rational coefficients consisting of brackets.*

We can recapture a previous result.

Corollary 1.7.9. *For any* X *and* $Y \in \mathfrak{g}$ *we have*

$$\lim_{n \to \infty} (\exp(\tfrac{1}{n}X)\exp(\tfrac{1}{n}Y))^n = \exp(X+Y).$$

In fact, more generally, if $X_n \to X$ *and* $Y_n \to Y$, *then*

$$\lim_{n \to \infty} (\exp(\tfrac{1}{n}X_n)\exp(\tfrac{1}{n}Y_n))^n = \exp(X+Y).$$

Using analyticity we now derive the functional equation for the exponential map of a connected real Lie group G. Namely, if X and $Y \in \mathfrak{g}$ commute then

$$\exp X \cdot \exp Y = \exp(X+Y). \tag{1.14}$$

To see this, apply the BCH formula.

$$\exp(X) \cdot \exp(Y) = \exp(X + Y + \tfrac{1}{2}[X,Y] + \ldots), \tag{1.15}$$

where the right side is an absolutely convergent series, involving higher-order brackets of X and Y, and (1.15) is valid for all *small* X and Y.

Definition 1.7.10. Let G be a Lie group and L be a manifold containing a neighborhood U of 1 in G. Suppose $VV^{-1} \subseteq U$ where V is itself a neighborhood of 1 in G such that the map on $V \times V \to U$ sending $(g,h) \mapsto gh^{-1}$ is C^∞ (equivalently $(g,h) \mapsto gh$ and $g \mapsto g^{-1}$ take values in U and are C^∞). Then L is called a *local Lie subgroup* of G and U is called a *germ* of L.

Proposition 1.7.11. *Let L be a local Lie subgroup of G and U a germ of L. If U is connected then there is a unique connected Lie subgroup H of G in which U is a neighborhood of 1. If (L', U') is another such pair, then $H' = H$ if and only if $U' \cap U$ is open in both U and U'.*

Proof. Let H be the subgroup of the abstract group G generated by U *i.e.* H is the set of all finite products of elements of U together with their inverses. It is easy to see that H is a submanifold of G and in fact a Lie subgroup of G (see p. 45-46 of [31] for details). We show that H is connected. Now H_0, its identity component, contains U. Hence U is a neighborhood of 1 in H_0. Therefore, as a connected Lie group, $H_0 = \bigcup_{n \geq 1} (U \cap U^{-1})^n$. But by definition this is H. H is clearly determined by U. In fact, since H is connected if W were any other connected neighborhood of 1 in G (say $W \subseteq U$), then since H is connected W generates H. Finally, if $H' = H$, then U' and U are both open in H and hence $U' \cap U$ is open in H and therefore in both U and U'. Conversely, if $U' \cap U$ is open in both U and U', then since they each contain 1 and L is a manifold, the identity component $(U' \cap U)_0$, is open in L and by the remark above generates both H and H'. So $H = H'$. $\qquad\square$

Our next result, due to Sophus Lie, was proved earlier by the Frobenius' Theorem. Here we observe that it follows from the BCH formula.

Corollary 1.7.12. *Let G be a Lie group and $\mathfrak{g}$ its Lie algebra. If $\mathfrak{h}$ is any Lie subalgebra of $\mathfrak{g}$, then there is a unique connected subgroup H of G with Lie algebra $\mathfrak{h}$. Since H is uniquely determined by U we have a bijective correspondence between connected subgroups of G and Lie subalgebras of $\mathfrak{g}$. In particular, if we knew that any Lie algebra had a faithful linear representation, then taking $G = \mathrm{GL}(n, R)$ we see that any Lie algebra over $\mathbb{R}$ is the Lie algebra of some real Lie group.*

Remark 1.7.13. In general, H need not be closed in G, but will be if G is simply connected. In particular, this applies to $[G, N]$ in the result below.

Proof. It is sufficient by the above to show that there is a local Lie subgroup L of G with its germ based on a connected neighborhood U of

1 in L. Let L be $\exp \mathfrak{h}$ and V be $\mathfrak{h} \cap W$, where W is a *sufficiently small* spherical canonical neighborhood of 0 in $\mathfrak{g}$. Then V is open, contains 0 and is connected, so $U = \exp(V)$ and this is what we want.

Since V is symmetric, $X \mapsto -X$ is smooth and $\exp(X)^{-1} = \exp(-X)$. Therefore, inversion is no problem. Now in V the group multiplication is for X and $Y \in V$ given by

$$\log(\exp(X) \cdot \exp(Y)) = X \circ Y = X + Y + \frac{1}{2}[X, Y] + \ldots$$

which is an analytic function of X and Y. These are called *local logarithmic* coordinates. Since $\mathfrak{h}$ is closed under $[\cdot, \cdot]$ and all the terms in the BCH formula involve brackets of various orders of X's and Y's, each term and therefore each partial sum is in $\mathfrak{h}$. But $\mathfrak{h}$ is a closed subspace of $\mathfrak{g}$ so $X \circ Y \in L$. If X and Y are sufficiently small $X \circ Y \in V$ and so $\exp(X \circ Y) \in U$. This proves that L is a local Lie subgroup of G with germ U and therefore the existence of H. The uniqueness of H also follows from Proposition 1.7.11. ∎

We shall now derive some consequences of the BCH formula. To do this we need the following result which itself requires BCH.

Proposition 1.7.14. *Let G be a Lie group, $X_1, \ldots, X_n$ be a basis for its Lie algebra $\mathfrak{g}$ and $\phi_i(t)$ be a family of smooth curves in G such that $\phi_i(0) = 1$ and $\frac{d}{dt}\phi_i(t)_{t=0} = X_i$. In particular, we could take $\phi_i(t) = \exp(tX_i)$. Since each element of $\mathfrak{g}$ can be written uniquely in the form $\sum_i t_i X_i$, we show that a small enough neighborhood of 1 in G can be parameterized by*

$$\phi\left(\sum_i t_i X_i\right) = \phi(t_1, \ldots, t_n) = \log(\phi_1(t_1) \ldots \phi_n(t_n)).$$

That is, for small $(t_1, \ldots, t_n)$, ϕ is an analytic map of a neighborhood of 0 in $\mathfrak{g} \to \log G \subseteq \mathfrak{g}$, and in fact is a local diffeomorphism at 0.

We say that ϕ is a set of *canonical coordinates of the 2nd kind.*

Proof. Since $\phi_i(t_i) = \exp(t_i X_i + $ higher order terms$)$, we see by BCH $\log(\phi_1(t_1) \ldots \phi_m(t_m)) = \sum_i t_i X_i + \ldots$. For purposes of calculating $d_0 \phi$

we may assume the higher order terms is not present and hence that $\phi(\sum_i t_i X_i) = \sum_i t_i X_i$. From this it follows that ϕ is the identity map near 0 so $d_0\phi = I$, and in particular is nonsingular. By the inverse function theorem ϕ is a local diffeomorphism at 0.   □

A direct consequence of our next result is the fact that for a connected Lie group the notions of nilpotence and solvability for the group and its Lie algebra coincide.

Another consequence of the BCH formula and other facts is

Proposition 1.7.15. *Let G be a connected Lie group and H and N connected Lie subgroups of G with N normal. Then $G = HN$ if and only if $\mathfrak{g} = \mathfrak{h} + \mathfrak{n}$, where $\mathfrak{g}$, $\mathfrak{h}$ and $\mathfrak{n}$ are the corresponding Lie algebras of G, H and N respectively.*

Proof. Suppose $G = HN$. Let $H \times N$ act on G by $(h, n)g = hgn^{-1}$. Since $\mathcal{O}_{H \times N}(1) = HN = G$, this action is transitive and hence by Theorem 0.4.5 G is $H \times N$ equivariantly homeomorphic with $H \times N/\operatorname{Stab}_{H \times N}(1)$. In particular, the multiplication map $H \times N \to G$ is open. Let U be a canonical neighborhood of 1 in G and V small enough so that $V^2 \subseteq U$. Let $V_H = H \cap V$ and $V_N = N \cap V$. Then these are canonical neighborhoods in H and N, respectively, and by the above $V_H V_N$ contains a neighborhood W of 1 in G which is canonical since $W \subseteq V^2 \subseteq U$. If $g = \exp X$ is in W then $g = hn$, where $h \in V_H$ and $n \in V_N$. Hence $\exp X = \exp Y \exp Z$ where $Y \in \mathfrak{h}$ and $Z \in \mathfrak{n}$. But the latter is

$$\exp(Y + Z + \frac{1}{2}[Y, Z] + \ldots) = \exp(Y + Z')$$

where $Z' \in \mathfrak{n}$ since $\mathfrak{n}$ is an ideal. By taking Y and Z small enough, $\exp(Y + Z') \in U$. It follows that $X = Y + Z'$. This proves the claim for small X. By scaling we see that $\mathfrak{g} = \mathfrak{h} + \mathfrak{n}$.

Conversely, suppose $\mathfrak{g} = \mathfrak{h} + \mathfrak{n}$ and $g \in U$. Then $g = \exp X$, where X is near 0. By assumption $X = Y + Z$, where $Y \in \mathfrak{h}$ and $Z \in \mathfrak{n}$ are near enough to 0 for the BCH series to converge. Accordingly, $\exp(-Y)g = \exp(-Y)\exp(Y + Z) = \exp(Z + \frac{1}{2}[-Y, Y + Z] + \ldots)$. Now $[-Y, Y + Z] = [Z, Y] \in \mathfrak{n}$ and, similarly, all subsequent terms are in

$\mathfrak{n}$, since $\mathfrak{n}$ is an ideal. Thus $\exp(-Y)g = \exp(Z + Z')$, where $Z' \in \mathfrak{n}$. This means $g = \exp Y \exp(Z + Z') \in HN$ for each $g \in U$. Now since U generates G and N is normal, $G = HN$.   □

Remark 1.7.16. An example of the use of Proposition 1.7.15 is its application to compact connected Lie groups $G = Z(G)_0[G, G]$ where $[G, G]$ is compact and semisimple.

We now turn to some results of Zassenhaus and Margulis concerning discrete subgroups, Γ, of a Lie group G. These were proved by Margulis using the BCH formula. The original proof, due to Zassenhaus, which we give here depends on the following lemma involving elementary matrix inequalities, and seems clearer. As usual, $\|\cdot\|$ denotes the operator norm on $M_n(\mathbb{C})$.

Lemma 1.7.17. *Let $A = I + \alpha \in \mathrm{GL}(n, \mathbb{C})$ where $\|\alpha\| < 1$. Then $I + \sum_{n \geq 1}(-1)^n \alpha^n$ converges in $M_n(\mathbb{C})$ and, equals A^{-1}. Moreover, for any $X \in M_n(\mathbb{C})$, $\|A^{-1}X\| \leq \frac{\|X\|}{1-\|\alpha\|}$ and similarly, $\|XA^{-1}\| \leq \frac{\|X\|}{1-\|\alpha\|}$. Finally, if $A = I + \alpha$ and $B = I + \beta \in \mathrm{GL}(n, \mathbb{C})$, where $\|\alpha\|$ and $\|\beta\| < 1$, then $\|[A, B] - I\| \leq 2\frac{\|\alpha\|\|\beta\|}{(1-\|\alpha\|)(1-\|\beta\|)}$.*

Proof. That $A^{-1} = I + \sum_{n \geq 1}(-1)^n \alpha^n$ is just a convergent geometric series. To see that $\|A^{-1}X\| \leq \frac{\|X\|}{1-\|\alpha\|}$, simply estimate $\|A^{-1}\|$ by the geometric series and then apply the fact that $M_n(\mathbb{C})$ is a Banach algebra. Finally, turning to our last inequality, we have $ABA^{-1}B^{-1} - I = (AB - BA)A^{-1}B^{-1} = [\alpha, \beta]A^{-1}B^{-1}$. Hence, applying the previous inequality twice and the fact that $\|[\alpha, \beta]\| \leq 2\|\alpha\|\|\beta\|$, yields the result.   □

We now turn to a result which is usually called the Margulis Lemma.

Theorem 1.7.18. *Any Lie group G has a neighborhood Ω of 1 such that for any sequence $\{g_n\} \in \Omega$, the sequence given by $h_1 = g_1$, $h_2 = [g_2, g_1]$, $h_3 = [g_3, [g_2, g_1]], \ldots$ converges to 1 in G.*

Proof. Since G is a Lie group G_0, its identity component, is open in G so we may assume that G is connected and, as this is a local question and any connected Lie group is locally isomorphic to a Lie subgroup of $GL(n, \mathbb{C})$, we may also assume that G is itself a linear group. Choose a neighborhood Ω of I so that for all $g \in \Omega$, $\|g - I\| < \epsilon$, where $0 < \epsilon < \frac{1}{3}$. We will prove by induction that, for all n, if C is an n-fold commutator, then $\|C - I\| < \epsilon(3\epsilon)^n$. To see this let $C = [A, B]$, where $A = I + \alpha$, $B = I + \beta$ and $\|\alpha\| < \epsilon$ and $\|\beta\| < \epsilon(3\epsilon)^{n-1}$. Since both $\|\alpha\|$ and $\|\beta\| < \epsilon$, then $\frac{1}{(1-\|\alpha\|)}$ and $\frac{1}{(1-\|\beta\|)}$ are each $< \frac{1}{(1-\epsilon)}$. Hence, since $\|[\alpha, \beta]\| \leq 2\|\alpha\|\|\beta\|$, we see by Lemma 1.7.17 that,

$$\|[A, B] - I\| \leq \frac{2\epsilon\epsilon(3\epsilon)^{n-1}}{(1 - \epsilon)^2}.$$

But $\epsilon < \frac{1}{3} < 1 - \sqrt{\frac{2}{3}}$, it follows that $\frac{1}{(1-\epsilon)^2} < \frac{3}{2}$ and therefore, that $\|C - I\| \leq \epsilon(3\epsilon)^n$, thereby proving the inductive statement. Now, for each n, h^n is an n-fold commutator and since $\epsilon < 1$, we see that $\|h^n - 1\| < (3\epsilon)^n$. As $3\epsilon < 1$, $(3\epsilon)^n$ converges to 0 and so $h^n \to 1$. $\square$

In the following corollary what is important is that k may depend on Γ, but Ω depends only on G.

Corollary 1.7.19. *Let G be a Lie group, Γ be any discrete subgroup of G and Ω as above. Then there is a fixed integer k such that for any finite set $g_1, g_2, \ldots, g_k \in \Omega \cap \Gamma$ we have $[g_k, [g_{k-1}, \ldots, g_1] \ldots] = 1$.*

Proof. Let $\{g_n\}$ be any sequence $\Omega \cap \Gamma$. By the Margulis lemma g_1, $[g_2, g_1]$, $[g_3, [g_2, g_1]], \ldots$ converges to 1. But this sequence lies in Γ which is discrete. Hence it is identically 1 from some term on. Thus there is an integer k such that $[g_k, [g_{k-1}, \ldots, g_1] \ldots] = 1$. $\square$

We shall always denote by k the smallest such integer. The discrete part of our next result is also usually called the Margulis Lemma. However this result was first proven by Zassenhaus in the 1930s.

Theorem 1.7.20. *In any Lie group G there exists a neighborhood Ω of 1 such that for any discrete subgroup Γ of G, $\Omega \cap \Gamma$ generates a discrete nilpotent subgroup N of G. In fact, N is contained in a connected nilpotent Lie subgroup of G.*

Proof. Choose Ω smaller than the one in Theorem 1.7.18 and symmetric. Then $\Omega \cap \Gamma$ is also symmetric. Since $N \subset \Gamma$ it is discrete. There is a fixed integer k so that for any choice of $g_1, g_2, \ldots, g_k \in \Omega \cap \Gamma$, $[g_k, [g_{k-1}, \ldots g_1] \ldots] = 1$. For each integer $j \geq 2$, let N_j be the subgroup of G (actually of N) generated by the set of commutators C_j of length at least j with $g_i \in \Omega \cap \Gamma$. Since $C_{j+1} \subseteq C_j$ it follows that $N_{j+1} \subseteq N_j \subseteq N$. We know that $N_j = \{1\}$ for $j \geq k$. We will prove by induction that each N_j is normal in N. For $j \geq k$ this is clearly so. Suppose inductively that N_{j+1} is normal in N where $j < k$. Consider the exact sequence

$$\{1\} \to N_{j+1} \to N \xrightarrow{\pi} N/N_{j+1} \to \{1\}.$$

Then,

$$[\pi(\Omega \cap \Gamma), \pi(C_j)] = \pi[\Omega \cap \Gamma, C_j] = \pi(C_{j+1}) \subseteq \pi(N_{j+1}) = \{1\}.$$

Since N is generated by $\Omega \cap \Gamma$, N/N_{j+1} is generated by $\pi(\Omega \cap \Gamma)$. Also, $\pi(N_j)$ is generated by $\pi(C_j)$. It follows that $\pi(N_j) = N_j/N_{j+1}$ is in the center of $\pi(N) = N/N_{j+1}$. Thus $[\pi(N), \pi(N_j)] = \pi[N, N_j] = \{1\}$ so

$$[N, N_j] \subseteq N_{j+1} \subseteq N_j.$$

This means that N_j is normal in N. Since we have also proven that for each j, $[N, N_j] \subseteq N_{j+1}$ (for $j \geq k$ this is also clearly so) we see that $N_j/N_{j+1} \subseteq Z(N/N_{j+1})$. Now $(N/N_{j+1})/(N_j/N_{j+1}) = N/N_j$ so if N/N_j were nilpotent then N/N_{j+1} would also be nilpotent. Since $[N, N] \subseteq N_2$, N/N_2 is abelian and therefore nilpotent. This shows by induction that N/N_j is nilpotent for all j and in particular $N/N_k = N$ is nilpotent.

We now strengthen our result by showing that N is actually contained in a connected nilpotent Lie subgroup of G. Let $\log$ be the inverse to $\exp$ on Ω. By taking Ω small enough we may assume in addition to its other properties that it has compact closure. Then since $\bar{\Omega}$ is compact, choose a neighborhood V of 0 in $\mathfrak{g}$ small enough so that $\Omega \subseteq \exp V$, and $\mathrm{Ad}\, y(V) \subseteq \log(\Omega)$ for all $y \in \Omega$. Let $\mathfrak{t} = \log(\exp V \cap \Gamma)$ (in other words, for this part of the argument we replace Ω by the smaller $\exp V$) and $\mathfrak{h}$ by the subalgebra of $\mathfrak{g}$ generated by $\mathfrak{t}$. We shall show by induction on dim G that $\mathfrak{h}$ is nilpotent. Then the corresponding connected Lie group H is also nilpotent. Since $\mathfrak{t} \subseteq \mathfrak{h}$, it follows that $\exp(\mathfrak{t}) = \exp V \cap \Gamma \subseteq H$. Since H is a group and $\exp V \cap \Gamma$ generates N, $N \subseteq H$. Now by the estimates of Lemma 1.7.17

$$C^{k-1} \subseteq (\exp V \cap \Gamma) \subseteq \exp V,$$

so each $g \in C^{k-1}$ is of the form $\exp X$ for some $X \in \mathfrak{h}$. Let

$$\mathfrak{n}^{k-1} = \{X \in \mathfrak{n} : \exp X \in C^{k-1}\}.$$

We show first that $[\mathfrak{n}, \mathfrak{n}^{k-1}] = \{0\}$. Let $y = \exp Y \in \mathfrak{n}$ and $x = \exp X \in \mathfrak{n}^{k-1}$. Since $y \in \Omega \cap \Gamma$ and $x \in C^{k-1}$, $[y, x] = 1$ so $yxy^{-1} = x$. But $yxy^{-1} = \exp \operatorname{Ad} y(X)$ so $\exp \operatorname{Ad} y(X) = x \in C^{k-1} \subseteq \Omega$. On the other hand, since $y \in \Omega$, $X \in \mathfrak{n} \subseteq V$ we know $\operatorname{Ad} y(X) \in \log(\Omega)$. But $\exp$ is one-to-one on $\log(\Omega)$, $X = \log x$ and $\operatorname{Ad} y(X) \in \log(\Omega)$. We conclude that $\operatorname{Ad} y(X) = X$. But $\operatorname{Ad} \exp Y = \operatorname{Exp} \operatorname{ad} Y$ so $\operatorname{Exp} \operatorname{ad} Y(X) = X$ and since $\operatorname{Exp} \operatorname{ad} Y$ is a linear operator on $\mathfrak{g}$, this means $\operatorname{Exp} \operatorname{ad} Y(tX) = tX$ for all t and hence that $\operatorname{ad} Y(X) = 0$. Thus $[\mathfrak{n}, \mathfrak{n}^{k-1}] = \{0\}$. In particular, $[\mathfrak{n}^{k-1}, \mathfrak{n}^{k-1}] = \{0\}$. Let $\mathfrak{a}$ be the abelian subalgebra of $\mathfrak{g}$ spanned by $\mathfrak{n}^{k-1}$ over $\mathbb{R}$, and A be the corresponding connected Lie subgroup, B its closure, and $\mathfrak{b}$ the Lie algebra of B. Now the centralizer, $\mathfrak{z} = \mathfrak{z}_\mathfrak{b}(\mathfrak{g})$ contains $\mathfrak{b}$ since B and therefore also $\mathfrak{b}$ is abelian. Let $Z = Z_B(G)_0 \supseteq B$ be the corresponding connected Lie subgroup of G. Z is evidently closed in G and so is a Lie group. Let $\pi : Z \to Z/B$ be the projection and $\pi' : \mathfrak{z} \to \mathfrak{z}/\mathfrak{z}_\mathfrak{b}(\mathfrak{g})$ be its differential. Since $\mathfrak{z}_\mathfrak{b}(\mathfrak{g})$ is central in $\mathfrak{z}$, B is central in Z. Because $[\mathfrak{n}, \mathfrak{n}^{k-1}] = \{0\}$ it follows that $\exp(\mathfrak{n})$ centralizes A and therefore also B. Hence $\mathfrak{n} \subseteq \mathfrak{z}$. Now $C^{k-1} = \exp \mathfrak{n}^{k-1}$ so since $C^{k-1} \neq \{1\}$ it follows that $\mathfrak{n}^{k-1} \neq \{0\}$. Hence $0 < \dim A \leq \dim B$ so $\dim Z/B < \dim Z \leq \dim G$. We see by induction that the subalgebra of $\mathfrak{z}/\mathfrak{b}$ generated by $\pi'(\mathfrak{z} \cap \mathfrak{b})$ is nilpotent. Because $\mathfrak{b}$ is central in $\mathfrak{z}$ the subalgebra of $\mathfrak{z}$ (and $\mathfrak{g}$) generated by $\mathfrak{z} \cap \mathfrak{b}$ is nilpotent. Since $\mathfrak{b} \subseteq \mathfrak{z}$ this is $\mathfrak{b}$. $\qquad\square$

Chapter 2

Haar Measure and its Applications

2.1   Haar Measure on a Locally Compact Group

Given a locally compact Hausdorff space X and a continuous real (or complex) valued function f we denote by $\mathrm{Supp}(f)$ the set $\overline{\{x : f(x) \neq 0\}}$. We shall denote by $C_0(X)$ the continuous real or complex valued functions on X with compact support and by $C_0^+(X)$ the ones with positive values. When X is a locally compact group G, $f \in C_0(G)$ and $g \in G$ we define the *left translate* of f by $g \in G$ to be $f_g(x) = f(g^{-1}x)$ for all $x \in G$.

On any locally compact topological group there is always a nontrivial and essentially unique *left (or right) invariant measure dx*, called *Haar measure* defined by $\mu(gE) = \mu(E)$ for every measurable set $E \subseteq G$ and $g \in G$. Alternatively, $\int_G f(g^{-1}x)dx = \int_G f(x)dx$, for all continuous functions f with compact support on G and $g \in G$. Here by a measure we shall always mean a nontrivial positive *regular measure*, that is one where the measure of a set E can be approximated by open sets containing E and by compact sets contained in E. Such measures are positive on nontrivial open sets and finite on compact sets. For the details regarding regular measures see [65]. Since an invariant measure can be modified by multiplying by a positive constant and still remain

89

nontrivial positive and invariant there can be no uniqueness to such a measure. However, if this is the worst that can happen we shall say the measure is *essentially unique*.

Theorem 2.1.1. *There is an essentially unique left (or right) invariant measure on any locally compact group G. This measure is called left (or right) Haar measure.*

We first deal with the existence of Haar measure.

Let f and g be nonzero functions in $C_0^+(G)$. Then for some positive integer n there are positive constants $c_1, \cdots, c_n$ and group elements $x_1, \cdots, x_n$ so that for all $x \in G$

$$f(x) \leq \sum_{i=1}^{n} c_i g(x_i x).$$

For example if M_f and M_g are the maximum values of f and g respectively, then $f(x) \leq (\frac{M_f}{M_g} + \epsilon) g_{x_i}(x)$ for any choice of n and the x_i. So consider the set of all possible such inequalities and let $(f : g)$ stand for the inf $\sum_{i=1}^{n} c_i$ over this set. Then evidently we have

(1) $(f_x : g) = (f : g)$, for every $x \in G$.
(2) $(f_1 + f_2 : g) \leq (f_1 : g) + (f_2 : g)$.
(3) $(cf : g) = c(f : g)$, for $c > 0$.
(4) If $f_1 \leq f_2$, then $(f_1 : g) \leq (f_2 : g)$.
(5) $(f : h) \leq (f : g)(g : h)$.
(6) $(f : g) \geq \frac{M_f}{M_g}$.

This gives us a relative idea of the size of f as compared to g. In order to have an absolute estimate of the size of f we must fix an $f_0 \in C_0^+(G)$ for which $(f_0 : g)$ is positive. So we define $I_g(f) = \frac{(f:g)}{(f_0:g)}$. Now the subadditivity of 2) will somehow have to be corrected to become close to additivity. This will be done by taking g with smaller and smaller support. Then we will take some kind of limit of the $I_g(f)$ for small g to get actual additivity of the integral $I(f)$. To do so we need the following lemma.

Lemma 2.1.2. *For f_1 and $f_2 \in C_0^+(G)$ and $\epsilon > 0$ there is a sufficiently small neighborhood U of 1 in G so that whenever $\mathrm{Supp}\, g \subseteq U$, we have*

$$I_g(f_1) + I_g(f_1) \le I_g(f_1 + f_1) + \epsilon.$$

Proof. Since $f_1 + f_2 \in C_0^+(G)$ we know $\mathrm{Supp}(f_1 + f_2)$ is compact. Choose $f' \in C_0^+(G)$ which is $\equiv 1$ on $\mathrm{Supp}(f_1 + f_2)$. Let δ and $\epsilon' > 0$, $f = f_1 + f_2 + \delta f'$ and for $i = 1, 2$ let $h_i = \frac{f_i}{f}$, it being understood that $h_i = 0$ whenever $f = 0$. Then $h_i \in C_0^+(G)$. By uniform continuity choose a neighborhood U of 1 in G so that if $x^{-1}y \in U$ and $i = 1, 2$ then $|h_i(x) - h_i(y)| < \epsilon'$. Now choose $g \in C_0^+(G)$ with $\mathrm{Supp}\, g \subseteq V$.

Suppose $f(x) \le \sum_{j=1}^{n} c_j g(s_j x)$. If some $g(s_j x) \ne 0$, then $|h_i(x) - h_i(s_j^{-1})| < \epsilon'$ for both i and

$$f_i(x) = f(x)h_i(x) \le \sum_{j=1}^{n} c_j g(s_j x) h_i(x) \le \sum_{j=1}^{n} c_j g(s_j x)(h_i(s_j^{-1}) + \epsilon').$$

Hence $(f_i : g) \le \sum_{j=1}^{n} c_j (h_i(s_j^{-1}) + \epsilon')$. So that $(f_1 : g) + (f_2 : g) \le \sum_{j=1}^{n} c_j(1 + 2\epsilon')$. Because $\sum_{j=1}^{n} c_j$ approximates $(f : g)$ we get $I_g(f_1) + I_g(f_1) \le I_g(f_1 + f_1 + \delta f')(1 + 2\epsilon') \le (I_g(f_1 + f_1) + \delta I_g(f'))(1 + 2\epsilon')$. Now choose first ϵ' and then δ small enough so that $2\epsilon'(f_1 + f_2 : f_0) + \delta(1 + 2\epsilon')(f' : f_0) < \epsilon$. $\square$

Now from 5) we see that $I_g(f)$ always lies in a closed interval

$$\frac{1}{(f_0 : f)} \le I_g(f) \le (f : f_0).$$

If we think of the space of functionals on $C_0(X)$ as a subspace of the product $\mathbb{R}^{C_0(X)}$ equipped with the product topology, then by the Tychonoff theorem I_g lies in the compact space which is the product of these intervals as f varies and the f component of I_g is $I_g(f)$. For each neighborhood U of 1 denote by K_U the closure of the set of all I_g where $\mathrm{Supp}\, g \subseteq U$. Now these closed sets K_U have the finite intersection property because for any finite number of U_i, $K_{U_1 \cap \dots \cap U_n} = \cap_{i=1}^{n} K_{U_i}$ which, by Urysohn's lemma, in non-empty. Hence we can find a point

$I \in \cap\{K_U\}$ the intersection of all of them. By the properties of the product topology, for any such U and any finite number of $f_1, \ldots f_n$, there is a g with $\mathrm{Supp}\, g \subseteq U$ so that for all i, $|I(f_i) - I_g(f_i)| < \epsilon$ and moreover

$$\frac{1}{(f_0 : f)} \leq I(f) \leq (f : f_0).$$

The lemma now shows that I is additive and, of course, invariant. Finally in the usual way one extends I from $C_0^+(G)$ to $C_0(G)$ itself by $I(f_1 - f_2) = I(f_1) - I(f_2)$ to get a left invariant Haar measure on G.

 We now turn to the essential uniqueness of Haar measure. This is very important because if one can find an invariant measure then it must be Haar measure, suitable normalized.

Proof. Let $I = dx$ and $J = dy$ be two positive left invariant measures on G and $f \in C_0^+(G)$. Let $C = \mathrm{Supp}\, f$ and choose an open set U about C with compact closure. By Urysohn's lemma choose a function $\phi \in C_0^+(G)$ which is identically 1 on U. Let $\epsilon > 0$ and V be a symmetric neighborhood of 1 in G such that $CV \cup VC \subseteq U$. Since $f \in C_0(G)$ it is uniformly continuous and therefore $|f(xy) - f(zx)| < \epsilon$, for all $x \in G$ and $y, z \in V$.

 Then $f(xy) = f(xy)\phi(x)$ and $f(yx) = f(yx)\phi(x)$ for $x \in G$ and $y \in V$. Hence for $y \in V$, $|f(xy) - f(yx)| < \epsilon\phi(x)$ everywhere on G. Now let h be any symmetric function in $C_0^+(G)$ supported on V. Then, by invariance and the Fubini theorem,

$$I(h)J(f) = \int\int f(x)h(y)dxdy = \int\int h(y)f(yx)dxdy.$$

and

$$J(h)I(f) = \int\int f(y)h(x)dxdy = \int\int h(y^{-1}x)f(y)dydx =$$
$$\int\int h(y)f(xy)dydx.$$

So that

$$|I(h)J(f) - J(h)I(f)| \leq \int\int h(y)|f(yx) - f(xy)|dxdy \leq$$

$$\epsilon \int\int h(y)\phi(x)dxdy = \epsilon I(h)J(\phi).$$

Similarly, if $g \in C_0^+(G)$ and h is symmetric and suitably chosen

$$|I(h)J(g) - J(h)I(g)| \leq \epsilon I(h)J(\psi).$$

Hence

$$|\frac{J(f)}{I(f)} - \frac{J(g)}{I(g)}| \leq \epsilon |\frac{J(\phi)}{I(f)} - \frac{J(\psi)}{I(g)}|.$$

Since ϵ is arbitrary we see the left side of this equation is zero and hence $\frac{J(f)}{I(f)} = \frac{J(g)}{I(g)}$ for any f and g satisfying the above conditions. Let g be fixed. Then there is a positive $c = \frac{J(g)}{I(g)}$ for which $J(f) = cI(f)$ for all $f \in C_0^+(G)$. $\square$

We now know that on any locally compact group G there is an essentially unique left invariant Haar measure. The same reasoning also shows that there is an essentially unique right invariant measure.

Of course since Haar measure is regular, compact groups have finite measure (which is usually normalized to have total mass 1). The converse is also true.

Corollary 2.1.3. *A locally compact group G has a finite Haar measure if and only if G is compact.*

Proof. Let U be a compact neighborhood of 1 in G and V be small enough so that $VV^{-1} \subseteq U$. We consider finite subsets g_iV which are pairwise disjoint. Now for any such subset $\mu(G) \geq \mu(\bigcup_{i=1}^n g_iV) = \sum_{i=1}^n \mu(g_iV) = n\mu(V)$. Therefore $n \leq \frac{\mu(G)}{\mu(V)}$. Since the number of such subsets is bounded there must be a maximum number of them which we again call n. Let $g \in G$, but $g \neq g_i$ for any $i = 1, \ldots, n$. Then gV must intersect one of the g_iV so that $g \in g_iVV^{-1} \subset g_iU$. Clearly each g_i is also in g_iU. Thus $G = \bigcup_{i=1}^n g_iU$ and the latter being a finite union of compact sets is compact. $\square$

Using the uniqueness an obvious example of Haar measure on $\mathbb{R}$ is Lebesgue measure since it is translation invariant. For the same reason Lebesgue measure on the circle $\mathbb{T} = S^1$ is Haar measure. Here since we have a compact group and therefore a finite measure it is customary to normalize and divide by 2π. Later we shall see that normalized Lebesgue measure on S^3 is Haar measure for $\mathrm{SU}(2, \mathbb{C})$ $(= S^3)$, but for more complicated reasons. Since on a finite direct product of groups evidently left Haar measure is the product of the left Haar measures on the components we know that product measure is Haar measure on T^n and $\mathbb{R}^n$. Evidently, counting measure is Haar measure on a discrete group.

In general in the case of a Lie group one can be somewhat more explicit concerning Haar measure. If the dimension of G is n we consider left invariant n forms on G. Such a form is determined on all of G by its value at 1. Also the space of all such forms has dimension 1. Since G is orientable choose a nonzero left invariant n-form ω consistent with the orientation of G and then for each $f \in C_0(G)$ define $I(f) = \int_G f(x)\omega(x)dx$, where dx is local Lebesgue measure on a coordinate patch. A partition of unity argument together with the change of variables formula for multiple integrals shows that I is well-defined. Although I depends on the ω chosen, ω is uniquely determined up to a positive constant. Therefore the same is true of I. Clearly I gives a measure on G. To see that it is left invariant observe $I(f) = \int_G f\omega = \int_G d(L_g)(f\omega)$ by the left invariance of the form. But by the change of variables formula for multiple integrals this is $\int_G f(L_g)\omega = I(f(L_g))$. Thus for the Lie group G we have an essentially unique left invariant Haar measure given by an invariant volume form. That is locally on a chart $U = (x_1, \ldots, x_n)$ we have $dx = \omega(x_1, \ldots, x_n)dx_1 \ldots dx_n$, where ω is a non-negative smooth function on U and $dx_1 \ldots dx_n$ is Lebesgue measure on U. Thus dx is absolutely continuous with respect to Lebesgue measure. For the Lie groups case (see [15]).

We will now see in a very explicit way how the change of variable formula can be used to identify Haar measure in many cases.

Proposition 2.1.4. *Let G be a Lie group modeled on some open subset* of some Euclidean space, $\mathbb{R}^n$ and dg be Lebesgue measure on G inherited from $\mathbb{R}^n$. Let L_g and R_g denote left and right translations on G by the element g and suppose for each $g \in G$, $|\det d(L_g)(x)|$ (respectively $|\det d(R_g)(x)|$) is independent of x. Then left Haar measure is $\frac{dg}{|\det d(L_g)|}$ (respectively right Haar measure is $\frac{dg}{|\det d(R_g)|}$). In particular, left and right Haar measures are absolutely continuous with respect to Lebesgue measure.

Proof. We prove this for left Haar measure. The case of right Haar measure is completely analogous. Since G is an open subset of Euclidean space we can apply the change of variables formula for multiple integrals.

$$\int_G f(x)dx = \int_G f(Tx)|\det(dT)x|dx,$$

where T is a smooth global change of variables, dT is its derivative, dx is Lebesgue measure on G and f is a continuous function with compact support on G.

We specialize this to $T = L_g$ for $g \in G$. By assumption $|\det d(L_g)(x)| = \phi(g)$ is independent of x. The function ϕ is positive everywhere on G. Hence for all $f \in C_0(G)$ and $g \in G$, $\int_G f(gx)\phi(g)dx = \int_G f(x)dx$. Now $\frac{f(x)}{\phi(x)}$ is again such a function. Applying the last equation to these functions shows

$$\int_G \frac{f(gx)}{\phi(gx)}\phi(g)dx = \int_G \frac{f(x)}{\phi(x)}dx.$$

Taking into account the chain rule and the fact that $|\det|$ is multiplicative we see that ϕ is a homomorphism on G and hence

$$\int_G f(gx)\frac{dx}{\phi(x)} = \int_G f(x)\frac{dx}{\phi(x)}.$$

Since f and g are arbitrary in this last equation, uniqueness tells us that left Haar measure is $\frac{dx}{\phi(x)}$. □

Now although left Haar measure is often right invariant, the next example shows this is not so in general.

Example 2.1.5. Let G be the affine group of the real line $\mathbb{R}$. G consists of all 2×2 real matrices

$$g = \begin{pmatrix} a & b \\ 0 & 1 \end{pmatrix}, \tag{2.1}$$

where $a \neq 0 \in \mathbb{R}$ and $b \in \mathbb{R}$. In this way G can be regarded as an open set in the (a, b) plane, $\mathbb{R}^2$. G is usually called the $ax + b$-group.

A direct calculation shows $d_{L_g} = aI$. Hence $|\det d_{L_g}| = a^2$, which is independent of the space variables (as well as b). A similar calculation shows $|\det d_{R_g}| = |a|$, which is also independent of the space variables (as well as b). Thus left Haar measure here is $\frac{dadb}{a^2}$ while right Haar measure is $\frac{dadb}{|a|}$. Clearly neither of these measures is a constant multiple of the other. In fact, they do not even have the same L^1 functions!

G is said to be *unimodular* if left invariant Haar measure is also right invariant. Of course abelian groups are unimodular, but as we just saw solvable ones need not be. Clearly discrete groups are unimodular.

Exercise 2.1.6. Use Proposition 2.1.4 to calculate Haar measure on the following examples which are all unimodular.

(1) Haar measure on $\mathrm{GL}(n, \mathbb{R})$ is $\frac{dx}{|\det x|^n}$, where dx is Lebesgue measure on $M_n(\mathbb{R})$. This is because $|\det L_g(x)| = |\det R_g(x)| = |\det g|^n$.

(2) Haar measure on $\mathrm{GL}(n, \mathbb{C})$ is $\frac{dx}{|\det x|^{2n}}$, where dx is Lebesgue measure on $M_n(\mathbb{C})$. This is because $|\det L_g(x)| = |\det R_g(x)| = |\det g|^{2n}$.

(3) Haar measure on $N_n(R)$, the $n \times n$ real unitriangular matrices, is just Lebesgue measure. This is because $|\det L_g(x)| = |\det R_g(x)| = 1$. This is a special case of the fact that nilpotent groups are always unimodular.

On the other hand as we know in the case of the affine group of the line, the group $\mathrm{GL}(n, \mathbb{R}) \times_\eta \mathbb{R}^n$ of all affine motions of $\mathbb{R}^n$ is not unimodular.

Indeed calculations similar to the ones we have made show that left Haar measure is $\frac{dxdy}{|\det x|^{n+1}}$, where dx is Lebesgue measure on $\mathrm{GL}(n,\mathbb{R})$ and dy is Lebesgue measure on $\mathbb{R}^n$. Thus $I(f) = \int_{\mathrm{GL}(n,\mathbb{R})} \int_{\mathbb{R}^n} \frac{f(x,y)dxdy}{|\det x|^{n+1}}$. This is because $|\det L_g(x)| = |\det g|$ together with what we know about $\mathrm{GL}(n,\mathbb{R})$ itself. Similarly, right Haar measure is $\frac{dxdy}{|\det x|^n}$, where dx is Lebesgue measure on $\mathrm{GL}(n,\mathbb{R})$ and dy is Lebesgue measure on $\mathbb{R}^n$.

Exercise 2.1.7. Generalize these facts concerning the group of affine motions to semi-direct products as follows:

Let $G \times_\eta H$ be a semidirect product of unimodular groups, where G acts on H and dg and dh are Haar measure on G and H respectively. Then right Haar measure on $G \times_\eta H$ is $dgdh$, while left Haar measure is $\frac{dgdh}{\Delta(\eta(g))}$, where $\Delta(\eta(g))$ is the amount that the automorphism $\eta(g)$ acting on H distorts Haar measure on H.

In particular, $G \times_\eta H$ is unimodular if and only if G acts on H by measure preserving automorphisms and in this case Haar measure is the product measure. So for example this is the case for $\mathrm{SL}(n,\mathbb{R}) \times_\eta \mathbb{R}^n$ or $\mathrm{O}(n,\mathbb{R}) \times_\eta \mathbb{R}^n$.

Exercise 2.1.8. We now consider the solvable, but not nilpotent, full triangular subgroup, B of $\mathrm{GL}(n,\mathbb{R})$. This is evidently an open set in a Euclidean space, X. Prove for $g = (g_{ij})$,

$$|\det d(L_g)(x)| = |g_{11}^n g_{22}^{n-1} \cdots g_{nn}^1|,$$

which is independent of x. Therefore

$$d(\mu_l) = \frac{dx}{|x_1^n x_2^{n-1} \cdots x_n^1|},$$

where dx is Lebesgue measure on X. Similarly,

$$d(\mu_r) = \frac{dx}{|x_n^n x_{n-1}^{n-1} \cdots x_1^1|}.$$

Therefore these groups are not unimodular.

Exercise 2.1.9. Let $n_1, \ldots, n_r$ be a partition of n. Thus each n_i is a positive integer and $\sum_{i=1}^{r} n_i = n$. Let P be the subgroup of $\mathrm{GL}(n, \mathbb{R})$ consisting of block triangular matrices with diagonal blocks g_i corresponding to the n_i. Notice that this includes $\mathrm{GL}(n, \mathbb{R})$ itself as well as B. Calculate Haar measure.

An example we have not seen before is that of Haar measure on a compact non-abelian group. We will consider the most important non-abelian compact group, namely $G = \mathrm{SU}(2, \mathbb{C})$. Since as we shall see in the next section, compact groups are unimodular we need only consider left invariant Haar measure. Once we determine normalized Haar measure μ on G this automatically gives normalized Haar measure ν on the quotient group, $\mathrm{SO}(3, \mathbb{R})$. For if π is the universal covering map and A is a Borel set in $\mathrm{SO}(3, \mathbb{R})$, then $\nu(A) = \mu(\pi^{-1})(A)$ is normalized and invariant.

As we saw (see Section 1.5) each $g \in \mathrm{SU}(2, \mathbb{C})$ is of the form

$$ g = \begin{pmatrix} \alpha & \beta \\ -\bar{\beta} & \bar{\alpha} \end{pmatrix}, $$

where $|\alpha|^2 + |\beta|^2 = 1$. In this way our group can be regarded as the 3-sphere, S^3. We go further and give a 4-dimensional real linear realization of the transformation group $G \times G \to G$ acting by left translation. This is actually a special case of the equivariant embedding theorem of Mostow-Palais (see [58]). Since

$$ \begin{pmatrix} \alpha & \beta \\ -\bar{\beta} & \bar{\alpha} \end{pmatrix} \begin{pmatrix} \gamma & \delta \\ -\bar{\delta} & \bar{\gamma} \end{pmatrix} = \begin{pmatrix} \alpha\gamma - \beta\bar{\delta} & \alpha\delta + \beta\bar{\gamma} \\ -\bar{\beta}\gamma + \bar{\alpha}\bar{\delta} & -\bar{\beta}\delta + \bar{\alpha}\bar{\gamma} \end{pmatrix}, $$

if we write $\alpha = \alpha_1 + i\alpha_2$ and similarly for β, γ and δ, this says

$$ X(\gamma_1, \gamma_2, \delta_1, \delta_2) = (Re(\alpha\gamma - \beta\bar{\delta}), Im(\alpha\gamma - \beta\bar{\delta}), Re(\alpha\delta + \beta\bar{\gamma}), Im(\alpha\delta + \beta\bar{\gamma})), $$

where $X = X(g)$ is the linear transformation on $\mathbb{R}^4$ given by

$$ \begin{pmatrix} \alpha_1 & -\alpha_2 & -\beta_1 & -\beta_2 \\ \alpha_2 & \alpha_1 & -\beta_2 & \beta_1 \\ \beta_1 & \beta_2 & \alpha_1 & -\alpha_2 \\ \beta_2 & -\beta_1 & \alpha_2 & \alpha_1 \end{pmatrix}. $$

Now by definition each $X(g)$ preserves S^3. Since $X(g)$ is linear it is therefore orthogonal and in particular it is invertible. The subset $\mathcal{X}$ of $O(4, \mathbb{R})$ consisting of these $X(g)$ when acting on the invariant set S^3 acts equivariantly with $G \times S^3 \to S^3$ under translation. The fact that the map $g \mapsto X(g)$ is a homomorphism and $\mathcal{X}$ is a subgroup of $O(4, \mathbb{R})$, is immaterial. In any case, since ordinary Lebesgue measure λ on S^3 is such that if dx is Lebesgue measure on $\mathbb{R}^4$ then $dx = r^3 dr d\lambda$ is invariant under all of $O(4, \mathbb{R})$ and by the proposition below $\lambda(S^3) = 2\pi^2$, it follows that $\mu = \frac{\lambda}{2\pi^2}$ is normalized Haar measure on G.

Proposition 2.1.10. *In $\mathbb{R}^4$ let $B_4(r)$ stand for the ball centered at the origin of radius $r > 0$ and $S^3(r)$ the surface of the corresponding sphere. We denote the Lebesgue measures on $\mathbb{R}^n$ and $S^{n-1}(r)$ by vol_n and vol_{n-1} respectively.*
Then $\mathrm{vol}_4(B_4(r)) = \frac{\pi^2}{2}r^4$ and $\mathrm{vol}_3(S^3(r)) = 2\pi^2 r^3$.

Proof. As is well known the improper integral, $\int_{-\infty}^{\infty} e^{-t^2} dt = \sqrt{\pi}$. Hence by Fubini's theorem and the functional equation for exp we get $\int \cdots \int_{\mathbb{R}^n} e^{-\|x\|^2} dx_1 \ldots dx_n = \pi^{\frac{n}{2}}$. Writing this integral in polar coordinates gives $\pi^{\frac{n}{2}} = \int_0^{\infty} e^{-\rho^2} \rho^{n-1} d\rho \int_{S^{n-1}} d\Theta$, where $d\rho$ is Lebesgue measure on $(0, \infty)$ and $d\Theta$ is vol_{n-1} on $S^{n-1}(1) = S^{n-1}$. Now considering the volumes of two concentric spherical balls of radius r and $r + dr$ centered at 0 we see that $\frac{d\,\mathrm{vol}_n}{dr} = \mathrm{vol}_{n-1}(r)$. Since $\mathrm{vol}_n(B_n(r)) = c_n r^n$, where c_n is some constant (to be determined) it follows that $\mathrm{vol}_{n-1}(S^{n-1}(r)) = nc_n r^{n-1}$ so that

$$c_n = \frac{\pi^{\frac{n}{2}}}{n \int_0^{\infty} e^{-\rho^2} \rho^{n-1} d\rho},$$

this latter integral being the value of the gamma function at some half integral point. We shall calculate this integral when $n = 4$ using integration by parts.

$$\int_0^{\infty} u\,dv = (uv)|_0^{\infty} - \int_0^{\infty} v\,du.$$

Letting $dv = e^{-\rho^2} d\rho$ and $u = \rho^2$ we get $du = 2\rho d\rho$ and $v = -\frac{1}{2}e^{-\rho^2}$. Since the evaluative term is zero we conclude $\int_0^\infty e^{-\rho^2} \rho^3 d\rho = \frac{1}{2}$ so that $c_4 = \frac{\pi^2}{2}$ and hence the conclusions.   □

Exercise 2.1.11. Show that $\lambda(S^3) = 2\pi^2$ as follows: $\int_{\mathbb{R}^4} e^{-||x||^2} dx = (\int_{-\infty}^\infty e^{-t^2} dt)^4 = \pi^2$. On the other hand this is $\int_{\mathbb{R}^4-0} e^{-||x||^2} dx = \int_0^\infty e^{-r^2} r^3 dr \int_{S^3} d\lambda$. Thus $\lambda(S^3) = \frac{\pi^2}{\int_0^\infty e^{-r^2} r^3 dr}$. Then calculate the denominator using integration by parts.

2.2   Properties of the Modular Function

In general left and right Haar measures on a group are connected by the *modular function* Δ_G. This is a continuous map $\Delta_G : G \to \mathbb{R}_+^\times$ which measures the deviation from right invariance of left Haar measure dg and is defined as follows

$$\Delta_G(g) \int_G f(x)dx = \int_G f(xg)dx \qquad (2.2)$$

for all $f \in C_c(G)$.

Lemma 2.2.1.   *(1)* $\Delta_G : G \to \mathbb{R}_+^\times$ *is a homomorphism.*
    (2)

$$\int_G f(x^{-1})\Delta(x^{-1})dx = \int_G f(x)dx.$$

Proof. The proof of 1) is a direct check of the definition of the modular function. For 2) note that $f \mapsto \int_G f(x^{-1})\Delta(x^{-1})dx$ define a left invariant measure, therefore by uniqueness has to be $\int_G f(x)dx$.   □

For example we see that the modular function of the affine group of the real line is given by $\Delta_G(\begin{pmatrix} a & b \\ 0 & 1 \end{pmatrix}) = \frac{1}{|a|}$.

From these properties of the modular function we can immediately see that certain groups must be unimodular:

Corollary 2.2.2. *Compact and semisimple groups are unimodular.*

This is because Δ_G is a continuous homomorphism from $G \to \mathbb{R}_+^\times$ and such groups have no nontrivial homomorphisms into $\mathbb{R}_+^\times$.

Suppose for example G has a compact invariant neighborhood of the identity, U. Then for each $x \in G$, $\mu(xUx^{-1}) = \mu(U)$. But by left invariance $\mu(xUx^{-1}) = \mu(Ux^{-1})$ while the latter is $\Delta_G(x)\mu(U)$. Thus $\Delta_G(x)\mu(U) = \mu(U)$. Since $\mu(U)$ is finite and positive we see that Δ_G is identically 1. In particular, compact, discrete and of course abelian groups are unimodular.

Another example of a class of unimodular groups are the connected nilpotent ones. Since this fact will have little bearing on our work we will just sketch the proof. In such a group the center $Z(G)$ is always nontrivial by Corollary 3.2.9. So by induction on the dimension $G/Z(G)$ is unimodular. Let $d\bar{\mu}$ be the left and right invariant measure on this quotient group and dz be (left and right) Haar measure on $Z(G)$. Then by Theorem 2.3.5 below, $I(f) = \int_{G/Z(G)} \int_Z f(zx) dz d\bar{\mu}(\bar{x})$ is both left and right invariant on G.

2.3 Invariant Measures on Homogeneous Spaces

A natural step after studying Haar measure is to find conditions that guarantee a homogeneous G-space has a G-invariant measure and particularly a finite G-invariant measure. That is the main purpose of this section. Of course if H is normal in G then since G/H is a group it has an G-invariant that is G/H-invariant measure, namely Haar measure.

Definition 2.3.1. Let G be a locally compact group acting continuously on a locally compact space X (all spaces considered being Hausdorff). We shall call a nontrivial positive (regular) measure $d\mu(x)$ on X *invariant* if for each $g \in G$, and each measurable set $E \subseteq X$, we have $\mu(gE) = \mu(E)$. Alternatively $\int f(g \cdot x) d\mu(x) = \int f(x) d\mu(x)$ for every continuous function f on X with compact support.

Just as with Haar measure there can be no uniqueness to invariant measures.

An important special case of the definition is when G acts transitively on X. When X is G itself and the action is by left translation we have Haar measure. When G merely acts transitively, we know $X = G/H$, where H is a closed subgroup of G and G operates on G/H by left translation. As we shall see G/H has an (essentially) unique *invariant measure* if and only if

$$\Delta_G|_H = \Delta_H.$$

So for example, if G is unimodular (and non-compact) and Γ is discrete then both sides of this equation would be identically one and so G/Γ would always have a G-invariant measure (which may be infinite). If G/H were compact where H is a closed subgroup and G/H had an invariant measure then by regularity μ would have to be finite. Thus μ is finite and invariant. Hence if $G = \mathrm{GL}(n, \mathbb{R})$ and B is the full triangular group then G/B can have no invariant measure, because it would have to be finite. This cannot happen because the modular functions do not agree as G is unimodular and B is not. As we shall see in Chapter 9.3 there are other reasons why this cannot happen. Another way to think of this situation without considering, the finiteness of the measure is that since G is unimodular and B is not the nontrivial character $\frac{1}{\Delta_B}$ which must extend to G to get an invariant measure cannot do it. Such an extension restricted to $\mathrm{SL}(n, \mathbb{R})$ must be trivial. Hence, $\frac{1}{\Delta_B}$ must be trivial on $B^* = B \cap \mathrm{SL}(n, \mathbb{R})$ which it clearly is not.

The following fact is basic. Let G be a locally compact group and H be a closed subgroup with respective left Haar measures dg and dh and let $\pi : G \to G/H$ be the natural map.

For $f \in C_0(G)$ and $g \in G$ consider $f_g|_H$, the left translate of f by g restricted to H. This is in $C_0(H)$ so $F(g) = \int_H f(gh)dh$ exists for each $g \in G$. Moreover, if $h_1 \in H$, then $\int_H f(gh_1 h)dh = \int_H f(gh)dh$ so that $F(gh_1) = F(g)$. Hence F is constant on left cosets and gives a function on G/H.

Lemma 2.3.2. $F \in C_0(G/H)$.

Proof. Let $\epsilon > 0$. Since f is uniformly continuous choose a neighborhood $U(\epsilon)$ of 1 in G so that if $xy^{-1} \in U(\epsilon)$, then $|f(x) - f(y)| < \epsilon$. Also let U_0 be a fixed neighborhood of 1 in G. If $g_\nu \to g$, then eventually $g_\nu \in (U_0 \cap U(\epsilon))g$. Now $|F(g_\nu) - F(g)| \leq \int_H |f(g_\nu h) - f(gh)| dh$. Since $g_\nu h(gh)^{-1} = g_\nu g^{-1} \in U(\epsilon)$ we see $|f(g_\nu h) - f(gh)| < \epsilon$. We will show that function on H is 0 whenever $h \in H$ is outside the fixed compact set $H \cap g^{-1} U_0 \operatorname{Supp} f$. This is because $f(g_\nu h) = 0$ if $h \notin g_\nu^{-1} \operatorname{Supp} f$ and similarly $f(gh) = 0$ if $h \notin g^{-1} \operatorname{Supp} f$. So if $h \notin g_\nu^{-1} \operatorname{Supp} \cup g^{-1} \operatorname{Supp} f$, then $|f(g_\nu h) - f(gh)| = 0$. But $g_\nu^{-1} \in g^{-1} U_0$ so if $h \notin g^{-1} U_0 \operatorname{Supp} f \cup g^{-1} \operatorname{Supp} f = g^{-1} U_0 \operatorname{Supp} f$ we get $|f(g_\nu h) - f(gh)| = 0$. Since this set has finite H measure and $|F(g_\nu) - F(g)| \leq \epsilon \mu_H (H \cap g^{-1} U_0 \operatorname{Supp} f)$ it follows that F is continuous at each $g \in G$. Thus $F \in C(G/H)$.

Finally, suppose $\bar{g} \notin \pi(\operatorname{Supp} f)$. Then $f(gh) = 0$ for all $h \in H$. Hence $\int_H f(gh) dh = F(\bar{g}) = 0$ so $\bar{g} \notin \operatorname{Supp} F$. Thus $\operatorname{Supp} F \subseteq \pi(\operatorname{Supp} f)$ and so is compact. ☐

This enables us to define $I : C_0(G) \to C_0(G/H)$ by $f \mapsto F$. Evidently I is a linear map taking positive functions to positive functions.

Lemma 2.3.3. *I is surjective.*

Proof. Let $v \in C_0(G/H)$ and denote its lift back to G by $\tilde{v}$. Then for any $\psi \in C_0(G)$ we have $I(\psi \cdot \tilde{v}) = I(\psi) \cdot v$. This is because $\int_H \phi(gh) \tilde{v}(gh) dh = \int_H \phi(gh) v(\bar{g}) dh = v(\bar{g}) \int_H \phi(gh)$. Now let $u \in C_0(G/H)$. Since u has compact support choose an open set Ω in G with compact closure so that u vanishes outside $\pi(\Omega)$. By Urysohn's lemma choose $\psi \in C_0(G)$ so that $\psi \geq 0$ and $\psi|_\Omega \equiv 1$. Since $\bar{g} = gH$ where $g \in \Omega$ we have $\psi(g) = 1$. Thus $\psi(xh) \geq 0$ for all h and $\psi(g1) > 0$. Hence for $\bar{g} \in \pi(\Omega)$, $I(\psi)(\bar{g}) > 0$.

Define $v \in C_0(G/H)$ by $v(\bar{g}) = \frac{u(\bar{g})}{I(\psi)(\bar{g})}$, if $g \in \pi(\Omega)$ and $v(\bar{g}) = 0$ otherwise. Then v has compact support and is continuous on the open set $\pi(\Omega)$. Since u vanishes outside $\pi(\Omega)$ and is continuous, v is also continuous on the boundary so $v \in C_0(G/H)$. Also $u(\bar{g}) = I(\psi) v(\bar{g})$ everywhere on G/H. Hence $u = I(\psi) v = I(\psi \cdot \tilde{v})$. ☐

We keep the same notation. Let the modular functions on G and H be Δ_G and Δ_H.

Lemma 2.3.4. *Suppose $\Delta_G|_H \equiv \Delta_H$. Let $f \in C_0(G)$. If $I(f) = 0$, then $\int_G f(g) dg = 0$.*

Proof. Let $\phi \in C_0(G)$ be a function such that $I(\phi) \equiv 1$ on Supp f. Now
$\int_H \int_G \phi(g)f(gh)dgdh = \int_G \phi(g)dg \int_H f(gh)dh = 0$.

Also since $\int_G |\phi(g)|dg \int_H |f(gh)|dh < \infty$, Fubini's theorem applies. Hence

$$0 = \int_H \int_G \phi(g)f(gh)dgdh = \int_H \int_G \phi(gh^{-1})f(g)\Delta_G(h)dgdh.$$

But this is

$$\int_G f(g) \int_H \phi(gh^{-1})\Delta_G(h)dhdg = \int_G f(g) \int_H \phi(gh)\Delta_H(h^{-1})\Delta_G(h)dhdg$$

$$= \int_G f(g) \int_H \phi(gh)dhdg = \int_G f(g)$$

by the choice of ϕ.   □

Theorem 2.3.5. *Let G be a locally compact group and H be a closed subgroup. Then there exists an essentially unique invariant measure $d\bar{g}$ on G/H satisfying*

$$\int_G f(g)dg = \int_{G/H} \int_H f_g(h)dhd\bar{g},$$

$f \in C_0(G)$ if and only if $\Delta_G|_H = \Delta_H$.

Proof. Suppose G/H has a G-invariant measure $d(\bar{g})$. Let $f \in C_0(G)$. Consider $F \in C_0(G/H)$ as above. Then $\int_{G/H} F(\bar{g})d(\bar{g}) = J(f)$ is a positive linear functional on $C_0(G)$.

$$J(f) = \int_{G/H} \int_H f(gh)dhd(\bar{g}).$$

For $g_1 \in G$,

$$J(f_{g_1}) = \int_{G/H} \int_H f(g_1gh)dhd(\bar{g}) = \int_{G/H} F(g_1g)d(\bar{g})$$

$$= \int_{G/H} F(g)d(\bar{g}) = J(f).$$

So that J is G-invariant. Since I is surjective, J is also non-trivial. This means by uniqueness of Haar measure that J must be Haar measure on G with some normalization. Any two such measures on G/H give Haar measure after normalization. Since I is surjective these measures must coincide. Furthermore $\int_G f(g)dg = \int_{G/H} \int_H f(gh)dhd(\bar{g})$. Let $h_1 \in H$, then $\int_G f(gh_1)dg = \int_G R_{h_1} f(g)dg = \Delta_G(h_1) \int_G f(g)dg$. On the other hand this is $\int_{G/H} \int_H R_{h_1} f(gh)dhd(\bar{g}) = \int_{G/H} \int_H \Delta_H(h_1)f(gh)dhd(\bar{g}) = \Delta_H(h_1) \int_G f(g)dg$. Hence for every $f \in C_0(G)$, $\Delta_G(h_1) \int_G f(g)dg = \Delta_H(h_1) \int_G f(g)dg$. This means $\Delta_G|_H \equiv \Delta_H$.

Conversely, suppose $\Delta_G|_H \equiv \Delta_H$. Then by Lemma 2.3.4 the linear form $f \mapsto \int_G f$ factors through I and defines an invariant measure on G/H. Uniqueness follows from that of Haar measure. □

Our next result is usually referred to as pushing a measure forward. Its proof is straightforward and is left to the reader.

Proposition 2.3.6. *Suppose X and Y are locally compact G-spaces, μ is a regular, G-invariant measure on X and $\pi : X \to Y$ is a continuous, surjective G-equivariant map, then for any measurable set $S \subseteq Y$ we define $\nu(S) = \mu(\pi^{-1}(S))$. It is easy to see that ν is a regular G-invariant measure on Y which is finite if μ is.*

We can apply some of this to calculate Haar measure on $\mathrm{SL}(n, \mathbb{R})$. (As with $\mathrm{SU}(2, \mathbb{C})$ this group is also not diffeomorphic to a single open set in some Euclidean space.) Write the Iwasawa decomposition of $\mathrm{SL}(n, \mathbb{R}) = G = KAN$, where here $K = \mathrm{SO}(n)$ and $AN = B^+$, the real triangular matrices all of whose eigenvalues are positive and of det $= 1$. Let dk, da^+, and dn be Haar measures on K, A^+ and N respectively. Because these groups are compact, abelian, or nilpotent they are all unimodular (for the compact case see Corollary 2.2.2. Here N is actually the nilpotent group of Exercise 2.1.6). Write $G = B^+K$. Hence the map $G/K \to B^+$ is a B^+-equivariant diffeomorphism. Since both G and K are unimodular G/K has a G-invariant and therefore B^+-invariant measure, which by Proposition 2.3.6 can be pushed forward to give left Haar measure db^+ on B^+. Hence by Theorem 2.3.5 $f \mapsto \int_{B^+} db^+ \int_K f(b^+k)dk$

is (left) Haar measure on the unimodular group G. So we are reduced to the question of what Haar measure is on B^+? Since B^+ is the semidirect product of A^+ and N using the semi-direct product result one gets $db^+ = \Pi_{i<j} \frac{a_{ii}}{a_{jj}} da^+ dn$ and so $dg = \Pi_{i<j} \frac{a_{ii}}{a_{jj}} dk \, da^+ dn$.

2.4   Compact or Finite Volume Quotients

Definition 2.4.1. We say that a closed subgroup H of G has *cofinite volume* in G if G/H has a finite G-invariant measure. We shall say that H is *cocompact* or a *uniform subgroup* of G if G/H is compact. In particular, if Γ is discrete and of cofinite volume then we say Γ is a *lattice* in G; if Γ is a discrete subgroup and G/Γ is compact then we say that Γ is a *uniform lattice* in G.

Notice that if Γ is a lattice or a uniform lattice in a connected Lie group, G, we can always pull this back to such a thing in $\tilde{G}$, its universal covering group. This means that in many situations we may as well assume G itself is simply connected.

Proposition 2.4.2. *If a locally compact group G admits a lattice it must be unimodular.*

Proof. Observe that $\Delta_G|_\Gamma = \Delta_\Gamma = 1$. Hence $\Gamma \subseteq \operatorname{Ker} \Delta_G$. So that the finite measure on G/Γ pushes forward to give a finite G-invariant measure on $G/\operatorname{Ker} \Delta_G \subseteq \mathbb{R}_+^\times$. As $\operatorname{Ker} \Delta_G$ is normal in G, $G/\operatorname{Ker} \Delta_G$ is a group so by uniqueness this must be left Haar measure. Since the measure is finite $G/\operatorname{Ker} \Delta_G$ is compact. On the other hand $\mathbb{R}_+^\times$ has no nontrivial compact subgroups and so G is unimodular. $\qquad \square$

Thus for example the group of affine motions of the real line not being unimodular cannot have lattices. There are also other necessary conditions, but there are no known necessary and sufficient conditions, in general, for a locally compact group or even a Lie group to possess a lattice. However, Borel has shown [7] that any connected semisimple Lie group of non-compact type has both uniform and non-uniform lattices. It is a theorem of Mostow [71] that in a connected solvable Lie group G

and a closed subgroup H, then G/H is compact if and only if it carries a finite invariant measure. In particular, this holds for nilpotent groups and discrete subgroups. A theorem of Malcev [71] tells us a simply connected nilpotent Lie group has a lattice if and only if the Lie algebra has a basis in which all structure constants are rational. Proposition 3.1.69 will show that there are simply connected 2-step nilpotent groups which have no lattices. The next few Propositions will be useful.

Proposition 2.4.3. *Let G be a locally compact group, Γ a discrete subgroup. If Ω is a measurable set in G of finite Haar measurable satisfying $\Omega\Gamma = G$, then Γ is a lattice in G. That is, G/Γ has a finite invariant measure.*

Proof. Choose measures dg, $d\bar{g}$ and $d\gamma$, with $d\bar{g}$ invariant as in Theorem 2.3.5 so that

$$\int_G dg = \int_{G/\Gamma} \int_\Gamma d\gamma d\bar{g}$$

and apply this to χ_Ω, the characteristic function of Ω. Then

$$\int_{G/\Gamma} \int_\Gamma \chi_\Omega(g\gamma) d\gamma d\bar{g} = \mu(\Omega) < \infty.$$

Since each $g \in G$ is of the form $g = \omega_1\gamma_1$ and $d\gamma$ is left invariant we see that because Γ is discrete

$$\int_\Gamma \chi_\Omega(g\gamma)d\gamma = \int_\Gamma \chi_\Omega(\omega_1\gamma)d\gamma = \sum_{\gamma \in \Gamma} \chi_\Omega(\omega_1\gamma).$$

Now this last term is everywhere ≥ 1. If not, $\chi_\Omega(\omega_1\gamma) = 0$ for all $\gamma \in \Gamma$; that is for all γ, $\omega_1\gamma$ lies outside of Ω. Thus $\Gamma \cap \omega_1^{-1}\Omega$ is empty. This is impossible since $1 \in \Gamma$ and $1 = \omega_1^{-1}\omega_1$. Thus $\infty > \mu(\Omega) \geq \int_{G/\Gamma} 1d\bar{g}$. An application of Proposition 2.3.6 completes the proof. $\square$

Conversely we have

Proposition 2.4.4. *Let G be a Lie group and Γ a lattice. Then there exists a measurable set Ω in G of finite measure satisfying $\Omega\Gamma = G$.*

Actually, more is true as we shall see in Chapter 8. For any discrete subgroup Γ, there exists an open set $\Omega \subseteq G$ satisfying

(1) For any two distinct γ_1 and $\gamma_2 \in \Gamma$ the sets $\gamma_1 \Omega$ and $\gamma_2 \bar{\Omega}$ are disjoint.

(2) $\bigcup_{\gamma \in \Gamma} \gamma \bar{\Omega} = G$.

Here we need the fact that the space on which Γ acts is a complete Riemannian manifold.

In particular from 2), taking inverses, we have $G = (\Omega)^{-1} \Gamma$ so $\pi : \Omega^{-1} \to G/\Gamma$ is surjective and injective. Since G/Γ has finite volume with respect to the push forward measure therefore Ω^{-1} has a finite measure and hence also Ω.

The sister result to Propositions 2.4.3 and 2.4.4 above is the following:

Proposition 2.4.5. *Let G be a locally compact group and H be a closed subgroup. Then G/H is compact if and only if there is a compact symmetric neighborhood Ω of 1 in G satisfying $\Omega H = G$.*

In particular, if H and G/H are both compact, then so is G. An important special case of this is the situation where G is a compact connected Lie group and $\tilde{G}$ is its universal cover. Then $\tilde{G}$ is compact if and only if $\pi_1(G)$ is finite.

Proof. If there is such an Ω, then $\pi : G \to G/H$ is surjective when restricted to Ω. Since π is continuous and Ω is compact so is G/H.

Conversely, choose a compact neighborhood U of 1 in G. Then since π is both continuous and open $\pi(U)$ is a compact neighborhood of $\pi(1)$ in G/H. Cover G/H by a finite number of its G translates $g_i \pi(U)$ where $i = 1, \ldots, n$. Then $\Omega = \bigcup_{i=1}^{n} g_i(U)$ is compact and $\Omega H = G$. Since, we can always include 1 as one of the translates, Ω is a neighborhood of 1. By replacing Ω by $\Omega \cup \Omega^{-1}$ we may assume Ω is symmetric. $\square$

Proposition 2.4.6. *If G/H is compact and dg and dh are the respective left Haar measures then there is a non-negative function ω in $C_0^+(G)$ such that $\int_H \omega_g|_H \equiv 1$. Hence if f the lift back to G of a continuous function $\bar{f}$ on G/H, then $\int_G \omega(g) f(g) dg = \int_{G/H} \bar{f}(\bar{g}) d\bar{g}$.*

Proof. Since G/H is compact and I is surjective by Lemma 2.3.3 the constant function 1 has an inverse image. For the second statement,

$$\int_G \omega(g)f(g)dg = \int_{G/H}\int_H \omega(gh)f(gh)dhd\bar{g}.$$

But this last term is $\int_{G/H}\bar{f}(\bar{g})(\int_H \omega(hg)dh)d\bar{g} = \int_{G/H}\bar{f}(\bar{g})d\bar{g}.$ ☐

We now formulate two general propositions concerning subgroups of cofinite volume which are analogous to the second and third isomorphism theorems for topological groups (see Corollaries 1.4.9 and 1.4.10).

Proposition 2.4.7. *Let G be a locally compact σ-compact group and L and H closed subgroups with H normalizing L and HL closed in G. Then HL/H has a finite HL-invariant measure if and only if $L/H \cap L$ has finite L-invariant measure.*

Proof. Consider the natural map $L/H \cap L \to HL/H$. As we saw there this map is a homeomorphism which intertwines the actions L on the first and HL on the second. By Proposition 2.3.6 (perhaps slightly generalized to two different groups acting) applied to the inverse of the map if HL/H has a finite HL-invariant measure then $L/H \cap L$ has finite L-invariant measure.

Conversely, if $L/H \cap L$ has finite L-invariant measure then by the same reasoning as above we see that HL/H also has a finite L-invariant measure μ. For $h \in H$ let ν_h be the measure on HL/H defined by $\mu_h(E) = \mu(hE)$, where E is a measurable set in HL/H. Now for $l \in L$, we have by L-invariance,

$$\mu_h(lE) = \mu(hlE) = \mu(hlh^{-1}hE) = \mu(l'hE) = \mu(hE) = \mu_h(E).$$

Thus each μ_h is also an L-invariant measure on HL/H. By uniqueness $\mu_h = \lambda(h)\mu$, where $\lambda : H \to \mathbb{R}_+^{\times}$ is a character. But μ is a finite measure. Therefore letting $E = HL/H$ we see $\lambda(h)\mu(E) = \mu_h(E) = \mu(hE) = \mu(E)$ so that $\lambda(h) \equiv 1$ and $\mu_h = \mu$ for all h and μ is H-invariant. This means μ is HL-invariant. ☐

Proposition 2.4.8. *Let G be a locally compact group and H_1 and H_2 closed subgroups with $H_1 \supseteq H_2$. Then G/H_2 has a finite G-invariant measure if and only if G/H_1 and H_1/H_2 each have a finite G-invariant measure.*

Proof. If G/H_2 has a finite G-invariant measure then so does G/H_1 since $\pi : G/H_2 \to G/H_1$ can be used to push this measure forward (see Proposition 2.3.6). Since both G/H_2 and G/H_1 carry invariant measures we know from Theorem 2.3.5 that $\Delta_G|H_i = \Delta_{H_i}$ for $i = 1, 2$. Hence $\Delta_{H_1}|H_2 = \Delta_{H_2}$ and therefore again by Theorem 2.3.5 H_1/H_2 also carries an H_1-invariant measure. Let dx, dy and dz be these measures. Consider the linear functional I defined on $C_0(G/H_2)$ by

$$I(f) = \int_{G/H_1} \left(\int_{H_1/H_2} f(ghH_2)dz \right)dy. \tag{2.3}$$

This is a G invariant measure on G/H_2 so by uniqueness it is dx after normalization. Applying (2.3) to the constant function 1 tells $\int_{G/H_1} dy \int_{H_1/H_2} dz = 1$. By Fubini's theorem $\int_{H_1/H_2} dz < \infty$.

Conversely if G/H_1 and H_1/H_2 each carries a finite G-invariant measure, then (2.3) defines a finite G-invariant measure on G/H_2. $\quad\square$

Next we come to a general result which is useful in distinguishing uniform from non-uniform lattices. Refinements of this result play an important role in arithmetic groups.

Theorem 2.4.9. *Let G be a connected Lie group, Γ be a lattice in G and $\pi : G \to G/\Gamma$ the natural projection. For a sequence x_n in G, $\pi(x_n)$ has no convergent subsequence if and only if there exists a sequence $\{\gamma_n\} \neq 1$ in Γ so that $\{x_n \gamma_n x_n^{-1}\}$ converges to 1.*

So for example if Γ is a uniform lattice, then given any sequence $x_n \in G$, the only sequence $\gamma_n \in \Gamma$ for which $x_n \gamma_n x_n^{-1} \to 1$ is one where eventually all $\gamma_n = 1$.

Proof. Since G is connected and locally compact it is σ-compact and because G has a lattice it is unimodular Proposition 2.4.2. As a σ-compact group choose an increasing sequence F_n of compact subsets which fill out G. Let μ be Haar measure on G and $\bar{\mu}$ a finite invariant measure on G/Γ. Since π is surjective $\pi(\cup F_n) = G/\Gamma$. Now $\pi(F_n)$ is compact and measurable and $\bar{\mu}$ is finite so by Ergoroff's theorem $\bar{\mu}(\pi(F_n)) \uparrow \bar{\mu}(G/\Gamma)$. Letting $\epsilon_n = \bar{\mu}(G/\Gamma) - \bar{\mu}(\pi(F_n))$ it follows that $\epsilon_n \downarrow 0$. Since G obeys the first axiom of countability, choose a fundamental sequence $\{V_n\}$ of compact neighborhoods of 1 in G with $\mu(V_n) > \epsilon_n$. This can be done by considering balls $B(r)$ of radius $r > 0$ centered at 0 in $\mathfrak{g}$, the Lie algebra. Now $r \mapsto \mu(\exp(B(r)))$ is a continuous function on some neighborhood $0 < r < \delta$ of 0 and takes on all positive values in some interval. Therefore, there is a sequence $B(r_n)$ with $r_n \downarrow 0$, with $\mu(\exp(B(r_n))) = 2\epsilon_n$ for each n. Letting $V_n = \exp(B(r_n))$ gives such a sequence.

Now $V_n V_n^{-1}$ is also a fundamental sequence of compact neighborhoods of 1 in G. Suppose $\pi(x_n)$ has no limit point in G/Γ. Since for each n, $\pi(V_n V_n^{-1} F_n)$ is compact for any n there must be an integer k_n so that $\pi(x_m) \notin \pi(V_n V_n^{-1} F_n)$ if $m \geq k_n$. As a consequence $\pi(V_n x_m) \cap \pi(V_n F_n) = \emptyset$ for all $m \geq k_n$ because if for some γ and $m \geq k_n$, $v_n x_m \gamma = v_n' f_n \gamma'$, then $x_m = v_n^{-1} v_n' f_n \gamma''$ so $\pi(x_m) \in \pi(V_n^{-1} V_n' F_n)$, a contradiction.

Therefore $\pi(V_n x_m) \subseteq G/\Gamma - \pi(V_n F_n) \subseteq G/\Gamma - \pi(F_n)$ and so $\bar{\mu}(\pi(V_n x_m)) \leq \bar{\mu}(G/\Gamma - \pi(F_n)) = \epsilon_n$. But $\mu(V_n x_m) = \mu(V_n) > \epsilon_n$. Since Γ is discrete so measure on Γ is given by counting, therefore for a measurable set $S \subset G$ which intersects Γ in at most one point we have $\mu(S) = \bar{\mu}(\pi(S))$. Since V_n is a neighborhood basis at 1 this is a contradiction unless (taking $V_n x_m$ for S) there is a γ_m and $\gamma_m' \in \Gamma$, where $\gamma_m \neq \gamma_m'$ so that $v x_m \gamma_m = v' x_m \gamma_m'$. But then $x_m \gamma_m'' x_m^{-1} = v^{-1} v'$. Therefore for each n there exist a large enough m and a $\gamma_m'' \in \Gamma$ such that $x_m \gamma_m'' x_m^{-1} \in V_n V_n^{-1}$; hence $x_m \gamma_m'' x_m^{-1}$ converges to 1.

Conversely, let $x_n \in G$ be a sequence and suppose there is a sequence $\gamma_n \in \Gamma$ eventually $\neq 1$ with $x_n \gamma_n x_n^{-1}$ converging to 1 as $n \to \infty$. We show that the image π of such a sequence cannot have a limit point. For suppose it did, say $\pi(x_n) \to \pi(x)$. Since Γ is discrete, π is a local homeomorphism. So in some neighborhood of x in G we can find $\theta_n \in \Gamma$ so that $x_n \theta_n \to x$. But then $x_n \gamma_n x_n^{-1} = x_n \theta_n \theta_n^{-1} \gamma_n \theta_n \theta_n^{-1} x_n^{-1}$ so the limiting value of this is $x \theta_n^{-1} \gamma_n \theta_n x^{-1} \to 1$. But then $\theta_n^{-1} \gamma_n \theta_n$ must also converge to 1. Since Γ is discrete $\theta_n^{-1}\gamma_n\theta_n$ must eventually be 1. Hence γ_n is also eventually 1, a contradiction. $\qquad\square$

We now apply Theorem 2.4.9 to the following question. Let G be a Lie group H a closed subgroup and Γ a lattice in G. When is $\Gamma \cap H$ a lattice in H? As we shall see this is a rare occurrence.

Corollary 2.4.10. *Let G be a connected Lie group, H be a closed subgroup and Γ be a lattice in G. If $H \cap \Gamma$ is a lattice in H if and only if $H\Gamma$ is closed in G. Equivalently the injection $\iota : H/H \cap \Gamma \to G/\Gamma$ is proper.*

If H is normal then these conditions are equivalent by Proposition 2.4.7.

Proof. Suppose $H \cap \Gamma$ is a lattice in H and $\pi_H : H \to H/H \cap \Gamma$ and $\pi_G : G \to G/\Gamma$ be the natural projections. To show that ι is proper it is enough to prove that for a sequence $h_n \in H$, $\pi_H(h_n)$ has a limit point if and only if $\pi_G(h_n)$ has one. If $\pi_H(h_n)$ converges so does $\pi_G(h_n)$ because ι is continuous. Suppose $\pi_G(h_n)$ converges, but $\pi_H(h_n)$ has no limit point. Then by Theorem 2.4.9 there are elements $\gamma_n \in H \cap \Gamma$ such that $h_n\gamma_nh_n^{-1}$ converges to 1. Then by Theorem 2.4.9 again $\pi_G(h_n)$ has no limit point, a contradiction. Consider the commutative diagram

$$
\begin{array}{ccc}
H & \longrightarrow & G \\
\downarrow{\scriptstyle \pi_H} & & \downarrow{\scriptstyle \pi_G} \\
H/H \cap \Gamma & \stackrel{\iota}{\longrightarrow} & G/\Gamma
\end{array}
\qquad (2.4)
$$

We have $H\Gamma = \pi_G^{-1}(\iota(H/\Gamma \cap H))$, by the argument above ι sends the closed sets to closed sets which shows $\iota(H/\Gamma \cap H)$ is closed and since π_G is continuous therefore $H\Gamma$ is closed. $\qquad\square$

2.5   Applications

In this section we shall give a number of applications of Haar measure on a compact group to derive various algebraic and geometric properties of compact Lie groups.

A first application is the fact that compact linear groups are completely reducible.

Theorem 2.5.1. $\rho : G \to \mathrm{GL}(V)$ *be a finite dimensional continuous representation of a compact group on a real or complex linear space V. Then any invariant subspace W of V has an invariant complement.*

Proof. By replacing G by $\rho(G)$ we may assume the compact group is a subgroup of $\mathrm{GL}(V)$. Let $(\cdot, \cdot)$ be any positive definite symmetric (respectively Hermitian) form on V and let dg be normalized right Haar measure on G. Then the form $\langle \cdot, \cdot \rangle$ is defined on V as follows:

$$\langle v, w \rangle = \int_G (gv, gw) dg.$$

Since $\int_G$ is linear and $(\cdot, \cdot)$ is bilinear symmetric (respectively Hermitian conjugate linear) it follows $\langle \cdot, \cdot \rangle$ is also bilinear symmetric (respectively Hermitian conjugate linear). In addition $\langle \cdot, \cdot \rangle$ is positive definite. For $\langle v, v \rangle = \int_G (gv, gv) dg \geq 0$ because the integrand $(gv, gv) \geq 0$ everywhere since $(\cdot, \cdot)$ itself is positive definite. If $\langle v, v \rangle = 0$ then $v = 0$. This is because the integrand $(gv, gv) \geq 0$ and is positive at $g = 1$ unless of course v itself is 0. Finally, $\langle \cdot, \cdot \rangle$ is G-invariant. For any $h \in G$, because dg is right invariant

$$\langle hv, hw \rangle = \int_G (ghv, ghw) dg = \int_G (gv, gw) dg = \langle v, w \rangle.$$

Thus V has a positive definite invariant symmetric (respectively Hermitian) form. It follows from this that $W^{\perp}$ with respect to this form is also G-invariant. For if $w^{\perp} \in W^{\perp}$, $g \in G$ and $w \in W$ $\langle (gw^{\perp}), w \rangle = \langle w^{\perp}, g^{-1}w \rangle$. Since W is G-invariant and $w^{\perp} \in W^{\perp}$ this last term is zero so that $gw^{\perp} \in W^{\perp}$. Thus $W^{\perp}$ is G-invariant. Choosing an orthonormal basis for V by putting together two such forms W and $W^{\perp}$ shows G acts completely reducibly. $\square$

Proposition 2.5.2. *Let G be a compactly generated locally compact group and H a closed subgroup with G/H compact. Then H is also compactly generated.*

Proof. By Proposition 2.4.5 choose a compact symmetric neighborhood U_0 of 1 in G with $G = U_0 H$ and large enough so that it generates G. Then U_0^2 is compact and is contained in $G = U_0 H$. Therefore $U_0^2 \subseteq \bigcup_{i=1}^{n} U_0 h_i$, where $h_i \in H$. Let $F = \{h_1, \ldots, h_n\}$ and $\langle F \rangle$ be the (finitely generated) subgroup of H generated by F. Since $U_0^2 \subseteq U_0 F \subseteq U_0 \langle F \rangle$ we see that $U_0^3 \subseteq U_0^2 \langle F \rangle = U_0 \langle F \rangle^2 = U_0 \langle F \rangle$. Continuing in this way it follows that $U_0^n \subseteq U_0 \langle F \rangle$ for every $n \geq 1$. Since U_0 generates G we get $G \subseteq U_0 \langle F \rangle$ and in particular $H \subseteq U_0 \langle F \rangle$. Let $U_{0,H} = U_0 \cap H$. Then this is a compact neighborhood of 1 in H and $H = U_{0,H} \langle F \rangle$. Thus H is compactly generated. $\qquad\square$

This has as a consequence

Corollary 2.5.3.   *(1) Let G be a connected locally compact group and Γ a discrete cocompact subgroup. Then Γ is finitely generated.*

(2) If X is a compact space on which a connected Lie group acts transitively, then $\pi_1(X)$ is finitely generated.

(3) If G be a compact connected Lie group, then $\pi_1(G)$ is finitely generated.

Proof.   (1) The first statement follows from the fact that a connected locally compact group is compactly generated.

(2) Let $X = G/H$ and $(\hat{G}, \pi)$ be the universal covering group of G. Then $\hat{G}/\pi^{-1}(H)$ is homeomorphic and $\hat{G}$-equivariantly equivalent with $G/H = X$. So these spaces have the same fundamental groups. But $\pi_1(\hat{G}/\pi^{-1}(H)) = \pi^{-1}(H)/\pi^{-1}(H)_0$. Since $\hat{G}/\pi^{-1}(H)$ is compact and $\hat{G}$ is connected $\pi^{-1}(H)$ is compactly generated. Hence so is $\pi^{-1}(H)/\pi^{-1}(H)_0$. It is also discrete because $\pi^{-1}(H)$ is closed in $\hat{G}$ and therefore is also a Lie group. Hence $\pi^{-1}(H)/\pi^{-1}(H)_0$ is discrete and therefore finitely generated.

(3) This follows from 2 by letting $G = X$ and $H = \{1\}$. $\qquad\square$

Proposition 2.5.4. *The fundamental group of a connected Lie group is abelian.*

Proof. Let $(\hat{G}, \pi)$ be the universal covering group of G. Then $\operatorname{Ker}\pi$ is the fundamental group $\pi_1(G)$ and it is normal discrete subgroup of $\hat{G}$. Thus all we need to know is that a discrete normal subgroup of a connected group is central and therefore abelian and for that see Corollary 0.3.7.   □

The following is a modification of an important observation of Pierre Cartier . We first remark that if H is a locally compact group and dh is right Haar measure then we can uniquely extend dh to the vector valued functions on H *i.e.* to $C_0(H, V)$ where V is a finite dimensional real vector space as follows: For $\lambda \in V^*$, the dual space and $\phi \in C_0(H, V)$ we define $\int \phi dh$ by

$$\lambda(\int_H \phi dh) = \int_H \lambda \phi dh.$$

Thus we integrate in each coordinate. When we do this we again get an H-invariant linear vector valued function. Now if we take as our vector space $\operatorname{End}_{\mathbb{R}}(V)$ and if T is a fixed linear operator on V, then for a continuous $\phi : H \to \operatorname{End}_{\mathbb{R}}(V)$ with compact support we have

$$T(\int_H \phi dh) = \int_H T\phi dh.$$

Theorem 2.5.5. *Let G be a locally compact group and H a closed normal subgroup with G/H compact and ρ be a continuous finite dimensional representation of G on the real vector space V whose restriction to H is trivial. If $f : H \to V$ is a group homomorphism i.e.*

$$f(hh') = f(h) + f(h')$$

satisfying the invariance condition

$$\rho(g)(f(h)) = f(ghg^{-1})$$

for all $g \in G$ and $h \in H$, then f extends to a continuous $f^* : G \to V$ satisfying[1] *$f^*(xy) = \rho(x)(f^*(y)) + f^*(x)$ for all $x, y \in G$.*

---

[1]This is 1-cocycle condition and leads to a degree one class in the cohomology of group of G.

Proof. Since G/H is compact by Proposition 2.4.6 there exists a weighting function $\omega \in C_0(G, \mathbb{R})$ such that for all $x \in G$

$$\int_H \omega(hx)dh \equiv 1.$$

Since $f(h'h) = f(h') + f(h)$ for $h, h' \in H$ we have,

$$\int_H \omega(h'x)f(h'h)dh' - \int_H \omega(h'x)f(h')dh' = f(h)\int_H \omega(h'h)dh' = f(h).$$

Now translate $h' \mapsto h'h^{-1}$ and apply right invariance giving

$$\int_H \omega(h'h^{-1}x)f(h')dh' - \int_H \omega(h'h^{-1}x)f(h'h^{-1})dh' = f(h).$$

So that

$$\int_H (h^{-1}x)\omega \cdot f\,dh' - \int_H (x)\omega \cdot f\,dh' = f(h), \qquad (2.5)$$

here $y.\omega$ means the right translation action. Define $f_1 : G \to V$ by

$$f_1(x) = \int_H x\omega \cdot f\,dh' - x \cdot \int_H \omega \cdot f\,dh',$$

for $x \in G$. Since ρ is continuous f_1 is a continuous V-valued function and

$$f_1(h^{-1}x) - f_1(x) = \int_H (h^{-1}x\cdot\omega)f - h^{-1}x\int_H (\omega)f - \left(\int_H (x\cdot\omega)f - x\cdot\int_H \omega f\right).$$

By (2.5) get $f_1(h^{-1}x) - f_1(x) = f(h) - h^{-1}x\int_H \omega f + x\int_H \omega f$. Let $v_0 = x\int_H \omega f \in V$. Then for all $h \in H$ and $x \in G$, $f_1(h^{-1}x) - f_1(x) = f(h) - h^{-1}v_0 + v_0$. But since H acts trivially on V we get

$$f_1(h^{-1}x) - f_1(x) = f(h). \qquad (2.6)$$

For each $x \in G$ let $g_x : G \to V$ be defined by

$$g_x(t) = t^{-1}(f_1(t) - f_1(tx)),$$

for $t \in G$. Since ρ is continuous and right translation, also each g_x is a continuous function of t. We prove that g_x is constant on right cosets of H in G. We have $g_x(ht) = (ht)^{-1}(f_1(ht) - f_1(htx))$. By (2.6) we know $f_1(htx) - f_1(tx) = f(h^{-1})$ and $f_1(ht) - f_1(t) = f(h^{-1})$, therefore

$$f_1(ht) - f_1(htx) = f_1(t) - f_1(xt).$$

Since H acts trivially we have $g_x(ht) = t^{-1}h^{-1}(f_1(t) - f_1(xt)) = g_x(t)$. Let $\bar{g}_x$ be the induced function on cosets and then consider the V-valued integral

$$f^*(x) = \int_{G/H} \bar{g}_x(\bar{t})d\bar{t},$$

where $d\bar{t}$ is the normalized Haar measure on G/H. Then $f^* : G \to V$ is continuous. For $h \in H$ we have $g_h(t) = t^{-1}(f_1(t) - f_1(th)) = t^{-1}(f_1(t) - f_1(tht^{-1}t))$. Since $h' = tht^{-1} \in H$, using (2.6) we get $-t^{-1}(f_1(h't) - f_1(t)) = -t^{-1}f((tht^{-1})^{-1}) = -t^{-1}f(th^{-1}t^{-1}) = -f(h^{-1}) = f(h)$. Thus for all $h \in H$ and $t \in G$ we get $g_h(t) = f(h)$. This means $f^*(h) = \int_{G/H} \bar{g}_h(\bar{t})d\bar{t} = \int_{G/H} f(h)d\bar{t} = f(h)$, so that f^* is an extension of f.

For $x, t, u \in G$, a direct calculation similar to one just above tells us

$$u^{-1}((ux)g_t(ux)) = u^{-1}(f_1(ux) - f_1(uxt))$$

and so $x \cdot g_t(ux) = u^{-1}(f_1(ux) - f_1(uxt))$. This means $g_{xt}(u) = g_x(u) + x \cdot g_t(ux)$ and therefore. $\bar{g}_{xt} = \bar{g}_x + x \cdot \bar{g}_t(R_x(u))$ (right translation). Integrating this last equation over G/H yields

$$f^*(xt) = f^*(x) + x \cdot \int_{G/H} \bar{g}_t(R_x(\bar{u}))d\bar{u}.$$

But since $\int_{G/H} \bar{g}_t(R_x(\bar{u}))d\bar{u} = \int_{G/H} \bar{g}_t(\bar{u})d\bar{u}$ by invariance of the integral we get

$$f^*(xt) = f^*(x) + x \cdot f^*(t).$$

$\square$

Corollary 2.5.6. *Let G be a locally compact group and V a normal vector subgroup with G/V compact. Then G is a semidirect product of a compact subgroup K of G with V.*

Proof. Now conjugation on G leaves V stable and gives a continuous representation $g \to \alpha_g|_V$ which we denote by α of G on V. Since V is abelian α restricted to V is trivial. Evidently i is a continuous homomorphism $V \to V$ satisfying the invariance condition with respect to α. Hence i extends to a continuous map $i^* : G \to V$ as in Theorem 2.5.5 satisfying

$$i^*(xy) = xi^*(y)x^{-1} \cdot i^*(x),$$

here we write the group V multiplicatively. We show the sequence

$$1 \to V \xrightarrow{i} G \xrightarrow{\pi} G/V \to 1$$

has a continuous global cross section.

For $x \in G$ let $\Phi(x) = i^*(x)^{-1}x$. Then Φ is a continuous map $G \to G$. Moreover

$$\Phi(xy) = i^*(xy)^{-1}xy = (xi^*(y)x^{-1}i^*(x))^{-1}xy = i^*(x)^{-1}xi^*(y)^{-1}x^{-1}xy$$
$$= i^*(x)^{-1}xi^*(y)^{-1}y = \Phi(x)\Phi(y).$$

Thus $\Phi : G \to G$ is a continuous homomorphism. If $v \in V$ then $\Phi(v) = i^*(v)v^{-1} = 1$ since i^* extends i. Hence $V \subseteq \mathrm{Ker}\,\Phi$ and therefore Φ induces a continuous homomorphism $\bar{\Phi} : G/V \to G$ with $\bar{\Phi}(G/V) = \Phi(G) = K$, a compact subgroup of G. Finally we show $\pi \circ \bar{\Phi} = id_{G/V}$. Since V is normal $g^{-1}i^*(g)^{-1}g \in V$ for all $g \in G$. Thus $gV = i^*(g)^{-1}gV$ from which $\pi \circ \bar{\Phi} = id_{G/V}$ follows. $\square$

Using Corollary 2.5.6 and induction on the length of the derived series together with the fact that a connected solvable Lie group is simply connected if and only if it has no nontrivial compact subgroup, one can extend this result to arbitrary simply connected solvable subgroups as follows. We leave this as an exercise for the reader.

Corollary 2.5.7. *Let G be a locally compact group and S a normal simply connected solvable subgroup with G/S compact. Then G is a semidirect product of a compact subgroup K of G with S.*

We now turn to a fundamental result, namely *Weyl's finiteness theorem.*

Corollary 2.5.8. *If G is a connected compact semisimple Lie group then $\pi_1(G)$ is finite. Alternatively $\tilde{G}$ or indeed any group locally isomorphic to G is compact.*

Proof. The property we need concerning semisimple groups L is that $\overline{[L, L]} = L$. (Actually, $[L, L] = L$.) We prove that if D is a discrete central subgroup of H with H/D compact and $\overline{[H/D, H/D]} = H/D$, then D is finite.

Let $\tilde{G} = H$ be the universal covering of G and D be its fundamental group. Then $H/D = G$ is compact and D is a finitely generated, discrete, abelian group. Hence $D = \mathbb{Z}^r \times F$, where F is finite and $r \geq 0$ is an integer. If D is not finite then $r \geq 1$ so there is a surjective continuous homomorphism $\phi : D \to \mathbb{Z}$. Injecting $\mathbb{Z} \to \mathbb{R}$ and composing gives a nontrivial homomorphism $f : D \to \mathbb{R}$. Consider the extension f^*, where ρ is the 1-dimensional trivial representation of H on $\mathbb{R}$. Then evidently all the requirements of Theorem 2.5.5 are satisfied. If f^* is the extension to $H \to \mathbb{R}$ we see that f^* is actually a continuous homomorphism and $f^*(D) = f(D) = \mathbb{Z}$. Since f^* is continuous $f^*(H)$ is connected so this must be larger than $\mathbb{Z}$. Compose f^* with the projection $\pi : \mathbb{R} \to \mathbb{T}$. Then this is a nontrivial homomorphism $H \to \mathbb{T}$, but $\pi \circ f^*(D) = \pi(\mathbb{Z}) = \{1\}$. Therefore this drops down to a nontrivial continuous homomorphism $\chi : H/D \to \mathbb{T}$. This is impossible since $\chi(\overline{[H/D, H/D]}) = \{1\} = \chi(H/D)$. Hence D is finite so $\tilde{G}$ is compact. Therefore so are all locally isomorphic groups since they are covered by $\tilde{G}$. $\square$

As a further application of the use of Haar measure we prove the Bochner linearization theorem which says that when a compact group acts on a manifold, near a fixed point the action is locally linear (and, of course, orthogonal).

Theorem 2.5.9. *Let $G \times M \to M$ denote the smooth action of a compact Lie group G on a smooth manifold M and let p be a G-fixed point of M. Then there is a G-invariant neighborhood U of p in M and a G equivariant diffeomorphism $F : U \to B$, where B is an open ball about 0 in $T_p(M)$.*

Proof. First we show that around p there is a neighborhood basis consisting of G-invariant neighborhoods. This actually follows as the equicontinuity from a general Ascoli type theorem for group actions, but we will give a direct proof. Let U be a neighborhood of $p \in M$. By continuity of the action together with the fact that p is G-fixed, we can find for each $g \in G$ neighborhoods W_g of g and U_g of p so that $W_g U_g \subseteq U$. By compactness $G = \bigcup_{i=1}^n W_{g_i}$. Let $U_* = \cap_{i=1}^n U_{g_i}$. Then U_* is a neighborhood of $p \in M$ and $U_* \subseteq GU_* \subseteq U$. Hence GU is also neighborhood of $p \in M$. It is clearly G-invariant and since U was arbitrary we have a neighborhood basis about $p \in M$.

Now let U be a G-invariant neighborhood of $p \in M$ small enough to be in a chart f around p. Then f can be regarded as mapping U diffeomorphically to $T_p(M)$, taking p to 0. Hence $d_p f$ is invertible. By invariance of U, $u \mapsto f(g^{-1}u)$, also takes values in $T_p(M)$. Because p is G fixed and U is G invariant, since $T_p(M) = T_p(U)$, each $g \in G$ has derivative $d(g) \in GL(T_p(M))$. Hence $g \mapsto d(g)(f(g^{-1}u))$ also takes values in $T_p(M)$. It follows that the integral $\int_G d(g)(f(g^{-1}u))dg$ (normalized Haar measure) is a tangent vector and so defines a function $F : U \to T_p(M)$ which by differentiation under the integral is a smooth function since f is. Now $f(g^{-1}p) = f(p) = 0$ so $F(p) = \int_G d(g)(f(g^{-1}p))dg = 0$. To calculate $d_p F$ let $\epsilon > 0$. Since for u near p,

$$f(g^{-1}u) - f(p) = d_p f \|f(u) - f(p)\| + \epsilon \|f(u) - f(p)\|,$$

taking into account $f(p) = 0$, applying $d(g)$ and integrating we get

$$\int_G d(g)f(g^{-1}u)dg = d_p f \int_G d(g)\|f(u) - f(p)\|dg$$
$$+ \epsilon \int_G d(g)\|f(u) - f(p)\|dg.$$

Since we have normalized Haar measure

$$F(u) = F(u) - F(p) = (d_p f + \epsilon)\|f(u) - f(p)\|.$$

Hence $d_p F = d_p f \neq 0$. By the inverse function theorem F is a local diffeomorphism on some neighborhood of p within U with a ball about 0 in $T_p(M)$. Finally by invariance of Haar measure, for $h \in G$,

$$F(hu) = \int_G d(g)(f(g^{-1}hu))dg = \int_G d(hg)(f((hg)^{-1}hu))dg$$
$$= d(h) \int_G d(g)(f((g)^{-1}u))dg = d(h)F(u).$$

$\square$

2.6   Compact Linear Groups and Hilbert's 14th Problem

Here we will prove a theorem of Chevalley which states that a compact linear group is the set of real points of an algebraic group defined over $\mathbb{R}$. We shall do this by means of the study of invariant polynomials. This method leads in a natural way to the classical solution of Hilbert's 14th problem in the case of compact linear groups. Namely, that the algebra of invariants, $P(V)^G$, is finitely generated. (Something of these methods can also be made to work in the case of non-compact reductive groups, but that is another story.)

Suppose k is a field and V a vector space of dimension n over k. We shall call $k[x_1, \ldots, x_n]$, the k-algebra of polynomials in n indeterminants, $P(V)$. Elements $p \in P(V)$ are finite sums

$$p(x) = \Sigma a_{(e_1, \ldots, e_n)} x_1^{e_1} \ldots x_n^{e_n}$$

where the monomials are formed with coefficients from k and $(e_1, \ldots, e_n)$ are n-tuples of non-negative integers. The degree of the monomial is $\sum_i e_i$ and the degree of p is the maximum of the degrees of the monomials of which p is composed. A polynomial is called homogenous if all its monomials have the same degree. Now, $P(V)$ can be regarded as a k-space (in fact a k-algebra) of k-valued functions on V as follows. Choose a basis $\{v_1, \ldots, v_n\}$ of V. If $x = \sum x_i v_i$, then the value of $p(x)$ is given by the equation above. As a result, we get an action of $\mathrm{GL}(V)$ on $P(V)$ by left translation, namely, $(g, p) \rightarrow p_g$ where $p_g(x) = p(g^{-1}x)$. Clearly, $p_g \in P(V)$ and $\mathrm{GL}(V) \times P(V) \rightarrow P(V)$ is a k-linear (infinite dimensional) representation of $\mathrm{GL}(V)$ on $P(V)$. Note also that $\mathrm{GL}(V)$ acts by

k-algebra automorphisms; that is, $(pq)_g = p_g q_g$ for all p and $q \in P(V)$. Now, $\deg p_g = \deg p$ for all p and g. So, if we consider the filtration of $P(V) = \bigcup_n P(V)_m$ by degree where $P(V)_m = \{p \in P(V) : \deg p \le m\}$, then each $P(V)_m$ is a finite dimensional GL(V)-invariant subalgebra of $P(V)$. So, for each integer m, we get a finite dimensional representation on $P(V)_m$ as well as on $P(V)_m \setminus P(V)_{m-1}$, the space of homogenous polynomials of degree m. Now let $G \times V \to V$ be a linear representation of G on V. Then by restriction from GL(V) to the image of G under the representation, we get an action of G on $P(V)$, $P(V)_m$ etc. A G-invariant polynomial p is one for which $p_g = p$ for all $g \in G$. The set of G-invariant polynomials will be denoted by $P(V)^G$. Clearly, $P(V)^G$ is a G-invariant k-subalgebra of $P(V)$. It is called the *algebra of invariants*.

As an illustration, let $k = \mathbb{R}$ and G be any subgroup of GL$(n, \mathbb{R})$ which acts transitively on the unit sphere S^{n-1} of V, such as O$(n, \mathbb{R})$ or SO$(n, \mathbb{R})$. Then

$$P(V)^G = \{q(x_1^2 + \cdots + x_n^2) : q(t) \in \mathbb{R}[t]\},$$

i.e., the algebra of invariant polynomials has a single generator.

To see this let $x \ne 0 \in V$ and write $x = \|x\| \cdot v$, where $v \in S^{n-1}$. If p is homogeneous, then $p(x) = p(\|x\| \cdot v) = \|x\|^{\deg p} \cdot v$. Now if, in addition, p is G-invariant, then since G operates transitively on S^{n-1}, $p(v) = c$, a constant. Let $p \in P(V)^G$ and write $p = \sum p_i$ where each p_i is homogeneous and, by Lemma 2.6.1 below, is also G-invariant. By the above, $p_i(x) = \|x\|^i c_i$, for $i = 1, \ldots, \deg p$ and $x \ne 0 \in V$. Clearly, this also holds for $i = 0$ and for $x = 0$ since $p_i(0) = 0$ if $i > 0$ and $p_0(x)$ is constant. If any $c_i = 0$, then $p_i = 0$ on S^{n-1} and therefore in all of V, by homogeneity. Hence we may assume all $c_i \ne 0$. Thus

$$\frac{p_i(x)}{c_i} = (x_1^2 + \cdots + x_n^2)^{\frac{i}{2}}.$$

Now, the left side is a polynomial so i must be even. Let $q(t) = \sum c_{2j} t_j$ where $j = \frac{i}{2}$. Then $p(x) = \sum p_i(x) = q(x_1^2 + \cdots + x_n^2)$. Conversely, any $q(x_1^2 + \cdots + x_n^2) \in P(V)^G$.

We shall presently see that the algebra $P(V)^G$ always has a *finite number of generators*; where by this we mean that an algebra A has elements $\alpha_1, \ldots, \alpha_r$ such that

$$A = \{q(\alpha_1, \ldots, \alpha_r) : q \in k[x_1, \ldots, x_r]\}.$$

We require three lemmas.

Lemma 2.6.1. *Let the polynomial p be written $p(x) = \sum p_i(x)$ where p_i is homogenous of degree i. If p is G-invariant, then each p_i is also G-invariant.*

Proof. Clearly $p(x)$ can be expressed in terms of $p_i(x)$. Now, suppose q_i are homogenous polynomials of distinct degrees and $\sum c_i q_i = 0$ where $c_i \in k$. Then for each i, either $c_i = 0$ or $q_i = 0$. To see this, we may assume all $q_i \neq 0$ and show all $c_i = 0$. But this is clear since the distinct monomials are linearly independant over k. Now, if $p(x) = \sum p_i(x) = \sum q_i(x)$ where p_i and q_i are homogeneous to degree i, then $\sum 1(q_i - p_i) = 0$. Since $q_i - p_i$ is homogeneous of degree i, we see that $q_i - p_i = 0$ for all i, by the above. For $g \in G$ if $p(x) = \sum p_i(x)$, then $p(g^{-1}x) = \sum p_i(g^{-1}x) = p(x)$ since p is G-invariant. Now $p_i(g)$ is a homogeneous polynomial of degree i for each $g \in G$. By uniqueness, $p_i(x) = p_i(gx)$ for all i. $\qquad\square$

Lemma 2.6.2. *Let I^+ denote the homogeneous G-invariant polynomials of positive degree and I the ideal in $P(V)$ generated by I^+. Then I as an ideal, has a finite number of G-invariant homogeneous generators.*

Proof. By the Hilbert's basis theorem [82], $I = (p_1, \ldots, p_r)$. Hence, each $p_i = \sum q_i^j s_i^j$ where $q_i^j \in P(V)$ and $s_i^j \in I^+$. Now, the ideal $(s_i^j) \subseteq I$. If $p \in I$, then $p = \sum t_i p_i$ where $t_i \in P(V)$ so $p = \sum t_i \sum q_i^j s_i^j \in (s_i^j)$. Thus $I = (s_i^j)$. $\qquad\square$

Lemma 2.6.3. *Let G be a compact subgroup of $\mathrm{GL}(V)$. Then there exists a map $\# : P(V) \to P(V)^G$ such that*

(1) $\#$ is $\mathbb{R}$ (respectively $\mathbb{C}$) linear

(2) $p = p^{\#}$ *if and only if* $p \in P(V)^G$

(3) $(pg)^{\#} = p^{\#}q$ *if* $p \in P(V)$ *and* $q \in P(V)^G$.

Proof. Let dg be normalized Haar measure on G. For $p \in P(V)$, define $p^{\#}(x) = \int_G p(g^{-1}x)dg$. Then $p\# \in P(V)$ and $\#$ is k-linear. By invariance of dg, $p^{\#} \in P(V)^G$. Clearly, if $p \in P(V)^G$, then $p^{\#} = p$. Finally, $(pq)^{\#}(x) = \int p(g^{-1}x)q(g^{-1}x)dg = q(x) \int p(g^{-1}x)dx$, since q is G-invariant. Thus $(pq)^{\#} = p^{\#}q$.   □

We now come to Hilbert's 14th problem in the case of compact groups. Such results are also known as the *fundamental theorem of invariant theory*.

Theorem 2.6.4. *Let G be a compact group and $G \times V \to V$ be a continuous real or complex linear action of G on V. Then $P(V)^G$ is a finitely generated algebra over $\mathbb{R}$ or $\mathbb{C}$. In fact, $P(V)^G$ is generated as an algebra by a finite number of homogeneous, G-invariant polynomials.*

Proof. Let $J^+ = \{p \in P(V)^G : \deg p > 0\}$ and I^+, as above, be those elements in J^+ which are homogeneous. Let J and I be the respective ideals generated by J^+ and I^+. By Lemma 2.6.2, I has a finite number of ideal generators belonging to I^+. We shall show $J = I$. Since $1 \in P(V)$, $J^+ \subset J$, so from $I^+ \subset J^+$ we know $I \subset J$. Let $p \in J$. Then $p = \sum q_i p_i$ where $q_i \in P(V)$ and $p_i \in J^+$. By Lemma 2.6.1, each $p_i = \sum p_j^i$, where $p_j^i \in I^+$. Hence, $p = \sum q_i \sum p_j^i \in I$ so $J = I$ and J as an ideal has a finite number of G-invariant homogeneous generators; $J = (p_1, \ldots, p_r)$. Let $p \in J^+$; we show $p = q(p_1, \ldots, p_r)$ for some $q \in k[x_1, \ldots, x_r]$ by induction on $\deg p$. This would complete the proof, since if $p \in P(V)^G$, then either $p \in J^+$ or p is constant. In the latter case, we would take q to be the constant polynomial. Of course, conversely all $q(p_1, \ldots, p_r)$ are in $P(V)^G$. Now, take $p \in J^+$. As an element of J, $p = \sum q_i p_i$. By Lemma 2.6.3,

$$p = p^{\#} = (\sum q_i p_i)\# = \sum q_i^{\#} p_i.$$

It follows that $\deg p \geq \deg q_i^{\#} p_i = \deg q_i^{\#} + \deg p_i$ for each i. If $\deg p_i = 0$, then p_i is a constant and does not generate any more subalgebra than if it were not there at all. We can therefore assume all $\deg p_i > 0$. The

$\deg q_i^{\#} < \deg p$. If $\deg q_i^{\#} > 0$, then $\deg q_i^{\#} \in J^+$. By induction, $\deg q_i^{\#}$ is in the algebra generated by $\{p_1, \ldots, p_r\}$. If $\deg q_i^{\#} = 0$, then $q_i^{\#}$ is constant and is also in this algebra. This means $\sum q_i^{\#} p_i = p$ is in the algebra. $\square$

Now we shall see that there are sufficiently many G-invariant polynomials on V.

Theorem 2.6.5. *Let G be a compact subgroup of $\mathrm{GL}(V)$. Then the G-invariant polynomials on V (with real coefficients) separate the disjoint compact G-invariant subsets of V. In particular, they separate the G-orbits.*

Proof. Suppose A and B are disjoint compact G-invariant subsets of V. Let $\phi(x) = d(x, A) - d(x, B)$ where $x \in V$ and d is the distance function on V. Then ϕ is continuous and is < 0 on A and > 0 on B. By compactness, there is a $\delta > 0$ so that $\phi > \delta$ on B and $\phi < -\delta$ on A. By the Weierstrass approximation theorem, ϕ can be approximated, to within $\frac{\delta}{2}$, by a polynomial on the compact set $A \cup B$. Then $p^{\#} \in P(V)^G$ and is $p^{\#} > 0$ on B and < 0 on A. $\square$

Finally, we come to Chevalley's theorem.

Corollary 2.6.6. *A compact linear Lie group is the set of real points of an algebraic group defined over $\mathbb{R}$.*

Proof. G acts on $\mathrm{End}_{\mathbb{R}}(V)$ by $(g, T) \to g \cdot T$. This is a linear representation of G. Now, $G = G \cdot 1$, the G-orbit of 1. If $T \in \mathrm{End}_{\mathbb{R}}(V)$, but not in G, then there is a $p \in P(\mathrm{End}_{\mathbb{R}}(V))^G$ so that $p(T) \neq p(1)$. Thus, $G = \cap_{p \in P(\mathrm{End}_{\mathbb{R}}(V))^G} \{T \in \mathrm{End}_{\mathbb{R}}(V) : p(T) - p(1) = 0\}$. $\square$

Chapter 3

Elements of the Theory of Lie Algebras

3.1 Basics of Lie Algebras

3.1.1 Ideals and Related Concepts

Definition 3.1.1. A subspace $\mathfrak{h}$ of $\mathfrak{g}$ is called an *ideal* in $\mathfrak{g}$ if $[X,Y]$ is in $\mathfrak{h}$ whenever X is $\mathfrak{h}$. Evidently an ideal is a subalgebra.

Example 3.1.2. The *center* of $\mathfrak{g}$, $\mathfrak{z}(\mathfrak{g}) = \{X \in \mathfrak{g} : [X,Y] = 0, Y \in \mathfrak{g}\}$, is an ideal in $\mathfrak{g}$.

Example 3.1.3. In $\mathfrak{g} = \mathfrak{gl}(n,k)$ we consider the center, $\mathfrak{z}(\mathfrak{g})$ namely all linear operators commuting with $\mathfrak{g}$. This is evidently just the scalar matrices. Consider $\mathfrak{s} = \mathfrak{sl}(n,k)$ which is the set of matrices of trace zero. This is a subalgebra since the trace of any commutator is zero. In fact $\mathfrak{s}$ is an ideal in $\mathfrak{g}$. Also $\mathfrak{s} \cap \mathfrak{z}(\mathfrak{g}) = 0$ since the characteristic of k is zero. Since $\dim \mathfrak{z}(\mathfrak{g}) + \dim \mathfrak{s} = n^2 = \dim \mathfrak{g}$ this is a direct sum of ideals. For $X \in \mathfrak{g}$ then $X = \frac{\operatorname{tr} X}{n} I_{n \times n} + Y$, where Y is in $\mathfrak{s}$, implements the decomposition.

For the particular case of $n = 2$ we see that a basis of $\mathfrak{sl}(2,k)$ is given by $H = \begin{pmatrix} 1 & 0 \\ 0 & -1 \end{pmatrix}$, $X^+ = \begin{pmatrix} 0 & 1 \\ 0 & 0 \end{pmatrix}$ and $X^- = \begin{pmatrix} 0 & 0 \\ 1 & 0 \end{pmatrix}$. The relations (structure constânts) are $[H, X^+] = 2X^+$, $[H, X^-] = -2X^-$ and $[X^+, X^-] = H$. Notice that $\operatorname{ad} H$ with respect to the basis $\{X^+, H, X^-\}$ is diagonal with (distinct) eigenvalues $\{2, 0, -2\}$.

Exercise 3.1.4. Prove that $\mathfrak{sl}(2, \mathbb{C})$ has a basis $\{X, Y, Z\}$ such that $[X, Y] = Z$, $[Y, Z] = X$ and $[Z, X] = Y$.
Hint: Let, $X = \frac{H}{2i}$, $Y = \frac{X^+ + X^-}{2}$ and $Z = \frac{X^+ - X^-}{2i}$.

Exercise 3.1.5. Prove that $\mathfrak{so}(2, 1) \cong \mathfrak{sl}(2, \mathbb{R})$ as Lie algebras.

Let $\mathfrak{g}$ be a Lie algebra and $\mathfrak{h}$ an ideal in $\mathfrak{g}$; then we can equip $\mathfrak{g}/\mathfrak{h}$ with a Lie bracket, making it a Lie algebra. For $X \in \mathfrak{g}$, $\overline{X}$ denotes its image in $\mathfrak{g}/\mathfrak{h}$. The Lie bracket on $\mathfrak{g}/\mathfrak{h}$ is defined by setting

$$[\overline{X}, \overline{Y}] = \overline{[X, Y]}.$$

To see that it is well-defined, replace X by $X + X_1$ and $Y + Y_1$ where X_1 and Y_1 are in $\mathfrak{h}$. Then,

$$\begin{aligned}
[\overline{X + X_1}, \overline{Y + Y_1}] &= \overline{[X + X_1, Y + Y_1]} \\
&= \overline{[X, Y]} + \overline{[X_1, Y]} + \overline{[X, Y_1]} + \overline{[X_1, Y_1]} \\
&= \overline{[X, Y]}.
\end{aligned}$$

Here $\overline{[X_1, Y]}, \overline{[X, Y_1]}$ and $\overline{[X_1, Y_1]}$ are 0 because $\mathfrak{h}$ is an ideal. It follows immediately from the definition that the projection map $\pi : \mathfrak{g} \to \mathfrak{g}/\mathfrak{h}$ is a Lie algebra homomorphism with kernel $\mathfrak{h}$.

Proposition 3.1.6. (*The First Isomorphism Theorem*) *Let $f : \mathfrak{g} \to \mathfrak{h}$ be a Lie homomorphism, then $\operatorname{Ker} f$ is an ideal in $\mathfrak{g}$ and $f(\mathfrak{g})$ is subalgebra of $\mathfrak{h}$. Moreover, f induces a Lie isomorphism from $\mathfrak{g}/\operatorname{Ker} f$ to $f(\mathfrak{h})$ making in the following diagram commutative:*

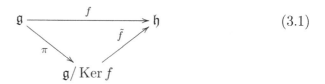

$$(3.1)$$

Proof. It is obvious that Ker f is a linear subspace. If $X \in \text{Ker} f$ and $Y \in \mathfrak{g}$ then we have

$$f([X,Y]) = [f(X), f(Y)] = [0, f(Y)] = 0.$$

Therefore $[X,Y] \in \text{Ker} f$ which means that Ker f is an ideal. Since $[f(X), f(Y)] = f([X,Y])$, $f(\mathfrak{g})$ is closed under bracketing so it is a sub-algebra of $\mathfrak{h}$. Then f induces a map $\tilde{f} : \mathfrak{g}/\text{Ker} f \to \mathfrak{h}$ defined as follows,

$$\tilde{f}(\overline{X}) = f(X)$$

where X is a representative of the equivalence class $\overline{X}$. One has to check that $\tilde{f}$ does not depend on the representative X, but if X and Y are two representatives for an equivalence class we have $X - Y = Z \in \text{Ker} f$ so $f(X) - f(Y) = f(Z) = 0$. This shows that $\tilde{f}$ is well-defined. $\tilde{f}$ is injective since if $\tilde{f}(\overline{X}) = f(X) = 0$ means that $X \in \text{Ker} f$ or in other words $\overline{X} = 0 \in \mathfrak{g}/\text{Ker} f$. It is obvious that $\tilde{f} : \mathfrak{g}/\text{Ker} f \to f(\mathfrak{g})$ is surjective since that $f(X) = \tilde{f}(\overline{X})$ for any $X \in \mathfrak{g}$. Therefore $\tilde{f}$ is an isomorphism. Also $f(X) = \tilde{f}(\overline{X})$ which means that the above diagram commutes. $\qquad\qquad \square$

Definition 3.1.7. Given two subalgebras $\mathfrak{a}$ and $\mathfrak{b}$ of a Lie algebra $\mathfrak{g}$ one can consider the subalgebra generated by $\mathfrak{a}$ and $\mathfrak{b}$. If $\mathfrak{a}$ and $\mathfrak{b}$ are ideals then $\mathfrak{a} + \mathfrak{b}$ is also an ideal.

Exercise 3.1.8. Prove that if $\mathfrak{a}$ and $\mathfrak{b}$ are ideals in $\mathfrak{g}$ then $\mathfrak{a} \cap \mathfrak{b}$, $[\mathfrak{a}, \mathfrak{b}]$ are ideals in $\mathfrak{a} + \mathfrak{b}$.

Definition 3.1.9. Let $\mathfrak{h}$ and $\mathfrak{l}$ be two Lie algebras and $\mathfrak{g} = \mathfrak{h} \oplus \mathfrak{l}$ be the *direct sum* of the two vector spaces. This vector space can be equipped with a Lie bracket such that $\mathfrak{h}$ and $\mathfrak{l}$ are subalgebras and $[\mathfrak{h}, \mathfrak{l}] = 0$. We call this Lie algebra the *external direct sum* of $\mathfrak{h}$ and $\mathfrak{l}$. Evidently $\mathfrak{h}$ and $\mathfrak{l}$ are ideals in $\mathfrak{g}$. A similar construction can be made with any finite number of factors.

Remark 3.1.10. Let $\mathfrak{g}$ be a Lie algebra and $\mathfrak{h}$ and $\mathfrak{l}$ two ideals in $\mathfrak{g}$ with trivial intersection. If $\mathfrak{g} = \mathfrak{h} + \mathfrak{l}$ then $\mathfrak{g} \simeq \mathfrak{h} \oplus \mathfrak{l}$.

Exercise 3.1.11. Let $\mathfrak{g}$ be a Lie algebra and $\mathfrak{h}$ and $\mathfrak{l}$ be two subalgebras with $\mathfrak{h}$ an ideal. Then,

(1) (The Second Isomorphism Theorem) The subalgebra $\mathfrak{h} + \mathfrak{l}$ generated by $\mathfrak{h}$ and $\mathfrak{l}$ contains $\mathfrak{h}$ as an ideal. Moreover $\mathfrak{h} \cap \mathfrak{l}$ is an ideal in $\mathfrak{l}$ and
$$\mathfrak{h} + \mathfrak{l}/\mathfrak{h} \simeq \mathfrak{l}/\mathfrak{h} \cap \mathfrak{l}.$$

(2) (The Third Isomorphism Theorem) If $\mathfrak{h} \subseteq \mathfrak{l} \subseteq \mathfrak{g}$ are ideals in $\mathfrak{g}$ then $\mathfrak{l}/\mathfrak{h}$ can be regarded as an ideal in $\mathfrak{g}/\mathfrak{h}$ (using the map induced by inclusion), and then
$$\frac{\mathfrak{g}/\mathfrak{h}}{\mathfrak{l}/\mathfrak{h}} \simeq \frac{\mathfrak{g}}{\mathfrak{l}}.$$

Clearly a subspace of the center $\mathfrak{z}(\mathfrak{g})$ is an ideal in $\mathfrak{g}$. Such an ideal is called a *central ideal*.

Definition 3.1.12. Let $\mathfrak{g}$ be Lie algebra and $\mathfrak{h}$ an ideal in $\mathfrak{g}$. We denote by $[\mathfrak{g}, \mathfrak{h}]$ the linear span of the set of all $[X, Y]$ where $X \in \mathfrak{g}$ and $Y \in \mathfrak{h}$. It is easy to see, using Jacobi identity, that $[\mathfrak{g}, \mathfrak{h}]$ is also an ideal. An important special case of this is $[\mathfrak{g}, \mathfrak{g}]$. This ideal is called the *derived subalgebra*.

Proposition 3.1.13. *If $\mathfrak{h}$ is a subalgebra of Lie algebra $\mathfrak{g}$ and $[\mathfrak{g}, \mathfrak{g}] \subseteq \mathfrak{h}$ then $\mathfrak{h}$ is an ideal.*

Proof. We have
$$[\mathfrak{g}, \mathfrak{h}] \subseteq [\mathfrak{g}, \mathfrak{g}] \subseteq \mathfrak{h}.$$

$\square$

Proposition 3.1.14. *For a Lie algebra $\mathfrak{g}$, $\mathfrak{g}/[\mathfrak{g}, \mathfrak{g}]$ is abelian. Moreover any ideal $\mathfrak{h} \subset \mathfrak{g}$ for which $\mathfrak{g}/\mathfrak{h}$ is abelian contains the derived subalgebra $[\mathfrak{g}, \mathfrak{g}]$.*

Proof. If $\overline{X}$ and $\overline{Y} \in \mathfrak{g}/[\mathfrak{g}, \mathfrak{g}]$ then $[\overline{X}, \overline{Y}] = \overline{[X, Y]} = 0 \in \mathfrak{g}/[\mathfrak{g}, \mathfrak{g}]$. So the derived subalgebra is abelian. Conversely if $\mathfrak{g}/\mathfrak{h}$ is abelian, $\overline{X}$ and $\overline{Y} \in \mathfrak{g}/\mathfrak{h}$ commute, for X and Y so
$$\overline{[X, Y]} = [\overline{X}, \overline{Y}] = 0 \in \mathfrak{g}/\mathfrak{h}.$$

This implies that $[X, Y] \in \mathfrak{h}$ and since this is true for all X and Y, we see that $\mathfrak{h}$ contains the derived subalgebra. $\square$

Definition 3.1.15. Let $\mathfrak{g}$ be a Lie algebra. A subspace $\mathfrak{a}$ of $\mathfrak{g}$ is said to be a *characteristic ideal* if it is invariant under every derivation of $\mathfrak{g}$.

A characteristic ideal is an ideal since it is invariant under $\operatorname{ad} X$ for all $X \in \mathfrak{g}$.

Example 3.1.16. Typical examples of characteristic ideals in a Lie algebra $\mathfrak{g}$ are the center $\mathfrak{z}(\mathfrak{g})$ and the derived subalgebra $[\mathfrak{g}, \mathfrak{g}]$.

Proposition 3.1.17. *Let* $\mathfrak{a}$ *and* $\mathfrak{b}$ *be two characteristic ideals of* $\mathfrak{g}$. *Then* $\mathfrak{a} + \mathfrak{b}$, $\mathfrak{a} \cap \mathfrak{b}$ *and* $[\mathfrak{a}, \mathfrak{b}]$ *are also characteristic ideals of* $\mathfrak{g}$.

Proof. We check this for $[\mathfrak{a}, \mathfrak{b}]$. For $X \in \mathfrak{a}$ and $Y \in \mathfrak{b}$ and $D \in \operatorname{Der}(\mathfrak{g})$ we have

$$D[X, Y] = [DX, Y] + [X, DY] \in [\mathfrak{a}, \mathfrak{b}]$$

since $\mathfrak{a}$ and $\mathfrak{b}$ are characteristic ideals. $\square$

Proposition 3.1.18. *Let* $\mathfrak{g}$ *be a Lie algebra,* $\mathfrak{h}$ *an ideal in* $\mathfrak{g}$, *and* $\mathfrak{l}$ *a characteristic ideal in* $\mathfrak{h}$. *Then* $\mathfrak{l}$ *is an ideal of* $\mathfrak{g}$.

Proof. If $X \in \mathfrak{g}$ then $\operatorname{ad} X$ is a derivation of $\mathfrak{h}$, therefore $\mathfrak{l}$ is stable under $\operatorname{ad} X$ and this means that $\mathfrak{l}$ is an ideal in $\mathfrak{g}$. $\square$

Definition 3.1.19. Let $\mathfrak{g}$ and $\mathfrak{h}$ be two Lie algebras and $\eta : \mathfrak{g} \to \operatorname{Der}(\mathfrak{h})$ be a Lie homomorphism. The *semi direct sum* $\mathfrak{g} \oplus_\eta \mathfrak{h}$ which is the vector space $\mathfrak{g} \oplus \mathfrak{h}$ equipped with the Lie bracket is

$$[(X, Y), (X', Y')] = ([X, X'], \eta_X(Y') - \eta_{X'}(Y) + [Y, Y']).$$

Making the obvious identifications $\mathfrak{h}$ is an ideal and $\mathfrak{g}$ is a subalgebra of $\mathfrak{g} \oplus_\eta \mathfrak{h}$.

Alternatively, let $\mathfrak{l}$ be a Lie algebra, $\mathfrak{h}$ an ideal and $\mathfrak{g}$ a subalgebra of $\mathfrak{l}$ such that $\mathfrak{l} = \mathfrak{h} \oplus \mathfrak{g}$ as vector spaces. For $X \in \mathfrak{g}$ let $\eta_X = \operatorname{ad} X|_{\mathfrak{h}}$. Then η is a Lie algebra homomorphism $\mathfrak{g} \to \operatorname{Der}(\mathfrak{h})$ and the resulting semi direct product is isomorphic to $\mathfrak{l}$.

Example 3.1.20. Let $\mathfrak{h}_n$ be the space of matrices in $\mathfrak{gl}(n+2,k)$ of the form

$$(\mathbf{X},\mathbf{Y},z) = \begin{pmatrix} 0 & x_1 & \dots & x_n & z \\ 0 & 0 & \dots & 0 & y_1 \\ \vdots & \vdots & & & \vdots \\ 0 & 0 & \dots & 0 & y_n \\ 0 & 0 & \dots & 0 & 0 \end{pmatrix}$$

where $\mathbf{X} = (x_1, x_2, \dots, x_n)$ and $\mathbf{Y} = (y_1, y_2, \dots, y_n)$ and $n \geq 1$. In this notation

$$[(\mathbf{X},\mathbf{Y},z),(\mathbf{X}',\mathbf{Y}',z')] = (0,0,\mathbf{X}\mathbf{Y}'^t - \mathbf{X}'\mathbf{Y}^t).$$

In particular this shows that $\mathfrak{h}_n$ is a Lie algebra called the *Heisenberg Lie algebra*. Here $[\mathfrak{h}_n,\mathfrak{h}_n] = \mathfrak{z}(\mathfrak{h}_n)$ which has dimension one.

It is of interest to identify the derivation algebra of $\mathfrak{h}_n$. Consider the basis

$$X_i = ((0,0,\dots,1,\dots,0),(0,0,\dots,0),0), 1 \leq i \leq n$$
$$Y_i = ((0,0,\dots,0),(0,0,\dots,0,1,0,\dots,0),0), 1 \leq i \leq n$$
$$Z = ((0,0,\dots,0),(0,0,\dots,0),1)$$

for $\mathfrak{h}_n$ and let D be a derivation for $\mathfrak{h}_n$. Then

$$DX_i = \sum a_{ij}X_j + \sum b_{ij}Y_j + \lambda_i Z,$$
$$DY_i = \sum c_{ij}X_j + \sum d_{ij}Y_j + \mu_i Z, \qquad (3.2)$$
$$DZ = \lambda Z.$$

The last equation follows because the center is a characteristic ideal. By applying D to the identities

$$[X_i,Y_j] = \delta_{ij}Z$$
$$[X_i,X_j] = 0$$
$$[Y_i,Y_j] = 0$$

and inserting the values from (3.2) we get,

$$a_{ij} + d_{ji} = \delta_{ij}\lambda$$
$$b_{ij} + b_{ji} = 0$$
$$c_{ij} + c_{ji} = 0.$$

In particular we have $\lambda = \frac{1}{n}(\sum a_{ii} + \sum d_{ii})$. Letting $A = (a_{ij})$, $B = (b_{ij})$, $C = (c_{ij})$ and $D = (d_{ij})$, we have

$$A + D^t = \lambda I_n$$
$$B = -B^t$$
$$C = -C^t.$$

The matrix representation of D is

$$D = \left(\begin{array}{cc|c} A & C & 0 \\ B & D & \vdots \\ \hline \lambda_1 \ldots \lambda_n \; \mu_1 \ldots \mu_n & & \lambda \end{array} \right)$$

When $\lambda = 0$ or equivalently $\operatorname{tr} D = 0$, we get

$$D = \left(\begin{array}{cc|c} A & C & 0 \\ B & -A^t & \vdots \\ \hline \lambda_1 \ldots \lambda_n \; \mu_1 \ldots \mu_n & & 0 \end{array} \right).$$

Notice that $H = \begin{pmatrix} A & C \\ B & -A^t \end{pmatrix}$ lies in $\operatorname{Sp}(n, k)$. Indeed, the set of derivations of $\mathfrak{h}_n$ of trace zero is isomorphic to the semidirect sum $\operatorname{Sp}(n, k) \oplus_\eta k^{2n}$ where η is the inclusion of $\operatorname{Sp}(n, k)$ in $\operatorname{GL}(2n, k)$ and k^{2n} is regarded as an abelian Lie algebra. Each such derivation

$$D = \begin{pmatrix} A & C & 0 \\ B & -A^t & 0 \\ \lambda_1 \ldots \lambda_n \; \mu_1 \ldots \mu_n & & 0 \end{pmatrix}$$

is mapped to $(\begin{pmatrix} A & C \\ B & -A^t \end{pmatrix}, (\lambda_1 \ldots \lambda_n, \mu_1 \ldots \mu_n))$.

We now turn to the concepts of nilpotence and solvability.

Definition 3.1.21. Let $\mathfrak{g}$ be a Lie algebra. We say that $\mathfrak{g}$ is *nilpotent* if its *lower central series*, $\mathfrak{g}_0 = \mathfrak{g}$, $\mathfrak{g}_1 = [\mathfrak{g}, \mathfrak{g}], \ldots, \mathfrak{g}_k = [\mathfrak{g}, \mathfrak{g}_{k-1}] \ldots$, eventually hits 0. The first index where this occurs is called the *index of nilpotence* of $\mathfrak{g}$. We call an ideal of $\mathfrak{g}$ *nilpotent* if it is nilpotent as a Lie algebra.

Exercise 3.1.22. Show that each term in the lower central series is a characteristic ideal in $\mathfrak{g}$.

Example 3.1.23. The Heisenberg Lie algebra $\mathfrak{h}_n$, is a 2-step nilpotent Lie algebra. This follows from the fact that $[\mathfrak{h}_n, \mathfrak{h}_n] = \mathfrak{z}(\mathfrak{h}_n)$.

We now construct two other important examples of 2-step nilpotent Lie algebras. Regard $\mathfrak{g}_n(\mathbb{C}) = \mathbb{C}^n \oplus i\mathbb{R}$ as a real vector space of dimension $2n + 1$ and let $\langle \cdot, \cdot \rangle$ be the Hermitian form on $\mathbb{C}^n$ given by $\langle \mathbf{X}, \mathbf{Y} \rangle = \sum_{i=1}^{n} x_i \overline{y_i}$. Now we define the bracket $[\cdot, \cdot]$ by

$$[(\mathbf{X}, it), (\mathbf{Y}, is)] = (0, i\Im\langle \mathbf{X}, \mathbf{Y} \rangle).$$

We want the reader to check that this is a 2-step nilpotent Lie algebra over $\mathbb{R}$ with 1-dimensional center, and in fact is isomorphic to $\mathfrak{h}_n$.

Exercise 3.1.24. Prove that any 2-step nilpotent Lie algebra over a field k is isomorphic to $\mathfrak{h}_n(k)$ if its center is 1-dimensional.

We now consider the quaternionic analogue of the example just above. Let $\mathbb{H}$ be the quaternions and $\mathbf{X} \mapsto \overline{\mathbf{X}}$ be quaternionic conjugation. Then $\mathfrak{g}_n(\mathbb{H}) = \mathbb{H}^n \oplus \Im(\mathbb{H})$ where $\Im(\mathbb{H})$ is the 3-dimensional real vector space spanned by i, j and k. Then $\mathfrak{g}_n(\mathbb{H})$ is a real vector space of dimension $4n + 3$. We make this into a Lie algebra by

$$[(\mathbf{X}, \mathbf{V}), (\mathbf{Y}, \mathbf{W})] = (0, \Im\langle \mathbf{X}, \mathbf{Y} \rangle)$$

where $\langle \mathbf{X}, \mathbf{Y} \rangle = \sum_{i=1}^{n} x_i \overline{y_i}$ is the Hermitian form on $\mathbb{H}^n$ (using quoternionic conjugation) and $\Im\langle \mathbf{X}, \mathbf{Y} \rangle$ is the imaginary part of $\langle X, Y \rangle$. Direct calculations show that $\mathfrak{g}_n(\mathbb{H})$ is a 2-step nilpotent Lie algebra with a 3-dimensional center[1]. These Lie algebras play an important role in the study of rank one simple Lie algebras and groups which will be given in Chapter 6.

---

[1] Similar constructions can be made using the Cayley numbers.

Proposition 3.1.25. *The sum of two nilpotent ideals, $\mathfrak{a}$ and $\mathfrak{b}$, is a nilpotent ideal.*

Proof. Let $\mathfrak{a}_n$ (respectively $\mathfrak{b}_n$) be the nth term of the lower central series of $\mathfrak{a}$ (respectively $\mathfrak{b}$). Now for any sequence $\mathfrak{h}_1, \mathfrak{h}_2, \ldots, \mathfrak{h}_k$ of subalgebras of $\mathfrak{g}$ where at least n of them are equal to $\mathfrak{a}$ (respectively $\mathfrak{b}$), then any bracketing $[[[\ldots [\mathfrak{h}_{i_1}, \mathfrak{h}_{i_2}], \mathfrak{h}_{i_3}] \ldots]]$ of these $\mathfrak{h}_i$ is in $\mathfrak{a}_n$ (respectively $\mathfrak{b}_n$). Let $m = \max(m_1, m_2)$ where m_1 (respectively m_2) is the index of nilpotence of $\mathfrak{a}$ (respectively $\mathfrak{b}$). Then

$$(\mathfrak{a} + \mathfrak{b})_{2m} \subset \mathfrak{a}_m + \mathfrak{b}_m = 0.$$

Hence $\mathfrak{a} + \mathfrak{b}$ is nilpotent. We already know it's an ideal.   □

Definition 3.1.26. It follows from Proposition 3.1.25 and finite dimensionality that every Lie algebra $\mathfrak{g}$ has a unique maximal nilpotent ideal $\mathrm{nil}(\mathfrak{g})$, called *nilradical*, which contains any nilpotent ideal.

Definition 3.1.27. A Lie algebra $\mathfrak{g}$ is said to be solvable if the *derived series* of $\mathfrak{g}$, $\mathfrak{g}_0 = \mathfrak{g}$, $\mathfrak{g}^1 = [\mathfrak{g}, \mathfrak{g}], \ldots, \mathfrak{g}^k = [\mathfrak{g}^{(k-1)}, \mathfrak{g}^{(k-1)}], \ldots$ hits 0. The first index where this occurs is called the *index of solvability* of $\mathfrak{g}$. We call an ideal of $\mathfrak{g}$ solvable if it is solvable as a Lie algebra.

Exercise 3.1.28. Show that each term in the derived series is a characteristic ideal in $\mathfrak{g}$.

Evidently since $[\mathfrak{g}^{k-1}, \mathfrak{g}^{k-1}] \subseteq [\mathfrak{g}, \mathfrak{g}_{k-1}]$ for every k, a nilpotent algebra is always solvable. Also it is clear that a subalgebra or quotient algebra of a nilpotent (respectively solvable) algebra is nilpotent (respectively solvable).

Example 3.1.29. Let V be a finite dimensional vector space and let $\mathfrak{s}(V)$ be the set of upper triangular matrices of $\mathfrak{gl}(V)$ and $\mathfrak{n}(V)$ be the set of upper triangular matrices with all zero entries on the diagonal. The latter are *niltriangular* matrices. We leave as an exercise that these are both subalgebras of $\mathfrak{gl}(V)$ and that $[\mathfrak{s}(V), \mathfrak{s}(V)] = \mathfrak{n}(V)$. A direct calculation shows that the series $\mathfrak{n}_k(V) = [\mathfrak{n}(V), \mathfrak{n}_{k-1}(V)]$ introduces another row of off diagonal zeros. This means that $\mathfrak{n}(V)$ is nilpotent of index $\dim V - 1$ and also that $\mathfrak{s}(V)$ is solvable of index $\dim V$.

Remark 3.1.30. As we shall see later, by Lie's theorem, Theorem 3.2.18, a Lie algebra is solvable if and only if its derived subalgebra is nilpotent.

Remark 3.1.31. Notice that (5) in Section 0.5 tells us that $[\mathfrak{n}(V), \mathfrak{n}(V)^t]$ is contained in the diagonal subalgebra, where $\mathfrak{n}(V)^t$ consists of the transpose of the elements of $\mathfrak{n}(V)$.

Definition 3.1.32. Let $f : \mathfrak{g} \to \mathfrak{h}$ be a surjective Lie algebra homomorphism. If $\operatorname{Ker} f \subset \mathfrak{z}(\mathfrak{g})$, then $\mathfrak{g}$ is called a *central extension* of $\mathfrak{h}$.

Proposition 3.1.33. *Let $\mathfrak{g}$ be a Lie algebra and $\mathfrak{h}$ be an ideal and consider the short exact sequence,*

$$0 \to \mathfrak{h} \to \mathfrak{g} \xrightarrow{\pi} \mathfrak{g}/\mathfrak{h} \to 0$$

then

(1) $\mathfrak{g}$ is solvable if $\mathfrak{h}$ and $\mathfrak{g}/\mathfrak{h}$ are solvable.

(2) If $\mathfrak{h}$ is a central ideal and $\mathfrak{g}/\mathfrak{h}$ is nilpotent, then so is $\mathfrak{g}$.

Proof. Because π is a homomorphism it takes the derived series of $\mathfrak{g}$ to the derived series of $\mathfrak{g}/\mathfrak{h}$. Since $\mathfrak{g}/\mathfrak{h}$ is solvable, $[\mathfrak{g}^k, \mathfrak{g}^k] \subseteq \mathfrak{h}$ for some k. It follows from the solvability of $\mathfrak{h}$ that $\mathfrak{g}$ is solvable. For the second part we observe that since $\mathfrak{g}/\mathfrak{h}$ is nilpotent and π is a homomorphism $\mathfrak{g}_{k+1} = [\mathfrak{g}, \mathfrak{g}_k] \subseteq \mathfrak{h}$. Hence $[\mathfrak{g}, \mathfrak{g}_{k+1}] \subset [\mathfrak{g}, \mathfrak{h}] = (0)$. □

Proposition 3.1.34. *Let $\mathfrak{g}$ be a Lie algebra and $\mathfrak{a}$ and $\mathfrak{b}$ two solvable ideals of $\mathfrak{g}$, then $\mathfrak{a} + \mathfrak{b}$ is also solvable.*

Proof. By the second isomorphism theorem we have $(\mathfrak{a} + \mathfrak{b})/\mathfrak{a} \cong \mathfrak{b}/(\mathfrak{a} \cap \mathfrak{b})$. The latter is solvable since it is homomorphic image of a solvable algebra. Therefore $\mathfrak{a} + \mathfrak{b}$ is solvable by Proposition 3.1.33. □

Definition 3.1.35. By virtue of Proposition 3.1.34 and finite dimensionality of $\mathfrak{g}$, every Lie algebra has a unique maximal solvable ideal $\operatorname{rad}(\mathfrak{g})$, called the *radical*.

Notice that by their very definitions $\mathrm{rad}(\mathfrak{g})$ and $\mathrm{nil}(\mathfrak{g})$ are stable under all *automorphisms* of $\mathfrak{g}$. Let $\mathfrak{h}$ stand for either of these ideals. Now suppose we are over the real, or complex field and D is a derivation of $\mathfrak{g}$. Then for all real t, $\mathrm{Exp}\, tD$ is an automorphism of $\mathfrak{g}$ (1.4.29) and so $\mathrm{Exp}\, tD(\mathfrak{h}) = \mathfrak{h}$. Differentiating at $t = 0$ shows $D(\mathfrak{h}) = \mathfrak{h}$. Hence both $\mathrm{rad}(\mathfrak{g})$ and $\mathrm{nil}(\mathfrak{g})$ are characteristic ideals in $\mathfrak{g}$.

Remark 3.1.36. Evidently $\mathrm{rad}(\mathfrak{g}) \supset \mathrm{nil}(\mathfrak{g})$. (We shall see later that $\mathrm{rad}(\mathfrak{g})/\mathrm{nil}(\mathfrak{g})$ is abelian (see Corollary 3.2.19).)

Proposition 3.1.37. *For every Lie algebra* $\mathfrak{g}$,

$$\mathrm{nil}(\mathrm{rad}(\mathfrak{g})) = \mathrm{nil}(\mathfrak{g}).$$

Proof. Now since, as we saw, $\mathrm{nil}(\mathrm{rad}(\mathfrak{g}))$ is a characteristic ideal in $\mathrm{rad}(\mathfrak{g})$ which is itself an ideal in $\mathfrak{g}$, it follows that $\mathrm{nil}(\mathrm{rad}(\mathfrak{g}))$ is an ideal in $\mathfrak{g}$. Since it is nilpotent it is contained in the largest such ideal of $\mathfrak{g}$; $\mathrm{nil}(\mathrm{rad}(\mathfrak{g})) \subset \mathrm{nil}(\mathfrak{g})$. Conversely, let $\mathfrak{a}$ be any nilpotent ideal in $\mathfrak{g}$. Then it is also a solvable ideal so $\mathfrak{a} \subset \mathrm{nil}(\mathrm{rad}(\mathfrak{g}))$. Taking $\mathfrak{a} = \mathrm{nil}(\mathfrak{g}))$ we get $\mathrm{nil}(\mathfrak{g})) \subset \mathrm{nil}(\mathrm{rad}(\mathfrak{g}))$. $\square$

Example 3.1.38. Here we introduce the affine Lie algebra. Let V be a vector space of dimension n over k, A a linear operator on V and b a vector in V. We consider the linear Lie algebra $\mathfrak{g}$ consisting of matrices of order $n + 1$,

$$\begin{pmatrix} A & b \\ 0 & 0 \end{pmatrix}.$$

It is easy to check that the matrices with $A = 0$ form an ideal in $\mathfrak{g}$. The Lie bracket is given by

$$[(A, b), (A', b')] = ([A, A'], Ab' - A'b).$$

This Lie algebra is called the $ax + b$-Lie algebra and any of its subalgebras is called an *affine Lie algebra*. It is the semidirect sum of $\mathfrak{gl}(n, k)$ with V. This is a subalgebra of $\mathfrak{gl}(n + 1, k)$. It is only solvable if $n = 1$.

3.1.2   Semisimple Lie Algebras

Definition 3.1.39. Let $\mathfrak{g}$ be a Lie algebra.

(1) $\mathfrak{g}$ is said to be *simple* if it has no nontrivial proper ideal.

(2) $\mathfrak{g}$ is said to be *semisimple* if it has no nontrivial solvable ideal.

It is clear that:

Proposition 3.1.40. *A Lie algebra is semisimple if and only if its radical is trivial.*

Proposition 3.1.41. *A Lie algebra is semisimple if and only if it has no nontrivial abelian ideal. In particular the center of a semisimple Lie algebra is trivial.*

Proof. Clearly a semisimple Lie algebra has no nontrivial abelian ideal as abelian ideals are solvable. Conversely, consider the derived series of $\mathrm{rad}(\mathfrak{g})$. By assumption that has no nontrivial abelian ideal. This is because the ideals in this derived series are characteristic ideals in $\mathrm{rad}(\mathfrak{g})$ and therefore are ideals in $\mathfrak{g}$. But since $\mathrm{rad}(\mathfrak{g})$ is solvable and nontrivial the last term in this series is abelian and nontrivial. Therefore $\mathrm{rad}(\mathfrak{g})$ must be (0) and $\mathfrak{g}$ is semisimple. □

Remark 3.1.42. Let $\mathfrak{g}$ be a Lie algebra and $\mathrm{rad}(\mathfrak{g})$ its radical. Then $\mathfrak{g}/\mathrm{rad}(\mathfrak{g})$ is semisimple. This is an immediate consequence of Proposition 3.1.40.
Examining the following exact sequence,

$$0 \to \mathrm{rad}(\mathfrak{g}) \to \mathfrak{g} \to \mathfrak{g}/\mathrm{rad}(\mathfrak{g}) \to 0$$

we see that any Lie algebra $\mathfrak{g}$ is the middle term of a short exact sequence whose other two terms are solvable and semisimple. So it seems plausible that studying solvable Lie algebras and semisimple Lie algebras is an appropriate guide to study of Lie algebras in general.

Definition 3.1.43. For a Lie algebra $\mathfrak{g}$ and $\mathfrak{h}$ a subalgebra of it, we define the normalizer $\mathfrak{n}_\mathfrak{g}(\mathfrak{h})$ of $\mathfrak{h}$ in $\mathfrak{g}$ to be the set of all $X \in \mathfrak{g}$ such that $[X, \mathfrak{h}] \subseteq \mathfrak{h}$. It is easy to see that $\mathfrak{n}_\mathfrak{g}(\mathfrak{h})$ is a subalgebra of $\mathfrak{g}$ containing

$\mathfrak{h}$ as an ideal and is the largest such subalgebra. We also define the centralizer $\mathfrak{z}_\mathfrak{g}(\mathfrak{h})$ of $\mathfrak{h}$ in $\mathfrak{g}$ to be the set of all $X \in \mathfrak{g}$ such that $[X,\mathfrak{h}] = 0$. Evidently $\mathfrak{z}_\mathfrak{g}(\mathfrak{h})$ is a subalgebra of $\mathfrak{g}$. If $\mathfrak{h} = \mathfrak{g}$ we get the center. If $\mathfrak{h}$ is abelian then the centralizer contains $\mathfrak{h}$. In any case the centralizer is a subalgebra of the normalizer. If $X \in \mathfrak{g}$ then $\mathfrak{z}_\mathfrak{g}(X)$ will mean the centralizer of the subalgebra generated by X.

Proposition 3.1.44. *If $\mathfrak{h}$ is an ideal in $\mathfrak{g}$ then its centralizer is also an ideal.*

Proof. Let $X \in \mathfrak{z}_\mathfrak{g}(\mathfrak{h})$, $Y \in \mathfrak{g}$ and $H \in \mathfrak{h}$ then

$$[[Y,X],H] + [[H,Y],X] + [[X,H],Y] = 0.$$

Since $[X,H] = 0$, the last term is trivial and since $\mathfrak{h}$ is an ideal $[H,Y] \in \mathfrak{h}$ and therefore the middle term is also zero. Hence the first term is zero, so $[Y,X] \in \mathfrak{z}_\mathfrak{g}(\mathfrak{h})$. $\qquad\square$

3.1.3 Complete Lie Algebras

Definition 3.1.45. A Lie algebra $\mathfrak{g}$ is called *complete* if its center is trivial and every derivation is inner.

Some examples of complete Lie algebras are semisimple Lie algebras as we shall see in the next section, and the affine Lie algebra of dimension 2 given just above.

Proposition 3.1.46. *The $ax + b$ Lie algebra $\mathfrak{g}$ is complete.*

Proof. Let X,Y be a basis of $\mathfrak{g}$ with $[X,Y] = Y$ and let $cX + dY$ be a generic element of the Lie algebra. If $cX + dY$ is in the center then it must bracket trivially with both X and Y. It follows that $c = d = 0$ and hence $\mathfrak{z}(\mathfrak{g}) = 0$.

Now let D be a derivation, so that

$$DY = D[X,Y] = [DX,Y] + [X,DY]. \tag{3.3}$$

Let $DX = aX + bY$ and $DY = cX + dY$. By substituting in (3.3) we get $c = a = 0$. Hence $DX = bY$ and $DY = dY$. Then it can be easily checked that $D = \operatorname{ad} U$ where $U = dX - bY$.   □

Proposition 3.1.47. *Let $\mathfrak{g}$ be a Lie algebra and $\mathfrak{h}$ an ideal. If $\mathfrak{h}$ is complete then it is a direct summand. In fact $\mathfrak{g} = \mathfrak{h} \oplus \mathfrak{z}_\mathfrak{g}(\mathfrak{h})$.*

Proof. Notice that $\mathfrak{h} \cap \mathfrak{z}_\mathfrak{g}(\mathfrak{h}) = \mathfrak{z}(\mathfrak{h}) = 0$. If $X \in \mathfrak{g}$ then $\operatorname{ad} X$ restricted to $\mathfrak{h}$ is a derivation of $\mathfrak{h}$ and therefore it is an inner derivation of $\mathfrak{h}$,

$$\operatorname{ad} X|_\mathfrak{h} = \operatorname{ad} H$$

for some $H \in \mathfrak{h}$. Hence $X - H$ is in the centralizer of $\mathfrak{h}$, therefore

$$\mathfrak{g} = \mathfrak{h} \oplus \mathfrak{z}_\mathfrak{g}(\mathfrak{h}).$$

□

3.1.4   Lie Algebra Representations

Definition 3.1.48. Let $\rho : \mathfrak{g} \to \mathfrak{gl}(V)$ be Lie algebra representation of $\mathfrak{g}$. A subspace W of V is called an *invariant subspace* if $\rho_X(W) \subset W$ for every $X \in \mathfrak{g}$. We shall say that a representation is *reducible* if it has a nontrivial proper invariant subspace, otherwise we call it *irreducible*.

Definition 3.1.49. Let $\rho_1 : \mathfrak{g} \to \mathfrak{gl}(V_1)$ and $\rho_2 : \mathfrak{g} \to \mathfrak{gl}(V_2)$ be two representations of a Lie algebra $\mathfrak{g}$. A *intertwining operator* from (ρ_1, V_1) to (ρ_2, V_2) is a linear map $T : V_1 \to V_2$ such that for every $X \in \mathfrak{g}$,

$$T \circ \rho_1(X) = \rho_2(X) \circ T.$$

Two representations are *equivalent* if there exists an invertible intertwining operator between them.

On can do certain operations on the space of representations.

Definition 3.1.50. Let $\rho_1 : \mathfrak{g} \to \mathfrak{gl}(V_1)$ and $\rho_2 : \mathfrak{g} \to \mathfrak{gl}(V_2)$ be two Lie representations of a Lie algebra $\mathfrak{g}$. Then one can consider the map

$\rho_1 \oplus \rho_2 : \mathfrak{g} \to \mathfrak{gl}(V_1) \oplus \mathfrak{gl}(V_2) \subset \mathfrak{gl}(V_1 \oplus V_2)$ defined as follows: For every $X \in \mathfrak{g}$,

$$(\rho_1 \oplus \rho_2)(X) = (\rho_1(X), \rho_2(X)) : V_1 \oplus V_2 \to V_1 \oplus V_2.$$

We call $(\rho_1 \oplus \rho_2, V_1 \oplus V_2)$ the *direct sum* of ρ_1 and ρ_2. Similarly, one can work with any finite number of representations.

A Lie representation is said to be *completely reducible* if it is isomorphic to a direct sum of irreducible representations.

Definition 3.1.51. A family of operators $\Upsilon \subset \mathfrak{gl}(V)$ is called *irreducible* if there is no nontrivial subspace of V stable under all $X \in \Upsilon$.

Lemma 3.1.52. (*Schur's Lemma*) *Let $\Upsilon_i \subset \mathfrak{gl}(V_i)$ be an irreducible set of operators for $i = 1, 2$ and $T : V_1 \to V_2$ be an operator such that $T\Upsilon_1 = \Upsilon_2 T$. Then either $T = 0$ or $\dim V_1 = \dim V_2$ and T is an isomorphism.*

Proof. Ker T is a Υ_1 invariant subspace of V_1 and $T(V_1)$ is a Υ_2 invariant subspace of V_2. Hence Ker $T = V_1$ or 0. If Ker $T = V_1$ then $T = 0$. Otherwise T is injective. Also $T(V_1) = V_2$ or 0. Since we have disposed of the case $T = 0$ we see that T is an isomorphism. $\square$

Corollary 3.1.53. (*Schur's Lemma over an algebraically closed field*) *Let $\Upsilon \subset \mathfrak{gl}(V)$ be an irreducible set of operators and $T : V \to V$ be an operator such that $T\Upsilon = \Upsilon T$. Then $T = cI$. In particular this is so if $TX = XT$ for all $X \in \Upsilon$.*

Proof. Let c be an eigenvalue of T. Since $T\Upsilon = \Upsilon T$ it follows that $(T - cI)\Upsilon = \Upsilon(T - cI)$. But $\det(T - cI) = 0$ so $T = cI$ by Schur's lemma. $\square$

The converse of the Schur's lemma is also true and does not depend on the field being algebraically closed.

Proposition 3.1.54. *Let ρ be a completely reducible representation of $\mathfrak{g}$ on a finite dimensional vector space V over k. If the intertwiners consist only of scalars, then ρ is irreducible.*

Proof. Let W be an invariant subspace of V. By complete reducibility choose an invariant complementary and subspace U to W. Let P_W be the projection of V onto W. Since both W and U are invariant an easy calculation shows that for each $X \in \mathfrak{g}$, $P_W \rho(X) = \rho(X)P_W$ when restricted to W and similarly for U. Hence they also agree on V. Thus P_W is an intertwiner. By hypothesis P_W is a scalar. But the eigenvalues of a projection consist of 0 and 1. Therefore $P_W = 0$ or $P_W = I$. Hence $W = 0$ or $W = V$ so ρ is irreducible. $\qquad\square$

3.1.5 The Irreducible Representations of $\mathfrak{sl}(2, k)$

We now consider the "simplest" non solvable Lie algebra, namely $\mathfrak{sl}(2, k)$ (see 3.1.3 for the definition). This algebra is of dimension 3 over k.

Proposition 3.1.55. $\mathfrak{sl}(2, k)$ *is a simple Lie algebra.*

Proof. If not, then it has a proper ideal which would be of dimension 1 or 2. But all Lie algebras of dimension ≤ 2 are solvable (see Example 3.1.38) so by Proposition 3.1.33 it would follow that $\mathfrak{g} = \mathfrak{sl}(2, k)$ is itself solvable. On the other hand, from the relations (see Example 3.1.3) we see that $[\mathfrak{g}, \mathfrak{g}] = \mathfrak{g}$ so $\mathfrak{g}$ cannot be solvable. $\qquad\square$

Corollary 3.1.56. *If ρ is a representation of $\mathfrak{sl}(2, k)$ on V and the intertwiners consist of scalars, then ρ is irreducible.*

Proof. Since $\mathfrak{sl}(2, k)$ is simple by Proposition 3.1.55 it is therefore semisimple. This means that ρ is completely reducible by Weyl's theorem so the proposition above applies. Here we refer to Weyl's theorem, Theorem 3.4.3, out of order. The proof we give of Weyl's theorem is independent of facts concerning $\mathfrak{sl}(2, k)$. $\qquad\square$

Corollary 3.1.57. *Any representation $\rho : \mathfrak{sl}(2, k) \rightarrow \mathfrak{gl}(n, k)$ is either trivial or faithful. In particular, any nontrivial representation of $\mathfrak{g}$ takes a basis of $\mathfrak{sl}(2, k)$ to a linearly independent family of operators in $\mathfrak{gl}(n, k)$.*

Proof. Let ρ be a nontrivial representation of $\mathfrak{g}$. Then $\mathrm{Ker}\,\rho$ is a proper ideal in $\mathfrak{g}$. Hence $\mathrm{Ker}\,\rho = \{0\}$ so ρ is faithful. $\qquad\square$

We shall now find all equivalence classes of finite dimensional irreducible representations of $\mathfrak{g} = \mathfrak{sl}(2, k)$, where k is algebraically closed. The set of all equivalence classes of finite dimensional representations of $\mathfrak{g}$ is called the finite dimensional irreducible dual.

Let V_{n+1} be a vector space of dimension $n + 1$ over k with basis $\{v_0, \ldots, v_n\}$ and $n \geq 1$. For each n and $X \in \mathfrak{g}$ we define operators $\rho(X)$ on $V = V_{n+1}$ as follows (following the notation in Example 3.1.3):

(1) $\rho(X^+)v_i = i(n - i + 1)v_{i-1}$ for $i = 1, \ldots, n$ and $\rho(X^+)v_0 = 0$.

(2) $\rho(H)v_i = (n - 2i)v_i$ for $i = 0, \ldots, n$.

(3) $\rho(X^-)v_i = v_{i+1}$ for $i = 0, \ldots, n - 1$ and $\rho(X^-)v_n = 0$.

Extending by linearity, this gives a well-defined linear operator $\rho(X)$ on V for each $X \in \mathfrak{g}$. With respect to the given basis $\rho(X^+)$ is lower triangular with integer entries, $\rho(X^-)$ is upper triangular with integer entries, and $\rho(H)$ diagonal with integer entries symmetric about 0.

To see that ρ is a representation, by linearity it is sufficient to verify that

$$[\rho(H), \rho(X^+)] = \rho([H, X^+]),$$
$$[\rho(X^+), \rho(X^-)] = \rho([X^+, X^-]),$$
$$[\rho(H), \rho(X^-)] = \rho([H, X^-]).$$

We leave this to the reader to check by using the definitions. These representations are all inequivalent since they all have different degrees. Thus for each integer n we have a representation of $\mathfrak{sl}(2, k)$ of degree $n + 1$.

We will now show that ρ is an irreducible representation of $\mathfrak{g}$. If T is an intertwining operator on V, then in particular $T\rho(H) = \rho(H)T$. Since $\rho(H)$ has $n+1$ distinct eigenvalues this means T is diagonal. Then applying $T\rho(X^+) = \rho(X^+)T$ tells us that all these diagonal entries are equal. Since k is algebraically closed it follows from the converse of Schur's Lemma 3.1.54 that ρ is irreducible.

We now show that if k is algebraically closed, then up to equivalence we have identified all finite dimensional irreducible representations of $\mathfrak{sl}(2, k)$.

Proof. To see this let σ be such a representation of degree $n+1$. Suppose $\sigma(h)v_0 = \lambda v_0$, where $\lambda \in k$ and $v_0 \neq 0$ in V. Since σ is a representation,

$$\sigma(H)\sigma(X^+)v_0 = \sigma(X^+)\sigma(H)v_0 + \sigma([X^+, H])v_0 = (\lambda + 2)\sigma(X^+)v_0.$$

But then λ, $\lambda + 2$, $\lambda + 4$ etc. is an infinite sequence (of distinct) eigenvalues of $\sigma(h)$ and $\dim_k V < \infty$. It follows that by replacing λ by one of the succeeding terms in the sequence we get $\sigma(X^+)v_0 = 0$ as well as $\sigma(H)v_0 = \lambda v_0$ for some $v_0 \neq 0$.

Define $v_i = \sigma(X^-)^i v_0$, for $i \geq 0$. Then arguing as above, $\sigma(H)\sigma(X^-)v_0 = (\lambda - 2)\sigma(X^-)v_0$, so that $\sigma(H)v_1 = (\lambda - 2)v_1$. We show by induction that $\sigma(H)v_i = (\lambda - 2i)v_i$, the case $i = 1$ having just been done. Indeed suppose $\sigma(H)v_{i-1} = (\lambda - 2(i-1))v_{i-1}$. Then

$$\begin{aligned}
\sigma(H)\sigma(X^-)v_{i-1} &= \sigma(X^-)\sigma(H)v_{i-1} + \sigma([H, X^-])v_{i-1} \\
&= \sigma(X^-)(\lambda - 2(i-1))v_{i-1} - 2\sigma(X^-)v_{i-1} \\
&= \sigma(X^-)(\lambda - 2i)v_{i-1}.
\end{aligned}$$

That is, $\sigma(H)v_i = (\lambda - 2i)v_i$. It follows from $\sigma(H)\sigma(X^-)v_{i-1} = (\lambda - 2i)\sigma(X^-)v_{i-1}$ that each $\sigma(X^-)^i v_0$ is an eigenvalue of $\sigma(H)$, or is zero. But just as before since the set of λ, $\lambda - 2$, $\lambda - 4$ etc. is an infinite sequence of distinct eigenvalues of $\sigma(H)$, $\sigma(X^-)^{n+1} = 0$ for some $n + 1$, where $n + 1$ is the smallest such integer. Now $\{v_0, \ldots, v_n\}$ is a linearly independent set since $v_0, \ldots, v_n$ are eigenvectors of $\sigma(H)$ with distinct eigenvalues. To summarize, $\sigma(H)v_i = (\lambda - 2i)v_i$, $\sigma(X^+)v_0 = 0$ and $\sigma(X^-)v_i = v_{i+1}$, $\sigma(X^-)v_{n+1} = 0$. We next prove $\{v_0, \ldots, v_n\}$ spans V. To see that this is so, we prove $\{v_0, \ldots, v_n\}$ is stable under $\sigma(X^+)$. Then since it is stable under $\sigma(H)$ and $\sigma(X^-)$ and these generate $\mathfrak{g}$, the linear span of $\{v_0, \ldots, v_n\}$ would be a nontrivial invariant subspace and because σ is irreducible this must be all of V. We prove our claim by actually showing $\sigma(X^+)v_i = i(\lambda - i + 1)v_{i-1}$ for each i.

First observe that

$$\begin{aligned}
\sigma(X^+)v_{i+1} &= \sigma(X^+)\sigma(X^-)v_i = \sigma([X^+, X^-])v_i + \sigma(X^-)\sigma(X^+)v_i \\
&= \sigma(H)v_i + \sigma(X^-)\sigma(X^+)v_i = (\lambda - 2i)v_i + \sigma(X^-)\sigma(X^+)v_i.
\end{aligned}$$

By inductive hypothesis this is

$$(\lambda - 2i)v_i + i(\lambda - i + 1)\sigma(X^-)v_{i-1} = (\lambda - 2i)v_i + i(\lambda - i + 1)v_i$$
$$= (i+1)(\lambda - i)v_i.$$

Thus the v_i's form a basis for V and in addition to the other relations, we have $\sigma(X^+)v_{i+1} = (i+1)(\lambda - i)v_i$ for each i.

We conclude the proof showing σ is equivalent to ρ_n, the representation defined above of degree $n+1$, by proving $\lambda = n$ and hence the relations are the same. But

$$\mathrm{tr}(\sigma(H)) = \mathrm{tr}(\sigma(X^+)\sigma(X^-) - \sigma(X^-)\sigma(X^+)) = 0.$$

This means $\sum_{i=0}^{n}(\lambda - 2i) = 0$ so that $(n+1)\lambda = 2\sum_{i=0}^{n} i = n(n+1)$. Thus $\lambda = n$. $\qquad\qquad\qquad\qquad\qquad\qquad\qquad\qquad\qquad\qquad\qquad\Box$

We remark that when $n = 0$ we get the trivial representation, when $n = 1$ we get the identity representation and when $n = 2$ we get the adjoint representation. We leave this as an exercise for the reader.

3.1.6   Invariant Forms

One can consider further structures on a Lie algebra, for example bilinear forms. But we must tie these to the Lie algebra structure (otherwise we are merely talking about the vector space). Therefore we require an invariance condition to be described below.

Definition 3.1.58. Let $\mathfrak{g}$ be a Lie algebra over a field k. A bilinear form $\beta : \mathfrak{g} \times \mathfrak{g} \to k$ is said to be *invariant* if for any triple X, Y, and Z in $\mathfrak{g}$,

$$\beta([X,Y],Z) = \beta(X,[Y,Z]).$$

Example 3.1.59. $\mathfrak{gl}(V)$ is equipped with a natural invariant bilinear form defined as follows:

$$\beta(X,Y) = \mathrm{tr}(XY).$$

It is easy to see that β is invariant,

$$
\begin{aligned}
\beta([X,Y],Z) = \operatorname{tr}([X,Y]Z) &= \operatorname{tr}(XYZ - YXZ) \\
&= \operatorname{tr}(XYZ) - \operatorname{tr}(YXZ) \\
&= \operatorname{tr}(XYZ) - \operatorname{tr}(XZY) \\
&= \operatorname{tr}(XYZ - XZY) = \operatorname{tr}(X(YZ - ZY)) \\
&= \operatorname{tr}(X[Y,Z]) \\
&= \beta(X,[Y,Z]).
\end{aligned}
$$

Here we use the fact that $\operatorname{tr}(AB) = \operatorname{tr}(BA)$, taking $A = Y$, $B = XZ$.

Example 3.1.60. Let $\phi : \mathfrak{g} \to \mathfrak{h}$ be a Lie homomorphism, and β an invariant form on $\mathfrak{h}$ then one can pull β back to $\mathfrak{g}$ using ϕ in the following manner,

$$
\beta_\phi(X,Y) = \beta(\phi(X),\phi(Y)) \text{ for } X, Y \in \mathfrak{g}.
$$

Since ϕ is a Lie homomorphism and β is an invariant form on $\mathfrak{h}$ it follows that β_ϕ is an invariant form on $\mathfrak{g}$.

Let $\rho : \mathfrak{g} \to \mathfrak{gl}(V)$ be a Lie representation. It follows from Examples 3.1.59 and 3.1.60 that the trace on $\mathfrak{gl}(V)$ induces an invariant form on $\mathfrak{g}$ given by the following formula,

$$
\beta_\rho(X,Y) = \operatorname{tr}(\rho(X)\rho(Y)).
$$

If ρ is adjoint representation this construction gives rise to an invariant which is called the *Killing form*.

Remark 3.1.61. There are invariant forms which are not the trace forms of any representation.

Lemma 3.1.62. *Let $\mathfrak{g}$ be a Lie algebra, $\mathfrak{h}$ an ideal and β the Killing form on $\mathfrak{g}$. Then the restriction of β to $\mathfrak{h} \times \mathfrak{h}$ is the Killing form of $\mathfrak{h}$.*

Proof. Let $X \in \mathfrak{h}$ then since $[X,\mathfrak{g}] \subset \mathfrak{h}$, $\operatorname{ad} X$ is of the form

$$
\operatorname{ad} X = \left(\begin{array}{c|c} \operatorname{ad}_\mathfrak{h} & * \\ \hline 0 & 0 \end{array} \right) \tag{3.4}
$$

Hence for X and $Y \in \mathfrak{h}$ it follows that

$$\operatorname{tr}(\operatorname{ad} X \operatorname{ad} Y) = \operatorname{tr}_\mathfrak{h}(\operatorname{ad}_\mathfrak{h}(X) \operatorname{ad}_\mathfrak{h}(Y)).$$

$\square$

Let $\mathfrak{g}$ be a Lie algebra endowed with an invariant form β and let W be a subset of $\mathfrak{g}$. One can consider the orthocomplement of W with respect to β,

$$W^\perp = \{X \in \mathfrak{g} \mid \ \beta(X, Y) = 0, \quad \forall \ Y \in W\}.$$

Corollary 3.1.63. *Let $\mathfrak{g}$ be a Lie algebra with an invariant form β and let $\mathfrak{h}$ be an ideal in $\mathfrak{g}$. Then $\mathfrak{h}^\perp$ is also an ideal.*

Proof. If $X \in \mathfrak{h}^\perp$, $Y \in \mathfrak{g}$ and $Z \in \mathfrak{h}$, then

$$\beta([X, Y], Z) = \beta(X, [Y, Z]) = 0$$

since $[Y, Z] \in \mathfrak{h}$. Therefore $[X, Y] \in \mathfrak{h}^\perp$ and $\mathfrak{h}^\perp$ is an ideal. $\square$

3.1.7   Complex, Real and Rational Lie Algebras

Let $\mathfrak{g}$ be a Lie algebra over a field k and k' be a field extension of k. The tensor product $\mathfrak{g}_{k'} = k' \otimes_k \mathfrak{g}$, a vector space over the field k', can be equipped with a Lie algebra structure induced by that of $\mathfrak{g}$. The Lie bracket on the generators of $\mathfrak{g}_{k'}$ is defined as follows,

$$[a \otimes X, b \otimes Y]_{\mathfrak{g}_{k'}} = ab \otimes [X, Y]_\mathfrak{g}$$

extending to all of $\mathfrak{g}_{k'}$ by linearity.

In other words $\mathfrak{g}_{k'}$ has the same structure constants as $\mathfrak{g}_k$.

In particular when $k = \mathbb{R}$ and $k' = \mathbb{C}$, we call $\mathfrak{g}_\mathbb{C}$ *complexification* of $\mathfrak{g}$. It is obvious that:

Proposition 3.1.64. *Let $\mathfrak{g}$, k and k' be as above. If $\mathfrak{h}$ is a subalgebra (respectively ideal) of $\mathfrak{g}$ then $\mathfrak{h}_{k'}$ is also subalgebra (respectively ideal) of $\mathfrak{g}_{k'}$.*

Proposition 3.1.65. *Let $\mathfrak{g}$ be a Lie algebra over k and k' a field extension of k. Then for any two ideals $\mathfrak{a}$, $\mathfrak{b}$ in $\mathfrak{g}$,*

$$[\mathfrak{a}, \mathfrak{b}]_{k'} = [\mathfrak{a}_{k'}, \mathfrak{b}_{k'}].$$

Proof is left to the reader as an exercise.

Proposition 3.1.66. *Let $\mathfrak{g}$, k and k' be as above. Then*

(1) $\mathfrak{g}$ is nilpotent if and only if $\mathfrak{g}_{k'}$ is nilpotent.

(2) $\mathfrak{g}$ is solvable if and only if $\mathfrak{g}_{k'}$ is solvable.

(3) $\mathfrak{g}$ is semisimple if and only if $\mathfrak{g}_{k'}$ is semisimple.

Proof. (i) and (ii) follow from the fact that lower central series and derived series of $\mathfrak{g}_{k'}$ are $\{\mathfrak{g}_i \otimes k'\}_{i=1,2..}$ and $\{\mathfrak{g}^i \otimes k'\}_{i=1,2..}$; part (iii) follows from Cartan's criterion which we prove later in the chapter. □

Proposition 3.1.67. *Let $\mathfrak{g}$ be a Lie algebra over a field k and let k' be a field extension of k. If β_k denotes the Killing form of $\mathfrak{g}$ and $\beta_{k'}$ the Killing form of $\mathfrak{g}_{k'}$, then*

$$\beta_{k'}|_{\mathfrak{g} \otimes \mathfrak{g}} = \beta_k.$$

Proof. This follows from the simple fact that the trace is independent of field extension. □

Let $\mathfrak{g}$ be a Lie algebra over k and let k_0 be a subfield of k. We say $\mathfrak{g}$ has a k_0-rational form if there is a basis for $\mathfrak{g}$ whose structure constants lie in k_0. If $k = \mathbb{R}$ and $k_0 = \mathbb{Q}$ then we say that $\mathfrak{g}$ has a *rational form*.

Lemma 3.1.68. *A Lie algebra $\mathfrak{g}$ over k has a k_0-rational form if and only if there is a Lie algebra $\mathfrak{h}$ over k_0 such that $\mathfrak{g} = \mathfrak{h}_k$.*

Proof. If $\mathfrak{g} = \mathfrak{h}_k$ then any basis for $\mathfrak{h}$ over k_0 is a basis for $\mathfrak{g}$ over k and obviously the structure constants are independent of the field extension.

To prove the converse, let $\{v_i\}_{i \in I}$ be a basis for $\mathfrak{g}$ such that the structure constants with respect to this basis are rational. Consider $\mathfrak{h}$, the vector space over k_0 spanned by $\{v_i\}_{i \in I}$. The Lie bracket of $\mathfrak{g}$ induces a Lie bracket on $\mathfrak{h}$, since the structure constants of the bracket of $\mathfrak{g}$ with respect to $\{v_i\}_{i \in I}$ are in k_0. Now it is clear that $\mathfrak{g} = \mathfrak{h}_k$ as Lie algebras over k. □

Lie algebras do not always have rational forms. In fact,

Proposition 3.1.69. *There exists a real (2-step) nilpotent Lie algebra without a rational form.*

As we shall see in Chapter 8, this fact will prove the well-known result of Malcev [39] that even simply connected 2-step nilpotent groups do not always contain lattices.

We shall need a lemma about families of surjective linear maps.

Lemma 3.1.70. *Let V and W be vector spaces of dimension n and m respectively where $n \geq m$, over a field k. Suppose that $\mathcal{T}$ is the subset of maps in $\mathrm{Hom}_k(V, W)$ consisting of the linear maps which are surjective. Then $\mathcal{T}$ is an open set in $\mathrm{Hom}_k(V, W)$.*

Proof. $\mathcal{T}$ is the subset of $\mathrm{Hom}_k(V, W) = \mathrm{Mat}_{n,m}(k)$ consisting of operators of maximal rank. That is to say the T's with some $m \times m$-minor with nonzero determinant. Therefore the complement of $\mathcal{T}$ consists of those matrices all of whose minors have determinant zero. This is the intersection of a finite number of (Zariski) closed sets which are Euclidean closed. Hence $\mathcal{T}$ is open. $\square$

Proof of Proposition 3.1.69: We want to construct a Lie algebra $\mathfrak{g} = E \oplus V$ where E and V are real vector spaces such that $[\mathfrak{g}, \mathfrak{g}] = V$ and $[\mathfrak{g}, V] = 0$. With these relations the Jacobi identity is automatically satisfied. Such a Lie algebra will be 2-step nilpotent and there is bijection between the set of such Lie algebra structures Φ and the surjective $\mathbb{R}$-linear maps $\phi : \bigwedge^2 E \to V$; this is a manifold by the previous lemma and hence has a dimension. Let $n = \dim E$ and $m = \dim V$; then $\dim \Phi$ is $m \cdot \frac{n(n-1)}{2}$. Existence of a rational structure, ϕ^0, on $\mathfrak{g}$ means that there is a basis $e_1^0, \ldots, e_n^0$ for E and a basis $v_1^0, \ldots, v_m^0$ for V such that the matrix of ϕ^0 has all rational entries. Note that $e_i^0 \wedge e_j^0$, $i < j$ is a basis for $\bigwedge^2 E$. Suppose that $e_1, \ldots, e_n$ and $v_1, \ldots, v_m$ are another such basis. Let $T_E : E \to E$ be an element of $\mathrm{GL}(E)$ taking the one basis of E to the other and let $T_{\wedge E} : \bigwedge^2 E \to \bigwedge^2 E$ be the map induced by T_E. Similarly $T_V : V \to V$ is an element of $\mathrm{GL}(V)$ taking one basis of V to the other. Then $\Phi_{\mathbb{Q}}$, the space of all rational structures on $\mathfrak{g}$, equals the set of all maps of the form $T_V \circ \phi^0 \circ T_{\wedge E}^{-1}$. There are countably many such ϕ^0 and the dimension of the set of all $T_V \circ \phi^0 \circ T_{\wedge E}$ for a fixed ϕ^0 is $n^2 + m^2$. For $m \geq 3$ and sufficiently large n, this is $< m \cdot \frac{n(n-1)}{2}$ which is the dimension of the ambient space. Hence each of these sets has Lebesgue measure zero. By countable subadditivity the union also has measure zero. Thus the complement is very large. In fact it is dense.

3.2   Engel and Lie's Theorems

Here we shall deal with certain results concerning representations of solvable and nilpotent Lie algebras. As we shall see each of these is a generalization of a familiar theorem of linear algebra concerning a single operator. These results are also important for non-solvable Lie algebras because those more general algebras have interesting solvable and nilpotent subalgebras. Later we shall consider the Lie group analogues of these results.

3.2.1   Engel's Theorem

Definition 3.2.1. An operator T on a finite dimensional vector space V over a field k is called *nilpotent* if $T^j = 0$ for some integer j.

Lemma 3.2.2. *Let N_1 and N_2 be commuting nilpotent operators on a vector space V of dimension n. Then $N_1 \pm N_2$ is nilpotent.*

Proof. By the binomial formula

$$(N_1 \pm N_2)^{2n} = \sum_{i=0}^{2n} (\pm 1)^i \binom{2n}{i} N_1^i N_2^{2n-i}.$$

Thus $(N_1 \pm N_2)^{2n} = 0$ since $N_1^i = 0$, if $i \geq n$ and $N_2^{2n-i} = 0$, if $i \leq n$. $\square$

Corollary 3.2.3. *Let $X \in \mathfrak{gl}(n,k)$ be nilpotent. Then $\mathrm{ad}\, X \in \mathfrak{gl}(\mathfrak{gl}(n,k))$ is also nilpotent.*

Proof. For any $X \in \mathfrak{g}(n,k)$ define the left and right translation operators, l_X and r_X by $l_X(T) = XT$ and $r_X(T) = TX$. These operators commute because of the associativity of multiplication in $\mathfrak{gl}(n,k)$ and also $l_X - r_X = \operatorname{ad} X$. Hence by Lemma 3.2.2, $\operatorname{ad} X$ is nilpotent. $\square$

Proposition 3.2.4. *Let $\mathfrak{g}$ be a subalgebra consisting of nilpotent operators in the Lie algebra $\mathfrak{gl}(n,k)$. Then there exists a $v_0 \neq 0 \in V$ such that $X(v_0) = 0$ for all $X \in \mathfrak{g}$.*

Such a vector is called an *invariant vector* of $\mathfrak{g}$.

Proof. We prove this by induction on the dimension of $\mathfrak{g}$. If the dimension is one we are really just talking about the line through X and so everything is determined by X itself. As is well known from linear algebra a nilpotent operator always annihilates some nonzero vector.

Now suppose inductively that the proposition holds for *all linear Lie algebras* of dimension strictly lower than $\dim_k(\mathfrak{g})$. Let $\mathfrak{h}$ be a proper subalgebra of $\mathfrak{g}$ of maximal dimension. Such an $\mathfrak{h}$ exists since 1-dimensional subspaces are (abelian) subalgebras. Let $H \in \mathfrak{h}$. By 3.2.3 $\operatorname{ad} H$ acting on $\mathfrak{gl}(V)$ is nilpotent. Hence $\operatorname{ad}_{\mathfrak{gl}(V)} H$ restricted to $\mathfrak{g}$, which is just $\operatorname{ad}_{\mathfrak{g}} H$ is also nilpotent. This operator stabilizes the subalgebra $\mathfrak{h}$ and induces a linear endomorphism $\widetilde{\operatorname{ad}_{\mathfrak{g}}}$ on $\mathfrak{g}/\mathfrak{h}$, which is also nilpotent, defined by $\widetilde{\operatorname{ad}_{\mathfrak{g}}}(X + \mathfrak{h}) = \operatorname{ad}_{\mathfrak{g}} H(X) + \mathfrak{h} = [X, H] + \mathfrak{h}$. Now these operators form a Lie subalgebra of $\mathfrak{gl}(\mathfrak{g}/\mathfrak{h})$ and since $\mathfrak{h}$ is a proper subalgebra, we get

$$\dim \operatorname{ad}_{\mathfrak{g}} \mathfrak{h} = \dim(\mathfrak{h} + \mathfrak{z}(\mathfrak{g})/\mathfrak{z}(\mathfrak{g})) = \dim(\mathfrak{h}/\mathfrak{h} \cap \mathfrak{z}(\mathfrak{g})) \leq \dim \mathfrak{h} < \dim(\mathfrak{g}).$$

Therefore by induction there is an $X \in \mathfrak{g} \setminus \mathfrak{h}$ such that $[X, H] \in \mathfrak{h}$ for all $H \in \mathfrak{h}$. This means that the linear span of $\mathfrak{h}$ and X is a subalgebra of $\mathfrak{g}$ strictly containing $\mathfrak{h}$. By maximality this subalgebra is $\mathfrak{g}$. But since $[X, \mathfrak{h}] \subseteq \mathfrak{h}$ we see that $\mathfrak{h}$ is actually an ideal in $\mathfrak{g}$. Let $W = \{w \in V : Hw = 0 \text{ for all } H \in \mathfrak{h}\}$. W is clearly a subspace of V. Now $\mathfrak{h}$ is a subalgebra of nilpotent operators on V of dimension strictly lower than that of $\mathfrak{g}$. Hence there is some $w \neq 0 \in W$ with $Hw = 0$ for all $H \in \mathfrak{h}$ and in particular, $W \neq 0$. For $w \in W$ and $H \in \mathfrak{h}$ we have $HX(w) = [H, X]w + XH(w)$. Hence $HX(w) = 0$ for all $H \in \mathfrak{h}$. Thus

W is an X stable subspace of V, and since X is nilpotent on V it is nilpotent on W. By the theory of a single nilpotent operator there is some $w_0 \neq 0 \in W$ such that $X(w_0) = 0$. Since $\mathfrak{h}$ kills everything in W and X kills w_0 it follows that $\mathfrak{g}$ kills w_0.   □

We now come to Engel's theorem itself which is just an amplification of the previous result. This is easily proved by induction on the dimension of V using the previous proposition and is left to the reader.

Theorem 3.2.5. *Let $\mathfrak{g}$ be a Lie subalgebra of the Lie algebra $\mathfrak{gl}(n, k)$ consisting of nilpotent operators. Then there exists a basis $v_1, \ldots v_n$ of V with respect to which $\mathfrak{g}$ is simultaneously nil-triangular.*

Corollary 3.2.6. *A subalgebra $\mathfrak{g}$ of $\mathfrak{gl}(n, k)$ consisting of nilpotent operators is a nilpotent Lie algebra.*

This is so because the full algebra of nil-triangular operators is nilpotent and hence so is any subalgebra. For similar reasons we get

Corollary 3.2.7. *A Lie subalgebra $\mathfrak{g}$ of $\mathfrak{gl}(n, k)$ consisting of nilpotent operators has the property that the associative product of any n elements is zero.*

The following variant is sometimes also called Engel's theorem.

Corollary 3.2.8. *A Lie algebra $\mathfrak{g}$ is nilpotent if and only if $\operatorname{ad} X$ is a nilpotent operator on $\mathfrak{g}$ for all $X \in \mathfrak{g}$.*

Proof. If $\mathfrak{g}$ is nilpotent, let $n = 1 + $index of nilpotence. Then $[X_1, [X_2, [X_3 \ldots]]] = 0$. Taking all $X_1 = X_2 \ldots = X_{n-1} = X$ and $X_n = Y$ we get $(\operatorname{ad} X)^{n-1}(Y) = 0$. Since Y is arbitrary each $\operatorname{ad} X$ is nilpotent. On the other hand if each $\operatorname{ad} X$ is nilpotent, then by a corollary to Engel's theorem the linear Lie algebra $\operatorname{ad} \mathfrak{g}$ is nilpotent. But then so is $\mathfrak{g}$ since it is a central extension of $\operatorname{ad} \mathfrak{g}$.   □

Corollary 3.2.9. *If $\mathfrak{g}$ is a nilpotent Lie algebra and $\mathfrak{h}$ is an ideal then $\mathfrak{h} \cap \mathfrak{z}(\mathfrak{g})$ is nonzero. In particular the center $\mathfrak{z}(\mathfrak{g})$ itself is nonzero.*

Proof. ad $\mathfrak{g}$ is a Lie subalgebra of $\mathfrak{gl}(\mathfrak{g})$ consisting of nilpotent operators and $\mathfrak{h}$ is an invariant subspace under this subalgebra. Therefore ad $\mathfrak{g}|_\mathfrak{h}$ is a Lie subalgebra of $\mathfrak{gl}(\mathfrak{h})$ consisting of nilpotent operators. Hence, there is a nonzero $H \in \mathfrak{h}$ such that ad $X(H) = 0$ for all $X \in \mathfrak{g}$ and thus $H \in \mathfrak{h} \cap \mathfrak{z}$. $\qquad\square$

Engel's theorem can be formulated in terms of representations as follows:

Theorem 3.2.10. *Let $\mathfrak{g}$ be a Lie algebra and ρ a representation of $\mathfrak{g}$ on V. If $\rho(\mathfrak{g})$ consists of nilpotent operators, then there exists an operator $P_0 \in GL(V)$ such that $P_0(\rho(\mathfrak{g}))P_0^{-1}$ is in nil-triangular form.*

Corollary 3.2.11. *If $\mathfrak{g}$ is a nilpotent Lie algebra and $\mathfrak{h}$ is a proper subalgebra of $\mathfrak{g}$, then $\mathfrak{h} \subsetneq \mathfrak{n}_\mathfrak{g}(\mathfrak{h})$.*

Proof. Consider the adjoint representation of $\mathfrak{h}$ on $\mathfrak{g}$. Since $\mathfrak{g}$ is nilpotent these operators are nilpotent. Because $\mathfrak{h}$ is a subalgebra this induces an action of $\mathfrak{h}$ on $\mathfrak{g}/\mathfrak{h}$ by nilpotent operators. Proposition 3.2.4 tells us that there has to be $X \in \mathfrak{g} \setminus \mathfrak{h}$ such that takes $[X, \mathfrak{h}] \subseteq \mathfrak{h}$. Therefore $X \in \mathfrak{n}_\mathfrak{g}(\mathfrak{h}) \setminus \mathfrak{h}$. $\qquad\square$

3.2.2   Lie's Theorem

Before turning to Lie's theorem we make some convenient definitions amplifying the notion of invariant in the previous section.

Definition 3.2.12. Let $\mathfrak{g}$ be a Lie algebra and ρ a representation of $\mathfrak{g}$ on V. If there is a nonzero vector $v \in V$ and a function $\chi : \mathfrak{g} \to k$ such that $\rho_X(v) = \chi(X)v$ for all $X \in \mathfrak{g}$ we shall call χ a *semi-invariant* or a *weight* of ρ and v a *weight vector*.

Evidently a semi-invariant is k-linear and the $\chi(X)$'s are simultaneously eigenvalues of the ρ_X's. When χ is identically zero we obtain invariants as before. Also, if $\rho = \mathrm{ad}$, we shall say that χ is a *root* and v a *root vector*. Since each semi-invariant χ kills the derived subalgebra, $\chi([\mathfrak{g}, \mathfrak{g}]) = \{0\}$, there will not be any nontrivial semi-invariants at all if $\mathfrak{g} = [\mathfrak{g}, \mathfrak{g}]$. This is the situation, for example, when $\mathfrak{g}$ is a semisimple Lie algebra. At the opposite extreme is the case of a solvable Lie algebra.

Theorem 3.2.13. *Let $\mathfrak{g}$ be a solvable Lie algebra and ρ a representation of $\mathfrak{g}$ on V over an algebraically closed field of characteristic zero. Then there exists an operator in $P_0 \in \mathrm{GL}(V)$ such that $P_0(\rho(\mathfrak{g}))P_0^{-1}$ is in triangular form.*

Before turning to the proof of Lie's theorem we make a few remarks. This result can be stated in several different ways. For example, it asserts that there exists a basis $v_1, \ldots v_n$ of V with respect to which $\rho(\mathfrak{g})$ is in simultaneously upper triangular form. Evidently, Lie's theorem generalizes the fact that, over an algebraically closed field (of characteristic zero), any operator can be put in triangular form. Indeed, in the case of a 1-dimensional (abelian) Lie algebra this is the content of Lie's theorem. It should also be remarked that even in the 1-dimensional case, the result fails if the field is not algebraically closed. For example, if $k = \mathbb{R}$, $\mathfrak{g}$ is the 1-dimensional abelian Lie algebra of 2×2 skew symmetric matrices and ρ is the identity representation, then there are no simultaneous eigenvectors in $\mathbb{R}$. Indeed, they are all in $i\mathbb{R}$.

It should also be mentioned that Lie's theorem is a generalization of the following result. Let $\mathcal{T}$ be a family of commuting operators on a vector space V over an algebraically closed field of characteristic zero, then these operators can be simultaneously put in triangular form. This is because the linear span of $\mathcal{T}$ is also a commuting family and therefore an abelian and hence solvable Lie algebra of operators on V.

Finally just as in the previous section it is sufficient to prove the following result by making an induction on $\dim_k V$.

Proposition 3.2.14. *Let $\mathfrak{g}$ be a solvable Lie algebra and ρ a representation of $\mathfrak{g}$ on V over an algebraically closed field of characteristic zero. Then there exists a nonzero weight χ and weight vector v.*

Proof. We shall argue by induction on $\dim_k(\mathfrak{g})$. Since $[\mathfrak{g}, \mathfrak{g}] \subseteq \mathfrak{g}$ choose a subspace $\mathfrak{h}$ of $\mathfrak{g}$ such that $[\mathfrak{g}, \mathfrak{g}] \subseteq \mathfrak{h} \subsetneq \mathfrak{g}$ with $\dim(\mathfrak{g}/\mathfrak{h}) = 1$, *i.e.* $\mathfrak{h}$ is maximal. Then $\mathfrak{h}$ is an ideal in $\mathfrak{g}$ ($[\mathfrak{g}, \mathfrak{h}] \subseteq [\mathfrak{g}, \mathfrak{g}] \subseteq \mathfrak{h}$). In particular, it is a subalgebra and hence solvable. Consider the restriction of ρ to $\mathfrak{h}$. By inductive hypothesis there is a $v \neq 0 \in V$ satisfying $\rho_H v = \chi(H)v$ for all $H \in \mathfrak{h}$. Let $X_0 \in \mathfrak{g} \setminus \mathfrak{h}$. Then $\mathfrak{g} = \mathfrak{h} + \{cX_0; c \in k\}$. We complete the proof using the following lemma.

Lemma 3.2.15. *Let $\rho : \mathfrak{g} \to \mathfrak{gl}(V)$ be a representation of a Lie algebra $\mathfrak{g}$ with an ideal $\mathfrak{h}$. Suppose there is a vector $v \neq 0 \in V$ and $\chi : \mathfrak{h} \to k$ such that $\rho(H)v = \chi(H)v$ for all $H \in \mathfrak{h}$. Then $\chi([X_0, H]) = 0$ for all $X_0 \in \mathfrak{g}$ and $H \in \mathfrak{h}$.*

Proof. Let V_i be the subspace of V generated by $\{v, \rho_{X_0}v, \ldots, \rho_{X_0}^{i-1}v\}$, with $V_0 = \{0\}$. We get an increasing sequence of subspaces of V. By finite dimensionality there must be a place where $V_i = V_{i+1}$. Let n be the smallest such index. But because $V_i = V_{i+1}$ if and only if $\rho_{X_0}^i v$ is a linear combination of $\{v, \rho_{X_0}v, \ldots, \rho_{X_0}^{i-1}v\}$, we see that $\dim V_i = i$, for $i = 0, \ldots, n$. In particular, $\dim V_n = n$. Now $\rho_{X_0}(V_n)$ is spanned by $\{\rho_{X_0}v, \ldots, \rho_{X_0}^n v\} \subseteq V_{n+1} = V_n$. Hence $\rho_{X_0}(V_n) \subseteq V_n$. From this it also follows that $V_n = V_i$ for all $i \geq n$.

Next we will prove by induction on i that for all $H \in \mathfrak{h}$,

$$\rho_H \rho_{X_0}^i v \equiv \chi(H)\rho_{X_0}^i v \bmod V_i.$$

When $i = 0$ this is just the statement $\rho_H v = \chi(H)v$ for all H. In the inductive step

$$
\begin{aligned}
\rho_H \rho_{X_0}^{i+1} v &= \rho_H \rho_{X_0} \rho_{X_0}^i (v) \\
&= \rho_{X_0} \rho_H \rho_{X_0}^i (v) + \rho_{[H,X_0]} \rho_{X_0}^i v \\
&= \rho_{X_0}(\chi(H)\rho_{X_0}^i(v) + v_i) + \chi([H, X_0])\rho_{X_0}^i(v) + v_i',
\end{aligned}
\tag{3.5}
$$

where v_i and $v_i' \in V_i$.

But this in turn is $\chi(H)\rho_{X_0}^{i+1}(v) + \rho_{X_0}(v_i) + \chi([H, X_0])\rho_{X_0}^i(v) + v_i'$. Since each of these terms is in V_{i+1} we see that indeed $\rho_H \rho_{X_0}^{i+1} v \equiv \chi(H)\rho_{X_0}^{i+1} v \bmod V_{i+1}$. Thus with respect to a basis compatible with our flag on V_n, for each $H \in \mathfrak{h}$, ρ_H is in simultaneous triangular form, with all diagonal terms equal to $\chi(H)$. Hence $\operatorname{tr}_{V_n}(\rho_H) = (\dim_k V_n)\chi(H)$, for all $H \in \mathfrak{h}$. In particular, $\operatorname{tr}_{V_n}(\rho_{[H,X_0]}) = 0 = \dim_k V_n \chi([H, X_0])$ and since k has characteristic zero it follows that $\chi([H, X_0]) = 0$ for all $H \in \mathfrak{h}$. $\qquad\square$

Let $W = \bigcap_{H \in \mathfrak{h}} \{\operatorname{Ker}(\rho_H - \chi(H)I)\}$. Then W is a nonzero subspace of

V since $v \in W$, and W is clearly $\mathfrak{h}$-invariant. For $H \in \mathfrak{h}$,

$$\rho_H \rho_{X_0}(w) = \rho_{X_0} \rho_H(w) + \rho_{[H,X_0]}(w) = \rho_{H_0} \chi(H)(w) + \chi([H, X_0])(w)$$
$$= \chi(H) \rho_{X_0}(w).$$

Since H is arbitrary we see that W is ρ_{X_0}-invariant and hence $\mathfrak{g}$ invariant. Choose an eigenvector w_0 for ρ_{X_0} in W with eigenvalue λ. Then since $\rho_X = \rho_H + c(X)\rho_{X_0}$ everywhere on V we see that $\rho_X(w_0) = (\chi(H) + \lambda c(X))w_0$. $\square$

Our next corollary is actually equivalent to Lie's theorem. Notice also that if all finite dimensional irreducible representations are 1-dimensional, this also implies solvability. This is because, as just remarked, $\rho(\mathfrak{g})$ is a subalgebra of the full triangular algebra and hence is solvable. In particular, ad $\mathfrak{g}$ is solvable. But then so is $\mathfrak{g}$ itself.

Corollary 3.2.16. *In a solvable Lie algebra each finite dimensional irreducible representation over an algebraically closed field of characteristic zero is 1-dimensional.*

Proof. Suppose ρ is the irreducible representation. By the proposition above there is a weight χ and a weight vector v in V, the representation space. The line through v is an invariant subspace and, by irreducibility, this must be all of V. $\square$

Applying Lie's theorem to the adjoint representation we get,

Corollary 3.2.17. *In a solvable Lie algebra over an algebraically closed field of characteristic zero there is always a flag of ideals.*

Here is a version of Lie's theorem which does not require the field to be algebraically closed.

Corollary 3.2.18. *Let $\mathfrak{g}$ be a Lie algebra over a field of characteristic zero. $\mathfrak{g}$ is solvable if and only if $[\mathfrak{g}, \mathfrak{g}]$ is nilpotent.*

Proof. If $[\mathfrak{g}, \mathfrak{g}]$ is nilpotent and hence solvable, then since $\mathfrak{g}/[\mathfrak{g}, \mathfrak{g}]$ is abelian and therefore also solvable so is $\mathfrak{g}$. Conversely, suppose $\mathfrak{g}$ is solvable and the field is algebraically closed of characteristic zero. Applying Lie's theorem to the adjoint representation we see that $\operatorname{ad}\mathfrak{g}$ is a subalgebra of upper triangular operators. Hence its derived algebra, $[\operatorname{ad}\mathfrak{g},\operatorname{ad}\mathfrak{g}] = \operatorname{ad}[\mathfrak{g},\mathfrak{g}]$ consists of nil-triangular operators and is therefore nilpotent. Since $\operatorname{Ker}\operatorname{ad} = \mathfrak{z}(\mathfrak{g})$, it follows that $[\mathfrak{g},\mathfrak{g}]$ itself is nilpotent. This completes the proof when k is algebraically closed. In general we proceed by complexifying everything. Now suppose we have a solvable Lie algebra over a field k of characteristic zero. Then, by Proposition 3.1.66 $[\mathfrak{g}_{\overline{k}},\mathfrak{g}_{\overline{k}}]$ is nilpotent, where $\overline{k}$ is algebraic closure of k. Therefore $\mathfrak{g}_{\overline{k}}$ is solvable and then by Proposition 3.1.66, $\mathfrak{g}$ is solvable. $\qquad\square$

From this follows:

Corollary 3.2.19. *For a Lie algebra* $\mathfrak{g}$,

$$[\operatorname{rad}(\mathfrak{g}),\operatorname{rad}(\mathfrak{g})] \subset \operatorname{nil}(\mathfrak{g}).$$

Corollary 3.2.20. *Let* $\mathfrak{g}$ *be a solvable real Lie algebra and* ρ *a real representation of* $\mathfrak{g}$ *on* V. *Then there exists an increasing family of invariant subspaces which are each of codimension 1 or 2. Choosing a natural basis for these quotients puts* $\rho(\mathfrak{g})$ *in simultaneous block triangular form over* $\mathbb{R}$, *where the* 2×2 *blocks are of the type.*

$$\begin{pmatrix} a_j(X) & b_j(X) \\ -b_j(X) & a_j(X) \end{pmatrix}$$

for $X \in \mathfrak{g}$.

Notice that when $a_j(X) = 0$ this is the form of the skew symmetric matrices mentioned earlier.

3.3 Cartan's Criterion and Semisimple Lie Algebras

3.3.1 Some Algebra

Definition 3.3.1. An operator T on a finite dimensional vector space V over k is called *semisimple* if T is diagonalizable over the algebraic closure of k.

Theorem 3.3.2. (Jordan Decomposition) *Let V be a finite dimensional space over an algebraically closed field and $T \in \operatorname{End}_k(V)$. Then $T = S + N$ where S is diagonalizable (semisimple), N is nilpotent and they commute. These conditions uniquely characterize S and N. Moreover there exist polynomials p and q without constant term in $k[x]$ such that $S = p(T)$ and $N = q(T)$. Hence not only do S and N commute with T but they commute with any operator which commutes with T. If $A \subset B$ are subspaces of V and $T(B) \subset A$ then $S(B) \subset A$ and $N(B) \subset A$. In particular if A is a T invariant subspace then it is S and N invariant. If $T(B) = 0$ then $S(B) = 0$ and $N(B) = 0$.*

The Jordan form of T consists of blocks each of which is the sum of a scalar and a nilpotent operator. This proves the first statement. Regarding the uniqueness, we first note that

Lemma 3.3.3. *An operator S is semisimple if and only if every S-invariant W space has a S-invariant complement.*

Proof. Suppose that S is semisimple and k is algebraically closed, then we can write $V = \bigoplus_{\alpha \in A} V_\alpha$ where V_α are 1-dimensional S-invariant vector spaces and A is a finite set. Consider the family of sets of the form $\bigcup_{\beta \in B}\{V_\beta\} \cup \{W\}$ where $B \subset A$ and V_β's and W are linearly independent. This family is nonempty as it contains $\{N\}$. Since it is finite, it has a maximal $K = \bigcup_{\beta \in B}\{V_\beta\} \cup \{W\}$. Let $M' = \bigoplus_{\beta \in B} V_\beta \oplus W$. We prove that $M' = M$, otherwise there exists $\alpha \in A$ such that $V_\alpha \not\subset M'$. Since V_α is 1-dimensional therefore $V_\alpha \cap M' = (0)$. Therefore $L = \bigcup_{\beta \in B}\{V_\beta\} \cup \{W\} \cup \{V_\alpha\}$ is in the family and $K \subset L$ which contradicts the maximality of K, hence $M' = M$. Now take $W' = \bigoplus_{\alpha \in A} V_\alpha$. Then $M = W \oplus W'$ and W' is S invariant.

The converse can be proved by an induction, and by noticing that there is always an eigenvector for S over an algebraically closed field.   □

Lemma 3.3.4. *The restriction of a semisimple operator S to an invariant subspace W is still a semisimple operator.*

Proof. Let U be an S-invariant subspace of W. Then U is an S-invariant subspace of V. By the previous lemma there is an S-invariant subspace

U' of V which complements U in V. Then $U' \cap W$ is an S-invariant subspace of W which complements U in W.   □

Lemma 3.3.5. *If S and S' are diagonalizable and commute, then $S \pm S'$ is diagonalizable. If N and N' are nilpotent and commute, then $N \pm N'$ is nilpotent.*

Proof. For $\alpha \in \mathrm{Spec}(S)$ let V_α be the eigenspace of α. Then V is the direct sum of the V_α. If $v \in V_\alpha$ then $S(v) = \alpha v$ so $S'S(v) = \alpha S'(v) = SS'(v)$ so that $S'(V_\alpha) \subset V_\alpha$. Now the restriction of S' to each V_α is still semisimple. Choose a basis in each V_α in which the restriction is diagonal and in this way get a basis of V. Since, on each V_α, $S = \alpha I$, it follows that $S + S'$ is diagonal. Moreover, $N \pm N'$ is nilpotent by Lemma 3.2.2.   □

Now suppose $S + N = T = S' + N'$, where $[S, N] = 0 = [S', N']$. Then S' and N' each commutes with T. Hence each commutes with S and N. In particular, S and S' and N and N' commute. But then by the lemma $S' - S$ is semisimple and $N - N'$ is nilpotent. Since $S' - S = N - N'$ each of these is zero. This proves the uniqueness part of Theorem 3.3.2.

Completion of proof of 3.3.2: Let $\chi_T(x) = \prod (x - \alpha_i)^{n_i}$ be the factorization of the characteristic polynomial of T into distinct linear factors over k. Since the α_i are distinct, the $(x - \alpha_i)^{n_i}$ are pairwise relatively prime. Consider the following

$$p(x) \equiv \alpha_i \bmod (x - \alpha_i)^{n_i}$$
$$p(x) \equiv 0 \bmod \quad (x)$$

If no $\alpha_i = 0$ then x together with the $(x - \alpha_i)^{n_i}$ is also relatively prime. If some $\alpha_i = 0$ then the last congruence follows from the others. In either case by the Chinese Remainder Theorem there is a polynomial p satisfying them. Then $p(T) - \alpha_i I = \phi_i(T)(T - \alpha_i I)^{n_i}$. So, on each V_{α_i}, $p(T) = \alpha_i I$. This is equal to S on V_{α_i} so $S = p(T)$. Taking $q(x) = x - p(x)$ we see that $q(0) = 0$ and $q(T) = T - p(T) = T - S = N$. Suppose $A \subset B$ are subspaces of V and $T(B) \subset A$. Since $S = \sum \alpha_i T^i$, we get $S(B) \subset A$ if we can show that $T^i(B) \subset A$ for $i \geq 1$. Now $T^2(B) = T(T(B)) \subset T(A) \subset T(B) \subset A$ and proceed by induction.

Proposition 3.3.6. (*Lagrange Interpolation Theorem*) *Let* $c_0, c_1, ..., c_n$ *be distinct elements of a field,* k, *and* $a_0, a_1, ..., a_n$ *lie in* k. *Then there is a unique polynomial* p *in* $k[x]$ *of degree* n *such that* $p(c_i) = a_i$ *for all* i.

Proof. For each $i = 1, ..., n$ let $\phi_i(x) = \prod_{j \neq i}(x - c_j)$. Then ϕ_i is a polynomial in $k[x]$ of degree n, $\phi_i(c_i) \neq 0$ and if $k \neq i, \phi_i(c_k) = 0$. Now $f_i = \phi_i/\phi_i(c_i)$ is also a polynomial of degree n and $f_i(c_j) = \delta_{ij}$ for all i, j. Let $a_0, a_1, ..., a_n$ be given and $p = \sum_0^n a_i f_i$. Then $p(c_j) = \sum a_i f_i(c_j) = \sum a_i \delta_{ij} = a_j$ for all j. If q is another such polynomial of degree $\leq n$ then $(p - q)(c_j) = 0$. Since $p - q$ has degree $\leq n$ and has $n + 1$ distinct roots $p - q = 0$ so $p = q$.   □

Corollary 3.3.7. *Let* $a_1, ..., a_n$ *be in a field of characteristic* 0, *and* $E = l.s._\mathbb{Q}\{a_1, ..., a_n\}$. *If* $f \in E^*$, *the* $\mathbb{Q}$-*dual space of* E, *then there is a polynomial* p *in* $k[x]$, *without constant term, such that* $p(a_i - a_j) = f(a_i) - f(a_j)$ *for all pairs* i, j.

Proof. Let S and T denote the following finite sets $S = \{a_i - a_j : i, j = 1, ..., n\}$ and $T = f(a_i) - f(a_j)$ for $i, j = 1, ..., n$. Consider the map $S \to T$ given by $a_i - a_j \to f(a_i) - f(a_j)$. This is well-defined since if $a_i - a_j = a_k - a_l$ then $f(a_i - a_j) = f(a_k - a_l)$ and since f is $\mathbb{Q}$-linear, $f(a_i) - f(a_j) = f(a_k) - f(a_l)$. If $a_i = a_j$ then $f(a_i) = f(a_j)$ so this map takes 0 to 0. By Lagrange interpolation there is a polynomial p such that $p(a_i - a_j) = f(a_i) - f(a_j)$ for all pairs i, j and in particular $p(0) = 0$.   □

In what follows k is an algebraically closed field of characteristic 0, $\mathfrak{gl}(V)$ stands for $\mathrm{End}_k(V)$ and $\{E_{ij} : i, j = 1, ..., n\}$ denotes its matrix units.

Lemma 3.3.8. *If* X *is a diagonal matrix with entries* $\{a_1, ..., a_n\}$ *then* $\mathrm{ad}_X(E_{ij}) = (a_i - a_j)(E_{ij})$. *Thus if* X *is semisimple on* V *then* $\mathrm{ad}\, X$ *is semisimple on* $\mathrm{End}_k(V)$.

Proof. We have $X = \sum_k a_k E_{kk}$ so

$$
\begin{aligned}
\operatorname{ad} X(E_{ij}) &= \sum_k a_k [E_{kk}, E_{ij}] = \sum_k a_k (E_{kk} E_{ij} - E_{ij} E_{kk}) \\
&= \sum_k a_k \delta_{ik} E_{kj} - \sum_k a_k E_{ik} \delta_{kj} \\
&= a_i E_{ij} - a_j E_{ij} = (a_i - a_j) E_{ij}.
\end{aligned}
$$

$\square$

Lemma 3.3.9. *Let $X \in \mathfrak{gl}(V)$ and let $X = S + N$ be its Jordan decomposition. Then $\operatorname{ad} X = \operatorname{ad} S + \operatorname{ad} N$ is the Jordan decomposition of $\operatorname{ad} X$.*

Proof. Since ad is linear $\operatorname{ad} X = \operatorname{ad} S + \operatorname{ad} N$. Moreover $[\operatorname{ad} S, \operatorname{ad} N] = \operatorname{ad}[S, N] = \operatorname{ad} 0 = 0$ so that $\operatorname{ad} S$ and $\operatorname{ad} N$ commute. By Lemma 3.3.8, $\operatorname{ad} S$ is semisimple. By Lemma 3.2.3, $\operatorname{ad} N$ is nilpotent. The uniqueness of the Jordan decomposition gives the result. $\square$

Lemma 3.3.10. *Let p be a polynomial without constant term and let $T, S \in \mathfrak{gl}(V)$ for some V. Suppose that, relative to a basis $\{v_1, ..., v_n\}$, T and S are diagonal with entries $\{a_1, ..., a_n\}$ and $\{p(a_1), ..., p(a_n)\}$ respectively. Then clearly $S = p(T)$. In particular, by the argument in the Jordan decomposition theorem, if $A \subset B$ are subspaces of V and $T(B) \subset A$ then $S(B) \subset A$.*

In order to formulate and prove Cartan's criterion for an arbitrary field of characteristic zero, it will be necessary to deal with certain finite dimensional rational vector spaces in the following lemma.

Lemma 3.3.11. *Let $A \subset B$ be subspaces of $\mathfrak{gl}(V)$ and $M = \{X \in \mathfrak{gl}(V) : [X, B] \subset A\}$. If $X \in M$ and $\operatorname{tr}(XY) = 0$ for all $Y \in M$ then X is nilpotent.*

Proof. Let $X = S + N$ be the Jordan decomposition of X and $v_1, ..., v_n$ be a basis such that S is diagonal with entries $\{a_1, ..., a_n\}$. Let E be the subspace of k spanned by $\{a_1, ..., a_n\}$ as a $\mathbb{Q}$-vector space. If $E = 0$

then all $a_i = 0$ and hence $S = 0$ and X is nilpotent. Since E is finite dimensional space to see $E = 0$ it suffices to show its dual E^*
over $\mathbb{Q}$, is trivial. If $f \in E^*$ let Y be the diagonal matrix with entries
$\{f(a_1), ..., f(a_n)\}$ with respect to $\{v_1, ..., v_n\}$. By a previous lemma,
$\operatorname{ad} S(E_{ij}) = (a_i - a_j)E_{ij}$ and $\operatorname{ad} Y(E_{ij}) = (f(a_i) - f(a_j))E_{ij}$. By Corollary 3.3.7, there is a polynomial p such that $p(0) = 0$ and $p(a_i - a_j) =$
$f(a_i) - f(a_j)$ for all pairs i, j. This means $\operatorname{ad} Y(E_{ij}) = p(a_i - a_j)E_{ij}$.
Since $X \in M$, $\operatorname{ad} X(B) \subset A$. As above $\operatorname{ad} S$ is a polynomial in $\operatorname{ad} X$
without constant term; it follows that $\operatorname{ad} S(B) \subset A$. By the last lemma
$\operatorname{ad} Y(B) \subset A$ and hence $Y \in M$. Now XY is triangular with diagonal
entries $\{a_1 f(a_1), ..., a_n f(a_n)\}$ and so $\operatorname{tr}(XY) = \sum a_i f(a_i) = 0$. This is a
$\mathbb{Q}$-linear combination of elements of E. Applying f yields $\sum f(a_i)^2 = 0$.
Since the $f(a_i)$ are in $\mathbb{Q}$ they are all 0. We conclude that f kills the
generators for E so $f = 0$ and since f is arbitrary $E^* = (0)$.   □

3.3.2   Cartan's Solvability Criterion

Cartan's solvability criterion is the following:

Theorem 3.3.12. (*Cartan's Criterion*) *Let* $\mathfrak{g}$ *be a subalgebra of* $\mathfrak{gl}(V)$
over any field of characteristic zero such that $\operatorname{tr}(XY) = 0$ *for all* $X \in$
$[\mathfrak{g}, \mathfrak{g}]$ *and* $Y \in \mathfrak{g}$. *Then* $\mathfrak{g}$ *is solvable.*

Proof. By a corollary to Lie's theorem it suffices to show that $[\mathfrak{g}, \mathfrak{g}]$ is
nilpotent; and by Engel's theorem that each element of $[\mathfrak{g}, \mathfrak{g}]$ is nilpotent.
Let $A = [\mathfrak{g}, \mathfrak{g}]$, $B = \mathfrak{g}$, X and $Y \in \mathfrak{g}$ and $Z \in M$ where M is defined as
in Lemma 3.3.11. Then $[Z, \mathfrak{g}] \subset [\mathfrak{g}, \mathfrak{g}]$. In particular $[Z, Y] \in [\mathfrak{g}, \mathfrak{g}]$ so our
hypothesis tells us that $\operatorname{tr}(X[Y, Z]) = 0$ or by invariance $\operatorname{tr}([X, Y]Z) =$
0. By linearity $\operatorname{tr}(UZ) = 0$ for any $U \in [\mathfrak{g}, \mathfrak{g}]$. Since $\mathfrak{g} \subset M$ Lemma
3.3.11 tells us that each such U is nilpotent.   □

Corollary 3.3.13. (*Also called Cartan's criterion*) *Let* $\mathfrak{g}$ *be a Lie alge-
bra* (*any field of characteristic zero*) *such that* $\operatorname{tr}(\operatorname{ad} X \operatorname{ad} Y) = 0$ *for all*
$X \in [\mathfrak{g}, \mathfrak{g}]$ *and* $Y \in \mathfrak{g}$ *then* $\mathfrak{g}$ *is solvable.*

Proof. By Cartan's criterion $\operatorname{ad} \mathfrak{g}$ is solvable. Since $\operatorname{Ker} \operatorname{ad} = \mathfrak{z}(\mathfrak{g})$ is also
solvable so is $\mathfrak{g}$.   □

Let $\rho : \mathfrak{g} \to \mathfrak{gl}(V)$ be a representation of a Lie algebra $\mathfrak{g}$ on V (any field of characteristic zero). Then, as above, the trace form

$$\beta_\rho : \mathfrak{g} \times \mathfrak{g} \to k$$

is given by $\beta_\rho(X,Y) = \mathrm{tr}(\rho(X)\rho(Y))$ and β_ρ is a symmetric bilinear form on $\mathfrak{g}$. In particular in the case of the Killing form we have an invariant symmetric bilinear form β on $\mathfrak{g}$.

Another way of expressing Cartan's criterion is the first half of the following:

Corollary 3.3.14. *Let $\mathfrak{g}$ be Lie algebra over any field k such that $[\mathfrak{g},\mathfrak{g}]$ is orthogonal to $\mathfrak{g}$ with respect to the Killing form then $\mathfrak{g}$ is solvable. Conversely if $\mathfrak{g}$ is a solvable then $[\mathfrak{g},\mathfrak{g}]$ is orthogonal to $\mathfrak{g}$ with respect to Killing form.*

Proof of Converse: Let $\bar{k}$ be the algebraic closure of k and regard ad $\mathfrak{g}$ as a subalgebra of $\mathfrak{gl}(\mathfrak{g}_{\bar{k}})$ rather than of $\mathfrak{gl}(\mathfrak{g})$. Since ad $\mathfrak{g}$ is solvable by Lie's theorem we know that ad $\mathfrak{g}$ can be triangularized over $\bar{k}$. Hence its derived algebra has zeros on the diagonal and so $\mathrm{tr}_{\mathfrak{g}_{\mathbb{C}}}(\mathrm{ad}\, X\, \mathrm{ad}\, Y) = 0$ for all $X \in [\mathfrak{g},\mathfrak{g}]$ and $Y \in \mathfrak{g}$. Since the trace is independent of field extensions, $[\mathfrak{g},\mathfrak{g}]$ is orthogonal to $\mathfrak{g}$ with respect to the Killing form.

Remark 3.3.15. The same argument shows that even without considering field extensions, the Killing form is identically zero if $\mathfrak{g}$ is nilpotent. For then ad $\mathfrak{g}$ can be nil-triangularized, therefore ad X ad Y is niltriangular and so has trace 0. Exercise: What about the converse?

Corollary 3.3.16. *A Lie algebra $\mathfrak{g}$ is semisimple if and only if its Killing form is nondegenerate. In fact (since the adjoint representation of a semisimple Lie algebra is faithful) more generally if $\mathfrak{g}$ is semisimple and ρ is any faithful representation of $\mathfrak{g}$ then β_ρ is nondegenerate.*

Proof. Suppose $\mathfrak{g}$ is semisimple. Since β_ρ is invariant, $\mathfrak{h} = \mathfrak{g}^\perp$ is an ideal. Because $\mathfrak{h}$ is orthogonal to $\mathfrak{g}$ it is orthogonal to $[\mathfrak{h},\mathfrak{h}]$. Hence by Cartan's criterion $\rho(\mathfrak{h})$ is solvable and since ρ is faithful this means $\mathfrak{h}$ is itself solvable and so is trivial. Thus β_ρ is nondegenerate. Conversely suppose the Killing form β is nondegenerate and $\mathfrak{h}$ is an abelian ideal. Let $X \in \mathfrak{h}$ and $Y \in \mathfrak{g}$. Then $\operatorname{ad} X \operatorname{ad} Y(\mathfrak{g}) \subset \mathfrak{h}$. So that $(\operatorname{ad} X \operatorname{ad} Y)^2(\mathfrak{g}) \subset \operatorname{ad} X \operatorname{ad} Y(\mathfrak{h})$. Since $\mathfrak{h}$ is an ideal, $\operatorname{ad} X \operatorname{ad} Y(\mathfrak{h}) \subset [X, \mathfrak{h}] = 0$ and $\mathfrak{h}$ is abelian. This means $\operatorname{ad} X \operatorname{ad} Y$ is nilpotent and so has trace 0. As β is nondegenerate we must have $X = 0$ so $\mathfrak{h} = 0$. $\qquad\square$

Proposition 3.3.17. *If $\mathfrak{h}$ is an ideal in a semisimple Lie algebra $\mathfrak{g}$ and $\mathfrak{a}$ is an ideal in $\mathfrak{h}$ then $\mathfrak{a}$ is an ideal in $\mathfrak{g}$.*

Proof. Let $\mathfrak{h}^\perp$ be the orthocomplement of $\mathfrak{h}$ in $\mathfrak{g}$ with respect to Killing form. Since Killing form is nondegenerate we have $\mathfrak{h} \oplus \mathfrak{h}^\perp = \mathfrak{g}$ and $\mathfrak{h}^\perp$ is also an ideal in $\mathfrak{g}$. We have $[\mathfrak{a}, \mathfrak{h}^\perp] \subset [\mathfrak{h}, \mathfrak{h}^\perp] \subset \mathfrak{h} \cap \mathfrak{h}^\perp = \{0\}$, therefore $[\mathfrak{a}, \mathfrak{g}] \subset [\mathfrak{a}, \mathfrak{h}] \subset \mathfrak{a}$. $\qquad\square$

Lemma 3.3.18. *Let V be a finite dimensional k-vector space and let $\beta : V \times V \to k$ be a bilinear form. Given a subspace W of V write $W^\perp = \{v \in V : \beta(v, W) = 0\}$. Then $W^\perp$ is a subspace of V and $\dim W + \dim W^\perp \geq \dim V$. If β is nondegenerate then $\dim W + \dim W^\perp = \dim V$.*

Proof. Choose a basis $\{v_1, ..., v_n\}$ of V so that $\{v_1, ..., v_k\}$ is a basis of W. For $x \in V$ and $i = 1, ..., k$ let $\alpha_i(x) = \beta(v_i, x)$ and $S = \{\alpha_i : i = 1, ..., k\}$. Then each $\alpha_i \in V^*$ (the dual space of V) and $W^\perp = \bigcap_S \operatorname{Ker} \alpha_i$. If j is the maximum number of linearly independent elements in the span of S then $j = k$ and $\dim W^\perp = n - j$ so $k + \dim W^\perp \geq n$. If β is nondegenerate and $\sum c_i \alpha_i = 0$ is a dependence relation among the elements of S then $\beta(\sum c_i v_i, x) = 0$, for all $x \in V$ and hence $\sum c_i v_i = 0$, and so each $c_i = 0$. This means that $j = k$ and so $\dim W + \dim W^\perp = \dim V$. $\qquad\square$

Corollary 3.3.19. *If $\mathfrak{g}$ is semisimple then adjoint representation is completely reducible i.e. $\mathfrak{g}$ is the direct sum of simple ideals $\mathfrak{g}_i$. Furthermore every simple ideal in $\mathfrak{g}$ coincides with one of the $\mathfrak{g}_i$. In particular the simple ideals are absolutely unique (not just up to equivalence of representations). Conversely, a direct sum of simple (or semisimple) algebras is semisimple.*

Proof. If $\mathfrak{h}$ is an ideal in $\mathfrak{g}$ then $\mathfrak{h}$ is an ad-invariant subspace. But $\mathfrak{h}^{\perp}$ is also an ideal. Hence $\mathfrak{h} \cap \mathfrak{h}^{\perp}$ is an ideal and β restricted to $\mathfrak{h} \cap \mathfrak{h}^{\perp}$ is identically 0. But β restricted to $\mathfrak{h} \cap \mathfrak{h}^{\perp}$ coincides with the Killing form of $\mathfrak{h} \cap \mathfrak{h}^{\perp}$. By Cartan's criterion $\mathfrak{h} \cap \mathfrak{h}^{\perp}$ is solvable and so is trivial. Since β is nondegenerate $\dim \mathfrak{h} + \dim \mathfrak{h}^{\perp} = \dim \mathfrak{g}$. Hence $\mathfrak{h} \oplus \mathfrak{h}^{\perp} = \mathfrak{g}$. Thus by induction on dimension ad is completely reducible and $\mathfrak{g} = \oplus \mathfrak{g}_i$, the direct sum of simple ideals. If $\mathfrak{h}$ is any simple ideal of $\mathfrak{g}$ then $[\mathfrak{h}, \mathfrak{g}]$ is an ideal in $\mathfrak{h}$ and therefore is either trivial or $\mathfrak{h}$. In the former case $\mathfrak{h} \subset \mathfrak{z}(\mathfrak{g}) = 0$. In the latter, $\mathfrak{h} = [\mathfrak{h}, \mathfrak{g}] = \oplus [\mathfrak{h}, \mathfrak{g}_i]$ a direct sum of ideals. Since $\mathfrak{h}$ is simple $\mathfrak{h} = [\mathfrak{h}, \mathfrak{g}_i]$ for some i. But $[\mathfrak{h}, \mathfrak{g}_i] \subset \mathfrak{g}_i$. Since $\mathfrak{h} \subset \mathfrak{g}_i$ and the latter is also simple, and $\mathfrak{h}$ is not the zero ideal we must have $\mathfrak{h} = \mathfrak{g}_i$.

For the converse it is sufficient to show that a direct sum of two semisimple algebras is semisimple. Let $\mathfrak{g} = \mathfrak{h} \oplus \mathfrak{l}$. If $\mathfrak{a}$ is an abelian ideal in $\mathfrak{g}$ then the $\pi(\mathfrak{a})$ is an abelian ideal in $\mathfrak{h}$ where π is the projection on $\mathfrak{h}$. Therefore $\pi(\mathfrak{a}) = 0$ and $\mathfrak{a} \subset \mathfrak{l}$. Similarly $\mathfrak{a} \subset \mathfrak{h}$. Hence $\mathfrak{a} = 0$. □

Corollary 3.3.20. *If $\mathfrak{g}$ is semisimple then $[\mathfrak{g}, \mathfrak{g}] = \mathfrak{g}$.*

Proof. Since $\mathfrak{g} = \oplus \mathfrak{g}_i$, a direct sum of simple ideals, we see that $[\mathfrak{g}, \mathfrak{g}] = \sum_{i,j} [\mathfrak{g}_i, \mathfrak{g}_j]$. If $i \neq j$ then $[\mathfrak{g}_i, \mathfrak{g}_j] \subset \mathfrak{g}_i \cap \mathfrak{g}_j = 0$ and if $i = j$ then $[\mathfrak{g}_i, \mathfrak{g}_i]$ is a nontrivial ideal in the simple algebra $\mathfrak{g}_i$. Hence $[\mathfrak{g}_i, \mathfrak{g}_i] = \mathfrak{g}_i$ and so $[\mathfrak{g}, \mathfrak{g}] = \mathfrak{g}$. □

Corollary 3.3.21. *If $\mathfrak{g}$ is semisimple Lie algebra so is any ideal of $\mathfrak{g}$, as is any homomorphic image of $\mathfrak{g}$. Any ideal in $\mathfrak{g}$ is the direct sum of some of the $\mathfrak{g}_i$.*

Proof. If $\mathfrak{h}$ is an ideal in $\mathfrak{g}$ then $\mathfrak{g} = \mathfrak{h} \oplus \mathfrak{h}'$. Continue decomposing these two summands. Then $\mathfrak{h}$ (and $\mathfrak{h}'$) is the direct sum of certain of the $\mathfrak{g}_i$. In particular $\mathfrak{h}$ is semisimple. Since $\mathfrak{g}/\mathfrak{h} \cong \mathfrak{h}'$ and $\mathfrak{h}'$ is semisimple this completes the proof. □

Corollary 3.3.22. *If $\mathfrak{g}$ is an arbitrary Lie algebra and $\mathfrak{h}$ is an ideal in $\mathfrak{g}$ which as a Lie algebra is semisimple then $\mathfrak{h}$ is a direct summand.*

Proof. Let β be the Killing form of $\mathfrak{g}$ and γ the restriction of β to $\mathfrak{h} \times \mathfrak{h}$. Then since $\mathfrak{h}$ is an ideal, γ is the Killing form of $\mathfrak{h}$ and is nondegenerate since $\mathfrak{h}$ is semisimple. If $X \in \mathfrak{h} \cap \mathfrak{h}^\perp$ then $\beta(X, H) = 0$ for all $H \in \mathfrak{h}$, but since X is itself in $\mathfrak{h}$ this means $\gamma(X, H) = 0$ and so $X = 0$. Thus $\mathfrak{h} \cap \mathfrak{h}^\perp = 0$. Since $\dim \mathfrak{h} + \dim \mathfrak{h}^\perp \geq \dim \mathfrak{g}$ and both $\mathfrak{h}$ and $\mathfrak{h}^\perp$ are ideals in $\mathfrak{g}$ this completes the proof. $\qquad\square$

Corollary 3.3.23. *In a semisimple algebra each derivation is inner.*

Proof. $\operatorname{ad}\mathfrak{g}$ is a subalgebra of $\operatorname{Der}(\mathfrak{g})$; in fact if $X \in \mathfrak{g}$ and $D \in \operatorname{Der}(\mathfrak{g})$, then $[D, \operatorname{ad} X] = \operatorname{ad} D(X)$, hence $\operatorname{ad}\mathfrak{g}$ is an ideal in $\operatorname{Der}(\mathfrak{g})$. Being a homomorphic image of a semisimple algebra $\operatorname{ad}\mathfrak{g}$ is semisimple. Hence it is a direct summand: $\operatorname{Der}(\mathfrak{g}) = \operatorname{ad}\mathfrak{g} \oplus \mathfrak{h}$. Let $D \in \mathfrak{h}$. Then for all X, $[D, \operatorname{ad} X] = 0 = \operatorname{ad} D(X)$. Since $\mathfrak{g}$ is semisimple ad is faithful and so $D = 0$ and $\operatorname{Der}(\mathfrak{g}) = \operatorname{ad}\mathfrak{g}$. $\qquad\square$

Corollary 3.3.24. *Let $\mathfrak{g}$ and $\mathfrak{g}'$ be an arbitrary Lie algebras with radicals $\mathfrak{r}$ and $\mathfrak{r}'$ respectively and let $f : \mathfrak{g} \to \mathfrak{g}'$ be a Lie algebra epimorphism. Then $f(\mathfrak{r}) = \mathfrak{r}'$.*

Proof. Clearly $f(\mathfrak{r}) \subset \mathfrak{r}'$. If $\tilde{f} : \mathfrak{g}/\mathfrak{r} \to \mathfrak{g}'/f(\mathfrak{r})$ denotes the induced Lie algebra epimorphism then since $\mathfrak{g}/\mathfrak{r}$ is semisimple so is $\mathfrak{g}'/f(\mathfrak{r})$. Since $f(\mathfrak{r})$ is an ideal in $\mathfrak{g}'$, we have $\mathfrak{r}' \subset f(\mathfrak{r})$. $\qquad\square$

3.3.3   Explicit Computations of Killing Form

We now compute the Killing form of certain Lie algebras explicitly. Although we do this over $\mathbb{R}$ or $\mathbb{C}$ our method works without change over fields k of characteristic 0. In order to do this we realize $\mathfrak{gl}(V)$ in another way. Let $\{v_1, ..., v_n\}$ be a basis of V. Then $\{v_i \otimes v_j : i, j = 1, ..., n\}$ is a basis of $V \otimes V$ and so this space has dimension n^2. This means that the k-linear map $\phi : \mathfrak{gl}(V) \to V \otimes V$ given by $Y \to \sum_{i,j} y_{ij} v_i \otimes v_j$ is a k-linear isomorphism. If $X \in \mathfrak{gl}(V)$ then $X(v_i) = \sum_{k,i} x_{ki} v_k$ and $X^t(v_j) = \sum_{j,l} x_{jl} v_l$.

The question is, what does ad look like on $V \otimes V$? For each $X \in \mathfrak{gl}(V)$

the diagram

$$
\begin{array}{ccc}
\mathfrak{gl}(V) & \xrightarrow{\ \phi\ } & V \otimes V \\
\downarrow {\scriptstyle \operatorname{ad} X} & & \downarrow {\scriptstyle X \otimes I - I \otimes X^t} \\
\mathfrak{gl}(V) & \xrightarrow{\ \phi\ } & V \otimes V
\end{array}
\qquad (3.6)
$$

is commutative. We have

$$
\phi(\operatorname{ad} X(Y)) = \phi([X,Y]) = \sum_{i,j}(XY - YX)_{ij} v_i \otimes v_j,
$$

whereas

$$
\begin{aligned}
(X \otimes I - I \otimes X^t)\phi(Y) &= (X \otimes I - I \otimes X^t)(\sum_{i,j} y_{ij} v_i \otimes v_j) \\
&= \sum_{i,j} x_{ij}(X \otimes I - I \otimes X^t)(v_i \otimes v_j) \\
&= \sum_{i,j} x_{ij}(X(v_i) \otimes v_j - v_i \otimes X^t(v_j)) \\
&= \sum_{i,j} x_{ij}(\sum_{k} x_{ki} v_k \otimes v_i - \sum_{1} x_{jl} v_i \otimes v_l) \\
&= \sum_{k,j,i} x_{ki} y_{ij} v_k \otimes v_j - \sum_{l,i,j} y_{ij} x_{jl} v_i \otimes v_l \\
&= \sum_{k,j} (XY)_{kj} v_k \otimes v_j - \sum_{i,l} (YX)_{il} v_i \otimes v_l \\
&= \sum_{s,t} (XY - YX)_{st} v_s \otimes v_t.
\end{aligned}
$$

Since this holds for all $Y \in \mathfrak{gl}(V)$, we conclude

$$
\phi(\operatorname{ad} X) = (X \otimes I - I \otimes X^t)\phi
$$

for all $X \in \mathfrak{gl}(V)$. Hence

$$
\operatorname{ad} X = \phi^{-1}(X \otimes I - I \otimes X^t)\phi
$$

and so
$$\mathrm{tr}_{\mathfrak{gl}(V)}(\mathrm{ad}\, X^2) = \mathrm{tr}_{V\otimes V}((X\otimes I - I\otimes X^t)^2).$$
But $(X\otimes I - I\otimes X^t)^2 = (X^2\otimes I - 2(X\otimes X^t) + I\otimes (X^t)^2)$. Since $\mathrm{tr}(X) = \mathrm{tr}(X^t)$ and $\mathrm{tr}(X^2) = \mathrm{tr}((X^t)^2)$ we see that
$$\mathrm{tr}_{V\otimes V}((X\otimes I - I\otimes X^t)^2) = 2n\,\mathrm{tr}(X^2) - 2\,\mathrm{tr}(X)^2.$$

Lemma 3.3.25. (*Polarization Lemma*) *Let α and β be symmetric bilinear forms $W \times W \to k$ where char $k \neq 2$. If $\alpha(X,X) = \beta(X,X)$ for all $X \in W$ then $\alpha = \beta$.*

Proof. $\alpha(X+Y,X+Y) = \alpha(X,X) + 2\alpha(X,Y) + \alpha(Y,Y)$ and similarly for β. Therefore $\alpha(X,Y) = \beta(X,Y)$. □

Since the Killing form satisfies $\beta(X,X) = 2n\,\mathrm{tr}(X^2) - 2\,\mathrm{tr}(X)^2$ and $\alpha(X,Y) = 2n\,\mathrm{tr}(XY) - 2\,\mathrm{tr}(X)\,\mathrm{tr}(Y)$ is a symmetric bilinear form on $\mathfrak{gl}(V)$ it follows, by polarization, that for $\mathfrak{gl}(V)$
$$\beta(X,Y) = 2n\,\mathrm{tr}(XY) - 2\,\mathrm{tr}(X)\,\mathrm{tr}(Y).$$

Corollary 3.3.26. *For $\mathfrak{sl}(V)$ (any k of char $\neq 2$) the Killing form β is given by $\beta(X,Y) = 2n\,\mathrm{tr}(XY)$.*

Proof. $\mathfrak{sl}(V)$ is an ideal in $\mathfrak{gl}(V)$. □

Corollary 3.3.27. *For $k = \mathbb{R}$ or $\mathbb{C}$ and $n \geq 2$, $\mathfrak{gl}(V)$ is not semisimple whereas $\mathfrak{sl}(V)$ is semisimple.*

Proof. If $\mathfrak{g} = \mathfrak{gl}(V)$ and $X = \alpha I$ then $\beta(\alpha I, Y) = 0$ so β is degenerate.

For $\mathfrak{sl}(V)$, $\beta(X,Y) = 2n\,\mathrm{tr}(XY)$ so if $\beta(X,Y) = 0$ for all $Y \in \mathfrak{g}$ then $\beta(X,X^t) = 0$ because $X^t \in \mathfrak{sl}(V)$. Therefore $\sum_{i,j}|x_{ij}|^2 = 0$ where x_{ij} are the entries of X. This means $X = 0$.

□

Let
$$X = \begin{pmatrix} 0 & * & * & \ldots & * \\ 0 & 0 & * & \ldots\ldots & * \\ 0 & 0 & 0 & * \ldots & * \\ \ldots & \ldots & \ldots & \ldots & \ldots & \ldots \\ \ldots & \ldots & \ldots & \ldots & \ldots & \ldots \\ 0 & 0 & 0 & 0 & 0 & 0 \end{pmatrix} \qquad Y = \begin{pmatrix} 0 & 1 & 0 & \ldots\ldots & 0 \\ -1 & 0 & 0 & \ldots\ldots & 0 \\ 0 & 0 & 0 & \ldots\ldots & 0 \\ \ldots & \ldots & \ldots & \ldots & \ldots & \ldots \\ \ldots & \ldots & \ldots & \ldots & \ldots & \ldots \\ 0 & 0 & 0 & 0 & 0 & 0 \end{pmatrix}$$

$$H = \begin{pmatrix} -1 & 0 & 0 & \dots & \dots & 0 \\ 0 & 1 & 0 & \dots & \dots & 0 \\ 0 & 0 & 0 & 0 & \dots & 0 \\ \dots & \dots & \dots & \dots & \dots & \dots \\ \dots & \dots & \dots & \dots & \dots & \dots \\ 0 & 0 & 0 & 0 & 0 & 0 \end{pmatrix}$$

Then X, H, $Y \in \mathfrak{sl}(n, \mathbb{R})$ and $2n \operatorname{tr}(X^2) = 0$, $2n \operatorname{tr}(H^2) = 4n$, $2n \operatorname{tr}(Y^2) = -4n$. Thus β is neither positive nor negative definite. Whereas for $\mathfrak{so}(n, \mathbb{R})$ we know β is negative definite (see Section 3.9). This shows that $\mathfrak{sl}(2, \mathbb{R})$ is not isomorphic to $\mathfrak{so}(3, \mathbb{R})$.

Now let $k = \mathbb{R}$ or $\mathbb{C}$, $n \geq 2$ and $\mathfrak{so}(V) = \{X \in \mathfrak{gl}(V) : X^t = -X\}$. We wish to compute the Killing form of $\mathfrak{h} = \mathfrak{so}(V)$. Let $S = \{X \in \mathfrak{gl}(V) : X^t = X\}$ and $\sigma : V \otimes V \to V \otimes V$ be defined by $\sigma(v \otimes w) = w \otimes v$. Then $\mathfrak{h}$ is a subalgebra of $\mathfrak{gl}(V)$ and $\mathfrak{gl}(V)$ is the direct sum of $\mathfrak{h}$ and S as k-spaces.

Lemma 3.3.28. $\operatorname{tr}_{V \otimes V}(\sigma(A \otimes B)) = \operatorname{tr}(AB)$.

Proof. Both sides are bilinear maps $\mathfrak{gl}(V) \times \mathfrak{gl}(V) \to k$. By polarization it suffices to show that $\operatorname{tr}_{V \otimes V}(\sigma(A \otimes A)) = \operatorname{tr}_V(A^2)$. If A is a matrix then

$$\sigma(A \otimes A)(v_i \otimes v_j) = A(v_j) \otimes A(v_i) = \sum_k a_{jk} v_k \otimes \sum_l a_{il} v_l$$

$$= \sum_{k,l} a_{jk} a_{il} v_k \otimes v_l.$$

Therefore $\operatorname{tr}_{V \otimes V}(\sigma(A \otimes A)) = \sum_{i,j} a_{ji} a_{ij} = \operatorname{tr}_V(A^2)$.   □

Decompose $\mathfrak{gl}(V) = \mathfrak{h} \oplus S$ under $Y \to (Y - Y^t)/2 + (Y + Y^t)/2$. Apply ϕ and get $V \otimes V = \phi(\mathfrak{h}) \oplus \phi(S)$ where $\phi(\mathfrak{h}) = \{\sum_{i,j} y_{ij} v_i \otimes v_j : y_{ij} = -y_{ji}\}$ and $\phi(S) = \{\sum_{i,j} y_{ij} v_i \otimes v_j : y_{ij} = y_{ji}\}$. Let $\pi : V \otimes V \to \phi(\mathfrak{h})$ be the projection onto $\phi(\mathfrak{h})$. Then clearly $\pi = (I - \sigma)/2$. Also notice that if $X \in \mathfrak{h}$, $Y \in \mathfrak{h}$ and $Z \in S$ then $[X, Y] \in \mathfrak{h}$ ($\mathfrak{h}$ is a subalgebra) and $[X, Z] \in S$; that is for $X \in \mathfrak{h}$, $\operatorname{ad} X$ leaves both $\mathfrak{h}$ and S stable. We are interested in

$$\text{tr}_{\mathfrak{h}}(\text{ad }X^2|_{\mathfrak{h}}) = \text{tr}((X^2 \otimes I - 2X \otimes X^t + I \otimes (X^t)^2)|_{\phi(\mathfrak{h})})$$

$$= \text{tr}_{V \otimes V}(\tfrac{1}{2}(I - \sigma)(X^2 \otimes I - 2X \otimes X^t + I \otimes (X^t)^2)).$$

But since $X \in \mathfrak{h}$, this is $\text{tr}_{V \otimes V}(\tfrac{1}{2}(I - \sigma)(X^2 \otimes I + 2(X \otimes X) + I \otimes X^2))$. Applying the lemma above to calculate this, we get $(n-2)\,\text{tr}(X^2)$.

Corollary 3.3.29. *For $k = \mathbb{R}$ or $\mathbb{C}$ the Killing form β of $\mathfrak{so}(V)$ is given by $\beta(X, Y) = (n-2)\,\text{tr}_V(XY)$.*

Proof. Apply the polarization lemma. □

Corollary 3.3.30. *For $k = \mathbb{R}$, or $\mathbb{C}$ and $n \geq 3$, $\mathfrak{so}(V)$ is semisimple.*

Proof. We will show β is nondegenerate. Let $X \in \mathfrak{h}$ and suppose $\text{tr}(XY) = 0$ for all $Y \in \mathfrak{h}$. If $Z \in \mathfrak{gl}(V)$ and $Y = Z - Z^t$. Then $Y \in \mathfrak{h}$ and so $0 = \text{tr}(X(Z - Z^t))$. Since $\text{tr}(XZ^t) = \text{tr}(ZX^t) = \text{tr}(X^tZ)$, we see that $0 = \text{tr}((X - X^t)Z)$. But Z is arbitrary and $\text{tr}(XY)$ is nondegenerate on $\mathfrak{gl}(V)$. Therefore $X \in S$ and since X is also in $\mathfrak{h}$, $X = 0$. □

3.3.4   Further Results on Jordan Decomposition

Corollary 3.3.31. *Let $\mathfrak{g}$ be a Lie algebra over an algebraically closed field k of characteristic 0, and let $D \in \text{Der}(\mathfrak{g})$. Then the Jordan components of D are also derivations.*

Proof. If $D = S + N$ it suffices to show that $S \in \text{Der}(\mathfrak{g})$. For $\alpha \in k$ let $\mathfrak{g}_\alpha = \{X \in \mathfrak{g} : (D - \alpha I)^i X = 0 \text{ for some integer } i\}$. Then $\text{Spec}(D) = \{\alpha : \mathfrak{g}_\alpha \neq 0\}$ and the $\mathfrak{g}_\alpha$'s are D (and S) invariant subspaces of $\mathfrak{g}$ on which S acts as scalars by αI; hence $\mathfrak{g} = \oplus\{\mathfrak{g}_\alpha | \alpha \in \text{Spec } D\}$. An inductive calculation similar to the one made earlier shows that for all n

$$(D - (\alpha + \beta)I)^n[X, Y] = \sum_{i \geq 0}^{n} \binom{n}{i} [(D - \alpha I)^{n-i}X, (D - \beta I)^i Y].$$

In particular if α and $\beta \in \operatorname{Spec} D$ then $[\mathfrak{g}_\alpha, \mathfrak{g}_\beta] \subset \mathfrak{g}_{\alpha+\beta}$, and if $\alpha + \beta$ is not in $\operatorname{Spec} D$ then $[\mathfrak{g}_\alpha, \mathfrak{g}_\beta] = 0$. Let $X = \sum X_\alpha$ and $Y = \sum Y_\beta$. Then $[X, Y] = \sum_{\alpha,\beta}[X_\alpha, Y_\beta]$, so

$$S[X,Y] = \sum_{\alpha,\beta} S[X_\alpha, Y_\beta] = \sum_{\alpha,\beta}(\alpha + \beta)[X_\alpha, Y_\beta].$$

On the other hand $[SX, Y] + [X, SY]$ clearly also equals $\sum_{\alpha,\beta}(\alpha + \beta)[X_\alpha, Y_\beta]$. □

In general, if k is an algebraically closed field of characteristic zero, for a derivation of a Lie algebra over k, the Jordan components of Lie derivation are also derivations. In a semisimple algebra $\mathfrak{g}$ each derivation is inner. Hence for $X \in \mathfrak{g}$, $\operatorname{ad} X = \operatorname{ad} Y + \operatorname{ad} Z$ where $\operatorname{ad} Y$ is semisimple and $\operatorname{ad} Z$ is nilpotent. We now derive a result which implies this in a stronger form.

Theorem 3.3.32. *Let $\mathfrak{g} \subset \mathfrak{gl}(V)$ be a semisimple Lie algebra over an algebraically closed field k of characteristic 0. Then $\mathfrak{g}$ contains the semisimple and nilpotent parts of each of its elements. In particular, for each $X \in \mathfrak{g}$, we have $X = S + N$, and hence we know that $\operatorname{ad} X = \operatorname{ad} S + \operatorname{ad} N$ is the Jordan decomposition of $\operatorname{ad} X \in \mathfrak{gl}(\mathfrak{gl}(V))$.*

Proof. For a subspace W of V let

$$\mathfrak{g}_W = \{X \in \mathfrak{gl}(V) : X(W) \subset W \text{ and } \operatorname{tr}(X|_W) = 0\}.$$

For each W, $\mathfrak{g}_W$ is a subalgebra of $\mathfrak{gl}(V)$. Now in general $\mathfrak{n}_{\mathfrak{gl}(V)}(\mathfrak{g})$ is a subalgebra of $\mathfrak{gl}(V)$ containing $\mathfrak{g}$. (If $[X, \mathfrak{g}] \subset \mathfrak{g}$ and $[Y, \mathfrak{g}] \subset \mathfrak{g}$ then $\operatorname{ad}[X, Y]$ stabilizes $\mathfrak{g}$ by Jacobi). Let $\mathfrak{s} = \{W : W \text{ is } \mathfrak{g}\text{-stable}\}$ and $\mathfrak{g}^\# = \bigcap_{W \in \mathfrak{s}} \mathfrak{g}_W \cap \mathfrak{n}_{\mathfrak{gl}(V)}(\mathfrak{g})$. If $W \in \mathfrak{s}$ then $\mathfrak{g}(W) \subset W$. The map $X \to X|_W$ is clearly a Lie algebra homomorphism so $\{X|_W : X \in \mathfrak{g}\}$ is semisimple. In particular if $X \in \mathfrak{g}$, $\operatorname{tr}(X|_W) = 0$ so $X \in \mathfrak{g}_W$. This means that $\mathfrak{g} \subset \bigcap_{W \in \mathfrak{s}} \mathfrak{g}_W$ so that $\mathfrak{g}^\#$ is a subalgebra of $\mathfrak{gl}(V)$ containing $\mathfrak{g}$ as an ideal.

We show $\mathfrak{g}^{\#} = \mathfrak{g}$: Since $\mathfrak{g}$ is a semisimple ideal in $\mathfrak{g}^{\#}$ then by Corollary 3.3.21 it is a direct summand so $\mathfrak{g}^{\#} = \mathfrak{g} \oplus \mathfrak{h}$. Let $H \in \mathfrak{h}$ and let W_0 be a minimal $\mathfrak{g}$-stable subspace of V. Then $W_0 \in \mathfrak{s}$ so $\mathfrak{g}^{\#} \subset \mathfrak{g}_{W_0}$ and each element of $\mathfrak{g}^{\#}$ leaves W_0 stable. In particular $H(W_0) \subset W_0$. Since $[\mathfrak{g}, \mathfrak{h}] = 0$ on V, and therefore also on W_0, each H is an intertwining operator on the irreducible subspace W_0 so $H = cI$ on W_0 by Schur's lemma. Hence $\operatorname{tr} H = c \dim W_0 = 0$ so $c = 0$ and $H = 0$ on W_0. By Weyl's theorem, to be proven in Section 3.4, V is the direct sum of irreducible $\mathfrak{g}$-subspaces so $H = 0$ on V. Thus $\mathfrak{h} = 0$ and $\mathfrak{g}^{\#} = \mathfrak{g}$.

Now let $X \in \mathfrak{g}$ and $X = S + N$ be its Jordan decomposition. Then $\operatorname{ad} X = \operatorname{ad} S + \operatorname{ad} N$ is the Jordan decomposition of $\operatorname{ad} X$ in $\mathfrak{gl}(\mathfrak{gl}(V))$. In particular, $\operatorname{ad} S = s(\operatorname{ad} X)$ where s is a polynomial without constant term and since $\operatorname{ad} X$ stabilizes $\mathfrak{g}$ so does $\operatorname{ad} S$. On the other hand $S = p(X)$ where p is also a polynomial without constant term. We prove $p(X) \in \mathfrak{g}^{\#}(= \mathfrak{g})$; hence also $N \in \mathfrak{g}$. Since $\operatorname{ad} S \in N_{\mathfrak{gl}(V)}(\mathfrak{g})$ we must show that $p(X) \in \mathfrak{g}_W$ for all $W \in \mathfrak{s}$. But for $W \in \mathfrak{s}$, $X(W) \subset W$ and hence $p(X)(W) \subset W$. Similarly N stabilizes W. Since $X \in \mathfrak{g}$ and N is nilpotent so is its restriction to W; hence $\operatorname{tr}(N|_W) = 0$. But $X|_W = N|_W + S|_W$ so $\operatorname{tr}(X|_W) = \operatorname{tr}(N|_W) + \operatorname{tr}(S|_W) = \operatorname{tr}(S|_W)$. Then since $\mathfrak{g} \subset \bigcap_{W \in \mathfrak{s}} \mathfrak{g}_W$ and $X \in \mathfrak{g}$, $\operatorname{tr}(X|_W) = 0$ we get $S \in \mathfrak{g}_W$.   □

Corollary 3.3.33. *Let $\mathfrak{g} \subset \mathfrak{gl}(V)$ be a semisimple Lie algebra over an algebraically closed field and $X \in \mathfrak{g}$. Then X is semisimple (respectively nilpotent) if and only if $\operatorname{ad} X$ is semisimple (respectively nilpotent). In fact, $\operatorname{ad}_{\mathfrak{g}} X = \operatorname{ad}_{\mathfrak{g}} S + \operatorname{ad}_{\mathfrak{g}} N$ is the Jordan decomposition of $\operatorname{ad}_{\mathfrak{g}} X$ in $\mathfrak{gl}(\mathfrak{g})$.*

Proof. Let S and N be the semisimple and nilpotent parts of X. By the theorem we know S and $N \in \mathfrak{g}$. We also know that $\operatorname{ad} X = \operatorname{ad} S + \operatorname{ad} N$ is the Jordan decomposition of $\operatorname{ad} X \in \mathfrak{gl}(\mathfrak{gl}(V))$. Since the restriction of a semisimple or nilpotent operator to an invariant subspace is respectively semisimple or nilpotent we see that $\operatorname{ad}_{\mathfrak{g}} S$ is the semisimple part of $\operatorname{ad}_{\mathfrak{g}} X$ and $\operatorname{ad}_{\mathfrak{g}} N$ is the nilpotent part of $\operatorname{ad}_{\mathfrak{g}} X$. In particular if X is semisimple (respectively nilpotent) then $\operatorname{ad}_{\mathfrak{g}} X$ is semisimple (respectively nilpotent). Conversely if $\operatorname{ad}_{\mathfrak{g}} X$ is semisimple *i.e.* $\operatorname{ad}_{\mathfrak{g}} X = \operatorname{ad}_{\mathfrak{g}} S$ then $X = S$ since ad is faithful, similarly for nilpotent case.   □

3.4 Weyl's Theorem on Complete Reducibility

We prove Weyl's complete reducibility theorem. Our first step is to define the Casimir element associated with a representation.

Lemma 3.4.1. *Let V be a finite dimensional vector space over a field k and $\beta : V \times V \to k$ be a nondegenerate symmetric bilinear form. If $\{X_1, \ldots, X_n\}$ is a basis of V then there exists a dual basis $\{Y_1, \ldots, Y_n\}$ that satisfies $\beta(X_i, Y_j) = \delta_{ij}$.*

Proof. For fixed X, the map $Y \mapsto \beta(X, Y)$ is in V^*, so we get a map μ from $V \to V^*$ which, by assumption is injective. Comparing dimensions we see that μ is an isomorphism. If $\{X_1, \ldots, X_n\}$ is a basis of V there is a corresponding dual basis $\{X_1^*, \ldots, X_n^*\}$ of V^* satisfying $X_j^*(X_i) = \delta_{ij}$. Taking $\{Y_1, \ldots, Y_n\}$ as the μ pre-image of $\{X_1^*, \ldots, X_n^*\}$ yields the result. $\qquad\square$

In particular, if $\mathfrak{g}$ is a semisimple Lie algebra, $\rho : \mathfrak{g} \to \mathfrak{gl}(V_\rho)$ is a faithful representation of $\mathfrak{g}$ on V_ρ and $\beta_\rho : \mathfrak{g} \times \mathfrak{g} \to k$ is the trace form of ρ then β_ρ is nondegenerate and for each basis $\{X_1, \ldots, X_n\}$ of $\mathfrak{g}$ there is a basis $\{Y_1, \ldots, Y_n\}$ satisfying $\beta_\rho(X_i, Y_j) = \delta_{ij}$.

We now define the Casimir operator C_ρ of ρ to be the element of the associative algebra $\mathrm{End}_k(V_\rho)$ given by

$$C_\rho = \sum_i \rho(X_i) \cdot \rho(Y_i).$$

Our next result shows that the Casimir operator is an invariant of the particular representation, $(\mathfrak{g}, \rho, V)$.

Proposition 3.4.2. *Let $\mathfrak{g}$ is a semisimple Lie algebra and $\rho : \mathfrak{g} \to \mathfrak{gl}(V)$ is a faithful representation of $\mathfrak{g}$.*

(1) *The operator C_ρ, is independent of the choice of the basis $\{X_1, \ldots, X_n\}$.*

(2) *If $\{X_1, \ldots, X_n\}$ is a basis and $\{Y_1, \ldots, Y_n\}$ is the dual basis $X \in \mathfrak{g}$ we write $[X, X_i] = \sum_j a_{ij}(X)X_j$ and $[X, Y_i] = \sum_j b_{ij}(X)Y_j$. Then $a_{ik}(X) = -b_{ki}(X)$ for all $i, k \leq \dim \mathfrak{g}$ and $X \in \mathfrak{g}$.*

(3) $\operatorname{tr} C_\rho = \dim \mathfrak{g}$.

(4) C_ρ is an intertwining operator on V. In particular, if ρ is irreducible and k is algebraically closed then $C_\rho = \frac{\dim \mathfrak{g}}{\dim V} \cdot I$.

Proof of 1: Let $\{X'_1, \ldots, X'_n\}$ be another basis and $\{Y'_1, \ldots, Y'_n\}$ its dual basis. Then $X'_i = \sum_j \alpha_{ij} X_j$ and $Y'_k = \sum_l \beta_{kl} Y_l$. This means

$$\delta_{ik} = \beta_\rho(X'_i, Y'_k) = \sum_{j,l} \alpha_{ij} \beta_{kl} \beta_\rho(X_j, Y_l).$$

But this is $\sum_{j,l} \delta_{jl} \alpha_{ij} \beta_{kl} = \sum_j \alpha_{ij} \beta_{kj}$, so $\alpha \cdot \beta^t = I$. Taking transposes, $\alpha^t \cdot \beta = I$. Now

$$\begin{aligned}
\sum_i \rho(X'_i) \cdot \rho(Y'_i) &= \sum_i \left(\sum_j \alpha_{ij} \rho(X_j)\right)\left(\sum_l \beta_{il} \rho(Y_l)\right) \\
&= \sum_i \sum_{j,l} \alpha_{ij} \beta_{il} \rho(X_j) \rho(Y_l) \qquad (3.7) \\
&= \sum_{j,l} \sum_i \alpha^t_{ji} \beta_{il} \rho(X_j) \rho(Y_l).
\end{aligned}$$

Since $\beta \cdot \alpha^t = I$ the last term is $\sum_{j,l} \delta_{jl} \rho(X_j) \rho(Y_l) = C_\rho$.

Proof of 2: $a_{ik}(X) = \sum_j a_{ij}(X) \beta_\rho(X_j, Y_k) = \beta_\rho(\sum_j a_{ij}(X) X_j, Y_k)$. But this is $\beta_\rho([X, X_i], Y_k) = -\beta_\rho([X_i, X], Y_k) = -\beta_\rho(X_i, [X, Y_k])$, which in turn equals

$$-\beta_\rho(X_i, \sum_j b_{kj}(X) Y_j) = -\sum_j b_{kj}(X) \beta_\rho(X_i, Y_j) = -\sum_j b_{kj}(X) \delta_{ij}$$

$$= -b_{ki}(X).$$

Proof of 3:

$$\operatorname{tr}(C_\rho) = \sum_i \operatorname{tr}(\rho(X_i) \rho(Y_i)) = \sum_i \beta_\rho(X_i, Y_i) = \sum_i \delta_{ii} = \dim \mathfrak{g}.$$

Proof of 4: For $X \in \mathfrak{g}$, we have

$$[\rho(X), C_\rho] = [\rho(X), \sum_i \rho(X_i)\rho(Y_i)] = \sum_i [\rho(X), \rho(X_i)\rho(Y_i)].$$

But this is $\sum_i \{[\rho(X), \rho(X_i)]\rho(Y_i) + \rho(X_i)[\rho(X), \rho(Y_i)]\}$ (we invoke the matrix identity $[A, BC] = [A, B]C + B[A, C]$). This last expression can be rewritten as

$$\sum_i \{\rho([X, X_i])\rho(Y_i) + \rho(X_i)\rho([X, Y_i])\}$$

$$= \sum_i \{\rho(\sum_j a_{ij}(X)X_j)\rho(Y_i) + \rho(X_i)\rho(\sum_j b_{ij}(X)Y_j)\}$$

$$= \sum_i \{\sum_j a_{ij}\rho(X_j)\rho(Y_i) + b_{ij}\rho(X_i)\rho(Y_j)\}.$$

This is zero $a_{ij}(X) = -b_{ji}(X)$ for all i, j. If ρ is irreducible and k is algebraically closed then by Schur's lemma $C_\rho = cI$. But, $\mathrm{tr}(C_\rho) = c \dim V = \dim \mathfrak{g}$. Hence $C_\rho = \frac{\dim \mathfrak{g}}{\dim V} \cdot I$.   □

Let ρ and σ be representations over k of $\mathfrak{g}$ on V_ρ and V_σ, respectively. We define a new representation of $\mathfrak{g}$ on the k space $\mathrm{Hom}_k(V_\rho, V_\sigma)$ as follows. For $X \in \mathfrak{g}$ and $T \in \mathrm{Hom}_k(V_\rho, V_\sigma)$ take $X(T) = T \circ \rho_X - \sigma_X \circ T$. One checks immediately that this is a Lie representation of $\mathfrak{g}$.

Theorem 3.4.3. (*Weyl*) *If* $\rho : \mathfrak{g} \to \mathfrak{gl}(V)$ *is a representation of a semisimple Lie algebra over* k, *then* ρ *is completely reducible.*

Proof. We may assume that $\mathfrak{g}$ acts faithfully since this cannot affect complete reducibility and still keeps semisimplicity. Using the remarks above we may also assume the field k is algebraically closed. We first deal with the case in which there is a $\mathfrak{g}$-invariant subspace W of V of codimension 1. Our proof, in this case, goes by induction on $\dim V$. If W has a proper $\mathfrak{g}$-invariant subspace then it has a minimal proper one, say W_0. Then we have an exact sequence of $\mathfrak{g}$ modules

$$0 \to W/W_0 \to V/W_0 \to V/W \to 0$$

Since W/W_0 is a submodule of codimension 1 of V/W_0 and $\dim V/W_0 < \dim V$, there is a complementary 1-dimensional $\mathfrak{g}$-invariant subspace U/W_0 in V/W_0 to W/W_0. But since U/W_0 is $\tilde{\rho}$-invariant and 1-dimensional, and $\mathfrak{g}$ is semisimple, $\tilde{\rho}$ acts trivially on U/W_0. Thus, $\tilde{\rho}_X(U/W_0) = 0$ for all $X \in \mathfrak{g}$. Hence $\rho_X(U) \subset W_0$, and since $W_0 \subseteq U$, this means that U is ρ-invariant. But $\dim U = 1 + \dim W_0$ and if W is reducible this is $< 1 + \dim W = \dim V$. Hence, again by induction, there is a ρ-invariant 1-dimensional subspace of U complementary to W_0. Let $u_0 \in U$ generate this line $\mathcal{L}$ over k. Then $u_0 \notin W$ and $U + W = V$, so $\mathcal{L} + W = V$. Since $\mathcal{L} \cap W = \{0\}$ and $\dim V/W = 1$, $\mathcal{L} \oplus W = V$ and W has a complementary 1-dimensional $\mathfrak{g}$-invariant subspace. This means we may assume that W is irreducible.

Because W is a $\mathfrak{g}$-invariant subspace of V of codimension 1, ρ induces $\tilde{\rho}$ of $\mathfrak{g}$ on V/W which, as above, is trivial. So $\rho_X(V) \subset W$ for all $X \in \mathfrak{g}$. If C_ρ is the Casimir operator, then $C_\rho(v) = \sum_i \rho(X_i)\rho(Y_i)(v)$ so $C_\rho(V) \subset W$. In particular W is C_ρ-invariant. Since W is irreducible, C_ρ restricted to W is cI. If this $c = 0$ then $C_\rho^2 = 0$. But $\operatorname{tr}_V C_\rho = \dim \mathfrak{g} = 0$, this is a contradiction. Now ρ is a faithful representation of a semisimple Lie algebra, so we also know C_ρ is an intertwining operator on V. This means that $\operatorname{Ker} C_\rho$ is a $\mathfrak{g}$-invariant subspace. Suppose $w \in \operatorname{Ker} C_\rho \cap W$. Then $C_\rho(w) = cw = 0$ and since $c \neq 0$, $w = 0$ so that $\operatorname{Ker} C_\rho \cap W = \{0\}$. On the other hand, $\dim \operatorname{Ker} C_\rho + \dim C_\rho(V) = \dim V$ and $\dim C_\rho(V) \leq \dim W$ so $\dim \operatorname{Ker} C_\rho \geq 1$. Thus $\dim \operatorname{Ker} C_\rho + \dim W \geq \dim V$. Together with disjointness from W this shows $\operatorname{Ker} C_\rho \oplus W = V$, completing the proof when W has codimension 1.

Finally let W be an arbitrary $\mathfrak{g}$-invariant subspace of V and consider the representation of $\mathfrak{g}$ on $\operatorname{Hom}_k(V, W)$ defined above where for σ we take ρ restricted to W. Let $\mathcal{V} = \{T \in \operatorname{Hom}_k(V, W) : T|_W = \lambda I_W\}$ and $\mathcal{W} = \{T \in \operatorname{Hom}_k(V, W) : T|_W = 0\}$.

Lemma 3.4.4. *$\mathcal{V}$ and $\mathcal{W}$ are subspaces of $\operatorname{Hom}_k(V, W)$; $\mathcal{W}$ is codimension 1 in $\mathcal{V}$ and $\mathfrak{g}(\mathcal{V}) \subseteq \mathcal{W}$. In particular $\mathcal{V}$ and $\mathcal{W}$ are $\mathfrak{g}$-invariant.*

Proof. Clearly $\mathcal{W} \subseteq \mathcal{V}$ and $\mathcal{V}$ and $\mathcal{W}$ are subspaces of $\operatorname{Hom}_k(V, W)$. Let $w_1, \ldots, w_k$ be a basis of W and extend this to a basis $\{w_1, \ldots, w_k, v_1, \ldots, v_j\}$ of V. Define $T_0 \in \operatorname{Hom}_k(V, W)$ by $T_0(w_i) = w_i$

and $T_0(v_i) = 0$. Then $T_0 \in \mathcal{V} - \mathcal{W}$. If $T \in \mathcal{V}$, then $T|_W = \lambda I_W$. So for $w \in W$, $T - \lambda T_0(w) = 0$ and $\dim \mathcal{V}/\mathcal{W} = 1$. For $X \in \mathfrak{g}$ and $T \in \mathrm{Hom}_k(V, W)$ we have

$$X(T) = \rho_X|_W \circ T - T \circ \rho_X.$$

If $T \in \mathcal{V}$ then since $T|_W = \lambda I_W$, and W is invariant we have

$$X(T)(w) = \rho_X \circ T(w) - T \circ \rho_X(w)$$
$$= \rho_X(\lambda w) - \lambda \rho_X(w) = 0.$$

This means $X(T) \in \mathcal{W}$, so $\mathfrak{g}(\mathcal{V}) \subseteq \mathcal{W}$. □

Continuing the proof, we see that by the lemma and the codimension 1 case there is some $T_0 \in \mathcal{V} \setminus \mathcal{W}$ such that $\mathcal{V} = \mathcal{W} + \{cT_0\}_{c \in k}$ as $\mathfrak{g}$-modules. Hence $T_0|_W = \lambda I_W$, where $\lambda \neq 0$. Normalizing, we may assume $T_0|_W = I$. Since $\mathfrak{g}$ is semisimple and the invariant subspace has codimension 1, $X \circ T_0 = 0$ for all $X \in \mathfrak{g}$. Thus T_0 is an intertwining operator and so $\mathrm{Ker}\, T_0$ is a $\mathfrak{g}$-invariant subspace of V. If $w \in W \cap \mathrm{Ker}\, T_0$, then $T_0(w) = w = 0$ so $W \cap \mathrm{Ker}\, T_0 = \{0\}$. Since T_0 maps onto W we have $\dim V = \dim W + \dim \mathrm{Ker}\, T_0$; it follows that $\mathrm{Ker}\, T_0$ is the desired complementary subspace. □

Here is a consequence for linear Lie groups.

Corollary 3.4.5. *A semisimple subgroup of* $\mathrm{GL}(n, \mathbb{C})$ *is closed.*

Proof. Let $H = \overline{G}$, the closure of G in $\mathrm{GL}(n, \mathbb{C}) = \mathrm{GL}(V)$ and $\mathfrak{h}$ and $\mathfrak{g}$ the respective Lie algebras. Because G normalizes itself, H also normalizes G. Therefore $\mathfrak{g}$ is an ideal in $\mathfrak{h}$. Since $\mathfrak{g}$ is semisimple it is a direct factor, so $\mathfrak{h} = \mathfrak{g} \oplus \mathfrak{l}$. Let L be the corresponding normal connected subgroup of H. Then $H = \mathrm{GL}$, and L commutes with G (and therefore also with H); thus $L \subseteq Z(H)_0$. By Weyl's theorem, the action of G on V is completely reducible, so $V = \sum_{i=1}^r V_i$ where each V_i is G invariant. Since G is semisimple each of these irreducible representations lies in $\mathrm{SL}(V_i)$ and since L commutes with G each $l \in L$ consists of diagonal operators which are scalars on each V_i. Hence all eigenvalues of each $l \in L$ are roots of unity with bounded order, namely the product $\prod_{i=1}^r \dim V_i$.

This is less than or equal to $n^r \leq n^n$, hence L is finite. Since it is also connected L is trivial and so $G = H$. $\square$

Corollary 3.4.6. $\mathfrak{sl}(n, \mathbb{C})$ *and* $\mathfrak{sl}(n, \mathbb{R})$ *are simple for* $n \geq 2$.

Proof. We first deal with the case of $\mathfrak{sl}(n, \mathbb{C}) = \mathfrak{sl}(V)$. We know from the calculation of the Killing form, or as a corollary of Lie's theorem, that $\mathfrak{sl}(n, \mathbb{C})$ is semisimple and so is the direct sum of simple ideals $\mathfrak{a}_j$. If $\mathfrak{a}_1$, for example, acts irreducibly on V, then since $\mathfrak{a}_j$ commutes with $\mathfrak{a}_1$ for each $j \geq 2$, each $\mathfrak{a}_j$ consists of scalars by Schur's lemma. But these scalars are of trace 0. Hence $\mathfrak{a}_j = 0$ for each $j \geq 2$ and $\mathfrak{sl}(V) = \mathfrak{a}_1$ is simple. Therefore we may assume each $\mathfrak{a}_j$ acts reducibly on V. Choose an $X \neq 0 \in \mathfrak{sl}(n, \mathbb{C})$ which is diagonal with distinct eigenvalues and write $X = \sum X_j$ according to the decomposition of $\mathfrak{g}$ given above. Relabeling the indices we may assume $X_1 \neq 0$. Clearly $[X, X_1] = 0$ and hence X_1 is also diagonal. Since $\mathfrak{a}_1$ does not act irreducibly there is a proper $\mathfrak{a}_1$-invariant subspace $W \subseteq V$. Let W' be a complementary $\mathfrak{a}_1$-invariant subspace as in Weyl's theorem. Since X_1 is diagonalizable, its restrictions to W and W' are also diagonalizable. Choose a basis of each so that both restrictions are diagonal. Together these form a basis $\{v_1, \ldots, v_n\}$ of V. If X_1 had only one eigenvalue it is a scalar of trace 0, and $X_1 = 0$, a contradiction. Let a_i and a_j, $i \neq j$ be distinct eigenvalues of X_1 corresponding to eigenvectors v_i and v_j. By interchanging the roles of W and W', if necessary, we may assume at least one of them, say v_i lies in W. Let T be a linear transformation defined by $T(v_i) = v_j$, and $T(v_k) = 0$ for all other k. Then $T \in \mathfrak{sl}(V)$ and $[T, X_1](v_i) = TX_1(v_i) - X_1T(v_i) = (a_i - a_j)v_j \neq 0$. Since $\mathfrak{a}_1$ is an ideal and W is invariant, we have a contradiction if $v_j \in W'$. The other possibility is that all eigenvalues of X_1 on W' have equal value, say a, and the v_j are in W. But then, a must differ from either a_i, or a_j, or both, and we argue as above replacing a_j by a.

Now since the complexification $\mathfrak{sl}(n, \mathbb{R})^{\mathbb{C}} = \mathfrak{sl}(n, \mathbb{C})$ is simple, to see that $\mathfrak{sl}(n, \mathbb{R})$ itself is simple we show in general that if $\mathfrak{g}$ is semisimple and $\mathfrak{g}^{\mathbb{C}}$ is simple then so is $\mathfrak{g}$. For let $\mathfrak{a}$ be a nonzero ideal in $\mathfrak{g}$. Then $\mathfrak{a}^{\mathbb{C}}$ is a nonzero ideal in $\mathfrak{g}^{\mathbb{C}}$ and so equals $\mathfrak{g}^{\mathbb{C}}$. Hence

$$\dim_{\mathbb{R}} \mathfrak{a} = \dim_{\mathbb{C}} \mathfrak{a}^{\mathbb{C}} = \dim_{\mathbb{C}} \mathfrak{g}^{\mathbb{C}} = \dim_{\mathbb{R}} \mathfrak{g},$$

and so $\mathfrak{a} = \mathfrak{g}$.   □

Another corollary of Weyl's complete reducibility theorem is White-head's lemma, which is actually equivalent to Weyl's theorem.

Lemma 3.4.7. (*Whitehead's lemma*) *Let* $\rho : \mathfrak{g} \to \mathfrak{gl}(V)$ *be a representation of the semisimple Lie algebra* $\mathfrak{g}$ *on* V *and* $\phi : \mathfrak{g} \to V$ *be a 1-cocycle, that is, a linear function satisfying*

$$\phi([X, Y]) = \rho_X(\phi(Y)) - \rho_Y(\phi(X)).$$

Then ϕ *is a coboundary. That is,* $\phi(X) = \rho_X(v_0)$ *for some* $v_0 \in V$.

Proof. Let $U = V \oplus (t)$ be a space of dimension 1 more than the dimension of V and let σ be defined by $\sigma_X(v, t) = (\rho_X(v) + t\phi(X), 0)$. Since ρ is a representation, one sees easily that σ is a representation on U if and only if ϕ is a 1-cocycle with values in V. Evidently $\sigma_X(U) \subseteq V$ for all $X \in \mathfrak{g}$, and in particular V is a σ-invariant subspace of U. By complete reducibility there is a vector $u_0 = (v_0, t_0) \in U - V$, with $t_0 \neq 0$ and $\sigma(u_0) = 0$. Normalizing by taking $t_0 = -1$, we see that $\phi(X) = \rho_X(v_0)$.   □

We remark that even if $\mathfrak{g}$ were not semisimple, but merely reductive (*i.e.* $\mathfrak{g}$ always acts completely reducibly in any finite dimensional linear representation ρ over k) then again $H^1(\mathfrak{g}, V, \rho) = \{0\}$. The argument is the same except now we get a vector $u_0 = (v_0, t_0) \in U - V$ with $t_0 \neq 0$, such that $(\rho_X(v_0) + t_0\phi(X), 0) = (\lambda_X v_0, \lambda_X t_0)$, where λ is a k-valued linear functional on $\mathfrak{g}$. Since $\lambda_X t_0 = 0$ for all X and $t_0 \neq 0$ it follows that $\lambda = 0$, so u_0 is killed by σ and we can proceed as above.

3.5   Levi-Malcev Decomposition

We now turn to the theorem of Levi-Malcev.

Theorem 3.5.1. *Let* $\pi : \mathfrak{g} \to \mathfrak{s}$ *be a Lie algebra homomorphism onto a semisimple Lie algebra* $\mathfrak{s}$. *Then there exists a Lie algebra homomorphism* ϵ *which gives a global cross section to* π. *That is to say,* $\epsilon : \mathfrak{s} \to \mathfrak{g}$ *and* $\pi \circ \epsilon = I_{\mathfrak{s}}$.

The uniqueness statement whose proof will be given below is due to Malcev. Before proving this theorem we give two of its corollaries.

Corollary 3.5.2. *Let $\mathfrak{r}$ be the radical of a Lie algebra $\mathfrak{g}$. Then there exists a semisimple subalgebra $\mathfrak{s}$ of $\mathfrak{g}$ such that $\mathfrak{g} = \mathfrak{r} \bigoplus \mathfrak{s}$. In fact, $\mathfrak{g}$ is the semidirect sum of the ideal $\mathfrak{r}$ with $\mathfrak{s}$. This is what is usually called Levi's splitting theorem. In addition, $\mathfrak{s}$ is unique up to conjugation by* $\mathrm{Exp}(\mathrm{ad}\, Y)$ *where $Y \in \mathfrak{n}$ and $\mathfrak{n}$ is the nilradical of $\mathfrak{r}$ (actually $Y \in [\mathfrak{g}, \mathfrak{r}]$). Notice that* $\mathrm{Exp}(\mathrm{ad}\, X)$ *is well-defined over any field of characteristic zero and that it is an inner automorphism of $\mathfrak{g}$.*

Proof. Let $\pi : \mathfrak{g} \to \mathfrak{g}/\mathfrak{r}$ be the projection mod $\mathfrak{r}$. Then $\mathfrak{g}/\mathfrak{r}$ is semisimple so there exists $\epsilon : \mathfrak{g}/\mathfrak{r} \to \mathfrak{g}$ such that $\pi \circ \epsilon = I_{\mathfrak{g}/\mathfrak{r}}$. In particular ϵ is injective and so $\epsilon(\mathfrak{g}/\mathfrak{r})$ is a semisimple Lie subalgebra $\mathfrak{s}$ of $\mathfrak{g}$ isomorphic with $\mathfrak{g}/\mathfrak{r}$ and hence is of dimension $\dim \mathfrak{g} - \dim \mathfrak{r}$. Thus $\dim \mathfrak{s} + \dim \mathfrak{r} = \dim \mathfrak{g}$. Now if $X \in \mathfrak{s} \cap \mathfrak{r}$, then $X = \epsilon(Y)$ for a unique $Y \in \mathfrak{g}/\mathfrak{r}$ and therefore $\pi(X) = \pi\epsilon(Y) = Y$. But since $\pi(X) = 0$ we see that $Y = 0$ and therefore $X = \epsilon(Y)$ is also 0. Thus $\mathfrak{s} \cap \mathfrak{r} = \{0\}$, and it follows that $\mathfrak{g} = \mathfrak{r} \bigoplus \mathfrak{s}$.   □

In general if $\mathfrak{g} = \mathfrak{a} \bigoplus \mathfrak{b}$, where $\mathfrak{a}$ is an ideal and $\mathfrak{b}$ is a subalgebra then $\mathfrak{g}$ is the semidirect sum of $\mathfrak{a}$ with $\mathfrak{b}$. For if A and $A' \in \mathfrak{a}$ and B and $B' \in \mathfrak{b}$ then, $[A + B, A' + B'] = [A, A'] + [B, A'] + [A, B'] + [B, B']$ and since $\mathfrak{a}$ is an ideal, the first three terms are $\in \mathfrak{a}$. By uniqueness of the decomposition we therefore get

$$[(A, B), (A', B')] = ([A, A'] + \mathrm{ad}\, B(A') - \mathrm{ad}\, B'(A), [B, B']).$$

Since $\mathrm{ad}\, \mathfrak{b}$ is an algebra of derivations of $\mathfrak{a}$, this is a semidirect sum.

Corollary 3.5.3. *Any finite dimensional Lie algebra $\mathfrak{g}$ that is not simple and not 1 dimensional is a semidirect sum of lower dimensional Lie subalgebras.*

Proof. We may clearly assume that $\mathfrak{g}$ is not semisimple for if it were we could decompose it as a sum of simple ideals. Thus we may assume its radical $\mathfrak{r} \neq 0$. If $\mathfrak{g}$ is not solvable then by the Levi theorem $\mathfrak{g}$ is the semidirect sum of the ideal $\mathfrak{r}$ with a Levi factor $\mathfrak{s}$. Thus we may assume $\mathfrak{g}$ is solvable. In particular, $\mathfrak{g} \neq [\mathfrak{g}, \mathfrak{g}]$ and we can find an ideal $\mathfrak{a}$ in $\mathfrak{g}$ of codimension 1. Thus $\mathfrak{g} = \mathfrak{a} \oplus kX_0$. Since kX_0 is a subspace this completes the proof. □

We now turn to the proof of Theorem 3.5.1. It clearly suffices to show that there exists a subalgebra $\mathfrak{t}$ of $\mathfrak{g}$ such that $\mathfrak{g} = \operatorname{Ker} \pi \oplus \mathfrak{t}$, as k-spaces. For if this were so π would give rise to a Lie algebra isomorphism $\tilde{\pi} : \mathfrak{t} \to \mathfrak{s}$ and we could then take $\epsilon = (\tilde{\pi})^{-1}$. Then $\epsilon : \mathfrak{s} \to \mathfrak{g}$, and $\pi \circ \epsilon = I_{\mathfrak{s}}$.

Now let $\mathfrak{a} = \operatorname{Ker} \pi$ and write $\mathfrak{s} = \mathfrak{g}/\mathfrak{a}$. We shall prove our result by induction on $\dim \mathfrak{a}$. Suppose there is a $\mathfrak{g}$-ideal $\mathfrak{a}_0$ lying within $\mathfrak{a}$. Then by inductive hypothesis we can find a supplementary subalgebra $\mathfrak{s}_0 = \mathfrak{g}_0/\mathfrak{a}_0$ to $\mathfrak{a}/\mathfrak{a}_0$ in $\mathfrak{g}/\mathfrak{a}_0$ and also a supplementary subalgebra $\mathfrak{s}'$ to $\mathfrak{a}_0$ in $\mathfrak{g}_0$. Then $\mathfrak{s}'$ is a supplementary subalgebra to $\mathfrak{a}$ in $\mathfrak{g}$, so we may assume that $\mathfrak{a}$ itself is a $\mathfrak{g}$-simple ideal. Now since $\mathfrak{s}$ is semisimple, $\mathfrak{a}$ must contain the radical $\mathfrak{r}$ of $\mathfrak{g}$. If $\mathfrak{r} = 0$, then $\mathfrak{g}$ would be semisimple and then we would be done, since the ideal $\mathfrak{a}$ would then be a direct summand. Otherwise, by irreducibility of $\mathfrak{a}$ under $\mathfrak{g}$, we have $\mathfrak{r} = \mathfrak{a}$. But then by solvability $[\mathfrak{a}, \mathfrak{a}]$ is a proper $\mathfrak{g}$-ideal in $\mathfrak{a}$ and hence is 0. Thus we may assume $\mathfrak{a}$ is an abelian ideal (on which $\mathfrak{g}$ and therefore also $\mathfrak{s}$ act irreducibly).

Lemma 3.5.4. *Let $\rho : \mathfrak{g} \to \mathfrak{gl}(V)$ be a representation of the Lie algebra $\mathfrak{g}$ over k, and suppose that there is a vector $v_0 \in V$ such that the map $X \mapsto \rho(X)v_0$ is a bijection of $\mathfrak{a}$ with the orbit $\rho(\mathfrak{a})v_0$. Assume also that $\rho(\mathfrak{g})v_0 = \rho(\mathfrak{a})v_0$. Then $\operatorname{Stab}_{\mathfrak{g}}(v_0) = \{X \in \mathfrak{g} : \rho(X)v_0 = 0\}$ is a subalgebra of $\mathfrak{g}$ and $\mathfrak{g} = \mathfrak{a} \oplus \operatorname{Stab}_{\mathfrak{g}}(v_0)$.*

Proof. If $\rho(X)v_0 = 0$ and $\rho(Y)v_0 = 0$, then $\rho([X, Y])v_0 = \rho(X)\rho(Y)v_0 - \rho(Y)\rho(X)v_0 = 0$ so $\operatorname{Stab}_{\mathfrak{g}}(v_0)$ is a subalgebra of $\mathfrak{g}$. By assumption, the map $X \mapsto \rho(X)v_0$ is a linear bijection of $\mathfrak{a}$ with the orbit $\mathfrak{g}(v_0)$. On the other hand the orbit $\rho(\mathfrak{g})(v_0)$ is isomorphic as a k-space with $\mathfrak{g}/\operatorname{Stab}_{\mathfrak{g}}(v_0)$. Thus $\dim \mathfrak{g} = \dim \mathfrak{a} + \dim \operatorname{Stab}_{\mathfrak{g}}(v_0)$. If $X \in \mathfrak{a}$ and $\rho(X)v_0 = 0$, then $X = 0$ (by assumption), so $\mathfrak{a} \cap \operatorname{Stab}_{\mathfrak{g}}(v_0) = \{0\}$. Hence $\mathfrak{g} = \mathfrak{a} \oplus \operatorname{Stab}_{\mathfrak{g}}(v_0)$. □

Continuing the proof of Theorem 3.5.1, let $V = \text{End}_k(\mathfrak{g})$ and ρ be the representation of $\mathfrak{g}$ on V defined by $\rho(X)T = [\text{ad}\,X, T]$. Define three subspaces of V as follows: $\mathcal{P} = \text{ad}_{\mathfrak{g}}(\mathfrak{a})$, $\mathcal{Q} = \{T \in \text{End}_k(\mathfrak{g}) : T(\mathfrak{g}) \subseteq \mathfrak{a}, T|_{\mathfrak{a}} = 0\}$, and finally, $\mathcal{R} = \{T \in \text{End}_k(\mathfrak{g}) : T(\mathfrak{g}) \subseteq \mathfrak{a}, T|_{\mathfrak{a}} = \lambda I_{\mathfrak{a}}\}$. Then $\mathcal{P} \subseteq \mathcal{Q} \subseteq \mathcal{R} \subseteq V$, and $\mathcal{Q}$ has codimension 1 in $\mathcal{R}$. We first show that $\mathcal{P}$, $\mathcal{Q}$ and $\mathcal{R}$ are sub $\mathfrak{g}$-modules of V. In fact, if $X \in \mathfrak{g}$ and $Y \in \mathfrak{a}$, then $[\text{ad}\,X, \text{ad}\,Y] = \text{ad}\,[X, Y] \in \mathcal{P}$, since $\mathfrak{a}$ is an ideal, thus $\mathcal{P}$ is a submodule of V. If $X \in \mathfrak{g}$ and $T(\mathfrak{g}) \subseteq \mathfrak{a}$, then for $Y \in \mathfrak{a}$, we have

$$[\text{ad}\,X, T](Y) = \text{ad}\,X \circ T(Y) - T \circ \text{ad}\,X(Y) \in \mathfrak{a}$$

since $\mathfrak{a}$ is an ideal. Finally, if $Y \in \mathfrak{a}$ and $T(\mathfrak{a})$ is a homothety, then $T(Y) = \lambda(Y)$ so

$$\text{ad}\,X \circ T(Y) - T \circ \text{ad}\,X(Y) = \lambda[X, Y] - \lambda[X, Y] = 0.$$

This shows that $\mathfrak{g}(\mathcal{R}) \subseteq \mathcal{Q}$; in particular, both $\mathcal{R}$ and $\mathcal{Q}$ are submodules of V. Since $\mathfrak{a}$ is an ideal in $\mathfrak{g}$, $\mathcal{P}$ is also a $\mathfrak{g}$-submodule of V.

Taking quotients by $\mathcal{P}$ we get an exact sequence of $\mathfrak{g}$ modules

$$0 \to \mathcal{Q}/\mathcal{P} \to \mathcal{R}/\mathcal{P} \to \mathcal{R}/\mathcal{Q} \to 0.$$

We show that $\mathfrak{a}$ acts trivially on $\mathcal{R}/\mathcal{P}$, that is $\mathfrak{a}(\mathcal{R}) \subseteq \mathcal{P}$. Let $Y \in \mathfrak{a}$, $T \in \mathcal{R}$ and $X \in \mathfrak{g}$. Then $\text{ad}\,Y T(X) = 0$ since $T \in \mathcal{R}$, $Y \in \mathfrak{a}$ and $\mathfrak{a}$ is abelian. On the other hand since $\mathfrak{a}$ is an ideal and $\text{ad}\,Y(X) \in \mathfrak{a}$, we have $T\,\text{ad}\,Y(X) = [\lambda Y, X]$. Thus $\rho_Y(T) = \text{ad}\,\lambda Y \in \mathcal{P}$ for each $Y \in \mathfrak{a}$.

Because $\mathfrak{a}$ acts trivially on $\mathcal{R}/\mathcal{P}$ these are actually all the representations of the semisimple algebra $\mathfrak{s}$, and this sequence splits by Weyl's theorem. Because $\mathcal{Q}$ has codimension 1 in $\mathcal{R}$ there exists $T_0 \in \mathcal{R} \setminus \mathcal{Q}$, such that $[\text{ad}\,\mathfrak{g}, T_0] \in \mathcal{P}$, that is $[\text{ad}\,\mathfrak{g}, T_0] \in \text{ad}_{\mathfrak{g}}(\mathfrak{a})$. This T_0 is our "v_0". We can even normalize T_0 so that the homothety has $\lambda = 1$.

We show that $\rho(A)T_0 = -\,\text{ad}_{\mathfrak{g}}\,A$. That is for every $X \in \mathfrak{g}$, $[\text{ad}\,A, T_0](X) = \text{ad}\,A \circ T_0(X) - T_0 \circ \text{ad}\,A(X)$. But since $T_0(X) \in \mathfrak{a}$, $\text{ad}\,A \circ T_0(X) = [A, T_0(X)] = 0$. Hence by our normalization of T_0, $\rho(A)T_0(X) = -T_0([A, X]) = -[A, X]$ and $\rho(A)T_0 = -\,\text{ad}_{\mathfrak{g}}(A)$.

For $A \in \mathfrak{a}$ the map $A \mapsto \rho(A)T_0$ is bijective (injective). Since this map is linear in A this means that if $\rho(A)T_0 = -\,\text{ad}_{\mathfrak{g}}(A) = 0$, then

$A = 0$. But this condition says that $[X, A] = 0$ for all $X \in \mathfrak{g}$. Thus A would be fixed under the original irreducible action, a contradiction unless $A = 0$. Finally, let $X \in \mathfrak{g}$. We must show that $\rho(X)T_0 = \rho(A)T_0$, for some $A \in \mathfrak{a}$ *i.e.* $\rho(X)T_0 = -\operatorname{ad}\mathfrak{g}(A)$ for that A. But $\rho(X)T_0 = [\operatorname{ad} X, T_0]$ is in $\operatorname{ad}\mathfrak{g}(\mathfrak{a})$. This completes the proof.

We now turn to the uniqueness part of the Levi theorem. For the convenience of the reader we recall the uniqueness statement.

Malcev uniqueness: If $\mathfrak{t}$ is any semisimple subalgebra of a finite dimensional Lie algebra $\mathfrak{g}$ there is an inner automorphism α such that $\alpha(\mathfrak{t}) \subset \mathfrak{s}$. In fact $\alpha = \operatorname{Exp}(\operatorname{ad} X)$ for some ad-nilpotent $X \in \operatorname{rad}(\mathfrak{g})$. In particular, a Levi factor of $\mathfrak{g}$ is unique up to inner automorphism of the form $\alpha = \operatorname{Exp}(\operatorname{ad} X)$, for some ad nilpotent $X \in \operatorname{rad}(\mathfrak{g})$.

Proof of uniqueness: By Levi splitting for $X \in \mathfrak{g}$ we can write $X = r(X) + s(X)$, the unique $\mathfrak{r}$ and $\mathfrak{s}$ components of X. If $Y = r(Y) + s(Y)$ is another such element, then since $\mathfrak{r}$ is an ideal,

$$r[X, Y] = [r(X), r(Y)] + [r(X), s(Y)] + [s(X), r(Y)]$$

and $s[X, Y] = [s(X), s(Y)]$. Now $[\mathfrak{r}, \mathfrak{r}]$ is an ideal in $\mathfrak{g}$ and hence also in $[\mathfrak{g}, \mathfrak{r}]$. Let π be the projection $[\mathfrak{g}, \mathfrak{r}] \to [\mathfrak{g}, \mathfrak{r}]/[\mathfrak{r}, \mathfrak{r}]$ and let $\phi = \pi \cdot r|_{\mathfrak{t}}$. Then ϕ is a linear map and if $\rho_X = (\widetilde{\operatorname{ad} X}|_{[\mathfrak{g},\mathfrak{r}]})$, where ~ means the induced map on $[\mathfrak{g}, \mathfrak{r}]/[\mathfrak{r}, \mathfrak{r}]$, then ρ is a representation of $\mathfrak{g}$ on $[\mathfrak{g}, \mathfrak{r}]/[\mathfrak{r}, \mathfrak{r}]$. We consider the restriction of the representation to $\mathfrak{t}$ which we again call ρ. We will show

$$\phi([H_1, H_2]) = \rho_{H_1}(\phi(H_2)) - \rho_{H_2}(\phi(H_1)),$$

that is, ρ is a 1 cocycle. One shows by direct calculation that for all $H_1, h_2 \in \mathfrak{t}$,

$$\phi([H_1, H_2]) - \rho_{H_1}(\phi(H_2)) + \rho_{H_2}(\phi(H_1)) \in [\mathfrak{r}, \mathfrak{r}]. \tag{3.8}$$

This means

$$r[H_1, H_2] - [H_1, r(H_2)] + [H_2, r(H_1)] \in [\mathfrak{r}, \mathfrak{r}].$$

But

$$r[H_1, H_2] = [r(H_1), r(H_2)] + [r(H_1), s(H_2)] - [r(H_2), s(H_1)],$$

while,

$$[H_1, r(H_2)] = [r(H_1), r(H_2)] + [s(H_1), r(H_2)],$$

and

$$[H_2, r(H_1)] = [r(H_2), r(H_1)] + [s(H_2), r(H_1)].$$

Hence (3.8) is just $[r(H_1), r(H_2)] \in [\mathfrak{r}, \mathfrak{r}]$. Since $\mathfrak{t}$ is semisimple, by Whitehead's lemma there is some $v_0 = X_0 + [\mathfrak{r}, \mathfrak{r}]$ where $X_0 \in [\mathfrak{g}, \mathfrak{r}]$ such that $\rho(H)v_0 = [H, X_0] + [\mathfrak{r}, \mathfrak{r}] = \phi(H)$. But since $\phi(H) = r(H) + [\mathfrak{r}, \mathfrak{r}]$ we see that $[H, X_0] - r(H) \in [\mathfrak{r}, \mathfrak{r}]$ for all $H \in \mathfrak{t}$. In other words, $H + [X_0, H] = H + \operatorname{ad} X_0(H) \in s(H) + [\mathfrak{r}, \mathfrak{r}]$. On the other hand, $[\mathfrak{r}, \mathfrak{r}]$ is an ideal in $\mathfrak{g}$ and therefore normalized by $\mathfrak{s}$. So $[\mathfrak{r}, \mathfrak{r}] \bigoplus \mathfrak{s} = \mathfrak{g}_1$ is a subalgebra of $\mathfrak{g}$, containing $[\mathfrak{r}, \mathfrak{r}]$ as a solvable ideal. Since $\mathfrak{s}$ is semisimple, its radical is $[\mathfrak{r}, \mathfrak{r}]$ and therefore $\mathfrak{s}$ is a Levi factor.

We show $\operatorname{Exp}(\operatorname{ad} X_0)(\mathfrak{t}) \subseteq \mathfrak{g}_1$. Now $(\operatorname{ad} X_0)^2(H) = [X_0, [X_0, H]]$ and since $X_0 \in [\mathfrak{g}, \mathfrak{r}]$ and this is an ideal, $[X_0, H] \in [\mathfrak{g}, \mathfrak{r}]$. Hence also $[X_0, [X_0, H]] \in [[\mathfrak{g}, \mathfrak{r}], [\mathfrak{g}, \mathfrak{r}]] \subseteq [\mathfrak{r}, \mathfrak{r}]$, since $[\mathfrak{g}, \mathfrak{r}] \subseteq \mathfrak{r}$. Thus $(\operatorname{ad} X_0)^2(H)/2! \in [\mathfrak{r}, \mathfrak{r}]$. Similarly, $(\operatorname{ad} X_0)^3(H) = [X_0, [X_0, [X_0, t]]]$ and since $[[\mathfrak{g}, \mathfrak{r}], [\mathfrak{r}, \mathfrak{r}]] \subseteq [[\mathfrak{g}, \mathfrak{r}], [\mathfrak{g}, \mathfrak{r}]] \subseteq [\mathfrak{r}, \mathfrak{r}]$, we see that by induction $(\operatorname{ad} X_0)^n(H)/n! \in [\mathfrak{r}, \mathfrak{r}]$ for all $n \geq 2$. Since $t + \operatorname{ad} X_0(H) \in \mathfrak{g}_1$, it follows that the automorphism $\operatorname{Exp}(\operatorname{ad} X_0)$ also takes $\mathfrak{t}$ into $\mathfrak{g}_1$ and hence $\operatorname{Exp}(\operatorname{ad} X_0)(\mathfrak{t})$ is a semisimple subalgebra of the latter. Since $[\mathfrak{r}, \mathfrak{r}] \neq \mathfrak{r}$, by solvability, we see by induction on the dimension of $\mathfrak{g}$ that there exists an $X_1 \in \operatorname{rad}(\mathfrak{g}_1)$, so that $\operatorname{Exp}(\operatorname{ad} X_1) \operatorname{Exp}(\operatorname{ad} X_0)(\mathfrak{t}) \subseteq \mathfrak{s}$. Thus if $\alpha = \operatorname{Exp}(\operatorname{ad} X_1) \cdot \operatorname{Exp}(\operatorname{ad} X_0)$, then $\alpha(\mathfrak{t}) \subseteq \mathfrak{s}$. Since $X_0 \in [\mathfrak{g}, \mathfrak{r}]$ and $X_1 \in [\mathfrak{r}, \mathfrak{r}] \subseteq [\mathfrak{g}, \mathfrak{r}]$, once we know that the operators $\operatorname{ad} X$ are nilpotent for $X \in [\mathfrak{g}, \mathfrak{r}]$ we would then argue as follows:

$$\operatorname{Exp}(\operatorname{ad} X_1) \cdot \operatorname{Exp}(\operatorname{ad} X_0) = \operatorname{Exp}(\operatorname{ad} X_1 + \operatorname{ad} X_0 + \frac{1}{2}[\operatorname{ad} X_1, \operatorname{ad} X_0] + \ldots).$$

But this is $\operatorname{Exp}(\operatorname{ad} Y)$ where

$$Y = X_1 + X_0 + \frac{1}{2}[X_1, X_0] + \ldots.$$

Since $[\mathfrak{g}, \mathfrak{r}]$ is a subalgebra and we have a finite sum, $Y \in [\mathfrak{g}, \mathfrak{r}]$. That concludes the uniqueness proof.

As a corollary of Malcev uniqueness we have:

Corollary 3.5.5. *In a finite dimensional Lie algebra $\mathfrak{g}$, a Levi factor is a maximal semisimple subalgebra, and conversely. In particular, any semisimple subalgebra is contained in some Levi factor.*

Proof. A Levi factor is a maximal semisimple subalgebra. For if it were properly contained in a larger semisimple subalgebra, the larger one would have to intersect the radical nontrivially in a solvable ideal in it, therefore violating semisimplicity. Conversely, if $\mathfrak{t}$ were a maximal semisimple subalgebra of $\mathfrak{g}$, then $\alpha(\mathfrak{t}) \subset \mathfrak{s}$. So that $\mathfrak{t} \subset \alpha^{-1}(\mathfrak{s})$. Since the latter is also semisimple, by maximality $\mathfrak{t} = \alpha^{-1}(\mathfrak{s})$, must be a Levi factor. $\square$

Proposition 1.7.15 enables one to transform decompositions of the Lie algebra usually gotten by linear algebra to the group. For example, in this way we get the Levi decomposition of a connected Lie group G with Lie algebra $\mathfrak{g}$. Let $\mathfrak{g} = \mathfrak{r} \oplus \mathfrak{s}$ be a Levi decomposition of $\mathfrak{g}$. Let R and S be the unique connected Lie subgroups of G corresponding to the Lie subalgebras $\mathfrak{r}$ and $\mathfrak{s}$ (see Theorem 1.3.3). Then $R \cap S$ is discrete, since R is normal in G, and $G = RS$. Here R is the radical of G and S is a Levi factor, a maximal semisimple connected subgroup. This is the Levi decomposition of G. For each such global decomposition there is a uniqueness statement concerning S. Notice $R \cap S$ is normal in S and since it is discrete and S is connected, $R \cap S$ is central in S.

We now make a few remarks about faithful representations of Lie groups. This works equally well in the real or complex cases and implies Ado's theorem, which we mentioned in Section 1.7. Let G be a connected Lie group and $G = RS$ be a Levi decomposition. A theorem of Hochschild and Mostow [33, 34] states that if R and S each have faithful representations, then G has a faithful representation and conversely. Any semisimple group always has a locally isomorphic group with a faithful representation namely the adjoint group and the universal covering group always have a faithful representation [33, 34]. Hence,

G is locally isomorphic to a faithfully represented group. It follows, therefore, that any connected Lie group is locally isomorphic to a faithfully represented Lie group. Taking the derivative and using Lie's third theorem (that Lie algebras of Lie groups comprise all Lie algebras), we get a proof of Ado's theorem for Lie algebras.

To inject a note of reality into our brief discussion of faithful representations we now give two examples of classes of connected Lie groups, one nilpotent and one semisimple which *have no faithful representations*. For the general situation see [48] and [49].

Let G be any simply connected 2-step nilpotent group, G (for example, G could be N_n, the Heisenberg group of dimension $(2n + 1)$). Since the center, $Z(G)$, is nontrivial abelian and simply connected, let D be a discrete subgroup of $Z(G)$ with $K = Z(G)/D$ compact. Then the locally isomorphic group $H = G/D$ has no faithful linear representation. We denote by $\pi : G \to H$ the canonical map. For suppose $\rho : H \to \mathrm{GL}(n, \mathbb{C})$ were such a representation. Then $\rho(K)$ is a compact and therefore a completely reducible subgroup of $\mathrm{GL}(n, \mathbb{C})$. On the other hand, since D is discrete $\pi(Z(G)) = Z(H)$, so $K = Z(H)$. Since G is 2-step nilpotent and H has the same Lie algebra H is also 2-step nilpotent. Therefore $[H, H] \subseteq Z(H) = K$. By Lie's theorem (see Theorem 3.2.13), $\rho(H)$ is contained in the triangular matrices and hence $\rho([H, H]) = [\rho(H), \rho(H)]$ acts by unipotent operators. But since $\rho([H, H])$ also completely reducible, $\rho([H, H]) = I$. Since ρ is faithful $[H, H] = 1$, a contradicion because H is non abelian being 2-step nilpotent.

Now consider $\widetilde{\mathrm{SL}(2, \mathbb{R})}$, or more generally $G = \widetilde{\mathrm{Sp}(n, \mathbb{R})}$. Since here $K = \mathrm{U}(n, \mathbb{C})$ is a maximal compact subgroup of G and $\pi_1(K) = \pi_1(G) = \mathbb{Z}$, we see that the center of G is infinite (it is actually $\mathbb{Z}$). Since ρ is faithful $\rho(G)$ also has infinite center. But by Theorem 7.5.17 a linear semisimple group must have a finite center, a contradiction.

Evidently this last example works whenever G is a non-compact semisimple group and the maximal compact subgroups are not semisimple (but are merely reductive).

The following is a useful result.

Theorem 3.5.6. *IF* $\mathfrak{g}$ *is a Lie algebra and* $\mathfrak{r}$ *is its radical then* $[\mathfrak{g}, \mathfrak{g}] \cap \mathfrak{r} = [\mathfrak{g}, \mathfrak{r}]$. *Moreover, if* $\rho : \mathfrak{g} \to V$ *is a representation of* $\mathfrak{g}$ *then* $[\mathfrak{g}, \mathfrak{r}]$ *acts on* V *by nilpotent operators.*

Proof. By the Levi decomposition, $\mathfrak{g} = \mathfrak{r} + \mathfrak{s}$ where $\mathfrak{s}$ is a semisimple subalgebra of $\mathfrak{g}$ and $\mathfrak{r} \cap \mathfrak{s} = \{0\}$ ($\mathfrak{s}$ being a Levi factor). Since $\mathfrak{r}$ is an ideal and $[\mathfrak{s}, \mathfrak{s}] = \mathfrak{s}$,

$$[\mathfrak{g}, \mathfrak{g}] = [\mathfrak{r}, \mathfrak{r}] + [\mathfrak{r}, \mathfrak{s}] + [\mathfrak{s}, \mathfrak{s}] \subseteq [\mathfrak{r}, \mathfrak{g}] + \mathfrak{s}.$$

This means that $[\mathfrak{g}, \mathfrak{g}] = [\mathfrak{r}, \mathfrak{g}] + \mathfrak{s}$ and hence that $[\mathfrak{g}, \mathfrak{g}] \cap \mathfrak{r} = [\mathfrak{g}, \mathfrak{r}]$. We know from Lie's theorem that $[\mathfrak{r}, \mathfrak{r}]$ acts on V by nilpotent operators. If $\mathfrak{m}$ is a maximal subspace of $[\mathfrak{g}, \mathfrak{r}]$ which acts nilpotently on V. We will show that $\mathfrak{m} = [\mathfrak{g}, \mathfrak{r}]$. Suppose this is not so and $\mathfrak{m}$ is a proper subspace. Then there exists an $X \in \mathfrak{g}$ and $R \in \mathfrak{r}$ so that $[X, R]$ does not act nilpotently on V. The subalgebra generated by X and $\mathfrak{r}$ consisting of $\{cX + R : c \in k, R \in \mathfrak{r}\}$ is clearly solvable since it contains a solvable ideal of codimension ≤ 1. Again by Lie's theorem, its derived also acts nilpotently on V. In particular, $[X, R]$ acts nilpotently on V, a contradiction. □

Corollary 3.5.7. *The subalgebra* $[\mathfrak{g}, \mathfrak{r}]$ *is in the nilradical* $\mathfrak{n}$ *of* $\mathfrak{g}$. *In particular, if* $\mathfrak{r}$ *is solvable then* $\mathfrak{r}/\mathfrak{n}$ *is abelian. In general, the radical of* $\mathfrak{g}/\mathfrak{n}$ *is abelian.*

Proof. Taking ρ to be the adjoint representation, we see by the present result together with Engel's theorem that $\mathrm{ad}[\mathfrak{g}, \mathfrak{r}] = [\mathrm{ad}\,\mathfrak{g}, \mathrm{ad}\,\mathfrak{r}]$ acts as a nilpotent Lie algebra. Hence $[\mathfrak{g}, \mathfrak{r}]$ is nilpotent ideal. Therefore $[\mathfrak{g}, \mathfrak{r}] \subseteq \mathfrak{n}$, the nilradical of $\mathfrak{g}$. Taking $\mathfrak{g}$ to be solvable we obtain the second statement. Finally, if $\mathfrak{g} = \mathfrak{r} \oplus \mathfrak{s}$ is the Levi decomposition of $\mathfrak{g}$, dividing by $\mathfrak{n} \subseteq \mathfrak{r}$ gives $\mathfrak{g}/\mathfrak{n} = \mathfrak{r}/\mathfrak{n} \oplus \mathfrak{s}$, a Levi decomposition of $\mathfrak{g}/\mathfrak{n}$. Hence $\mathfrak{r}/\mathfrak{n}$ is its radical which is abelian. □

3.6   Reductive Lie Algebras

Definition 3.6.1. We say a Lie algebra $\mathfrak{g}$ is *reductive* if it is non-abelian and the adjoint representation is completely reducible.

Lemma 3.6.2. *If $\rho : \mathfrak{g} \to \mathfrak{gl}(V)$ is a representation of a semisimple Lie algebra then $\rho(\mathfrak{g}) \subset \mathfrak{sl}(V)$. In particular $\mathfrak{g}$ acts trivially on any 1-dimensional space.*

Proof. $\mathfrak{g} = [\mathfrak{g}, \mathfrak{g}]$ so $\rho(\mathfrak{g}) = \rho[\mathfrak{g}, \mathfrak{g}] = [\rho(\mathfrak{g}), \rho(\mathfrak{g})] \subset [\mathfrak{gl}(V), \mathfrak{gl}(V)] \subset \mathfrak{sl}(V)$. $\qquad\qquad\square$

Proposition 3.6.3. *The following conditions are equivalent.*

(1) $\mathfrak{g}$ is reductive.

(2) $[\mathfrak{g}, \mathfrak{g}]$ is semisimple.

(3) $\mathfrak{g} = \mathfrak{z}(\mathfrak{g}) \oplus [\mathfrak{g}, \mathfrak{g}]$ where $[\mathfrak{g}, \mathfrak{g}]$ is semisimple.

(4) $\mathfrak{z}(\mathfrak{g}) = \mathrm{rad}(\mathfrak{g})$.

Proof. If the adjoint representation is completely reducible then, as above (even if $\mathfrak{g}$ is not semisimple) $\mathfrak{g} = \bigoplus \mathfrak{g}_i$, the direct sum of simple ideals, except that now some of them may be 1-dimensional abelian. Thus $\mathfrak{g}$ is the direct sum of an abelian and a semisimple algebra $\mathfrak{h}$. This means that $[\mathfrak{g}, \mathfrak{g}] = [\mathfrak{h}, \mathfrak{h}] = \mathfrak{h}$. If $[\mathfrak{g}, \mathfrak{g}]$ is semisimple then $\mathfrak{g} = [\mathfrak{g}, \mathfrak{g}] \oplus \mathfrak{h}$ for some ideal $\mathfrak{h}$. Since $\mathfrak{g}/[\mathfrak{g}, \mathfrak{g}] = \mathfrak{h}$ and the former is abelian this proves (iii) because $\mathfrak{z}([\mathfrak{g}, \mathfrak{g}]) = 0$. If $\mathfrak{g} = \mathfrak{z}(\mathfrak{g}) \oplus [\mathfrak{g}, \mathfrak{g}]$ where $[\mathfrak{g}, \mathfrak{g}]$ is semisimple then $\mathrm{rad}(\mathfrak{g}) = \mathrm{rad}(\mathfrak{z}(\mathfrak{g})) \oplus \mathrm{rad}([\mathfrak{g}, \mathfrak{g}]) = \mathfrak{z}(\mathfrak{g})$. Finally, suppose $\mathfrak{z}(\mathfrak{g}) = \mathrm{rad}(\mathfrak{g})$. Then ad induces a map $\mathfrak{g}/\mathfrak{z} = \mathfrak{g}/\mathrm{rad}(\mathfrak{g}) \to \mathrm{ad}\,\mathfrak{g}$. The algebra $\mathfrak{g}/\mathfrak{r}$ is semisimple, hence so is $\mathrm{ad}\,\mathfrak{g}$. By Weyl's theorem ad is completely reducible. $\qquad\qquad\square$

An example of a reductive Lie algebra is a Lie algebra of compact type. Another example is provided by

Corollary 3.6.4. *If $\mathfrak{g}$ is a Lie algebra and $\mathfrak{n}$ is its nil radical then $\mathfrak{g}/\mathfrak{n}$ is reductive.*

Proof. Let $\mathfrak{r} = \mathrm{rad}(\mathfrak{g})$ then $\mathfrak{r}/\mathfrak{n}$ is an ideal in $\mathfrak{g}/\mathfrak{n}$ and the quotient, $\mathfrak{g}/\mathfrak{r}$ is semisimple. By Corollary 3.2.19, $[\mathfrak{r}, \mathfrak{r}] \subset \mathfrak{n}$ so $\mathfrak{r}/\mathfrak{n}$ is abelian. By the Levi theorem $\mathfrak{g}/\mathfrak{n} = \mathfrak{r}/\mathfrak{n} \oplus \mathfrak{g}/\mathfrak{r}$ the semi-direct sum of an abelian ideal and a semisimple algebra. To see that this is a direct sum, *i.e.* that $\mathfrak{r}/\mathfrak{n}$ is central in $\mathfrak{g}/\mathfrak{n}$, we must show that $[\mathfrak{g}/\mathfrak{n}, \mathfrak{r}/\mathfrak{n}] = \bar{0}$ *i.e.* that $[\mathfrak{g}, \mathfrak{r}] \subset \mathfrak{n}$. This is so because $[\mathfrak{g}, \mathfrak{r}]$ is a nilpotent ideal. $\qquad\qquad\square$

Definition 3.6.5. Let $\mathfrak{g}$ be a subalgebra of $\mathfrak{gl}(V)$ where V is a finite dimensional k-vector space and k is a field of characteristic 0. We say that $\mathfrak{g}$ is to be *linearly reductive* if $\mathfrak{g}$ acts completely reducibly V.

We now study linearly reductive subalgebras of $\mathfrak{gl}(V)$.

Proposition 3.6.6. *A representation $\rho : \mathfrak{g} \rightarrow \mathfrak{gl}(V)$ of a Lie algebra $\mathfrak{g}$ is completely reducible if and only if V is the direct sum of nontrivial $\mathfrak{g}$-invariant subspaces W_i of V on each of which $\mathfrak{g}$ acts irreducibly.*

Proof. The restriction of a completely reducible representation to an invariant subspace W is still a completely reducible representation. For if U is a $\mathfrak{g}$-invariant subspace of W, then U is a $\mathfrak{g}$-invariant subspace of V. If U' is a complementary $\mathfrak{g}$-invariant subspace of V then $U' \cap W$ is a $\mathfrak{g}$-invariant subspace of W which complements U in W. Hence by induction on the finite dimension of V, if ρ is completely reducible then V is the direct sum of such subspaces W_i of V on which $\mathfrak{g}$ acts irreducibly. Conversely, suppose V is the direct sum of nontrivial $\mathfrak{g}$-invariant irreducible subspaces V_i. Let W be a $\mathfrak{g}$-invariant subspace of V. We argue by induction on the codimension of such W. Since $V_i \cap W$ is a $\mathfrak{g}$-invariant subspace and V_i is irreducible we have either $V_i \cap W = V_i$ or $V_i \cap W = 0$ for each i. If $V_i \cap W = V_i$ for all i then $W = V$. Otherwise choose i so that $V_i \cap W = 0$ and let $U = V_i + W$. Then U is a $\mathfrak{g}$-invariant subspace of V of codimension less than that of W. Hence there is a complementary $\mathfrak{g}$-invariant subspace U' of V, which means that $U' + V_i$ complements W. □

Proposition 3.6.7. *If a representation $\rho : \mathfrak{g} \rightarrow \mathfrak{gl}(V)$ of a Lie algebra $\mathfrak{g}$, and if V is the direct sum of nontrivial $\mathfrak{g}$-invariant subspaces W_i of V on which $\mathfrak{g}$ acts irreducibly, then this decomposition is unique up to equivalence. That is if $\rho = \sum_1^n n_i \rho_i$ where n_i and n are integers and the ρ_i's are irreducible, the equivalence of classes of ρ_i's and their multiplicities n_i's are uniquely determined by ρ.*

Proof. Apply the Jordan-Holder theorem. □

We now deal with some further questions about extension of the base field. If $\rho : \mathfrak{g} \rightarrow \mathfrak{gl}(V)$ is a representation over $\mathbb{R}$ of a real Lie algebra let

$\rho^{\mathbb{C}}$, $\mathfrak{g}_{\mathbb{C}}$, and $V^{\mathbb{C}}$ denote the respective complexifications. Then $\mathfrak{g}^{\mathbb{C}}$ is a complex Lie algebra, $V^{\mathbb{C}}$ is a complex vector space and $\rho^{\mathbb{C}} : \mathfrak{g}_{\mathbb{C}} \to \mathfrak{gl}(V^{\mathbb{C}})$ is a representation over $\mathbb{C}$ where $\rho^{\mathbb{C}}$ is defined by

$$\rho^{\mathbb{C}}(X + iY)(v + iw) = \rho_X(v) - \rho_Y(w) + i(\rho_X(w) + \rho_Y(v)) \qquad (3.9)$$

for $X + iY \in \mathfrak{g}_{\mathbb{C}}, v + iw \in V^{\mathbb{C}}$.

We now show that the trace form $\beta_{\rho\mathbb{C}} : \mathfrak{g}_{\mathbb{C}} \times \mathfrak{g}_{\mathbb{C}} \to \mathbb{C}$ of $\rho^{\mathbb{C}}$ is given by

$$\beta_{\rho\mathbb{C}}(X + iY, X' + iY') = \mathrm{tr}_{V\mathbb{C}}(\rho^{\mathbb{C}}(X + iY)\rho^{\mathbb{C}}(X' + iY')).$$

The latter term equals

$$\mathrm{tr}_V(\rho(X)\rho(X') - \rho(Y)\rho(Y')) + i(\mathrm{tr}_V(\rho(X)\rho(Y')) + \mathrm{tr}_V(\rho(Y)\rho(X'))).$$

In particular if Y and $Y' = 0$, we have $\beta_{\rho\mathbb{C}}(X, X') = \beta_\rho(X, X')$. Now β_ρ is nondegenerate if and only if there exists a $\mathbb{R}$-basis $\{X_1, ..., X_n\}$ of $\mathfrak{g}$ (also a $\mathbb{C}$-basis for $\mathfrak{g}_{\mathbb{C}}$) such that $\det \beta_\rho(X_i, X_j) \neq 0$. But then $\det \beta_{\rho\mathbb{C}}(X_i, X_j)$ is also $\neq 0$. We have proved that if β_ρ is nondegenerate so is $\beta_{\rho\mathbb{C}}$. Applying this to the adjoint representation of $\mathfrak{g}$ we see that if $\mathfrak{g}$ semisimple Lie algebra then $\mathfrak{g}_{\mathbb{C}}$ is also semisimple.

Conversely, suppose $\mathfrak{g}_{\mathbb{C}}$ is semisimple and $\mathfrak{h}$ is an abelian ideal in $\mathfrak{g}$. From the identity $[\mathfrak{h}, \mathfrak{k}]_{\mathbb{C}} = [\mathfrak{h}_{\mathbb{C}}, \mathfrak{k}_{\mathbb{C}}]$ we see that $\mathfrak{h}_{\mathbb{C}}$ is an abelian ideal in $\mathfrak{g}_{\mathbb{C}}$ and hence is trivial. But $\mathfrak{h} \subset \mathfrak{h}_{\mathbb{C}}$. Thus $\mathfrak{g}$ is also semisimple. We have proved:

Proposition 3.6.8. *A real Lie algebra $\mathfrak{g}$ semisimple if and only if $\mathfrak{g}_{\mathbb{C}}$ is also semisimple.*

Lemma 3.6.9.   *(1) Any $\mathbb{C}$-subspace S of $V^{\mathbb{C}}$ is of the form $S = W^{\mathbb{C}}$ for some $\mathbb{R}$-subspace W of V.*

(2) If W is a subspace of V then $W = W^{\mathbb{C}} \cap V$.

(3) If W is a subspace of V then W is $\mathfrak{g}$-invariant if and only if $W^{\mathbb{C}}$ is $\mathfrak{g}_{\mathbb{C}}$-invariant.

Proof. (i) is clear. Let $\{w_1, ..., w_j\}$ be a basis of W and extend it to a basis $\{w_1, ..., w_j, v_1, ..., v_k\}$ of V over $\mathbb{R}$ (also a basis of $W^{\mathbb{C}}$ over $\mathbb{C}$). Let $\sum_i c_i w_i \in W^{\mathbb{C}}$ (where $c_i \in \mathbb{C}$). If this also lies in V then $\sum_i c_i w_i =$

$\sum_i a_i w_i + \sum_j b_j v_j$. Hence $\sum_i (c_i - a_i) w_i + \sum_j b_j v_j = 0$. It follows that $c_i = a_i$ and $b_1 = 0$, and in particular $\sum_i c_i w_i \in W$. Since $W \subset W^{\mathbb{C}} \cap V$ this proves (ii).

As for (iii), if W is $\mathfrak{g}$-invariant then for $w + iw' \in W^{\mathbb{C}}$ we know $\rho_X(w) - \rho_Y(w')$ and $\rho_X(w') + \rho_Y(w) \in W$. By (3.9) $\rho^{\mathbb{C}}(X + iY)(w + iw') \in W^{\mathbb{C}}$. Conversely, if $\rho^{\mathbb{C}}(X + iY)(w + iw') \in W^{\mathbb{C}}$ for all X, Y, w and w' then by (3.9) $\rho_X(w) \in W^{\mathbb{C}} \cap V = W$ for all X and w.   □

Corollary 3.6.10. *A linear Lie algebra* $\mathfrak{g} \subset \mathfrak{gl}(V)$ *is completely reducible if and only if* $\mathfrak{g}_{\mathbb{C}} \subset \mathfrak{gl}(V^{\mathbb{C}})$ *is also completely reducible.*

Proof. If $\mathfrak{g}_{\mathbb{C}}$ is completely reducible and W is a $\mathfrak{g}$-invariant subspace of V then $W^{\mathbb{C}}$ is $\mathfrak{g}_{\mathbb{C}}$-invariant. Hence $V^{\mathbb{C}} = W^{\mathbb{C}} \oplus S$ where S is $\mathfrak{g}_{\mathbb{C}}$-invariant. But $S = U^{\mathbb{C}}$ for some U which is $\mathfrak{g}$-invariant by (ii). From $V^{\mathbb{C}} = W^{\mathbb{C}} \oplus U^{\mathbb{C}}$ it follows that $V = W \oplus U$. Now suppose $\mathfrak{g}$ is completely reducible and let S be a $\mathfrak{g}_{\mathbb{C}}$-invariant subspace of $V^{\mathbb{C}}$. Then $S = W^{\mathbb{C}}$ for some $W \subset V$ and W is $\mathfrak{g}$-invariant. Hence W has a complementary $\mathfrak{g}$-invariant subspace U and $V = W \oplus U$. But then $U^{\mathbb{C}}$ is $\mathfrak{g}_{\mathbb{C}}$-invariant and $V^{\mathbb{C}} = W^{\mathbb{C}} \oplus U^{\mathbb{C}}$.   □

Theorem 3.6.11. *A linear Lie algebra* $\mathfrak{g}$ *is linearly reductive if and only if*

(1) $\mathfrak{g}$ *is reductive and*

(2) the elements of $\mathfrak{z}(\mathfrak{g})$ *are simultaneously diagonalizable.*

Proof. Using the remarks above we may assume the field k is algebraically closed. Let $\mathfrak{r} = \mathrm{rad}(\mathfrak{g})$. By Lie's theorem there exists a semi-invariant $\chi \in \mathfrak{r}^*$ with nonzero semi-invariant vector w. Then $V_\chi = \{v \in V : Xv = \chi(X)v \text{ for all } X \in \mathfrak{r}\}$ is a nonzero subspace of V. By Lemma 3.2.15, which was used in the proof of Lie's theorem, $\chi([\mathfrak{g}, \mathfrak{r}]) = 0$. If $v \in V_\chi$, $Y \in \mathfrak{r}$ and $X \in \mathfrak{g}$ then $YXv = XYv + [Y, X]v = X(\chi(Y)v) + 0v = \chi(Y)X(v)$ so that V_χ is a $\mathfrak{g}$-invariant subspace of V. By complete reducibility (and the fact that a submodule of a completely reducible module is itself completely reducible) there is a finite set of χ such that $V = \bigoplus V_\chi$ and each V_χ is $\mathfrak{g}$-invariant. Choose a basis in each V_χ and put these together to get a basis of V. On each V_χ, $\mathfrak{r}$ acts by $Y \to \chi(Y)I$ so on V, $\mathfrak{r}$ acts diagonally and in particular $\mathfrak{z}$ acts diagonally, proving *(ii)*. Moreover for $v \in V_\chi$, $Y \in \mathfrak{r}$ and $X \in \mathfrak{g}$ we have $XYv = X(\chi(Y)v) = \chi(Y)(X(v))$, while $YX(v) = \chi(Y)(X(v))$ because V_χ is X-stable. Thus $YX = XY$ on each V_χ, $[X, Y]$ acts trivially V and $\mathfrak{r} \subset \mathfrak{z}$. Hence $\mathfrak{r} = \mathfrak{z}$, proving (i) by Proposition 3.6.3. Conversely suppose (i) and (ii) hold. By (ii) if $Z \in \mathfrak{z}(\mathfrak{g})$ we have

$$
Z = \begin{pmatrix}
\chi_1(Z) & 0 & 0 & 0 & 0 \\
0 & \chi_2(Z) & 0 & 0 & 0 \\
\ldots & \ldots & \ldots & \ldots & \ldots \\
\ldots & \ldots & \ldots & \ldots & \ldots \\
0 & 0 & 0 & 0 & \chi_n(Z)
\end{pmatrix}, \tag{3.10}
$$

where $\chi_i \in \mathfrak{z}(\mathfrak{g})^*$. Here we use the well-known fact from linear algebra that a commuting family of diagonalizable operators can be simultaneously diagonalized. Let $V_{\chi_i} = \{v \in V : Zv = \chi_i(Z)v \text{ for all } Z \in \mathfrak{z}(\mathfrak{g})\}$. Then each V_{χ_i} is $\mathfrak{g}$-stable. For if $X \in \mathfrak{g}$ and $Zv = \chi_i(Z)v$ for all $Z \in \mathfrak{z}(\mathfrak{g})$ then $ZXv = XZv = X(\chi_i(Z)v) = \chi_i(Z)Xv$. Since this holds for all $Z \in \mathfrak{z}(\mathfrak{g})$ we conclude that $Xv \in V_{\chi_i}$. To prove that $\mathfrak{g}$ acts completely reducibly on V we may assume $V = V_{\chi_i}$. Then $\mathfrak{z}(\mathfrak{g})$ acts by scalars so by (i) $\mathfrak{g}$-submodules are the same as $[\mathfrak{g}, \mathfrak{g}]$-submodules. By Weyl's theorem $[\mathfrak{g}, \mathfrak{g}]$ acts completely reducibly. $\square$

3.7   The Jacobson-Morozov Theorem

We recall (see Example 3.1.3) that if $\{X_+, H, X_-\}$ are the usual generators of $\mathfrak{sl}(2) = \mathfrak{sl}(2, \mathbb{C})$, then they form a basis for $\mathfrak{sl}(2)$ and $[H, X_+] = 2X_+$, $[H, X_-] = -2X_-$ and $H = [X_+, X_-]$.

Definition 3.7.1. Let $\mathfrak{g}$ be a Lie algebra containing linearly independent elements, X_+, H and X_+. We shall say $\{X_+, H, X_-\}$ is an $\mathfrak{sl}(2)$ triple if they satisfy the $\mathfrak{sl}(2)$ relations.

Our objective here is to prove the following result which provides us another criterion (see Theorem 3.6.11) for complete reducibility of a Lie algebra of operators in characteristic 0. Here k is algebraically closed of characteristic 0.

Theorem 3.7.2. *Let $\mathfrak{g}$ be a completely reducible Lie subalgebra of $\mathfrak{gl}(V)$ and let $N \neq 0 \in \mathfrak{g}$. If $\mathrm{ad}\, N$ is nilpotent then $\mathfrak{g}$ can be imbedded in an $\mathfrak{sl}(2)$ triple, in a way that $N = X_+$. We shall call this condition J-M. Moreover, $\mathfrak{g}$ contains the nilpotent and semisimple parts of each of its elements.*

Conversely, if J-M holds and $\mathfrak{g}$ contains the nilpotent and semisimple parts of each of its elements, then $\mathfrak{g}$ is completely reducible.

In particular, if J-M holds and $\mathfrak{g}$ has a trivial center, then $\mathfrak{g}$ is completely reducible. Of course if $\mathfrak{g}$ is semisimple, we already know this by Weyl's theorem, which suggests that Weyl's theorem will have to play a role in our proof. Moreover, by Weyl's theorem, if $\mathfrak{g}$ is a linear semisimple algebra then it contains the nilpotent and semisimple parts of each of its elements.

We begin with a result known as Morozov's lemma.

Lemma 3.7.3. *Let $\mathfrak{g}$ be a Lie algebra containing elements X_+ and H such that $[H, X_+] = 2X_+$ and $H = [Z, X_+]$, for some $Z \in \mathfrak{g}$. Then there exists an $X_- \in \mathfrak{g}$ such that $\{X_+, H, X_-\}$ form an $\mathfrak{sl}(2)$-triple.*

Proof. Let $X'_+ = \mathrm{ad}\, X_+$, $H' = \mathrm{ad}\, H$ and $Z' = \mathrm{ad}\, Z$. Since ad is a representation we know $[H', X'_+] = 2X'_+$ and $H' = [Z', X'_+]$. From the first of these relations we see by Lemma 3.7.6 that X'_+ is a nilpotent operator on $\mathfrak{g}$. Moreover, $[[Z, H] - 2Z, X_+] = [[Z, H], X_+] - 2[Z, X_+]$ which, by the Jacobi identity, equals $-[[X_+, Z], H] - [[H, X_+], Z] - 2[Z, X_+] = [H, H] - 2[X_+, Z] + 2[X_+, Z] = 0$. In other words, $[Z, H] - 2Z$ is in $\mathfrak{z}_{\mathfrak{g}}(X_+)$, the centralizer of X_+. We can therefore write $[Z, H] = 2Z + C$, where $C \in \mathfrak{z}_{\mathfrak{g}}(X_+)$.

Now let $U \in \mathfrak{z}_{\mathfrak{g}}(X_+)$. Because $[H', X'_+] = 2X'_+$ and $X'_+(U) = 0$,

$$X'_+ H'(U) = [X'_+, H'](U) + H' X'_+(U)$$
$$= -2X'_+(U) + H' X'_+(U) = 0,$$

and hence $H'(U) \in \mathfrak{z}_{\mathfrak{g}}(X_+)$. Thus H' leaves $\mathfrak{z}_{\mathfrak{g}}(X_+)$ stable. Moreover for a positive integer i, we have

$$[Z', X'^i_+] = X'^{i-1}_+[Z', X'_+] + X'^{i-2}_+[Z', X'_+]X'_+ + \cdots + [Z', X'_+]X'^{i-1}_+$$
$$= X'^{i-1}_+ H' + X'^{i-2}_+ H' X'_+ + \cdots + H' X'^{i-1}_+.$$

By an easy induction, we see that for every positive integer i, $H'X''^i_+ = X'^k_+ H' + 2iX''^i_+$,

$$[Z', X''^i_+] = (H' - 2(i-1)I + H' - 2(i-2)I + \cdots + H')X''^{i-1}_+$$
$$= i(H' - (i-1)I)X''^{i-1}_+.$$

Now suppose $U \in \mathfrak{z}_\mathfrak{g}(X_+) \cap X''^{i-1}_+(\mathfrak{g})$. Then $U = X''^{i-1}_+(V)$ for some V and so $X'_+(U) = X'^i_+(V) = 0$. But then,

$$i(H' - (i-1)I)X''^{i-1}_+(V) = Z'X''^i_+(V) - X''^i_+ Z'(V)$$
$$= -X''^i_+ Z'(V) \in X''^i_+(\mathfrak{g}),$$

so that $(H' - (i-1)I)(U) \in X''^i_+(\mathfrak{g})$. This means that for every i, $H' - (i-1)I$ sends $\mathfrak{z}_\mathfrak{g}(X_+) \cap X''^{i-1}_+(\mathfrak{g})$ to $X''^i_+(\mathfrak{g})$. But since X'_+ is nilpotent it follows that there is some m for which $(H' - mI) \cdots (H' - 2I)(H' - I)H'(U) = 0$ for all $U \in \mathfrak{z}_\mathfrak{g}(X_+)$.

Consider the restriction of H' to $\mathfrak{z}_\mathfrak{g}(X_+)$ and put it in upper triangular form over the algebraic closure of our field. From this and the above equation, it follows that each eigenvalue of this restriction is a nonpositive integer, so the restriction of $H' + 2I$ to $\mathfrak{z}_\mathfrak{g}(X_+)$ is invertible. Since $[Z, H] = 2Z + C$ where $C \in \mathfrak{z}_\mathfrak{g}(X_+)$, and since the restriction of $H + 2I$ to $\mathfrak{z}_\mathfrak{g}(X_+)$ is onto, there must be a $Y \in \mathfrak{z}_\mathfrak{g}(X_+)$ for which $(H' + 2I)(Y) = C$. But then, $[H, Y] = -2Y + C$. Letting $X_- = -(Y + Z)$, we see that

$$[H, X_-] = -[H, Y + Z] = -[H, Y] - [H, Z]$$
$$= -(-2Y + C) + (2Z + C) = 2(Y + Z) = -2X_-.$$

Since $Y \in \mathfrak{z}_\mathfrak{g}(X_+)$, we also get

$$[X_+, X_-] = [X_+, -(Y + Z)] = [X_+, -Z] = [Z, X_+] = H.$$

$\square$

Before turning to the J-M theorem itself we need the following lemma.

Lemma 3.7.4. *Suppose $\mathfrak{g}$ is a Lie subalgebra of $\mathfrak{gl}(V)$ with the property that every nonzero nilpotent element can be imbedded in an $\mathfrak{sl}(2)$ triple. Let $\mathfrak{h}$ be a subalgebra of $\mathfrak{g}$ satisfying*

(1) $\mathfrak{g} = \mathfrak{h} \oplus \mathfrak{l}$, where $\mathfrak{l}$ is a subspace of $\mathfrak{g}$.

(2) $[\mathfrak{h}, \mathfrak{l}] \subseteq \mathfrak{l}$.

Then $\mathfrak{h}$ also has the property that every nonzero nilpotent element can be imbedded in an $\mathfrak{sl}(2)$ triple lying in $\mathfrak{h}$.

Proof. Suppose F is a nonzero nilpotent operator in $\mathfrak{h}$. Choose E and $H \in \mathfrak{g}$ so that $\{E, H, F\}$ form an $\mathfrak{sl}(2)$ triple in $\mathfrak{g}$. Using the decomposition above write $H = H_\mathfrak{h} + H_\mathfrak{l}$ and $E = E_\mathfrak{h} + E_\mathfrak{l}$. Then $-2F = [F, H] = [F, H_\mathfrak{h}] + [F, H_\mathfrak{l}]$, where $[F, H_\mathfrak{h}] \in \mathfrak{h}$ and $[F, H_\mathfrak{l}] \in \mathfrak{l}$. Since we have a direct sum decomposition, $-2F = [F, H_\mathfrak{h}]$. Also, $H = [E, F] = [E_\mathfrak{h}, F] + [E_\mathfrak{l}, F]$ where the components belong to $\mathfrak{h}$ and $\mathfrak{l}$, respectively, hence $H_\mathfrak{h} = [E_\mathfrak{h}, F]$. Thus by Morozov's lemma applied to $H_\mathfrak{h}$ and F, both of which are in $\mathfrak{h}$, we get an $E^- \in \mathfrak{h}$ so that $\{E^-, H_\mathfrak{h}, F\}$ satisfy the relations of an $\mathfrak{sl}(2)$ triple. The subalgebra of $\mathfrak{h}$ generated by these elements is a homomorphic image of $\mathfrak{sl}(2)$ and is therefore a simple Lie algebra of dimension ≤ 3. Therefore, since k is algebraically closed of characteristic 0 it is either trivial or isomorphic to $\mathfrak{sl}(2)$ as Lie algebras of dimension 1 or 2 are always solvable. But since $F \neq 0$ this means it is isomorphic to $\mathfrak{sl}(2)$, that is, $\{E^-, H_\mathfrak{h}, F\}$ are linearly independent. $\qquad\qquad\square$

To prove the J-M theorem we need the following lemmas.

Lemma 3.7.5. *Let $T \in \mathrm{End}_k(V)$ where V is a k-vector space of dimension n. If $\mathrm{tr}(T^j) = 0$ for all $j = 1, ..., n$ then T is nilpotent.*

Proof. We may clearly assume k is algebraically closed. Hence T is triangular with diagonal entries $\alpha_1, ..., \alpha_n$. This means that for all j, T^j is triangular with diagonal entries $\alpha_1^j, ..., \alpha_n^j$. Thus our hypothesis says $\alpha_1^j + ... + \alpha_n^j = 0$ for $j = 1, ..., n$ and we must show that each $\alpha_i = 0$. If we knew one of the α_i say $\alpha_n = 0$ then we would have (by throwing away the last equation) a system of $n - 1$ equations, which by induction would have only the trivial solution. This would complete the proof.

Now let $\chi_T(x) = x^n - \mathrm{tr}(T)x^{n-1} + \ldots + \det(T)$ be the characteristic polynomial of T. By the Cayley-Hamilton theorem $T^n - \mathrm{tr}(T)T^{n-1} + \ldots + \det(T) = 0$. Taking traces and using our hypothesis we get $\det(T) = 0$. Thus one of the $\alpha_i = 0$. □

Lemma 3.7.6. *Let* $X \in \mathfrak{gl}(V)$ *and assume* $X = \sum_i [X_i, Y_i]$ *where* $[X, X_i] = 0$ *for all* i. *Then* X *is nilpotent.*

Proof. We first show that $X^j = \sum_i [X_i, X^{j-1} Y_i]$ for $j \geq 1$. Now $X^j = \sum_i X^{j-1}(X_i Y_i - Y_i X_i) = \sum_i X^{j-1} X_i Y_i - X^{j-1} Y_i X_i$. Since X commutes with all the X_i so does any power X^j. Hence this last term equals $\sum_i X_i X^{j-1} Y_i - X^{j-1} Y_i X_i = \sum_i [X_i, X^{j-1} Y_i]$. Since tr is linear and takes the value 0 on a commutator it follows that $\mathrm{tr}(X^j) = 0$ for all j. By Lemma 3.7.5, X is nilpotent. □

We now turn to the proof of the J-M theorem.

Proof. Let $\mathfrak{g}$ be a completely reducible Lie subalgebra of $\mathfrak{gl}(V)$ and $F \neq 0$ be a nilpotent element in it. Let $V = \bigoplus V_i$ be the decomposition into Jordan blocks relative to F, so in each V_i we have a basis, $\{v_0, \ldots v_{r_i}\}$ such that $Fv_i = v_{i+1}$, when $i < r_i$ and $Fv_{r_i} = 0$. We define H and E to be the linear transformations on V which leave each V_i-invariant and on V_i we define for $i = 0, \ldots r_i$, $Hv_i = (r_i - 2i)v_i$, $E(v_0) = 0$ and for $i > 0$, $E(v_i) = (-ir_i + i(i-1))v_{i+1}$. Then, $[E, H] = 2E$, $[F, H] = -2F$, $[E, F] = H$ and $\{E, H, F\}$ are linearly independent. Therefore they form a subalgebra of $\mathfrak{g}$ isomorphic to $\mathfrak{sl}(2)$ containing F.

Next let $X_0 \in \mathfrak{g}$ and $X_0 = N + S$ be its Jordan decomposition in $\mathfrak{gl}(V)$. We denote by ad the adjoint representation of $\mathfrak{gl}(V)$ on itself. Then $\mathrm{ad}\, X_0 = \mathrm{ad}\, N + \mathrm{ad}\, S$ and moreover $\mathrm{ad}\, N$ is nilpotent and $\mathrm{ad}\, S$ is semisimple (see Lemma 3.3.9) and they commute since N and S commute. Thus by uniqueness of the additive Jordan decomposition $\mathrm{ad}\, X_0 = \mathrm{ad}\, N + \mathrm{ad}\, S$ is the Jordan decomposition of $\mathrm{ad}\, X_0$. This means they are polynomials in $\mathrm{ad}\, X_0$ without constant term. Since $\mathrm{ad}\, X_0$ leaves $\mathfrak{g}$ stable, the same is true of $\mathrm{ad}\, N$ and $\mathrm{ad}\, S$. Hence the maps $\mathfrak{g} \to \mathfrak{g}$ given by $X \mapsto [X, N]$ and $X \mapsto [X, S]$, are both derivations of $\mathfrak{g}$. Since $\mathfrak{g}$ acts completely reducibly on V we know by Proposition 3.6.3 that $\mathfrak{g} = \mathfrak{s} \oplus \mathfrak{z}$, where $\mathfrak{s}$ is a Levi factor and $\mathfrak{z}$ is the center. But the derivations of a semisimple Lie algebra are all inner by Corollary 3.3.23, and hence any derivation of $\mathfrak{g}$ which maps $\mathfrak{z}$ to 0 is also an inner derivation determined by an element of $\mathfrak{s}$. Because $[X_0, Z] = 0$ for all $Z \in \mathfrak{z}$ and N is a polynomial without constant term in X_0 it follows that $[N, Z] = 0$ for all $Z \in \mathfrak{z}$, which means the derivation $X \mapsto [X, N]$ maps $\mathfrak{z}$ to zero. Thus the derivation $X \mapsto [X, N]$ of $\mathfrak{g}$ is inner and determined by an element, say N_1 in $\mathfrak{s}$. Thus $[X, N] = [X, N_1]$ for all $X \in \mathfrak{g}$, or alternatively, $\operatorname{ad} N$ coincides with $\operatorname{ad} N_1$ as operators on $\mathfrak{g}$. But $\operatorname{ad} N$ is nilpotent as an operator on $\mathfrak{gl}(V)$ and therefore is also nilpotent on $\mathfrak{g}$. Hence so is $\operatorname{ad} N_1$. Therefore the result proved just above applied to $\operatorname{ad} \mathfrak{s}$ shows that there is some $X_0 \in \mathfrak{s}$ so that $[\operatorname{ad} N_1, \operatorname{ad} X_0] = 2 \operatorname{ad} N_1$ (on $\mathfrak{s}$). But $\mathfrak{s}$ is semisimple and therefore centerless this means $[N_1, X] = 2N_1$. By Lemma 3.7.6, N_1 is a nilpotent operator on V. Since $[X_0, N] = 0$ we know $[X_0, N_1] = 0$ and hence $[N_1, N] = 0$. Because both N and N_1 are nilpotent operators $N - N_1$ is also nilpotent (see Lemma 3.2.2). But we also showed $[X, N] = [X, N_1]$ for all $X \in \mathfrak{g}$. Therefore $[X, N - N_1] = 0$ for all $X \in \mathfrak{g}$. Now consider the associative algebra $\mathfrak{g}^*$ generated by $\mathfrak{g}$ in $\mathfrak{gl}(V)$. It contains N as a polynomial in X_0. It also contains $N_1 \in \mathfrak{s}$, so $N - N_1 \in \mathfrak{g}^*$. On the other hand $\mathfrak{g}^* \supseteq \mathfrak{g}$ so it also acts completely reducibly. It follows that $N - N_1$ is diagonalizable. Since it is also nilpotent it must be zero and so $N = N_1$. Therefore $N \in \mathfrak{g}$. Since $S = X_0 - N$ it is also in $\mathfrak{g}$.

Conversely, suppose $\mathfrak{g} \subseteq \mathfrak{gl}(V)$ satisfies the J-M condition and $\mathfrak{z}$ contains the nilpotent and semisimple parts of each of its elements. We will show $\mathfrak{g}$ acts completely reducibly on V. Let $\mathfrak{r}$ be the radical of $\mathfrak{g}$. If $F \in [\mathfrak{g}, \mathfrak{r}]$, then by Theorem 3.5.6, F is nilpotent. If F were nonzero, it could be imbedded in a 3-dimensional simple subalgebra $\mathfrak{s}$ of $\mathfrak{g}$. Hence $\mathfrak{s} \cap \mathfrak{r} \neq \{0\}$. But $\mathfrak{s}$ is simple so $\mathfrak{s} \cap \mathfrak{r} = \{0\}$. Thus $F = 0$ and therefore $[\mathfrak{g}, \mathfrak{r}] = 0$. This means $\mathfrak{r} = \mathfrak{z}$, and therefore by Proposition 3.6.3 $\mathfrak{g}$ is reductive, and acts completely reducibly on V. $\square$

3.8   Low Dimensional Lie Algebras over $\mathbb{R}$ and $\mathbb{C}$

In this section we indicate a classification of Lie algebras over $\mathbb{R}$ and $\mathbb{C}$, up to dimension 3. Having done dimensions 1 and 2 already we now deal with dimension 3. The classification is done by dimension of the derived subalgebra. We consider four subcases corresponding to dimension of $[\mathfrak{g}, \mathfrak{g}]$ equal to $0, 1, 2$ and 3.

(a) $\dim_k[\mathfrak{g}, \mathfrak{g}] = 0$. Here $\mathfrak{g}$ is abelian so in the complex case $\mathfrak{g} = \mathbb{C}^3$ and in the real case $\mathfrak{g} = \mathbb{R}^3$.

(b) $\dim_k[\mathfrak{g}, \mathfrak{g}] = 1$. In the complex case $\mathfrak{g} = \mathfrak{h}(\mathbb{C}) \oplus \mathbb{C}$, where $\mathfrak{h}(\mathbb{C})$ is the $ax + b$-Lie algebra over $\mathbb{C}$, or $\mathfrak{g} = \mathfrak{n}_1(\mathbb{C})$, the complex Heisenberg (see Example 3.1.20). In the real case, $\mathfrak{g} = \mathfrak{h}(\mathbb{R}) \oplus \mathbb{R}$, where $\mathfrak{h}(\mathbb{R})$ is the real $ax + b$-Lie algebra, or $\mathfrak{g} = \mathfrak{n}_1(\mathbb{R})$ the real Heisenberg.

Proof. There are two possibilities depending on whether $[\mathfrak{g}, \mathfrak{g}] \subseteq \mathfrak{z}(\mathfrak{g})$ or not. In the former case let $Z \neq 0 \in [\mathfrak{g}, \mathfrak{g}] \cap \mathfrak{z}(\mathfrak{g})$ and extend this to a basis X, Y, Z of $\mathfrak{g}$. Then $[X, Y] = \lambda Z$. Now $\lambda \neq 0$ for if it were otherwise $[X, Y] = 0$ and since everything commutes with Z, $\mathfrak{g}$ would be abelian, whereas here we are assuming $[\mathfrak{g}, \mathfrak{g}]$ has exactly dimension 1. Since $\lambda \neq 0$ we can absorb it into X or Y and then $[X, Y] = Z$ and all other brackets are zero. This is the real or complex Heisenberg Lie algebra.

On the other hand suppose $[\mathfrak{g}, \mathfrak{g}]$ has dimension 1, but is not contained in the center. Let $X \neq 0 \in [\mathfrak{g}, \mathfrak{g}]$, X not in $\mathfrak{z}(\mathfrak{g})$. Then there is some $Y \in \mathfrak{g}$ with $[X, Y] \neq 0$ and so $[X, Y] = \lambda X$. Since $\lambda \neq 0$ we can absorb it into y and get $[X, Y] = X$, so X, Y generate the $ax + b$-Lie algebra, $\mathfrak{h}$. Since $\mathfrak{h} \supset [\mathfrak{g}, \mathfrak{g}]$ it is an ideal (see Proposition 3.1.13). By Propositions 3.1.46 and 3.1.47 $\mathfrak{h}$ is a direct summand $\mathfrak{g} = \mathfrak{h} \oplus \mathfrak{a}$, where $\mathfrak{a}$ is abelian since it has dimension 1. $\qquad\square$

In both cases there are only finitely many non-isomorphic Lie algebras and they are all solvable (both real and complex).

(c) $\dim_k[\mathfrak{g},\mathfrak{g}] = 2$ we shall see that we get two continuous families for the complex field and three continuous families over the real field.

Proof. Suppose $\dim_k[\mathfrak{g},\mathfrak{g}] = 2$. Then $[\mathfrak{g},\mathfrak{g}]$ cannot be the $ax + b$-algebra since it would then be a direct summand by Corollary 3.3.22. But if $\mathfrak{g} = \mathfrak{h} \oplus \mathfrak{a}$ where $\mathfrak{h} = [\mathfrak{g},\mathfrak{g}]$, then since $\mathfrak{a}$ is abelian we have $\mathfrak{h} = [\mathfrak{g},\mathfrak{g}] = [\mathfrak{h},\mathfrak{h}]$ and this is a contradiction. Thus $[\mathfrak{g},\mathfrak{g}]$ is abelian. Choose a basis X, Y for $[\mathfrak{g},\mathfrak{g}]$ and extend this to a basis X, Y, U of $\mathfrak{g}$. Then $[U, X] = aX + bY$, $[U, Y] = cX + dY$ and $[X, Y] = 0$, where $a, b, c, d \in k$. This gives a matrix

$$A = \begin{pmatrix} a & c \\ b & d \end{pmatrix}$$

which determines the Lie algebra. Since $[X, Y] = 0$, $[\mathfrak{g},\mathfrak{g}]$ is generated by $[U, X] = \operatorname{ad} U(X)$ and $[U, Y] = \operatorname{ad} U(Y)$. Hence $\operatorname{ad} u|_{[\mathfrak{g},\mathfrak{g}]}$ is one-to-one. This means that A is nonsingular and we may take any nonsingular A because the skew symmetry and Jacobi identity are automatically satisfied with these structure constants. Thus we get many different Lie algebras in this way. Now the question is exactly which of these are non-isomorphic? Evidently, we can change the basis X, Y of $[\mathfrak{g},\mathfrak{g}]$ and also change the U. The first results in changing A to a conjugate PAP^{-1} by $P \in \operatorname{GL}(2,k)$. We can also change the U to $\lambda U + W$, where $w \in [\mathfrak{g},\mathfrak{g}]$. But then $[\lambda U + W, X] = \lambda[U, X] + [W, X] = \lambda[U, X]$, since $[W, X] = 0$. Similarly, $[\lambda U + W, Y] = \lambda[U, Y]$. Thus the effect of this is to change A to λA. Thus the invariants are those of $A \mapsto \lambda PAP^{-1}$. If $k = \mathbb{C}$ we can choose

$$A = \begin{pmatrix} 1 & 0 \\ 0 & \alpha \end{pmatrix},$$

$\alpha \neq 0 \in \mathbb{C}$, or

$$A = \begin{pmatrix} 1 & \beta \\ 0 & 1 \end{pmatrix},$$

$\beta \neq 0 \in C$, denoting these Lie algebras $\mathfrak{g}_{3,\alpha}$ and $\mathfrak{g}_{\beta,3}$.

When $k = \mathbb{R}$ we get

$$A = \begin{pmatrix} 1 & 0 \\ 0 & \alpha \end{pmatrix},$$

$\alpha \neq 0 \in \mathbb{R}$, or

$$A = \begin{pmatrix} \alpha & \beta \\ -\beta & \alpha \end{pmatrix}$$

where both α and β are real and $\beta \neq 0$,
or

$$A = \begin{pmatrix} 1 & \beta \\ 0 & 1 \end{pmatrix}$$

$\beta \neq 0 \in \mathbb{R}$. These Lie algebras are denoted by $\mathfrak{g}_{3,\alpha}^{\mathbb{C}}$, $\mathfrak{g}_{\alpha,\beta}^{\mathbb{C}}$ and $\mathfrak{g}_{\beta,3}^{\mathbb{C}}$. $\quad\square$

(d) $\dim_k[\mathfrak{g}, \mathfrak{g}] = 3$. Here $\mathfrak{g}$ is simple and $\mathfrak{g} = \mathfrak{sl}(2, \mathbb{C})$ in the complex case, while in the real case $\mathfrak{g} = \mathfrak{sl}(2, \mathbb{R})$ or $\mathfrak{g} = \mathfrak{so}(3, \mathbb{R})$.

The proof of this requires two lemmas.

Lemma 3.8.1. *If $\mathfrak{g}$ is a 3-dimensional Lie algebra over any field k and $\mathfrak{g} = [\mathfrak{g}, \mathfrak{g}]$, then $\mathfrak{g}$ is simple.*

Proof. Let $\mathfrak{a}$ be a nontrivial abelian ideal in $\mathfrak{g}$ then $\dim \mathfrak{a} = 1, 2$ or 3. In the latter case $\mathfrak{g}$ is abelian and so solvable. If the dimension of $\mathfrak{a}$ is 1 or 2, then $\dim \mathfrak{g}/\mathfrak{a}$ is 2 or 1. Hence $\mathfrak{g}/\mathfrak{a}$ is solvable (see Example 3.1.38), therefore $\mathfrak{g}$ is also solvable by Proposition 3.1.33. Thus when $\mathfrak{a}$ is nontrivial $\mathfrak{g}$ is solvable so $[\mathfrak{g}, \mathfrak{g}] \neq \mathfrak{g}$, a contradiction. This shows $\mathfrak{g}$ is semisimple. Now the same argument as in the proof of Proposition 3.1.55 shows that $\mathfrak{g}$ is simple. $\quad\square$

Lemma 3.8.2. *Let $\mathfrak{g}$ be a simple Lie algebra of dimension 3. If $k = \mathbb{C}$ then $\mathfrak{g} = \mathfrak{sl}(2, \mathbb{C})$ and if $k = \mathbb{R}$ then $\mathfrak{g} = \mathfrak{sl}(2, \mathbb{R})$ or $\mathfrak{so}(3, \mathbb{R})$.*

Proof. We first consider the complex case. Let X, Y, Z be a basis for $\mathfrak{g}$. It is easy to see that we can choose this basis such that $[X, Y] = Z$. Then $[Z, Y] = aX + bY + cZ$ for some complex numbers a, b, c. Note that $a \neq 0$ otherwise the vector space generated by Y and Z would be a nontrivial ideal of $\mathfrak{g}$ contradicting the fact that $\mathfrak{g}$ is simple.

Let $d = \sqrt{a}$ then $[Y/d, Z/d] = X + a/d.Y/d + c/d.Z/d$, therefore the new basis $X' = X + a/d.Y/d$, $Y' = Y/d$ and $X' = Z/d$ satisfies the relations

$$[X', Y'] = Z'$$
$$[Y', Z'] = X' + tZ' \qquad (3.11)$$

where $t \in \mathbb{C}$. If $t = 0$ then it immediately follows from the Jacobi identity that $\mathfrak{g}$ has a basis X'', Y'', Z'' in which the Lie bracket is defined by

$$[X'', Y''] = Z''$$
$$[Y'', Z''] = X''$$
$$[Z'', X''] = Y'',$$

therefore by Exercise 3.1.4 $\mathfrak{g}$ is $\mathfrak{sl}(2, \mathbb{C})$. If $t \neq 0$ then it follows from (3.11),

$$-[[Z', X'], Y'] = t[Z', X'],$$

therefore by Lemma 3.7.6 $[Z', X']$ is nilpotent. Note that $[Z'', X''] \neq 0$ as $[\mathfrak{g}, \mathfrak{g}] = \mathfrak{g}$. Now that there is a nilpotent element in $\mathfrak{g}$, by Jacobson-Morozov there must be a $\mathfrak{sl}(2)$ triple within $\mathfrak{g}$. By dimension we see that $\mathfrak{g}$ is $\mathfrak{sl}(2, \mathbb{C})$. This takes care of the complex case.

Turning to the real case, consider the Killing form β of $\mathfrak{g}$. β is nondegenerate. If it is indefinite then its signature is either $(2, 1)$ or $(1, 2)$. These are really the same because one is the negative of the other, so we shall consider only the former. Now in any case since $\mathfrak{g}$ is semisimple it is a linear Lie algebra because the adjoint representation is faithful; also by semisimplicity $\operatorname{tr}(\mathfrak{g}) = 0$. If β is of type $(2, 1)$ then $\mathfrak{g}$ (or rather its adjoint algebra) preserves the Killing form, so we have $\mathfrak{g} \subseteq \mathfrak{o}(2, 1)$, and then because the trace is identically zero $\mathfrak{g} \subseteq \mathfrak{so}(2, 1)$. As both these Lie algebras are of dimension 3 they are equal: $\mathfrak{g} = \mathfrak{so}(2, 1)$ (see Exercise 3.1.5). The proof of this part can be completed by observing that $\mathfrak{so}(2, 1) = \mathfrak{sl}(2, \mathbb{R})$. On the other hand, if β is negative definite then $-\beta$ is positive definite and $\mathfrak{g}$ preserves these forms similarly, we have $\mathfrak{g} \subseteq \mathfrak{so}(3, \mathbb{R})$ and by dimension counting $\mathfrak{g} = \mathfrak{so}(3, \mathbb{R})$.

$\square$

	$\dim[\mathfrak{g}, \mathfrak{g}]$	$\mathbb{C}$	$\mathbb{R}$
a	0	$\mathbb{C}^3$	$\mathbb{R}^3$
b	1	$\mathfrak{h}(\mathbb{C}) \oplus \mathbb{C}, \mathfrak{n}_1(\mathbb{C})$	$\mathfrak{h}(\mathbb{R}) \oplus \mathbb{R}, \mathfrak{n}_1(\mathbb{R})$
c	2	$\mathfrak{g}_{3,\alpha}, \mathfrak{g}_{\beta,3}$	$\mathfrak{g}_{3,\alpha}^{\mathbb{C}}, \mathfrak{g}_{\alpha,\beta}^{\mathbb{C}}, \mathfrak{g}_{\beta,3}^{\mathbb{C}}$
d	3	$\mathfrak{sl}(2, \mathbb{C})$	$\mathfrak{sl}(2, \mathbb{R}), \mathfrak{so}(3, \mathbb{R})$

Table 3.1: The 3-dimensional Lie algebras over $\mathbb{R}$ and $\mathbb{C}$

Corollary 3.8.3. *A 3-dimensional simple Lie algebra over $\mathbb{R}$ or $\mathbb{C}$ must have rank one (Theorem 6.7.1 for the definition of rank).*

3.9   Real Lie Algebras of Compact Type

In this section we will find out which real Lie algebras are the Lie algebras of compact real Lie groups.

Proposition 3.9.1. *Let $\mathfrak{g}$ be a real Lie algebra and ρ any faithful (real) representation of $\mathfrak{g}$ on V such that the matrices $\rho(X)$ are skew symmetric with respect to some positive definite symmetric form $\langle \cdot, \cdot \rangle$ on V. Then the trace form, β_ρ, is negative definite on $\mathfrak{g}$.*

Proof. Relative to some orthonormal basis $\{v_1, \dots, v_n\}$ of V we have $\rho_{ij}(X) = -\rho_{ji}(X)$ where $(\rho_{ij}(X))$ is the matrix of $\rho(X)$ with respect to $\{v_1, \dots, v_n\}$. Hence for each i, $\rho(X)(v_i) = \sum_j \rho_{ji}(X)v_j$ and so

$$\rho(X)^2(v_i) = \sum_j \rho_{ji}(X)\rho(X)(v_j) = \sum_{j,k} \rho_{ji}(X)\rho_{kj}(X)v_k.$$

Thus

$$\mathrm{tr}(\rho(X)^2) = \sum_{i,j} \rho_{ij}(X)\rho_{ji}(X) = -\sum_{i,j} \rho_{ij}(X)^2 \leq 0.$$

Since ρ is faithful, β_ρ is negative definite.

$\square$

We recall the definition of an invariant form β on a Lie algebra $\mathfrak{g}$: for all X, Y and $Z \in \mathfrak{g}$,

$$\beta(\operatorname{ad} X(Y), Z) + \beta(X, \operatorname{ad} Y(Z)) = 0.$$

Definition 3.9.2. A real Lie algebra $\mathfrak{g}$ is said to be of *compact type* if it has a positive definite invariant form, β.

An obvious consequence of the definition is that subalgebras of Lie algebra of compact type are themselves of compact type.

Some examples of Lie algebras of compact type:
$\mathfrak{so}(n, \mathbb{R}) = \{X \in \mathfrak{gl}(V) : X^t = -X\}$ is of compact type since β_ρ, where ρ is the inclusion in $\mathfrak{gl}(V)$, is positive definite by the previous result.

Every abelian Lie algebra $\mathfrak{g}$ is of compact type since any form is automatically invariant. So here we may simply choose any positive definite form on $\mathfrak{g}$.

As a final example, let G be a compact connected Lie group and $\mathfrak{g}$ be its Lie algebra. Then $\mathfrak{g}$ is of compact type. To see this observe that by the proof of Theorem 2.5.1 there is an $\operatorname{Ad} G$-invariant inner product on $\mathfrak{g}$. Relative to this inner product the operators of $\operatorname{Ad} G$ are all orthogonal and hence those of $\operatorname{ad} \mathfrak{g}$ are skew symmetric. By Proposition 3.9.1, the negative of the trace form is bilinear, symmetric, invariant and positive definite.

Remark 3.9.3. Let $\mathfrak{g}$ be a semisimple Lie algebra over $\mathbb{R}$. Then the Killing form, β, is nondegenerate and invariant. Hence for each $Y \in \mathfrak{g}$, $\operatorname{ad} Y$ is skew symmetric with respect to β. This, however, does not mean that β is definite. The result above requires a positive definite form on $V \; (= \mathfrak{g})$ and in general (for non-compact semisimple Lie algebra) the Killing form is of mixed type. For example when $\mathfrak{g} = \mathfrak{sl}(2, \mathbb{R})$ it is a $(1, 2)$ form.

We now come to the following result which characterizes real Lie algebras of compact type.

Theorem 3.9.4. *If $\mathfrak{g}$ is of compact type then $\mathfrak{g} = \mathfrak{z}(\mathfrak{g}) \oplus [\mathfrak{g}, \mathfrak{g}]$ where $[\mathfrak{g}, \mathfrak{g}]$ is semisimple (and of compact type as remarked above). If $\mathfrak{g}$ is a semisimple Lie algebra of compact type then its Killing form is negative definite. Conversely, if $\mathfrak{g} = \mathfrak{z}(\mathfrak{g}) \oplus [\mathfrak{g}, \mathfrak{g}]$, where $[\mathfrak{g}, \mathfrak{g}]$ is semisimple and of compact type then $\mathfrak{g}$ is of compact type.*

Proof. Since $\mathfrak{g}$ has a positive definite invariant form β, the invariance tells us that for each $X \in \mathfrak{g}$, ad X is skew symmetric with respect to β. By the argument given for semisimple algebras, the orthocomplement of any ideal $\mathfrak{h}$ in $\mathfrak{g}$ is also an ideal and $\mathfrak{g}$ is the direct sum of these two ideals. In particular since $\mathfrak{z}(\mathfrak{g})$ is an ideal $\mathfrak{g} = \mathfrak{z}(\mathfrak{g}) \oplus \mathfrak{l}$. If $\mathfrak{a}$ is an abelian ideal in $\mathfrak{l}$, then since $\mathfrak{l}$ is of compact type, $\mathfrak{a}$ is a direct summand; $\mathfrak{l} = \mathfrak{a} \oplus \mathfrak{b}$. Therefore $\mathfrak{a}$ commutes with $\mathfrak{b}$. Since $\mathfrak{a}$ is itself abelian $\mathfrak{a}$ commutes with all of $\mathfrak{l}$. On the other hand $\mathfrak{a} \subseteq \mathfrak{l}$ so $\mathfrak{a}$ commutes with $\mathfrak{z}(\mathfrak{g})$. But then $\mathfrak{a}$ commutes with all of $\mathfrak{g}$. Therefore $\mathfrak{a} \subseteq \mathfrak{z}(\mathfrak{g})$. But $\mathfrak{a}$ is a subset of $\mathfrak{l}$ so $\mathfrak{a} = 0$ and $\mathfrak{l}$ is semisimple. Now from $\mathfrak{g} = \mathfrak{z}(\mathfrak{g}) \oplus \mathfrak{l}$ it follows directly that $[\mathfrak{g}, \mathfrak{g}] = [\mathfrak{l}, \mathfrak{l}]$ and since $\mathfrak{l}$ is semisimple, $[\mathfrak{l}, \mathfrak{l}] = \mathfrak{l}$. Because $\mathfrak{l}$ is semisimple, the adjoint representation is faithful and the Killing form is negative definite by Proposition 3.9.1.

Before turning to the converse we need the following lemma which shows that the direct sum of Lie algebras of compact type is again of compact type.

Lemma 3.9.5. *Suppose $\mathfrak{g} = \mathfrak{u} \oplus \mathfrak{v}$ is a direct sum of ideals each having a positive definite invariant forms $\langle \cdot, \cdot \rangle_\mathfrak{u}$ and $\langle \cdot, \cdot \rangle_\mathfrak{u}$. Then $\mathfrak{g}$ has a positive definite invariant form.*

Proof. Putting these two forms together we get a positive definite form on $\mathfrak{g}$ with the summands orthogonal.

$$\langle (U, V), (U', V') \rangle_\mathfrak{g} = \langle U, U' \rangle_\mathfrak{u} + \langle V, V' \rangle_\mathfrak{v}.$$

Now

$$\begin{aligned}
\langle \mathrm{ad}\, \mathfrak{g}(U, V)(U_1, V_1), (U_2, V_2) \rangle_\mathfrak{g} &= \langle (\mathrm{ad}\, U(U_1), \mathrm{ad}\, V(V_1)), (U_2, V_2) \rangle_\mathfrak{g} \\
&= \langle \mathrm{ad}\, U(U_1), U_2 \rangle_\mathfrak{u} + \langle \mathrm{ad}\, V(V_1), V_2 \rangle_\mathfrak{v} \\
&= -\langle U_1, \mathrm{ad}\, U(U_2) \rangle_\mathfrak{u} - \langle V_1, \mathrm{ad}(V)V_2 \rangle_\mathfrak{v}.
\end{aligned}$$

Since the forms on $\mathfrak{u}$ and $\mathfrak{v}$ are invariant, the form on $\mathfrak{g}$ is also invariant.

□

Now for the converse, if $\mathfrak{g} = \mathfrak{z}(\mathfrak{g}) \oplus [\mathfrak{g}, \mathfrak{g}]$ where $[\mathfrak{g}, \mathfrak{g}]$ is semisimple and of compact type, then the negative of the Killing form of $[\mathfrak{g}, \mathfrak{g}]$ is positive definite (and invariant) and hence $[\mathfrak{g}, \mathfrak{g}]$ is of compact type. As above the abelian Lie algebra $\mathfrak{z}(\mathfrak{g})$ is also of compact type and therefore so is $\mathfrak{g}$.

□

Corollary 3.9.6. *In a Lie algebra of compact type the Killing form is negative semidefinite. It is negative definite if and only if $\mathfrak{g}$ is semisimple.*

Corollary 3.9.7. *A Lie algebra $\mathfrak{g}$ is of compact type if and only if it is the Lie algebra of a compact connected Lie group.*

Proof. We know the Lie algebra of a compact connected Lie group is of compact type. Conversely let $\mathfrak{g}$ be a Lie algebra of compact type. Then $\mathfrak{g} = \mathfrak{z}(\mathfrak{g}) \oplus [\mathfrak{g}, \mathfrak{g}]$, where $[\mathfrak{g}, \mathfrak{g}]$ is semisimple and of compact type. Therefore $\mathrm{ad}\,\mathfrak{g}$ is a semisimple Lie algebra of compact type. Let $\tilde{G}$ be the simply connected real Lie group whose Lie algebra is $\mathfrak{g}$. Then $\tilde{G}$ is a direct product $\mathbb{R}^n \times H$, where H is a subgroup of $\tilde{G}$ whose Lie algebra is $[\mathfrak{g}, \mathfrak{g}]$. Clearly, $\mathbb{T}^n \times H$ is also a Lie group whose Lie algebra is $\mathfrak{g}$ which will be compact if and only if H is. Since the Lie algebra of H is compact and semisimple we may assume these properties for $\mathfrak{g}$. Thus we may assume $\mathfrak{g}$ is semisimple and of compact type. In particular, $\mathfrak{g}$ is isomorphic to $\mathrm{ad}\,\mathfrak{g}$. Since $\mathrm{Ad}\,G$ is a Lie group whose Lie algebra is $\mathrm{ad}\,\mathfrak{g}$ we see that its Lie algebra is $\mathfrak{g}$. We need only show that $\mathrm{Ad}\,G$ is compact. However since $\mathfrak{g}$ is semisimple $\mathrm{ad}\,\mathfrak{g} = \mathrm{Der}(\mathfrak{g})$. It follows that $\mathrm{Ad}\,G = \mathrm{Aut}_0(\mathfrak{g})$. Since the latter is an algebraic group $\mathrm{Ad}\,G$ it is a closed subgroup of $\mathrm{GL}(\mathfrak{g})$. On the other hand, since $\mathfrak{g}$ is semisimple and of compact type the negative of its Killing form is positive definite. Therefore by invariance of this form $\mathrm{ad}\,\mathfrak{g}$ consists of skew symmetric operators. This means, at least near the identity, $\mathrm{Ad}\,G$ consists of orthogonal operators. Because $\mathrm{Ad}\,G$ is a group it is a subgroup of the orthogonal group on $\mathfrak{g}$. In particular, $\mathrm{Ad}\,G$ is bounded. Since it is also closed, it is compact. □

Corollary 3.9.8. *Any compact connected Lie group G is isomorphic to*

$$(Z(G)_0 \times [G, G])/F,$$

the quotient group of a direct product of a torus, $Z(G)_0$, and a compact semisimple group by a finite central subgroup F.

Proof. If G is compact then, as we showed, its Lie algebra, $\mathfrak{g}$, is of compact type. By the argument just above $\tilde{G}$ is a direct product $\mathbb{R}^n \times H$, where H is as above. But since this H is locally isomorphic with $\mathrm{Ad}\, G$ (has the same Lie algebra) and the latter is compact, so is H, by Theorem 2.5.8. The covering $\pi : \tilde{G} \to G$ maps onto a compact group so its kernel must contain a lattice of maximal rank in $\mathbb{R}^n$. Hence G is covered by a direct product $T^n \times H$. Then $H = [H, H]$ since the Lie algebra of H is $[\mathfrak{g}, \mathfrak{g}]$, (see Theorem 3.9.4). Moreover this product group is compact since both factors are. We have shown G is covered by $\mathbb{T}^n \times [H, H]$. But the image of $[H, H]$ is clearly $[G, G]$. Therefore $G = (\mathbb{T}^n \times [G, G])/F$ where F is a discrete central subgroup, which must be finite since $\mathbb{T}^n \times [G, G]$ is compact. Since we have a direct product upstairs, G is the commuting product of $\mathbb{T}^n$ and $[G, G]$ downstairs. This implies its center $Z(G) = \mathbb{T}^n \cdot Z[G, G]$. As $[G, G]$ is compact and semisimple its center is finite so $\mathbb{T}^n = Z(G)_0$. $\qquad\square$

Chapter 4

The Structure of Compact Connected Lie Groups

4.1 Introduction

In this chapter we deal with the important role a maximal torus plays in the structure and representation theory of a compact connected Lie group.

As we know, if H is a connected abelian Lie group then it is isomorphic as a Lie group to $\mathbb{R}^m \times \mathbb{T}^n$. In particular, a compact connected abelian Lie group is isomorphic with $\mathbb{T}^n$. A *maximal connected abelian subgroup* H of a connected Lie group G means one which is contained in no strictly larger such subgroup. These clearly exist in any connected Lie group G for dimension reasons. Similarly, *maximal tori* exist in any connected Lie group. If H is a maximal connected abelian subgroup of a connected Lie group G, then H must be closed, for its closure is a possibly larger connected abelian Lie subgroup of G. If G is also compact then H, being closed, is also compact and hence is a torus. So in the compact case maximal abelian connected is the same as a maximal torus. Similarly, a maximal abelian subalgebra in a Lie algebra is an abelian subalgebra which is not properly contained in a larger abelian subalgebra. These also exist for dimension reasons.

For example suppose $G = U(n, \mathbb{C})$, the unitary group. This compact, connected Lie group is a good place to start. Consider the subgroup D of diagonal matrices which is clearly isomorphic to $\mathbb{T}^n$. Since it is compact D must be closed. Now D is actually a maximal torus. For if there were a strictly larger one, then there would be some element g which commutes with each point of D. But D contains elements with n distinct eigenvalues. Since g commutes with such an element, g is itself diagonal, a contradiction. At the same time this shows D is its own centralizer $Z_G(D) = D$. Another observation is that every point of G is conjugate to something in D. That is to say, $G = \bigcup_{g \in G} gDg^{-1}$. This follows by the finite dimensional spectral theorem, which says that any unitary operator is similar under a unitary operator to a diagonal unitary operator. We shall see that this holds for any compact connected Lie group G so that $G = \bigcup_{g \in G} gTg^{-1}$, where T is any maximal torus in G. However, this is a much deeper fact than the finite dimensional spectral theorem because of the profusion of compact connected Lie groups. As we shall see in the next chapter, such groups are linear, but they may be much smaller than the ambient unitary group. Hence the nature of the maximal torus may also be different because of the relations defining G. Here the example to keep in mind is the most important compact, connected, nonabelian, simply connected Lie group $G = \mathrm{SU}(2)$. As we have seen above, here the maximal torus has dimension 1.

4.2   Maximal Tori in Compact Lie Groups

The point of looking at maximal tori in a compact Lie group is to reduce a more complicated non-abelian situation to an abelian one. We first deal with some issues of abelian groups.

Abelian Lie groups

If an abelian topological group A has an element a_0 which generates a dense cyclic subgroup, we call a_0 a *quasi-generator* of A. By the Kronecker approximation theorem (Appendix B), a torus $\mathbb{T}^n$ has a quasi-generator.

Definition 4.2.1. Let G be a group and let k be a fixed positive integer. Consider the map $\pi_k : g \mapsto g^k$. We say G is *divisible* if π_k is surjective for all $k \in \mathbb{Z}^+$.

This notion is particularly important when G is abelian. For example, if $\mathbb{T}^n$ is a torus and $\pi : \mathbb{R}^n \to \mathbb{T}^n$ is its universal covering, then since $\mathbb{R}$ is a field it, and therefore also $\mathbb{R}^n$ is divisible. Hence by using the homomorphism π so is $\mathbb{T}^n$. As to the significance of divisibility in abelian groups we mention the following well-known fact (see [36]).

Proposition 4.2.2. *Let H be a subgroup of the abelian group G and $f : H \to D$ be a homomorphism where D is a divisible group. Then f extends to a homomorphism $\hat{f} : G \to D$.*

Corollary 4.2.3. *A divisible subgroup H of an abelian group G is a direct summand of G.*

Proof. Take the identity map $i : H \to H$ and extend it to a homomorphism, $\hat{i} : G \to H$. Then as is easily seen $G = H \oplus \operatorname{Ker} \hat{i}$.   □

Corollary 4.2.4. *A compact abelian Lie group G is the direct product of a torus and a finite group.*

Proof. Clearly the identity component G_0 being a compact connected abelian Lie group is a torus, $\mathbb{T}^n$ (which is normal). Because G is a Lie group, $\mathbb{T}^n$ is open in G. Hence $G/\mathbb{T}^n$ is discrete and compact and so finite. Now as remarked earlier $\mathbb{T}^n$ is divisible. Therefore the identity homomorphism $i : \mathbb{T}^n \to \mathbb{T}^n$ extends to a homomorphism $\hat{i} : G \to \mathbb{T}^n$. If $\hat{i}$ were continuous the proof of the previous results would show $G = \mathbb{T}^n \oplus G/\mathbb{T}^n$ as topological groups. But a homomorphism is continuous if it is so at the identity. Since $\mathbb{T}^n$ is an open subgroup containing the identity element and $\hat{i}|_{\mathbb{T}^n} = i$ which is continuous, $\hat{i}$ is indeed continuous on G.   □

It now makes sense to ask when does a compact abelian Lie group G have a quasi-generator?

Corollary 4.2.5. *A compact abelian Lie group G has a quasi-generator if and only if $G = \mathbb{T}^n \oplus \mathbb{Z}_m$. That is if and only if $G/\mathbb{T}^n$ is cyclic.*

Proof. Since the projection is a continuous homomorphism if G has a quasi-generator then so does G/H for any subgroup H. Because $G/\mathbb{T}^n$ is discrete this quotient must be cyclic. More generally let $\mathbb{T}^n$ be a toral subgroup and suppose $G/\mathbb{T}^n$ is a finite cyclic group, then G will have a quasi-generator.

To see this choose a quasi-generator of $t_0 \in \mathbb{T}^n$ and a $g_0 \in G$ which projects onto the generator of $G/\mathbb{T}^n = \mathbb{Z}_m$. Then $mg_0 \in \mathbb{T}^n$. Because $\mathbb{T}^n$ is divisible there is a $t \in \mathbb{T}^n$ satisfying $mt = t_0 - mg_0$, and then $g = t + g_0$ is a quasi-generator of G because $mg = m(t + g_0) = t_0$. Thus the m-fold powers of g which lie in $\mathbb{T}^n$, are dense in $\mathbb{T}^n$. As to the other powers, $ng = nt + ng_0$. Write $n = qm + r$, where $0 \le r < m$. Then $ng = (qm + r)t + rg_0$. This gives a dense set in the coset $\mathbb{T}^n + rg_0$. Since we have a finite coset decomposition, $G = \bigcup_{r=0}^{m-1} \mathbb{T}^n + rg_0$, we see that g is a quasi-generator for G. □

Another important feature of tori is the fact that their automorphism groups are discrete. When the group is an abelian Lie group it is easy to see this by duality, because $\mathrm{Aut}(G)$ is isomorphic to $\mathrm{Aut}(\hat{G})$ where $\hat{G}$ is character group of G. Since $\hat{G}$ is a finitely generated discrete group its automorphism group is surely discrete. Actually in the case of a torus $\mathbb{T}^n$ we can calculate $\mathrm{Aut}(G)$ explicitly: $\mathrm{Aut}(\mathbb{Z}^n) = \mathrm{GL}(n, \mathbb{Z})$.

Exercise 4.2.6. Work out the details of the paragraph above. In particular, show that $\mathrm{Aut}(G)$ is discrete for a compact abelian Lie group G. Show that $\mathrm{Aut}(G)$ is not discrete when $G = \mathrm{SO}(3, \mathbb{R})$.

4.3   Maximal Tori in Compact Connected Lie Groups

We now turn to the following basic result. This argument is due to G. Hunt [35].

Theorem 4.3.1. *Let G be a compact connected Lie group and $\mathfrak{g}$ its Lie algebra. Then*

(1) G has a maximal torus, T.

(2) $\mathfrak{g}$ has a maximal abelian subalgebra, $\mathfrak{t}$.

(3) If $\mathfrak{h}$ is a maximal abelian subalgebra, then the connected subgroup H of G with Lie algebra $\mathfrak{h}$ is a maximal torus of G.

(4) Any two maximal abelian subalgebras $\mathfrak{h}_1$ and $\mathfrak{h}_2$ are conjugate; $\mathrm{Ad}\, g(\mathfrak{h}_1) = \mathfrak{h}_2$ for some $g \in G$.

(5) Similarly, any two maximal tori T_1 and T_2 of G are conjugate; $g T_1 g^{-1} = T_2$.

In particular, any two maximal tori of G have the same dimension, and any two maximal abelian subalgebras of $\mathfrak{g}$ have the same dimension. This number, called the *rank* of G, is an important invariant. As we shall see later, an analogue of the conjugacy theorem does hold in the case of connected *complex* semisimple (or reductive) Lie groups.

Proof. We know G has a torus and each torus is contained in a maximal one; and similarly for abelian subalgebras of $\mathfrak{g}$. The correspondence between Lie subgroups and subalgebras, Theorem 1.3.3, proves the first 3 items and shows that 4 and 5 imply each other. We shall prove 4.

Let $X_1 \in \mathfrak{h}_1$ and $X_2 \in \mathfrak{h}_2$ be fixed and $f(g) = \langle \mathrm{Ad}\, g(X_1), X_2 \rangle$, where by compactness $\langle \cdot, \cdot \rangle$ can be taken to be an Ad-invariant inner product on $\mathfrak{g}$ (see Chapter 2). Again by compactness this smooth function has a minimum value at say $g_0 \in G$. Let $X \in \mathfrak{g}$. Then $\frac{d}{dt}\big|_{t=0} \langle \mathrm{Ad}\, \exp tX g_0 X_1, X_2 \rangle = 0$. But $\langle \mathrm{Ad}\, \exp tX g_0 X_1, X_2 \rangle = \langle \mathrm{Ad}\, \exp tX \, \mathrm{Ad}\, g_0 X_1, X_2 \rangle = \langle \mathrm{Exp}(\mathrm{ad}\, tX)\, \mathrm{Ad}\, g_0 X_1, X_2 \rangle$ so taking the derivative at $t = 0$ tells us $\langle \mathrm{ad}\, X \, \mathrm{Ad}\, g_0 X_1, X_2 \rangle = 0$. Now Ad leaves $\langle \cdot, \cdot \rangle$ invariant, ad leaves it infinitesimally invariant (*i.e.* operates by skew symmetric matrices). Therefore $\langle X, [\mathrm{Ad}\, g_0 X_1, X_2] \rangle = 0$ and since X is arbitrary and $\langle \cdot, \cdot \rangle$ is positive definite, we get $[\mathrm{Ad}\, g_0 X_1, X_2] = 0$. But $X_i \in \mathfrak{h}_i$ are also arbitrary hence $\mathrm{Ad}\, g_0(\mathfrak{h}_1)$ and $\mathfrak{h}_2$ commute pointwise. Since $\mathfrak{h}_1$ is a maximal abelian subalgebra and $\mathrm{Ad}\, g_0$ is an automorphism of $\mathfrak{g}$, so is $\mathrm{Ad}\, g_0(\mathfrak{h}_1)$. By maximality $\mathrm{Ad}\, g_0(\mathfrak{h}_1) + \mathfrak{h}_2 = \mathfrak{h}_2$ from which it follows easily that $\mathrm{Ad}\, g_0(\mathfrak{h}_1) = \mathfrak{h}_2$. $\qquad\square$

Exercise 4.3.2. Consider the compact connected Lie group $\mathrm{SO}(3, \mathbb{R})$. It contains a subgroup A (isomorphic to the Klein 4 group) consisting of the identity together with diagonal elements with two -1 and one 1

in the various possible places. Show that A is maximal abelian. This does not contradict Theorem 4.3.1 above because A is not connected; rather it is discrete.

We will illustrate the importance of the rank of G by finding the (non-abelian) compact connected Lie groups of rank 1.

Theorem 4.3.3. *Let G be a non-abelian compact connected Lie group of rank 1. Then G is either* $\mathrm{SU}(2,\mathbb{C})$ *or* $\mathrm{SO}(3,\mathbb{R})$.

Proof. Since G is compact we know that $\mathfrak{g}$ is the direct sum of its center and its derived subalgebras (Theorem 3.9.4). But $[G,G]$ has positive rank and G has rank 1 so this means the center is trivial and G is semisimple. Next we prove $\dim G = 3$. To do so we may assume G is simply connected. This is because G is semisimple so its universal covering group is compact (by Theorem 2.5.8). As they have the same Lie algebra Theorem 4.3.1 tells us the ranks are the same.

As above, let $\langle \cdot, \cdot \rangle$ be an Ad-invariant inner product on $\mathfrak{g}$ and let $X_0 \in \mathfrak{t}$ be a unit vector in the Lie algebra of a maximal torus T. Define $f(g) = \mathrm{Ad}\, g(X_0)$. This is a smooth function $f : G \to \mathfrak{g}$ which is constant on cosets. The induced map $\bar{f} : G/T \to \mathfrak{g}$ is injective. This is because if $\mathrm{Ad}\, g(X_0) = \mathrm{Ad}\, h(X_0)$, then $\mathrm{Ad}\, h^{-1} g(X_0) = X_0$ and since the group has rank is 1, $\mathrm{Ad}(h^{-1}g)$ fixes all of $\mathfrak{t}$. By maximality $h^{-1}g \in T$ so $hT = gT$. By compactness of G/T, $\bar{f}$ is a homeomorphism onto its image. What is this image? If $n = \dim G$, $\bar{f}(G/T) = f(G)$ is the orbit of X_0 so it is contained in S^{n-1}. On the other hand $\dim G/T = n - 1$ because G has rank 1. Since this is also the dimension of the sphere and everything is connected we see G/T and S^{n-1} are homeomorphic. Now apply the long exact homotopy sequence to the fibration $T \to G \to G/T = S^{n-1}$ to get $\pi_2(S^{n-1}) \to \pi_1(T) \to \pi_1(G)$. If $n \geq 3$ then $\pi_2(S^{n-1}) = \{1\}$, $\pi_1(T) \to \pi_1(G)$ is an injection. This is impossible since G is simply connected and T is not. Hence $\pi_2(S^{n-1}) \neq \{1\}$. Therefore $n - 1 = 2$ and $n = 3$.

It remains to prove G is either $\mathrm{SU}(2,\mathbb{C})$, or $\mathrm{SO}(3,\mathbb{R})$. We continue to assume G is simply connected (and semisimple). Then $Z(G)$ is finite, hence $\dim \mathrm{Ad}\, G = 3$. But by invariance of the form we have $\mathrm{Ad}\, G \subseteq \mathrm{SO}(3)$ and since both these are connected and $\mathrm{SO}(3)$ also has dimension

3, we conclude that $\mathrm{Ad}\, G = \mathrm{SO}(3)$. This means the Lie algebra is that of $\mathrm{SU}(2, \mathbb{C})$, proving the theorem (see Lemma 3.8.2). $\qquad\square$

Remark 4.3.4. As a corollary of the proof (since, as we saw above, $G = \mathrm{SU}(2, \mathbb{C})$ has rank 1) we know $G/T = S^2$. That is, $S^3/S^1 = S^2$. This is the Hopf fibration.

In order to deal with many of the properties of a compact connected Lie group we shall require the concept of an exponential Lie group.

Definition 4.3.5. Let G be connected Lie group. We say G is *exponential* if $\exp : \mathfrak{g} \to G$ is surjective. Alternatively, every point of G lies on a 1-parameter subgroup.

Remark 4.3.6. A few comments about this notion are in order. As we know, the exponential map is surjective on sufficiently small neighborhoods of 1. The question is, can this be extended to the entire group. Of course, an exponential group would have to be connected. That is why we restricted ourselves to connected groups in the definition above. Non-compact connected Lie groups are rarely exponential, for example $\mathrm{SL}(2, \mathbb{R})$ is not exponential. In fact, here the range of exp is not even dense in G.

We now prove a lemma due to H. Hopf. Because the theorem on the degree of a mapping requires compactness, the proof of this lemma breaks down if the group is not compact. In fact this statement, as well as the following two (or their analogues), are false for non-compact semisimple Lie groups.

Lemma 4.3.7. *Compact connected Lie groups are divisible.*

Proof. Our proof relies on the concept of the degree of a mapping. We refer the reader to [18] for the definition of the degree of a map. Let X be a smooth oriented *compact* connected manifold and $f : X \to X$ a smooth map. It follows from the very definition of degree that if the degree of f is different from zero, then f is surjective [18]. Here we take G itself as the manifold, and for the mapping, $\pi_k : g \mapsto g^k$ and we determine the sign $\deg(\pi_k)$. First we fix a left invariant orientation and volume form on G. Let R_g and L_g denote the right and the left translation by g. We shall prove that $\deg(\pi_k)$ is positive. For that we must determine the sign of the Jacobian of π_k, i.e. $J_a(\pi_k) = \det(d_a\pi_k)$ and for $a \in G$, we know that T_aG is generated by left invariant vector fields, so let X_a be the invariant vector field generated by $X \in \mathfrak{g}$. Note that $\pi_k \circ L_a$ is given by $g \mapsto a \cdot g \cdot a \cdots a \cdot g$ and its derivative at the identity element is

$$d_a\pi_k d_1 L_a = d_1(\pi_k L_a) = \sum_{i+j=k} dL_{a^i} dR_{a^j}$$

$$= dL_{a^k}(I_n + T + T^2 + \cdots + T^{k-1})$$

where $T = \mathrm{Ad}(a^{-1})$. Therefore

$$d_a\pi_k(X_a) = d_a\pi_k d_1 L_a(X) = d_1 L_{a^k}(I_n + T + T^2 + \cdots + T^{k-1})(X),$$

and since dL_{a^k} is an orientation preserving diffeomorphism of G, the sign of the determinant of $d\pi_k$ and of $T_a^k = Id + T + T^2 + \cdots + T^{k-1}$ are the same.

As G is compact, we can choose an Ad-invariant measure (see Chapter 2) so $\mathrm{Ad}\, G$ is contained in some unitary group $\mathrm{U}(n, \mathbb{C})$, hence the eigenvalues of T, we have absolute value 1. Now consider the characteristic polynomial of $\mathrm{Ad}(g^{-1})$,

$$P_g(t) = \det(tI_n - \mathrm{Ad}(g^{-1}))$$

which is positive for large enough t. Its roots have absolute value one, therefore $P_g(t)$ is positive for $t > 1$.

Let

$$T_a^k(t) = t^{k-1}I_n + t^{k-2}T + T^2 + \cdots + tT^{k-1},$$

then $T_a^k(1) = T_a^k$ so we have

$$(tI_n - T)T_a^k(t) = T_a^k(t)(tI_n - T) = t^k I_n - A^k,$$

and $P_a(t) \det T_a^k(t) = P_a(t^k)$. Since $P_a(t) > 0$ and $P_a(t^k) > 0$ for $t > 1$, we get

$$\det T_a^k(t) > 0$$

for $t > 1$ therefore $\det T_a^k = \det T_a^k(1) \geq 0$. One can calculate the degree of π_k by

$$\int_G \pi_k^*(dg) = \deg(\pi_k) \int_G dg,$$

where dg is the volume form on G. We have

$$\int_G \pi_k^*(dg) = \int_G J(\pi_k)(g)dg.$$

Since $J(\pi_k)(g) \geq 0$ for all $g \in G$, and $J(\pi_k)(1) = k^n$, we get $\int_G \pi_k^*(dg) > 0$ which proves that $\deg(\pi_k) > 0$. $\qquad\square$

We are now in a position to prove:

Theorem 4.3.8. *Compact connected Lie groups are exponential.*

Proof. Let $g \in G$, let U be a canonical neighborhood of 1 in G and consider the closed subgroup H generated by g. Since a compact Lie group is linear by the Peter-Weyl theorem, Chapter 5, it is also second countable, therefore $\{g^n : \mathbb{Z}^+\}$, has a convergent subsequence g^{n_i} converging to g_0, say. Hence $\lim_{i\to\infty} g^{n_i - n_{i-1}} = 1$. This means that some power g^k lies in U. By continuity of π_k, there is a neighborhood $V(g)$ of g so that $\pi_k(V(g)) \subseteq U$. Hence $G = \bigcup_{g \in G} V(g)$ and by compactness $G = \bigcup_{g_i \in S} V(g_i)$ for a finite number of g_i where $\pi_i(V(g_i)) \subset U$. Let m be the product of these i's. For each i, we have $i \cdot \hat{i} = m$, where $\hat{i}$ is the product of all the others. Then for all i, $\pi_i(V(g_i)) \subseteq U$. Hence for each i, $\pi_m(V(g_i)) \subseteq \pi_{\hat{i}}(U)$. But every point of U does lie on a 1-parameter group of G therefore $U \subseteq \pi_i(U)$ for each i. Applying $\pi_{\hat{i}}$ we get $\pi_{\hat{i}}(U) \subseteq \pi_m(U)$ and therefore $\pi_m(V(g_i)) \subseteq \pi_m(U)$. But the $V(g_i)$ cover G so $\pi_m(G) \subseteq \pi_m(U)$ and since G is divisible by Lemma 4.3.7 we finally get $G = \pi_m(U)$. Thus some power of each group element lies on a 1-parameter subgroup of G and this means the same is true of the group element itself. $\qquad\square$

Corollary 4.3.9. *For a compact connected Lie group, G, the conjugates of a maximal torus T fill out all of G:*

$$G = \bigcup_{g \in G} gTg^{-1}.$$

Proof. Since each point of G lies on a 1-parameter subgroup which is itself contained in a maximal torus we conclude that G is a union of its maximal tori. The result now follows from the fact that the maximal tori are all conjugate, Theorem 4.3.1.   □

We remark that actually the conclusion of Corollary 4.3.9 implies exponentiality which in turn implies divisibility. Thus all three notions are equivalent in the case of compact connected Lie groups. To see this, let $g \in G$. Then $g = g_1 t g_1^{-1}$ where $t \in T$. Since G is linear, because it is compact, we have $\exp = \mathrm{Exp}$ and therefore $\exp(PXP^{-1}) = P\exp(X)P^{-1}$. Since T is abelian here $\exp$ is the universal covering of T by $\mathbb{R}^n$, where $n = \dim T$. In particular $\exp|_T$ is surjective. Therefore $t = \exp X$ and $g = \exp(g_1 X g_1^{-1})$. Conversely if $g = \exp(X)$ and k is a positive integer, then $g = [\exp \frac{1}{k}X]^k$.

Corollary 4.3.10. *If G is a compact connected Lie group, then $Z(G)$ is the intersection of all maximal tori of G.*

Proof. Let $g \in Z(G)$. Then $g \in T$ for some maximal torus. Therefore $g = hgh^{-1} \in hTh^{-1}$. Hence g lies in every maximal torus of G. On the other hand suppose g lies in every maximal torus of G. Then g commutes pointwise with every maximal torus. Since G is the union of all maximal tori, g has to be in the center.   □

Corollary 4.3.11. *If $A \subseteq G$ is a connected abelian subgroup of G its centralizer $Z_G(A)$ is the union of the maximal tori in G containing A.*

Proof. Since A is a connected abelian subgroup of G so is its closure, $\bar{A}$. Therefore $\bar{A}$ is a torus. If g centralizes A, then it also centralizes $\bar{A}$ (and vice versa). Thus we can assume A is a torus and must show $Z_G(A) = \cup T'$, where $T' \supseteq A$. We have $T' \subseteq Z_G(A)$, for any torus containing A hence $\cup T' \subseteq Z_G(A)$.

Conversely, let $g \in Z_G(A)$ and let B be the closure of the (abelian) subgroup of G generated by g and A. In particular, $B \supseteq A$. Also, B is compact and abelian. Hence its identity component B_0 is a torus. Since $B_0 \supseteq A$ and $g \in B$, gB_0 generates B/B_0, which is finite and hence finite cyclic. By Corollary 4.2.5, since B_0 has a quasi-generator, so does B. Let b be a quasi-generator of $B \subseteq G$. Hence by Corollary 4.3.9 $b \in T'$, some maximal torus of G. Therefore B must also be contained in T' and $g \in T'$. Since $T' \supseteq A$, this completes the proof. □

We now come to a criterion for maximality which made an appearance in many previous examples.

Corollary 4.3.12. *Let G be a compact connected Lie group, T be a torus and $\mathfrak{g}$ and $\mathfrak{t}$ be the respective Lie algebras. Then T is maximal if and only if $Z_G(T) = T$.*

Proof. We know from the previous proposition that if T is maximal, then $Z_G(T) = T$. Suppose $Z_G(T) = T$ and T' is a torus in G containing T. Then $T' \subseteq Z_G(T)$. Therefore $T' = T$. □

4.4 The Weyl Group

In this section we introduce *Weyl group*, an invariant for compact connected Lie groups.

Proposition 4.4.1. *Let G be a compact connected Lie group and T maximal torus. Then G/T is simply connected.*

Proof. First assume G is compact semisimple. Consider the universal covers, $\tilde{G}$ and $\tilde{T}$ of G and T respectively. Since G is semisimple $\tilde{G}$ is compact and $\tilde{T}$ is a maximal torus in $\tilde{G}$. Now $\tilde{G}/\tilde{T} = G/T$ and since $\tilde{G}$ is simply connected we have $\pi_1(\tilde{G}/\tilde{T}) = \tilde{T}/\tilde{T}_0$ which is trivial because $\tilde{T}$ as a maximal torus is connected. Therefore G/T is simply connected. Now in general $G = Z(G)_0[G,G]$ where $[G,G]$ is semisimple. Since G is compact $Z(G)_0$ is a torus and hence $Z(G)_0T$ is a maximal torus of G where T is a maximal torus of $[G,G]$. Therefore by the second isomorphism theorem $G/Z(G)_0T = [G,G]/T$. □

Definition 4.4.2. We denote the *centralizer* and *normalizer* of T by $Z_G(T)$ and $N_G(T)$ respectively.

Proposition 4.4.3. *Let G be a compact connected Lie group and T a maximal torus. Then $N_G(T)$ contains T as a subgroup of finite index.*

Proof. Using joint continuity $\mathrm{Aut}(G) \times G \to G$ one sees easily that $N_G(T)$ is a closed subgroup of G, so $N_G(T)$ and $N_G(T)/N_G(T)_0$ both are compact. Moreover since $N_G(T)$ is a Lie group, $N_G(T)_0$ is open in $N_G(T)$ so $N_G(T)/N_G(T)_0$ is discrete. Therefore $N_G(T)/N_G(T)_0$ is finite. We complete the proof by showing $N_G(T)_0 = T$. Evidently $N_G(T)_0 \supseteq T$. Consider $\alpha_{n_0}|_T$ is in $\mathrm{Aut}(T)$, the conjugation by $n_0 \in N_g(T)$. This gives a connected subgroup of $\mathrm{Aut}(T)$. On the other hand, as we saw earlier, $\mathrm{Aut}(T)$ is discrete. Therefore this subgroup is trivial and n_0 centralizes T. Therefore by the previous paragraph $n_0 \in T$. □

Definition 4.4.4. Since $N_G(T)$ contains T as a closed normal subgroup of finite index, the quotient $N_G(T)/T$ is a finite group called the *Weyl group*, $\mathcal{W}(G)$.

In principle $\mathcal{W}$ could depend on the choice of maximal torus T. However, suppose $gTg^{-1} = T'$ were another maximal torus, with $\mathcal{W}' = N_G(gTg^{-1})/gTg^{-1}$. An easy calculation shows $gN_G(T)g^{-1} = N_G(gTg^{-1})$. This means that $\mathcal{W}'$ is naturally isomorphic with $\mathcal{W}$. Thus the Weyl group, and in particular its order, is another important invariant of G.

Exercise 4.4.5. Suppose $G = \mathrm{U}(n, \mathbb{C})$. Its Weyl group is the symmetric group $\mathcal{W} = S_n$. In particular, here $|\mathcal{W}| = n!$.

Now we turn to the mapping $\phi : G/T \times T \to G$ given by $(gT, t) \mapsto gtg^{-1}$ which is of some importance in the representation theory of compact connected Lie groups (see *e.g.* [1]), where G is a compact connected Lie group, and T is a maximal torus. The map is well-defined since if $h^{-1}g \in T$, then $hth^{-1} = gtg^{-1}$. It is also surjective by Corollary 4.3.9, below. Evidently here we have a smooth map between compact connected manifolds of the same dimension. What is the degeneracy of ϕ in some generic sense? Let t_0 be a quasi-generator of T. Then $\phi(gT, s) = gsg^{-1} = t_0$ if and only if g normalizes T. Therefore $|\phi^{-1}(t_0)| = |N_G(T)/T| = |\mathcal{W}|$. We want to see what this map looks like locally near $(T, 1)$, to do so, consider $\mathfrak{g}$ and its subalgebra $\mathfrak{t}$.

Now the significance of the Weyl group is that it operates on T by inner automorphisms $\mathcal{W} \times T \to T$, $(nT, t) \mapsto ntn^{-1}$. Since $n \in N_G(T)$ and $t \in T$, $ntn^{-1} \in T$. This action is effective, *i.e.* the map $\mathcal{W} \to \mathrm{Aut}(T)$ is $1:1$, since if $w = nT \in \mathcal{W}$ and $ntn^{-1} = t$, for all $t \in T$ then $n \in Z_G(T) = T$ so $nT = w$ is the identity. We shall see in a moment that the action of the Weyl group on T reflects exactly the action of G on itself by conjugation.

Lemma 4.4.6. *Let t_1 and t_2 lie in T, a maximal torus. Then there is a $g \in G$ with $gt_1 g^{-1} = t_2$ if and only if $w(t_1) = t_2$ for some $w \in \mathcal{W}$.*

Proof. Suppose $gt_1 g^{-1} = t_2$. A direct calculation shows $gZ_G(t_1)g^{-1} = Z_G(t_2)$. Since $T \subset Z_G(t_1)$ we get $gTg^{-1} \subseteq Z_G(t_2)$. Thus T and gTg^{-1} are maximal tori in $Z_G(t_2)_0$. They are conjugate in this compact connected Lie group, so there is some $h \in Z_G(t_2)_0$ satisfying $h(gTg^{-1})h^{-1} = T$. Hence $hg \in N_G(T)$. Also $hgt_1(hg)^{-1} = ht_2 h^{-1} = t_2$. Hence $hgT = w \in \mathcal{W}$ and $w(t_1) = t_2$. The converse statement is obvious. $\square$

Let $\mathcal{O}$ be the space of conjugacy classes of G with the quotient topology. Since G is compact one defines a compact Hausdorff topology on this orbit space. Let $T/\mathcal{W}$ be the orbit space of the action of $\mathcal{W}$ on T. Then there is a canonical homeomorphism $T/\mathcal{W} \to \mathcal{O}$ given by $\mathcal{W}(t) \mapsto \mathcal{O}(t)$, $t \in T$, which is surjective because of Corollary 4.3.9 and injective because of Lemma 4.4.6 above. Since $T/\mathcal{W}$ is also compact and Hausdorff we have a homeomorphism.

Let $C(X)$ be the complex valued continuous functions on the compact space, X. If a compact group G acts continuously on X, we recall that this gives rise to a linear representation of G on $C(X)$ via $f_g(x) = f(g^{-1}x)$. We denote the G-fixed functions by $C(X)^G$. For example, if G acts on itself by conjugation, then $C(G)^G$ indicates the continuous class functions on G. In particular conjugation also gives rise to an action of $\mathcal{W}$ on $C(T)$. Here $C(T)^{\mathcal{W}} = C(\mathcal{O}_{\mathcal{W}})$.

For a compact connected Lie group G and maximal torus T we consider the restriction map $C(G) \to C(T)$. Since conjugation gives rise to an action of G on G and $\mathcal{W}$ on T we see that the restriction map takes $C(G)^G \to C(T/\mathcal{W}) = C(T)^{\mathcal{W}}$. If $f|_T = f_1|_T$, where f_1 is another continuous class function, then since these are class functions we have $f(gtg^{-1}) = f_1(gtg^{-1})$ for all $g \in G$. If $h \in G$, then $h \in gTg^{-1}$ for some g so $f(h) = f_1(h)$, thus $f = f_1$ and the restriction map is $1 : 1$. This map is also onto. Suppose f is a continuous function on T invariant under the Weyl group. Define $\bar{f}$ on G by $\bar{f}(gtg^{-1}) = f(t)$. It is easy to see that $\bar{f}$ is well defined, continuous and a class function on G. This we leave as an exercise to the reader. Evidently $\bar{f}|_T = f$. The restriction map is a complex algebra homomorphism, hence an isomorphism.

Now let $K(G)$ denote the ring of isomorphism classes of representations of G. The basic operations are the tensor product and the direct sum. Note that $K(G)$ is not quite a ring but to obtain a ring we must take all formal finite $\mathbb{Z}$-linear combinations of representations (including those with possibly negative coefficients), then $(-n\rho) \oplus \sigma = -(n\rho) \oplus \sigma = -n(\rho \oplus \sigma)$ and $(-n\rho) \otimes \sigma = -n(\rho \otimes \sigma)$, etc. for $n \in \mathbb{Z}^+$. Now we really do have a ring with identity called the *ring of virtual representations*. Since the character of a finite dimensional representation determines the representation (see Chapter 5), $K(G)$ is basically the set of all the characters of the finite dimensional representations of G, together with 0 under pointwise operations. We recall that $\chi_{\rho \otimes \rho'} = \chi_\rho \chi_{\rho'}$ and $\chi_{\rho \oplus \rho'} = \chi_\rho + \chi_{\rho'}$.

Now we consider a character χ_ρ of the finite dimensional continuous representation ρ of G on V; ρ is determined by its character χ_ρ and since χ_ρ is a class function it is determined by its restriction to T.

Corollary 4.4.7. $K(G)$ *is an integral domain.*

Proof. The paragraph above tells us that restriction $K(G) \to K(T)^{\mathcal{W}}$ is an injective $\mathbb{C}$-algebra homomorphism. Suppose $\chi_\rho \chi_\sigma = 0$, or even if $\chi_\rho|_T \chi_\sigma|_T = 0$. We will show either $\chi_\rho = 0$ or $\chi_\sigma = 0$. Let t_0 be a quasi-generator of T and let $\chi_\rho = \sum_i n_i \chi_{\rho_i}$, $\chi_\sigma = \sum_j m_i \chi_{\sigma_j} \in K(G)$ where n_i and $m_j \in \mathbb{Z}$. Then since $\chi_\rho(t_0)\chi_\sigma(t_0) = 0$ and $\mathbb{C}$ is a field we get $\chi_\rho(t_0) = 0$, or $\chi_\sigma(t_0) = 0$. Let us assume it is the former. Ignoring, as we may, the zero coefficients, reorder the irreducibles of ρ so that $n_1, \ldots, n_k$ are positive and $n_{k+1}, \ldots, n_r$ are negative. Let $\rho^+ = n_1 \rho_1 \oplus \cdots \oplus n_k \rho_k$ and $\rho^- = (-n_{k+1}\rho_{k+1}) \oplus \cdots \oplus (-n_r\rho_r)$. Then ρ^+ and ρ^- are finite dimensional representations of G and $\chi_{\rho^+}(t_0) = \chi_{\rho^-}(t_0)$. Since t_0 is a quasi-generator of T, they agree on all of T. Hence χ_{ρ^+} and χ_{ρ^-} agree on T. Therefore they agree on all of G so that $\chi_\rho = 0$. $\qquad\square$

We conclude with an application of representation theory.

Proposition 4.4.8. *Let G be a compact connected Lie group and T maximal torus. Then G/T has even dimension.*

Proof. Consider the adjoint representation of G on its Lie algebra $\mathfrak{g}$ and let ρ be the restriction to a maximal torus T. Since T is compact and abelian the continuous representation ρ decomposes into the direct sum of irreducibles over $\mathbb{R}$. Thus we have some 1-dimensional representations of the form $\rho(t) = \lambda(t)X$, $t \in T$, where λ is a continuous homomorphism $T \to \mathbb{R}^\times$, or we have 2-dimensional real representations. In the former case since T is compact $\lambda(t) = \pm 1$, and since T is connected $\lambda(t) \equiv 1$. On the other hand, suppose we have a 2-dimensional invariant real subspace, V, of $\mathfrak{g}$. Then $\rho_t|V = R_t$ of rotations. Thus $\mathfrak{g}$ is the direct sum of $\mathfrak{g}_0$, the space on which ρ acts trivially, with $\sum_{j=1}^{k} \mathfrak{g}_j$, of rotations. Hence $\dim \mathfrak{g} = \dim \mathfrak{g}_0 + 2k$. But $\mathfrak{g}_0$ is actually the Lie algebra of T. Hence $\dim(G/T) = \dim G - \dim T = 2k$.

To complete the proof of even dimensionality we must show $\mathrm{Ad}\, t(X) = X$ for all $t \in T$ if and only if $\mathrm{Exp}(sX) \in T$ for all $s \in \mathbb{R}$. Suppose $\mathrm{Exp}(sX) \in T$ for all $s \in \mathbb{R}$. Then $\mathrm{Ad}\, t(\mathrm{Exp}(sX)) = t\,\mathrm{Exp}(sX)t^{-1} = \mathrm{Exp}(sX)$ for all s and $t \in \mathbb{R}$. Hence $\mathrm{Exp}(tsXt^{-1}) = \mathrm{Exp}(sX)$ for all s and $t \in \mathbb{R}$. Taking $\frac{d}{ds}|_{s=0}$ on both sides tells us $tXt^{-1} = \mathrm{Ad}\, t(X) = 0$. Conversely, if $tXt^{-1} = 0$, then reversing our argument we get $\mathrm{Ad}\, t(\mathrm{Exp}\, sX) = \mathrm{Exp}\, sX$ for all s and t so that for all s, $\mathrm{Exp}(sX) \in Z_G(T) = T$. $\qquad\square$

4.5   What Goes Wrong If G is Not Compact

Is there any hope of finding a connected abelian subgroup whose conjugates fill out the group. Let us take an important and familiar non-compact, but simple Lie group, namely $G = \mathrm{SL}(2, \mathbb{R})$. Write $G = KAN$

as in Section 1.6. We claim that both A and N are maximal connected abelian subgroups of G. First, they are both abelian connected Lie subgroups. Suppose H was such a subgroup of G containing A. Every $h \in H$ centralizes A. But a matrix commuting with a diagonal matrix that has distinct eigenvalues must itself be diagonal, and since H is connected the diagonal entries of h are positive. Therefore $h \in A$, which shows A is maximal. To see that N is also maximal, let $h \in H$ which is a connected abelian subgroup of G containing N. A direct calculation shows that $Z_{\mathrm{SL}(2,\mathbb{R})}(N)$ consists of $\pm N$. Since H is connected, $h \in N$. Thus N is also a maximal abelian subgroup of G.

Now is it possible that $G = \bigcup_{g \in G} gAg^{-1}$, or $G = \bigcup_{g \in G} gNg^{-1}$? The answer to each of these questions is no, both for the same reason. If either were true then each element of A would be conjugate to an element of N and vice versa. But this cannot be because $\operatorname{tr} h = a + \frac{1}{a}$, where $h \in A$, while if $h \in N$, then $\operatorname{tr} h = 2$. However, $a + \frac{1}{a} > 2$, unless of course $a = 1$.

Another possibility for a connected abelian subgroup (which like A acts completely reducibly) is K, which after all is a torus. Could $G = \bigcup_{g \in G} gKg^{-1}$? (Even if K were not a maximal torus it would be contained in one and so the conjugates of it would also fill out G and so we could argue as below.) If $G = \bigcup_{g \in G} gKg^{-1}$, then each element of G would be diaganolizable with eigenvalues on the unit circle. But A is diagonal with positive eigenvalues. This means $A = \{1\}$, a contradiction.

Moreover the conjugacy relation between maximal abelian subgroups is also false. For K cannot be conjugate to either A or N since K is compact and the other two are not. Nor can A and N be conjugate since the eigenvalues of elements of A give all positive reals while those of N are always 1. It is worth noting that these groups are not conjugate in spite of the fact that they are isomorphic.

Chapter 5

Representations of Compact Lie Groups

In this chapter we deal with the classical representation theory of a compact group. We shall see that compact Lie groups play a special role although the case of finite or abelian groups is also of great interest. The central object of study here is the set of all finite dimensional, continuous, irreducible unitary representations. Although we will not need this, actually every continuous, irreducible, unitary representation of a compact group on a Hilbert space is automatically finite dimensional. In Section 1 we introduce the players, in Section 2 we prove the Schur orthogonality relations. In Section 3 we develop what we need from functional analysis and in Section 4 we prove the Peter-Weyl theorem and its many consequences. Section 5 deals with characters and class functions. In our final section we study induced representations and the Frobenius reciprocity theorem as well as a number of related ideas which have proven to be quite useful in geometric questions (such as the Mostow-Palais equivariant embedding theorem) and spherical harmonics. It might also be mentioned that the results on representations and harmonic analysis have been generalized from compact to other classes of groups. The most direct analogies have been found in the case of *central groups*, those which are compact modulo their center (see [27] and [28]).

5.1   Introduction

Unless otherwise stated, throughout this chapter G will denote a compact topological group and dg will be the normalized Haar measure on G (recall G is unimodular by Corollary 2.2.2), $L_1(G)$ and $L_2(G)$ will denote the integrable, respectively square integrable, complex valued measurable functions on G with respect to the Haar measure and $C(G)$ the continuous complex valued functions on G. Likewise if G operates on a space X with a measure dx preserved by G we denote by $L_1(X)$ and $L_2(X)$ the integrable, respectively square integrable, complex valued measurable functions on X and $C(X)$ the continuous ones.

Even though compact groups have many interesting infinite dimensional representations, we shall also see why we concentrate on finite dimensional representations.

For the readers convenience we list the following definitions. These are fundamental and actually do not require compactness of G, but merely that the representations are continuous and finite dimensional.

Definition 5.1.1.   (1) Given two such representations of ρ and σ of G we shall call an operator $T : V_\rho \to V_\sigma$ an *intertwining operator* if $T\rho_g = \sigma_g T$, $g \in G$.

   (2) ρ and σ are said to be *equivalent* if there exists an invertible intertwining operator between them.

   (3) We shall say ρ is a *unitary representation* if $\rho(G) \subseteq \mathrm{U}(n,\mathbb{C})$ for some n.

   (4) A Hermitian inner product $\langle \cdot, \cdot \rangle$ on V_ρ is called *invariant* if $\langle \rho_g(v), \rho_g(w) \rangle = \langle v, w \rangle$ for all $g \in G$ and $v, w \in V_\rho$.

   (5) A subspace W of V_ρ is called an *invariant subspace* if $\rho_g(W) \subseteq W$ for all $g \in G$.

   (6) ρ is called *completely reducible* if every invariant subspace has a complementary invariant subspace.

   (7) ρ is called *irreducible* if it has no nontrivial invariant subspaces.

   (8) Finally, we denote by $\mathcal{R}(G)$ the *equivalence classes of finite dimensional, continuous, irreducible unitary representations* of G. As we shall see this set is quite interesting even when G is finite

or abelian.

The following proposition and its corollary was proved in Chapter 2 (see the proof of Theorem 2.5.1).

Proposition 5.1.2. *Any finite dimensional representation of a compact group G has a G-invariant inner product and hence is equivalent to a unitary representation.*

In particular,

Corollary 5.1.3. *Any finite dimensional continuous representation of a compact group G is completely reducible.*

We leave the following important exercise to the reader.

Exercise 5.1.4.   (1) Show that equivalence of representations is an equivalence relation.

(2) Given a single representation ρ, show the set of intertwining operators forms a subalgebra of $\mathrm{End}(V_\rho)$.

(3) Give an example to show that the proposition and corollary above is false if G is not compact e.g. consider a unipotent representation of $\mathbb{R}$.

(4) Show that two representations are equivalent if and only if the modules (G, V_ρ, ρ) and (G, V_σ, σ) are isomorphic. Thus they share all module theoretic properties such as a composition series for one corresponds to such a series for the other etc.

(5) Show that a finite dimensional continuous representation ρ is completely reducible if and only if the corresponding module is semisimple.

(6) Show that a finite dimensional continuous representation ρ is irreducible if and only if the corresponding module is simple.

(7) Define a unitary representation (not necessarily continuous) of a group (not necessarily compact) on a Hilbert space V and show that it is completely reducible in the sense that any closed G invariant subspace of V has a complimentary closed invariant subspace.

We conclude this section with an important example. We shall find all the finite dimensional irreducible unitary representation of $SU(2, \mathbb{C})$. Now we know all the complex irreducible representations of the Lie algebra $\mathfrak{sl}(2, \mathbb{C})$ (see Section 3.1.5). Since the Lie group $SL(2, \mathbb{C})$ is simply connected (Corollary 6.3.7) and has $\mathfrak{sl}(2, \mathbb{C})$ as its Lie algebra its irreducible representations are in bijective correspondence with those of the Lie algebra by $\rho \mapsto \rho'$ (Corollary 1.4.15). Similarly, $SU(2, \mathbb{C})$ is also simply connected (Corollary 1.5.2), so its real continuous (smooth) irreducible representations are in bijective correspondence with those of its Lie algebra, $\mathfrak{su}(2, \mathbb{C})$. Finally, as we will see in Chapter 7, $\mathfrak{su}(2, \mathbb{C})$ is a compact real form of $\mathfrak{sl}(2, \mathbb{C})$. Hence complex irreducibles of the latter bijectively correspond with the real irreducibles of the former.

Corollary 5.1.5. *There are an infinite number of continuous, finite dimensional, irreducible, unitary representations of* $SU(2, \mathbb{C})$, *one for each degree.*

Exercise 5.1.6. Show that, within these, the representations of $SO(3, \mathbb{R})$ are the ones of odd degree.

5.2   The Schur Orthogonality Relations

The Schur orthogonality relations are the following.

Theorem 5.2.1. *Let G be a compact group, dg be the normalized Haar measure and ρ and σ finite dimensional continuous irreducible unitary representations of G. Then*

    (1) If ρ and σ are inequivalent, then $\int_G \rho_{ij}(g)\overline{\sigma_{lk}(g)}dg = 0$ for all $i, j = 1, \ldots, d_\rho$ and $k, l = 1, \ldots, d_\sigma$.

    (2) $\int_G \rho_{ij}(g)\overline{\rho_{kl}(g)}dg = \frac{\delta_{ik}\delta_{jl}}{d_\rho}$.

Proof. Let V_ρ and V_σ be the respective representation spaces and $B(V_\sigma, V_\rho)$ be the (finite dimensional) $\mathbb{C}$-vector space of linear operators between them. Let $T \in B(V_\sigma, V_\rho)$ and consider the map $G \to B(V_\sigma, V_\rho)$ given by $g \mapsto \rho(g)T\sigma(g^{-1})$. This is a continuous operator valued function on G and so $\int_G \rho(g)T\sigma(g^{-1})dg$ is also an operator in $B(V_\sigma, V_\rho)$. For

$h \in G$ we have

$$\rho(h) \int_G \rho(g)T\sigma(g^{-1})dg\sigma(h)^{-1} = \int_G \rho(h)\rho(g)T\sigma(g^{-1})\sigma(h)^{-1}dg$$

$$= \int_G \rho(hg)T\sigma(hg)^{-1}dg$$

$$= \int_G \rho(g)T\sigma(g)^{-1}dg$$

Letting $T_0 = \int_G \rho(g)T\sigma(g)^{-1}dg$ we see $\rho(g)T_0 = T_0\sigma(g)$ for every $g \in G$. That is, T_0 is an intertwining operator. By Schur's lemma, Lemma 3.1.52, there are only two possibilities. Either ρ and σ are equivalent and T_0 is invertible (and implements the equivalence), or ρ and σ are inequivalent and $T_0 = 0$. In the latter case $\int_G \rho(g)T\sigma(g)^{-1}dg = 0$, where T is arbitrary. Let $T = (t_{jk})$. Then $(\rho(g)T\sigma(g)^{-1})_{il} = \sum_{jk} \rho_{ij}(g)t_{jk}\sigma_{kl}(g^{-1})$. Since (t_{jk}) are arbitrary we get $\int_G \rho_{ij}(g)\sigma_{kl}(g^{-1})dg = 0$ for all $i, j = 1, \ldots, d_\rho$ and $k, l = 1, \ldots, d_\sigma$. Because σ is a unitary representation $\sigma(g^{-1}) = \sigma(g)^{-1} = \sigma(g)^*$. Thus $\int_G \rho_{ij}(g)\overline{\sigma_{lk}(g)}dg = 0$ for all $i, j = 1, \ldots, d_\rho$ and $k, l = 1, \ldots, d_\sigma$.

We now consider the case when we have equivalence. Here we may as well just take σ to be ρ. In this case Schur's lemma tells us T_0 is a scalar multiple of the identity. Thus $\int_G \rho(g)T\rho(g)^{-1}dg = \lambda(T)I$. Taking the trace of each side yields

$$\mathrm{tr}(\int_G \rho(g)T\rho(g)^{-1}dg) = \int_G \mathrm{tr}(\rho(g)T\rho(g)^{-1})dg = \int_G \mathrm{tr}(T)dg = \mathrm{tr}(T),$$

while $\mathrm{tr}(\lambda(T)I) = \lambda(T)d_\rho$. We conclude $\lambda(T) = \frac{\mathrm{tr}(T)}{d_\rho}$ and so $\int_G \rho(g)T\rho(g)^{-1}dg = \frac{\mathrm{tr}(T)}{d_\rho}I$. Using reasoning similar to the earlier case one finds $\int_G \rho_{ij}(g)\overline{\rho_{lk}(g)}dg = 0$ for all $i, j, k, l = 1, \ldots, d_\rho$ whenever $i \neq l$, or $j \neq k$. Now we consider the case when $i = l$ and $j = k$. By taking T to be diagonal with all zero entries except for one we get $i, j = 1, \ldots, d_\rho$. Hence in general one has $\int_G \rho_{ij}(g)\overline{\rho_{kl}(g)}dg = \frac{\delta_{ik}\delta_{jl}}{d_\rho}$.   □

5.3   Compact Integral Operators on a Hilbert Space

Before proceeding further we must now prove the spectral theorem for compact self-adjoint operators on a Hilbert space. Then we will apply this result to compact self-adjoint integral (Fredholm) operators to conclude that the range of such an operator always has an eigenfunction expansion. It is this fact which is behind the Peter-Weyl theorem.

Definition 5.3.1. Let V and W be real or complex Hilbert spaces. A bounded linear operator $T : V \to W$ is called a *compact operator* if $T(B_1(0))$ is compact where $B_1(0)$ is the unit ball in V.

For an operator T, the *norm*, if it exists, is defined to

$$\|T\| = \sup\{\|Tv\| : \|v\| = 1\}.$$

Exercise 5.3.2. Evidently, if this were so it would be true of every ball about 0. In fact, T is compact if and only if it takes bounded sets to compact ones. Notice that when $V = W$ and is infinite dimensional, then the identity map I or, more generally λI, $\lambda \neq 0$ is not compact while if W is finite dimensional all bounded linear operators are compact. Such operators are called *finite rank operators*. Observe also that the restriction of a compact operator to a closed invariant subspace is again compact.

What would be a nontrivial example of a compact operator? Suppose $V = L_2(X)$, where X is a compact Hausdorff space, dx is a (finite which we may as well normalize to have total mass 1) regular measure on X and k is a continuous function on $X \times X$. We can use k to define an operator $T_k : V \to V$, by $T_k(f)(x) = \int_X k(x,y)f(y)dy$. T_k is well defined since by compactness k is bounded and by the Schwarz inequality together with compactness tells us $L_2(X) \subseteq L_1(X)$. T_k is evidently linear. Applying the Schwarz inequality again shows $\int_X |k(x,y)f(y)|^2 dy \leq \|k\|_{X \times X}^2 \int_X |f(y)|^2 dy \leq \|k\|_{X \times X}^2 \|f\|_2^2$. Hence T_k is a bounded operator (whose operator norm is $\leq \|k\|_{X \times X}^2$).

Definition 5.3.3. In this context the k above is called a *kernel function* and T_k a Fredholm operator.

So for example, we can let ϕ_i and ψ_i be two sets of n continuous functions on X, where n is any integer and $k(x,y) = \sum_{i=1}^n \phi_i(x)\psi_i(y)$. Then T_k is a compact operator for a very simple reason. $T_k(V) \subseteq W$, where W is the linear span of ϕ_i and hence is finite dimensional.

Now this conclusion actually holds for any jointly continuous k. We shall see that it will be sufficient for our purposes to understand Fredholm integral operators.

Theorem 5.3.4. *For jointly continuous k, T_k is a compact operator. Furthermore, $T_k(L_2(X)) \subseteq C(X)$.*

Proof. We first prove the second statement. Since $X \times X$ is compact, k is uniformly continuous. That is given $x_0 \in X$ and $\epsilon > 0$ there exists a neighborhood U_{x_0} of x_0 so that $|k(x,y) - k(x_0,y)| < \epsilon$, whenever $y \in X$ and $x \in U_{x_0}$. Therefore $|T_k(f)(x) - T_k(f)(x_0)| \leq \int_X |k(x,y) - k(x_0,y)|\|f(y)\|dy \leq \epsilon\|f\|_1$ and since $L_2 \subseteq L_1$, $\|f\|_1 < \infty$. Thus $T_k(f)$ is always a continuous function.

We now show that T_k is a compact operator. Notice that on $C(X)$ we have two norms, the sup norm $\|\cdot\|_X$ and the restricted L_2 norm. But since we have normalized the measure, $\|f\|_2 \leq \|f\|_X$. Let B be a bounded set in L_2. If we can prove $T_k(B)$ is compact in $C(X)$, then by continuity of the injection $C(X) \to L_2(X)$ we will be done. To do this we apply the Ascoli theorem. Now again by Schwarz, for $x \in X$, $|T_k f(x)| \leq \int_X |k(x,y)|\|f(y)|dy \leq \|k\|_{X \times X}^2\|f\|_2^2$. Thus $\|T_k f\|_X < \infty$ and $T_k(B)$ is uniformly bounded. Moreover $|T_k(f)(x) - T_k(f)(x_0)|^2 \leq \int_X |k(x,y) - k(x_0,y)|^2|f(y)|^2 dy \leq \epsilon^2\|f\|_2^2$, if $x \in U_{x_0}$. Thus $T_k(B)$ is equicontinuous at every point of X. By Ascoli, $T_k(B)$ has compact closure in $C(X)$. $\qquad\square$

Definition 5.3.5.   (1) A linear operator $T : V \to V$ on a Hilbert space is called *self adjoint* if for all $v, w \in V$, $\langle Tv, w \rangle = \langle v, Tw \rangle$.

  (2) For such an operator if $\lambda \in \mathbb{C}$, we define $V_\lambda = \{v \in V : Tv = \lambda v\}$. Here λ is called an eigenvalue of T and V_λ the corresponding eigenspace.

Exercise 5.3.6.   (1) So for example, a Fredholm operator T_k is self-adjoint if and only if $k(x,y) = \overline{k(y,x)}$ for all $x,y \in X$.

   (2) If a linear operator $T : V \to V$ on a Hilbert space is self adjoint, then all its eigenvalues are real.

Now we wish to prove the following *spectral theorem* for compact self adjoint operators on a Hilbert space.

Theorem 5.3.7. *Let T be a compact self-adjoint operator on a Hilbert space V. Then*

   (1) These are all real.

   (2) If $\lambda \neq \mu$ are distinct eigenvalues then V_λ and V_μ are orthogonal.

   (3) If $\lambda \neq 0$ then V_λ is finite dimensional.

   (4) T has at most a countable number of nonzero eigenvalues.

   (5) $\operatorname{Ker} T = V_0$.

   (6) $V = V_0 \oplus \overline{\left(\sum_{\lambda_i \neq 0} V_{\lambda_i}\right)}$ (orthogonal direct sum).

The main point being the last item which says, in particular, that the range of T can be expanded in a convergent series of eigenvectors *i.e.* given $\epsilon > 0$, for any $v \in V$, there exists a positive integer $n(v)$ so that $\|T(v) - \sum_{i=1}^{n} \lambda_i T(v_i)\| < \epsilon$.

Here are some consequences of the spectral theorem.

Exercise 5.3.8. Prove that:

$$\sum_{i=1}^{\infty} \lambda_i^2 \dim V_{\lambda_i} = \|T\|.$$

In particular, for any nonzero eigenvalue, $\lambda_i^2 \dim V_{\lambda_i} \leq \|T\|$. Hence $\dim V_{\lambda_i} \leq \frac{\|k\|_2^2}{\lambda_i^2}$.

Exercise 5.3.9. Notice that in the case of a finite dimensional operator this just amounts to the fact that a self-adjoint operator is unitarily diagonalizable with real eigenvalues.

Before turning to the proof of Theorem 5.3.7 we need some preparatory results.

Lemma 5.3.10. *Let* $T : V \to V$ *be a compact operator and* $\delta > 0$, *then the number of eigenvectors of norm 1 with eigenvalues* $\lambda > \delta$ *is finite.*

In particular the number of such distinct (*i.e.* orthogonal) eigenspaces is finite. In particular, the total number of such distinct eigenspaces associated with positive eigenvalues is countable (even if V has an uncountable orthonormal basis!). Moreover, (using the positive integers) if we order the positive eigenvalues, λ_n, in decreasing order, then $\lambda_n \to 0$.

Proof. Let λ and μ be distinct eigenvalues of T both bigger than δ and v and w be the respective eigenvectors of norm 1. Then since these eigenspaces are orthogonal (see item 2), $||Tv - Tw||^2 = ||\lambda v - \mu w||^2 = \lambda^2 + \mu^2 \geq 2\delta^2$. Thus $||Tv - Tw|| \geq \sqrt{2}\delta$. Clearly if there were an infinite number of such eigenvalues there could be no convergent subsequence contradicting the fact that T is compact. $\qquad\square$

Lemma 5.3.11. *Let* $T : V \to V$ *be a bounded self-adjoint operator and* W *a* T-*invariant subspace of* V. *Then* $W^{\perp}$ *is also* T-*invariant.*

In particular if T is compact self-adjoint and W is a closed T-invariant subspace then T restricted to $W^{\perp}$ is again a compact self adjoint operator.

Proof. Let $w \in W$ and $w^{\perp} \in W^{\perp}$. Then $\langle Tw^{\perp}, w \rangle = \langle w^{\perp}, Tw \rangle = 0$ since W is T-invariant. Thus since w is arbitrary $Tw^{\perp} \in W^{\perp}$. $\qquad\square$

Lemma 5.3.12. *For a self-adjoint operator* T,

$$||T|| = \sup\{|\langle T(v), v \rangle : ||v|| = 1\}.$$

Proof. Let $M = \sup\{|\langle T(v), v \rangle : ||v|| = 1\}$. By the Cauchy-Schwarz inequality it is obvious that

$$||\langle Tv, v \rangle|| \leq ||T(v)|| \cdot ||v|| \leq ||v|| = ||T||$$

if $||v|| = 1$, therefore M exists and $M \leq ||T||$. It remains to prove that $||T|| \leq M$ for which it suffices to show that $||T(v)|| \leq M$ if $||v|| = 1$. We assume that $T(v) \neq 0$ and let $w = Tv/\|Tv\|$. Then $\langle Tv, w \rangle = \langle v, Tw \rangle = \|Tv\|$ and

$$4\|Tv\| = \langle T(v+w), v+w \rangle - \langle T(v-w), v-w \rangle$$
$$\leq M\|v+w\|^2 + M\|v-w\|^2 = 4M.$$

$\square$

Proposition 5.3.13. *Let $T : V \to V$ be a compact self-adjoint operator on a Hilbert space. Then there is some w of norm 1 with $T(w) = \pm\|T\|w$.*

Proof. By previous result there is sequence of vectors v_n of norm 1 such that $\|T\| = \lim_{n\to\infty} |\langle T(v_n), v_n \rangle|$. By passing to a subsequence, we may assume that $\langle T(v_n), v_n \rangle$ converges to r which is $\|T\|$ or $-\|T\|$, and $T(v_n)$ converges to some vector v, as T is compact. Then

$$0 \leq \|T(v_n) - rv_n\| = \|T(v_n)\|^2 - 2r\langle T(v_n), v_n \rangle + r^2\|v_n\|^2$$
$$\leq 2r^2 - 2r\langle T(v_n), v_n \rangle.$$

As the right side of the inequality converges to zero therefore $\lim_{n\to\infty} \|T(v_n) - rv_n\| = 0$. On the other hand, $w = \lim_{n\to\infty} T(v_n)$ hence $\lim_{n\to\infty} rv_n = w$. For $w = r^{-1}v$, we have $T(w) = rw$.

$\square$

Proof of the spectral theorem.

(1) This is the exercise above.

(2) Suppose $Tv = \lambda v$ and $Tw = \mu w$. Then $\langle Tv, w \rangle = \lambda\langle v, w \rangle$. But its also $\langle v, Tw \rangle = \bar{\mu}\langle v, w \rangle$. Therefore either $\langle v, w \rangle = 0$ or $\lambda = \bar{\mu}$, but since μ is real this would mean $\lambda = \mu$.

If $\lambda \neq 0$, then since T acts on V_λ as a nonzero scalar multiple of the identity, V_λ is finite dimensional by a remark made earlier. By Lemma 5.3.10 T has at most a countable number of positive eigenvalues. Since $-T$ is also a compact operator, T must also have at most a countable number of negative eigenvalues hence a countable number of nonzero eigenvalues.

(3) Clearly $V_0 = \operatorname{Ker} T$.

(4) Finally, let $W = \overline{(\sum_{\lambda_i \neq 0} V_{\lambda_i})}$. Then we have $V = W \oplus W^\perp$. We prove that $V_0 = W^\perp$ which complete the proof. Since each $V_{\lambda_i} \subseteq V_0^\perp$, hence $V_0 \subseteq W^\perp$.

Since W is clearly T-invariant, the same is true of $W^\perp$ by Lemma 5.3.11 and moreover T restricted to $W^\perp$ is a compact self-adjoint operator. By Proposition 5.3.13 choose $w_0^\perp$ of norm 1 in $W^\perp$ so that $T(w_0^\perp) = \pm\|S\|w_0^\perp$, where S is the restriction of T to $W^\perp$. If $\|S\| > 0$, then $w_0^\perp \in V_{\lambda_i}$ for some i. Since $V_{\lambda_i} \subseteq W$ this means $w_0^\perp = 0$ which is impossible as its norm is 1. Thus $\|S\| = 0$ so T restricted to $W^\perp$ is zero or in other words $W^\perp \subseteq \operatorname{Ker} T = V_0$.

Since we were within $C(X)$ and estimated by the sup norm we get

Corollary 5.3.14. *The range, $T_k(L_2(X))$, can be expanded in a uniformly convergent series of eigenfunctions of T_k with $\lambda_i \neq 0$.*

$$T_k(f) = \sum_{i=1}^{\infty} f_{\lambda_i}.$$

Moreover each eigenfunction ϕ of T_k associated with a nonzero eigenvalue is continuous because $T(\phi) = \lambda\phi$ and $T(\phi)$ is continuous, hence so is ϕ.

This completes our study of compact operators. We remark that using the spectral theorem for compact integral operators proven above, one can also get the following theorem which is important in the study of compact Riemann surfaces or, more generally, compact hyperbolic manifolds of higher dimension. Here G is a non-compact simple Lie group and H is the (discrete) fundamental group of the compact manifold. For the details the reader is referred to Representation Theory and Automorphic Functions by I.M. Gelfand et al. The definition of induced representations is given in the last section of this chapter.

Theorem 5.3.15. *Let G be a locally compact group, H a closed subgroup with G/H compact and having a finite G-invariant measure. Let σ be a finite dimensional, continuous, unitary representation of H. Then the induced representation, $\operatorname{Ind}(H \uparrow G, \sigma)$, decomposes into a countable orthogonal direct sum of irreducible unitary representations, each of finite multiplicity (but usually of infinite dimension).*

5.4   The Peter-Weyl Theorem and its Consequences

In order to prove the Peter-Weyl theorem it will be necessary to study a certain infinite dimensional representation called the left regular representation L which is defined as follows.

The representation space of L is $L_2(G)$ and the action is given by left translation, $L_g(f)(x) = f(g^{-1}x)$, where $g, x \in G$ and $f \in L_2(G)$. We leave it to the reader to check that this is well-defined on L_2 and is a linear action. It is actually a unitary representation; that is each L_g is a unitary operator since $\langle L_g(f_1), L_g(f_2) \rangle = \langle f_1, f_2 \rangle$ for all $g \in G$. $f_1, f_2 \in L_2(G)$ because of the invariance of Haar measure.

L has another important feature. Namely it is continuous in the following sense (called strong continuity). If $g_\nu \to g$ in G and $f \in L_2(G)$, then $L_{g_\nu}(f) \to L_g(f)$. First let $f \in C(G)$. Then by compactness, f is uniformly continuous. So if $\epsilon > 0$ then $|f(g_\nu^{-1}x) - f(g^{-1}x)| < \epsilon$ whenever $g_\nu^{-1}x(g^{-1}x)^{-1} = g_\nu^{-1}xx^{-1}g = g_\nu^{-1}g \in U$, where U is a sufficiently small neighborhood of 1 in G. Hence $||L_{g_\nu}(f) - L_g(f)||_G \leq \epsilon$ and so also. $||L_{g_\nu}(f) - L_g(f)||_2 \leq \epsilon$. Now since Haar measure is regular, $C(G)$ is dense in L_2. So if $f \in L_2$ we can choose $f_1 \in C(G)$ with $||f - f_1|| < \epsilon$. Then

$$||L_{g_\nu}(f) - L_g(f)||_2 \leq ||L_{g_\nu}(f) - L_{g_\nu}(f_1)||_2 + ||L_{g_\nu}(f_1) - L_g(f_1)||_2$$
$$+ ||L_g(f_1) - L_g(f)||_2 \leq 3\epsilon$$

if $g_\nu^{-1}g \in U$.

Now let ϕ be a continuous non-negative function on G which is not identically zero. We can make it symmetric (that is $\phi(x) = \overline{\phi(x^{-1})}$) by replacing ϕ by $\phi(x) + \overline{\phi(x^{-1})}$. We can also have $\int \phi dx = 1$ by normalizing $i.e.$ replacing ϕ by $\frac{\phi}{\int \phi dx}$. Let $k(x, y) = \phi(x^{-1}y)$. Then k is continuous and since ϕ is symmetric $k(x, y) = \overline{k(y, x)}$. Because ϕ is real we see T_k is a compact self-adjoint Fredholm operator. In fact here $T_k(f)$ is called the convolution of ϕ and f and this is precisely the type of integral operator we are interested in. Let Ω be the set of all the eigenfunctions associated with nonzero eigenvalues of all such T_k. Then Ω and therefore also its complex linear span, $l.s._{\mathbb{C}}(\Omega)$, is contained in $C(G)$. We will now show that any continuous function on G is the uniform limit of some complex linear combination of such T_k. To do so requires a lemma, sometimes called the *approximate identity* lemma.

Lemma 5.4.1. *Let $f \in C(G)$, $\epsilon > 0$ and U be a symmetric neighborhood of 1 in G so that $|f(x) - f(y)| < \epsilon$, if $x^{-1}y \in U$. Let ϕ be a function as described above with a support contained in U. Then for all $x \in G$* $|f(x) - T_k(f)(x)| < \epsilon$.

Proof. $|f(x) - T_k(f)(x)| = |f(x) \int_G \phi(y) - \int_G \phi(x^{-1}y)f(y)|$. But by invariance of Haar measure and the fact that ϕ non-negative this is $|f(x) \int_G \phi(x^{-1}y)dy - \int_G \phi(x^{-1}y)f(y)dy| \leq \int_G \phi(x^{-1}y)|f(x) - f(y)|dy$. Now because $\mathrm{Supp}\,\phi_{x^{-1}} \subseteq xU$, we see $|f(x) - T_k(f)(x)| \leq \epsilon \int_{\mathrm{Supp}\,\phi_x} \phi(x^{-1}y) = \epsilon \int_G \phi(x^{-1}y) = \epsilon$. □

Proposition 5.4.2. *l.s.$_{\mathbb{C}}(\Omega)$ is dense in $C(G)$.*

Proof. Let $f \in C(G)$, $\epsilon > 0$ and U be a symmetric neighborhood of 1 in G sufficiently small so that $|f(x) - f(y)| < \epsilon$, if $x^{-1}y \in U$. Choose a neighborhood U_1 of 1 so that $\overline{U_1} \subseteq U$ and by Urysohn's lemma a continuous function $h : G \to [0,1]$ with $\mathrm{Supp}\,h \subseteq U$ and $h \equiv 1$ on U_1. Then this gives rise to a function ϕ as above with $\mathrm{Supp}\,\phi \subseteq U$. By Lemma 5.4.1 $|f(x) - T_k(f)(x)| < \epsilon$ for all $x \in G$ so that $||f - T_k(f)||_G \leq \epsilon$. This proves the proposition since by spectral theorem $T_k(f)$ itself is the uniform limit of a finite linear combination of nonzero eigenfunctions associated with T_k. □

Our next lemma shows that L restricted to V_λ gives a finite dimensional continuous unitary representation of G.

Lemma 5.4.3. *If V_λ is an eigenspace of such a T_k, where $\lambda \neq 0$. Then V_λ (which we know is a finite dimensional $\subseteq C(G) \subseteq L_2$) is invariant under L.*

Proof. We know $\int_G \phi(x^{-1}y)\psi(y)dy = \lambda\psi(x)$, $x \in G$, we apply L and replace x by $g^{-1}x$. Then

$$\int_G \phi((g^{-1}x)^{-1}y)\psi(y)dy = \lambda\psi(g^{-1}x).$$

That is

$$\int_G \phi(x^{-1}gy)\psi(y)dy = \lambda\psi(g^{-1}x).$$

Applying left invariance, the first term is

$$\int_G \phi(x^{-1}gy)\psi(y)dy = \int_G \phi(x^{-1}y)\psi(g^{-1}y)dy,$$

therefore

$$\int_G \phi(x^{-1}y)\psi_g(y)dy = \lambda\psi_g(x).$$

$\square$

Now let Δ be the set of all matrix coefficients of all finite dimensional continuous unitary representations of G and $R(G)$ be $l.s._{\mathbb{C}}\Delta$. $R(G)$ is called the space of *representative functions* on G. We leave it to the reader to check that $R(G)$ is intrinsic to G and does not depend on the choice of basis needed to get these matrices. Notice that $R(G)$ is stable under conjugation since if ρ is a finite dimensional continuous unitary representations of G so is $\bar{\rho}$, its conjugate. If ρ is irreducible so is $\bar{\rho}$.

Exercise 5.4.4. The reader should verify these statements. Particularly the irreducibility of $\bar{\rho}$. Hint use Schur's lemma.

We now show that $l.s._{\mathbb{C}}\Omega \subseteq R(G)$ and $R(G)$ is uniformly dense in $C(G)$. To do so only requires the following.

Lemma 5.4.5. $\Omega \subseteq R(G)$.

Proof. Let $f \in \Omega$. Then for some appropriate k, $T_k(f) = \lambda f$, where $\lambda \neq 0$. Choose an orthonormal basis, $\phi_1, \ldots, \phi_n$ for V_λ. Then $f = \sum_{i=1}^n c_i\phi_i$. Since V_λ is invariant under L we get $L_g(\phi_i) = \sum_{i=1}^n \rho_{ij}(g)\phi_j$. That is $\phi_i(g^{-1}x) = \sum_{i=1}^n \rho_{ij}(g)\phi_j(x)$ for all $g, x \in G$. Taking $x = 1$ and replacing g by its inverse tell us $\phi_i(g) = \sum_{i=1}^n \rho_{ij}(g^{-1})\phi_j(1)$ for all $g \in G$. But since $\rho(g^{-1}) = \rho(g)^{-1} = \rho(g)^*$ and $\rho_{ij}^* = \overline{\rho_{ji}}$ we see $\phi_i(g) = \sum_{i=1}^n \overline{\rho_{ji}(g)}\phi_j(1)$. Thus each $\phi_i \in R(G)$ and since this is a linear space so is f. $\square$

Corollary 5.4.6. *For a compact topological group G, $R(G)$ is uniformly dense in $C(G)$. Also $R(G)$ is dense in L_2 (with respect to $\|\cdot\|$ norm).*

This yields the following which is also called the *Peter-Weyl theorem*.

Corollary 5.4.7. *For a compact topological group G, $R(G)$ separates the points of G.*

Proof. Let $g \neq h \in G$ and suppose $r(g) = r(h)$ for all $r \in R(G)$. Choose a continuous real valued function f such that $f(g) \neq f(h)$. Since $R(G)$ is uniformly dense in $C(G)$ we can choose a representative function r so that $\|r - f\|_G < \frac{1}{2}|f(g) - f(h)|$. Since $|r(g) - f(g)|$ and $|r(h) - f(h)| \leq \frac{1}{2}|f(g) - f(h)|$, it follows that $|f(g) - f(h)| \leq |f(g) - r(g)| + |r(g) - r(h)| + |r(h) - f(h)| < |f(g) - f(h)|$, a contradiction. Now there must be an irreducible representation $\rho \in \mathcal{R}(G)$ satisfying $\rho(g) \neq \rho(h)$. For otherwise by complete reducibility all continuous finite dimensional unitary representations would take the same value on g and h. Hence $r(g) = r(h)$ for all $r \in R(G)$, a contradiction. $\square$

Corollary 5.4.8. *A compact topological group G is isomorphic to a closed subgroup of a product of unitary groups. Conversely, such a group is compact.*

Proof. For each $\rho \in \mathcal{R}(G)$ we get a unitary representation $\rho : G \to U_\rho$. Putting them together gives a continuous homomorphism $G \to \Pi_{\rho \in \mathcal{R}(G)} U_\rho$, a product of unitary groups. Since $R(G)$ separates the points of G this map is injective. By compactness G is homeomorphic (and isomorphic) to its image which is closed. The converse is obvious. $\square$

Corollary 5.4.9. *Given a compact topological group G and a neighborhood U of 1 there is a closed normal subgroup H_U of G contained in U such that G/H_U is isomorphic to a closed subgroup of some $U(n, \mathbb{C})$.*

Proof. Now $G \setminus U$ is closed and therefore compact. By Corollary 5.4.7 each $g \in G \backslash U$ has $\rho_g \in \mathbb{R}(G)$ so that $\rho_g(g) \neq I$. By continuity of ρ_g there is a neighborhood, V_g of g where ρ_g is never the identity anywhere on V_g. Since these V_g cover $G \setminus U$ we have by throwing in U an open covering of G itself. By compactness $G = U \cup V_{g_i}$, the union of a finite number of these. Consider the corresponding ρ_{g_i}. Let $H_U = \cap_i \operatorname{Ker} \rho_{g_i}$. Then H_U is a closed normal subgroup of G. Let $\rho = \oplus_i \rho_{g_i}$. Then ρ is a finite dimensional unitary representation, $\operatorname{Ker} \rho = H_U$ and $G/H_U = \rho(G)$ is a closed subgroup of some unitary group. Let $g \in H_U$. If g is not in U then $g \in V_{g_i}$ for some i. But then $\rho_{g_i}(g) \neq I_{\rho_{g_i}}$, On the other hand since $g \in H_U$, $\rho(g) = I_\rho$ and hence $\rho_{g_i}(g) = I_{\rho_{g_i}}$, a contradiction.   □

Corollary 5.4.10. *A compact Lie group G is isomorphic to a closed subgroup of some* $U(n, \mathbb{C})$ *and conversely.*

Proof. This follows immediately from Corollary 5.4.9 since G has no small subgroups. That is, there is some U containing only the subgroup $\{1\}$. Hence the H_U is trivial and so G is isomorphic to a closed subgroup of some $U(n, \mathbb{C})$. The converse follows from Cartan's theorem.   □

We make a few final remarks concerning the abelian case. Here the irreducibles are all 1-dimensional. That is, they are multiplicative characters $\chi : G \to \mathbb{T}$ and enables us to sharpen the conclusions of the Peter-Weyl theorem in this case.

The next corollary shows that if a compact group has only 1-dimensional irreducible representations it must be abelian since it is embedded in an abelian group.

Corollary 5.4.11. *For a compact abelian topological group G, the characters separate the points and the linear span of the characters is uniformly dense in $C(G)$. G is isomorphic to a closed subgroup of a product of tori and if G is a Lie group it is isomorphic to a closed subgroup of a torus.*

We can now study the (in general infinite dimensional) left regular representation, L.

Definition 5.4.12. If ρ is a finite dimensional irreducible representation of the compact group G we denote by $R(\rho)$ the linear span of the coefficients of ρ *i.e.* the representative functions associated with ρ.

$R(\rho)$ is a subspace of $R(G) \subseteq C(G) \subseteq L_2(G)$. By Section 5.2 dimension is d_ρ^2. If ρ and σ are distinct in $R(G)$ then $R(\rho)$ and $R(\sigma)$ are orthogonal (see Section 5.2). Now the linear span of all $R(\rho)$, as $\rho \in \mathcal{R}(G)$, is $R(G)$. Hence by the Peter-Weyl theorem we have

Corollary 5.4.13. $L_2(G) = \overline{\bigoplus_{\rho \in R(G)} R(\rho)}$ *with* $\{d_\rho^{\frac{1}{2}} \rho_{ij}\}$ *as an orthonormal basis.*

We can look at $R(\rho)$ in another way as follows.

Proposition 5.4.14. $R(\rho)$ *is both a left and right invariant subspace of* L_2. *In particular,* $R(\rho)$ *is also invariant under inner automorphisms.*

Proof. We prove left invariance. Right invariance is done similarly. Let $r(x) = \sum c_{ij}\rho_{ij}(x)$, $c_{ij} \in \mathbb{C}$, be a generic element of $R(\rho)$. Since L is a linear representation it is sufficient to show $L_g\rho_{ij} \in R(\rho)$. But $\rho(g^{-1}x) = \rho(g^{-1})\rho(x)$ so $\rho_{ij}(g^{-1}x) = \sum_k \rho_{ik}(g^{-1})\rho_{k,j}(x) \in R(\rho)$.   □

Thus we have decomposed L_2 into the orthogonal direct sum of perhaps a large number of finite dimensional (closed) *left invariant subspaces*. In order to completely analyze L we merely need to know which irreducibles occur in each of the $R(\rho)$.

Now if τ is a finite dimensional representation of G on V and ρ is an irreducible representation of G, then $[\tau : \rho]$, the multiplicity that ρ occurs in τ, is given in Corollary 5.5.5 below by $\langle \chi_\tau, \chi_\rho \rangle = \int_G \chi_\tau(x)\overline{\chi_\rho(x)}dx$. In our case $\tau = L|_{R(\rho)}$. It can be easily checked that $\chi_{L|_{R(\rho)}}(g) = d_{\bar\rho}\chi_{\bar\rho}$. We also saw that $\bar\rho \in \mathcal{R}(G)$ if ρ is. Thus the multiplicity of $\bar\rho$ in $L|_{R(\rho)}$ is $d_{\bar\rho} = d_\rho$. Since $\dim_{\mathbb{C}} R(\rho) = d_\rho^2$ it follows that $L|_{R(\rho)}$ contains only $\bar\rho$ with multiplicity $d_{\bar\rho}$ and nothing else. Since ρ then occurs in $L|_{R(\bar\rho)}$ with multiplicity $d_{\bar\rho} = d_\rho$ we have

Corollary 5.4.15. *Each irreducible of G occurs in L with a multiplicity equal to its degree.*

In particular, if the group is finite one has

Corollary 5.4.16. *A finite group has a finite number of inequivalent finite dimensional irreducible representation $\rho_1, \ldots, \rho_r$. These are constrained by the requirement $\sum_{i=1}^{r} d_{\rho_i}^2 = |G|$.*

Example 5.4.17. Let $G = S_3$, the symmetric group on 3 letters. This group has two 1-dimensional characters. These are the characters of $S_3/A_3 = \mathbb{Z}_2$ lifted to G. It has no others since $[S_3, S_3] = A_3$. S_3 must have an irreducible of degree $d > 1$ for otherwise it would be abelian see 1.4.20. Since $1^2 + 1^2 + 2^2 = 6$, the order of S_3, we see $|\mathcal{R}(S_3)| = 3$ and the higher dimensional representation has degree 2.

If we consider the two-sided regular representation of G on L_2 (the Haar measure is both left and right invariant and left and right translations commute) then a similar analysis shows that this representation on $R(\rho)$ is now actually irreducible and equivalent to $\bar{\rho} \otimes \rho$. This can be done by calculating the character of this representation (see beginning of the next section). Here $R(\rho)$ is identified with $V_{\bar{\rho}} \otimes V_\rho$ and $\chi_{\bar{\rho} \otimes \rho}(g, h) = \chi_{\bar{\rho}}(g)\chi_\rho(h)$, $g, h \in G$. Applying Proposition 5.5.4 shows that these representations are equivalent and Corollary 5.5.6 that they are irreducible. We leave this verification to the reader as an exercise.

We now turn to the *Plancherel theorem*. Let $\rho \in \mathcal{R}(G)$ and ϕ be an L_1 function. We define $T_\phi(\rho) = \int_G \phi(g)\rho(g)dg$. Thus $T_\phi(\rho)$ is a linear operator on V_ρ. It is called the *Fourier transform* of ϕ at ρ and so each fixed ϕ gives an operator valued function on $\mathcal{R}(G)$, but always taking its value in a different space of operators. Since $\phi \in L_1$ and the coefficients of ρ are bounded, $T_\phi(\rho)$ always exists.

Now let ϕ and $\psi \in L_2(G)$. We want to calculate $\langle \phi, \psi \rangle_{L_2}$ by Fourier analysis. This is exactly what the Plancherel theorem does

$$\langle \phi, \psi \rangle_{L_2} = \sum_{\rho \in R(G)} d_\rho \operatorname{tr}(T_\phi(\rho)T_\psi(\rho)^*),$$

where $*$ means the adjoint operator. Since $L_2 \subseteq L_1$ the Fourier transform applies. To prove this using polarization we may take $\psi = \phi$. Then we get the following formula involving the Hilbert Schmidt norm of an operator.

$$\|\phi\|_2^2 = \sum_{\rho \in R(G)} d_\rho \operatorname{tr}(T_\phi(\rho)T_\phi(\rho)^*).$$

Matrix calculations similar to those involved in the orthogonality relations themselves yield

$$d_\rho \operatorname{tr}(T_\phi(\rho) T_\phi(\rho)^*) = \sum_{i,j=1}^{d_\rho} |d_\rho^{\frac{1}{2}} \int_G \phi(g) \rho_{ij}(g) dg|^2.$$

Since by Corollary 5.4.13 $\{d_\rho^{\frac{1}{2}} \rho_{ij}\}$ form an orthonormal basis, the claim follows from the identity

$$||\phi||_2^2 = \sum \langle \phi, d_\rho^{\frac{1}{2}} \rho_{ij} \rangle^2.$$

Corollary 5.4.18. *Let G be a compact Lie group and ρ_0 be a faithful finite dimensional unitary representation as guaranteed by Corollary 5.4.10. Then each irreducible representation ρ of G is an irreducible component of $\otimes^n \rho_0 \otimes^m \bar{\rho}_0$ for some n and m non-negative integers.*

Proof. Consider the representative functions $\mathcal{F}$ associated with irreducible subrepresentations of $\otimes^n \rho_0 \otimes^m \bar{\rho}_0$ as n and m vary. Since ρ_0 is faithful, $\mathcal{F}$ separates the points. It is clearly stable under conjugation, contains the constants and is a subalgebra of $C(G)$. By the Stone-Weierstrass theorem $\mathcal{F}$ is dense in $C(G)$. Let $\sigma \in \mathcal{R}(G)$. If σ is not equivalent to some irreducible component of $\otimes^n \rho_0 \otimes^m \bar{\rho}_0$ for some n and m then $R(\sigma) \perp \mathcal{F}$. On the other hand given σ_{ij} we can choose $f_\nu \to \sigma_{ij}$ uniformly on G. Therefore $\langle f_\nu, \sigma_{ij} \rangle = 0 \to \langle \sigma_{ij}, \sigma_{ij} \rangle \neq 0$, a contradiction. $\qquad\square$

We conclude this section with the following result concerning infinite dimensional representations of a compact group.

Theorem 5.4.19. *Let γ be a strongly continuous unitary representation of a compact group, G, on a complex Hilbert space, V. Then γ is the direct sum of finite dimensional, continuous, irreducible unitary subrepresentations.*

Notice that here the multiplicities need not be finite. Also observe that it follows from Theorem 5.4.19 that irreducible unitary representations of a compact group on a Hilbert space are finite dimensional.

Before turning to the proof of this result we extend the definition of the Fourier transform to the case of a strongly continuous unitary representation, γ on a complex Hilbert space V. For $f \in L_1(G)$ define $T_f(\gamma) = \int_G f(x)\gamma_x dx$. This is the integral of an operator valued function. Hence if the integral exists, the result is an operator. This integral does indeed exist since $f \in L_1$ and the coefficients, $x \mapsto \langle \gamma_x(v), w \rangle$, $(v, w \in V)$ are all bounded. Hence this operator has the property that $\langle T_f(\gamma)(v), w \rangle = \int_G f(x)\langle \gamma_x(v), w \rangle dx$. Also, since γ is unitary and $T_f(\gamma)(v) = \int_G f(x)\gamma_x(v)dx$. It follows that for $v \in V$,

$$\|T_f(\gamma)(v)\| \leq \|f\|_1 \|v\|. \tag{5.1}$$

Lemma 5.4.20. *For each $f \in L_1$, T_f preserves all γ-invariant subspaces of V.*

Proof. Let W be an invariant subspace and $W^\perp$ be its orthocomplement. We want to prove that if $w \in W$, then $T_f(\gamma)(w) \in W$. That is, $\langle T_f(\gamma)(w), w^\perp \rangle = 0$ for all $w^\perp \in W^\perp$. But $\langle T_f(\gamma)(w), w^\perp \rangle = \int_G f(x)\langle \gamma_x(w), w^\perp \rangle dx$. Since $\gamma_x(w) \in W$ for all $x \in G$ and $w^\perp \in W^\perp$ the integrand is zero. $\square$

Proof of Theorem 5.4.19. Ordering the set of all orthonormal, finite dimensional, irreducible, γ-invariant subspaces of V by inclusion and applying Zorn's lemma shows there is such a maximal set. Let W be the closure of the subspace generated by the subspaces in this maximal set. We want to show that $W = V$. In any case W is a γ-invariant subspace. Hence since γ is unitary the orthocomplement $W^\perp$ is also a γ-invariant subspace (by the same argument we used for the finite dimensional case). Choose a family of functions (approximate identity), f_U, consisting of continuous non-negative functions on G with $\operatorname{Supp} f_U \subseteq U$ and $\int_G f_U dx = 1$, as in the beginning of Section 5.4. By Lemma 5.4.20 for each U, T_{f_U} preserves $W^\perp$. Hence if $v \in W^\perp$, $T_{f_U} v \in W^\perp$ for all U. Let us assume there is such a nonzero v.

Now let $v_1 \in V$. By the Schwarz inequality $|\langle T_{f_U}(v) - v, v_1 \rangle| \leq \int_G |f_U \gamma_x(v) - v| dx \|v_1\|$. Since f_U is non-negative, is supported on U and has integral 1 we get $|\langle T_{f_U} v - v, v_1 \rangle| \leq \sup_{x \in U} \|\gamma_x(v) - v\| \|v_1\|$.

Taking the sup over all v_1 with $||v_1|| \leq 1$ we conclude that $||T_{f_U}v - v|| \leq \sup_{x \in U} ||\gamma_x(v) - v||$. Hence by strong continuity of γ, $||T_{f_U}v - v|| \to 0$ as U shrinks to 1. (The reader will notice the similarity with Lemma 5.4.1.) Finally, since $v \neq 0$, $T_{f_U}v \neq 0$ for some small U. Next we apply the Peter-Weyl theorem to uniformly approximate f_U by a representative function $r \in \mathcal{R}(G)$ to within $\frac{\epsilon}{||v||}$. Since this is also an approximation in L_2 we see $||T_{f_U}v - T_r(v)|| = ||T_{f_U - r}v||$. The latter is $\leq ||f_U - r||_1 ||v|| \leq ||f_U - r||_2 ||v|| < \epsilon$. Thus also $T_r(v) \neq 0$.

Now the linear span of all left translates by $x \in G$ of r lies in a finite dimensional subspace $\mathcal{F}$ of $L_2(G)$ and so gives a finite dimensional continuous unitary representation of G. Let $f_1, \ldots, f_n$ be an orthonormal basis of $\mathcal{F}$. Then $L_g(f_i) = \sum_{j=1}^{n} c_{ij}(g)f_j$. These functions all being in $L_2(G)$ and hence also in $L_1(G)$. Therefore for $g \in G$ and $i = 1, \ldots, n$,

$$\gamma_g T_{f_i}(v) = \gamma_g \int_G f_i(x)\gamma_x(v)dx = \int_G f_i(x)\gamma_g\gamma_x(v)dx = \int_G f_i(x)\gamma_{gx}(v)dx.$$

Now by invariance of the integral under translation this is

$$\int_G f_i(g^{-1}x)\gamma_x(v)dx = T_{L_g(f_i)}(v) = T_{\sum_{j=1}^{n} c_{ij}(g)f_j}(v) = \sum_{j=1}^{n} c_{ij}(g)T_{f_j}(v).$$

Hence, $T_{\mathcal{F}}(v)$ is a finite dimensional γ-invariant subspace of V which lies in $W^{\perp}$ by Lemma 5.4.20. Moreover, it is nontrivial since $T_r(v) \neq 0$. This means it lies in W, a contradiction.

5.5   Characters and Central Functions

Definition 5.5.1. Let ρ be a finite dimensional, continuous, unitary representation of G. We shall call $\chi_\rho(g) = \text{tr}(\rho(g))$, $g \in G$ the *character* of ρ. Then $\chi_\rho : G \to \mathbb{C}$ is a bounded continuous function on G.

Exercise 5.5.2. Why is χ_ρ bounded? Where does it take its largest absolute value?

Corollary 5.5.3. *Let ρ and σ be a finite dimensional, continuous, irreducible, unitary representation of G. Then $\langle \chi_\rho, \chi_\sigma \rangle = 0$ if ρ and σ are inequivalent and $\langle \chi_\rho, \chi_\rho \rangle = 1$.*

Thus the set $\mathcal{X}(G)$ consisting of the characters of the irreducible unitary representations of G form an orthonormal family of functions in $L_2(G)$. In particular they are linearly independent.

Proof. We have,

$$\sum_{i=1}^{d_\rho} \rho_{ii}(g)(\sum_{j=1}^{d_\sigma} \overline{\sigma_{jj}(g)}) = \sum_{i=1}^{d_\rho}\sum_{j=1}^{d_\sigma} \rho_{ii}(g)\overline{\sigma_{jj}(g)}.$$

Hence

$$\int_G \chi_\rho\overline{\chi_\sigma}dg = \sum_{i=1}^{d_\rho}\sum_{j=1}^{d_\sigma} \int \rho_{ii}(g)\overline{\sigma_{jj}(g)}dg.$$

This is clearly 0 if ρ and σ are inequivalent. Now if $\sigma = \rho$, then

$$||\chi_\rho||_2^2 = \sum_{i=1}^{d_\rho}\sum_{j=1}^{d_\rho} \int_G \rho_{ii}(g)\overline{\rho_{jj}(g)}dg.$$

If $i \neq j$ we get 0. Hence $||\chi_\rho||_2^2 = \sum_{i=1}^{d_\rho} \int_G \rho_{ii}(g)\overline{\rho_{ii}(g)}dg = 1$ by Theorem 5.2.1.

$\square$

Proposition 5.5.4. *Let ρ and σ be finite dimensional, continuous, unitary representation of G. Then ρ and σ are equivalent if and only if they have the same character.*

Proof. Evidently, equivalent representations have the same character. We now suppose $\chi_\rho \equiv \chi_\sigma$. Decompose ρ and σ into irreducibles. $\rho = \sum_{i=1}^r n_i\rho_i$, $n_i > 0$ and $\sigma = \sum_{i=k}^s m_i\rho_i$, $m_i > 0$.

After renumbering we can consider the overlap to be from $\rho_k, \ldots, \rho_r$. Then

$$0 = \chi_\rho - \chi_\sigma = \sum_{i=1}^{k-1} n_i\chi_{\rho_i} + \sum_{i=k}^r (n_i - m_i)\chi_{\rho_i} + \sum_{i=r+1}^s -m_i\chi_{\rho_i}.$$

Since the χ_{ρ_i} are linearly independent we conclude $n_i = 0$ for $i = 1, \ldots, k-1$, $n_i = m_i$ for $i = k, \ldots, r$ and $-m_i = 0, i = r+1, \ldots, s$.

But since the n_i and m_i are positive $k = 1$, $r = s$ and $n_i = m_i$ for all $i = 1, \ldots, r$. That is, ρ and σ are equivalent. □

The next result follows from the orthonormality of $\mathcal{X}(G)$ in a similar manner. We leave its proof to the reader as an exercise.

Corollary 5.5.5. *Let ρ be a finite dimensional, continuous, unitary representation of G whose decomposition into irreducibles is $\rho = \sum_{i=1}^{r} n_i \rho_i$, $n_i > 0$. Then $||\chi_\rho||_2^2 = \sum_{i=1}^{r} n_i^2$. In particular ρ is irreducible if and only if $||\chi_\rho||_2^2 = 1$. Moreover, the multiplicity of an irreducible ρ_i in ρ is $\langle \chi_\rho, \chi_{\rho_i} \rangle$.*

We can use our irreducibility criterion to study tensor product representations. Let G and H be compact groups and ρ and σ are finite dimensional continuous representations of G and H respectively. Form the representation $\rho \otimes \sigma$ of $G \times H$ on $V_\rho \otimes V_\sigma$ by defining $\rho \otimes \sigma(g, h) = \rho_g \otimes \sigma_h$.

We leave it to the reader to check that this is a continuous finite dimensional representation of $G \times H$. It is actually unitary, but this does not matter since everything is equivalent to a unitary representation anyway.

Corollary 5.5.6. *If $\rho \in \mathcal{R}(G)$ and $\sigma \in \mathcal{R}(H)$, then $\rho \otimes \sigma \in \mathcal{R}(G \times H)$. Conversely, all irreducibles of $G \times H$ arise in this way.*

Proof. If dg and dh are normalized Haar measures on G and H respectively then $dgdh$ is normalized Haar measure on the compact group $G \times H$. Now $\chi_{\rho \otimes \sigma}(g, h) = \chi_\rho(g)\chi_\sigma(h)$. Hence

$$||\chi_{\rho \otimes \sigma}||_2^2 = \int_G \int_H \chi_\rho(g)\chi_\sigma(h)\overline{\chi_\rho(g)\chi_\sigma(h)}dgdh = ||\chi_\rho||_2^2||\chi_\sigma||_2^2 = 1 \cdot 1 = 1,$$

proving the irreducibility.

Now let $\tau \in \mathcal{R}(G \times H)$ and consider, as before, all $\rho \otimes \sigma$ where $\rho \in \mathcal{R}(G)$ and $\sigma \in \mathcal{R}(H)$. Let $f \in C(G \times H)$. By the Stone-Weierstrass theorem f can be uniformly approximates by the functions of the form $\sum_{i=1}^{n} g_i(x)h_i(y)$, where $g_i \in C(G)$ and $h_i \in C(H)$. But by the Peter-Weyl theorem these in turn can be uniformly approximated by $\phi(x,y) = \sum_{i=1}^{n} r_i(x)s_i(y)$, where $r_i \in R(G)$ and $s_i \in R(H)$. Hence Φ the collection of these ϕ's are the representative functions of the irreducible representations of $G \times H$ which are of the form $\rho \otimes \sigma$, form a uniformly dense linear subspace of $C(G \times H)$. If an irreducible representation τ is not of the form $\rho \otimes \sigma$, then its coefficients must be perpendicular to Φ and therefore to all of $L_2(G \times H)$. In particular it must be orthogonal to itself, a contradiction.   □

Definition 5.5.7. A function $f : G \to \mathbb{C}$ is called a *central function* or a *class function* if $f(xy) = f(yx)$ for all $x, y \in G$. Equivalently, f is a class function if and only if $f(gxg^{-1}) = f(x)$ for all $g, x \in G$. That is, f is constant on conjugacy classes of G. We denote the central functions by $C(G)^G$.

Exercise 5.5.8. Show that these two definitions are equivalent.

   Obvious examples of class functions are characters of finite dimensional, continuous representations ρ of G and since a linear combination of a class function is again such a function, the linear span of all characters is a class function. Pursuing this idea somewhat further, it is quite clear that the uniform limit (even the pointwise limit if the limiting function is continuous) of class functions is again a class function. Thus we know that the elements of $\overline{\mathcal{X}(G)}$ are class functions. It turns out that the converse is also true. Namely,

Theorem 5.5.9. *Every central function is a uniform limit of functions in $\mathcal{X}(G)$ and conversely.*

   Before turning to the proof we need a pair of lemmas.

Lemma 5.5.10. *Let $\rho \in \mathcal{R}(G)$ and $r \in R(\rho)$. If r is central, then $r = \lambda \chi_\rho$, where $\lambda \in \mathbb{C}$.*

Proof. Here $r(x) = \sum_{i,j=1}^{d_\rho} c_{ij} \rho_{ij}(x)$. Since $r(x) = r(gxg^{-1})$ we conclude upon substituting and taking into account the linear independence of ρ_{ij} that $C = \rho(g)^t C \overline{\rho(g)}$. Alternatively, $\rho(g)\bar{C} = \bar{C}\rho(g)$ for all $g \in G$, where C is the matrix of c_{ij}. By Schur's lemma $\bar{C}$ is a scalar multiple of the identity and hence so is C.   □

Lemma 5.5.11. *Suppose that $f \in R(G)$ is central then f is in the linear span of $\mathcal{X}(G)$.*

Proof. $f = \sum_{i=1}^{n} r_i$, where $r_i \in R(\rho_i)$ and the ρ_i are distinct in $\mathcal{R}(G)$. Applying the assumption $f(gxg^{-1}) = f(x)$ and taking into account the linear independence of the r_i and Proposition 5.4.14 tells us each r_i is itself a class function. Hence by Lemma 5.5.10 each $r_i = \lambda_i \chi_{\rho_i}$ and $f = \sum_{i=1}^{n} \lambda_i \chi_{\rho_i}$.   □

Proof of Theorem 5.5.9. We first define a projection operator $\#$: $C(G) \to C(G)^G$ via the formula $f^{\#}(x) = \int_G f(gxg^{-1})dg$. It is easy to see that this gives a continuous function $f^{\#}$, the operator is norm decreasing $\|f^{\#}\|_G \leq \|f\|_G$ and f is central if and only if $f = f^{\#}$. We leave these details to the reader to check.

   Let $f \in C(G)^G$. Then by the Peter-Weyl theorem f can be uniformly approximated on all of G by representative functions, ϕ, $\|f - \phi\|_G < \epsilon$. Apply the $\#$ operator and get $\|f^{\#} - \phi^{\#}\|_G = \|(f - \phi)^{\#}\|_G \leq \|f - \phi\|_G < \epsilon$. On the other hand $f = f^{\#}$ and $\phi^{\#}$ is a central representative function which by Lemma 5.5.11 is a linear combination of characters of $\mathcal{R}(G)$. Thus f is the uniform limit of a linear combination of irreducible characters.

Exercise 5.5.12. The reader should verify the various properties of $\#$ mentioned above as these are necessary to complete the proof of Theorem 5.5.9.

Corollary 5.5.13. *$\mathcal{X}(G)$ separates the conjugacy classes of G.*

Proof. Let C_x and C_y be disjoint conjugacy classes. Since these are disjoint compact sets Urysohn's lemma tells us there is $f \in C(G)$ with $f|_{C_x} = 0$ and $f|_{C_y} = 1$. Applying $\#$ yields $f^{\#}|_{C_x} = 0$ and $f^{\#}|_{C_y} = 1$. Now approximate $f^{\#}$ by a linear combination of characters to within $\frac{1}{2}$ by Theorem 5.5.9. If $\chi_\rho(x) = \chi_\rho(y)$ for every $\rho \in \mathcal{R}(G)$ this would give a contradiction. Hence the conclusion.   □

   Thus the irreducible representations of a compact group are in bijective correspondence with the irreducible characters and, if the group is finite, the characters are in bijective correspondence with the set of conjugacy classes. This is the basis of the so-called character tables of a finite group. Vertically the characters are listed and horizontally the conjugacy classes are listed. Then the table must be filled in with the value of that character on that particular conjugacy class. For example as we saw above, S_3 has exactly 3 characters and therefore also 3 conjugacy classes.

We conclude this section with the functional equation for a character of a representation in $\mathcal{R}(G)$.

Theorem 5.5.14. *Let $f = \chi_\rho$ be the character of a finite dimensional, continuous, irreducible, unitary representation ρ. Then for all x, $y \in G$,*

$$f(x)f(y) = f(1)\int_G f(gxg^{-1}y)dg.$$

Conversely, if f is a continuous function $G \to \mathbb{C}$, not identically zero satisfying this equation, then $\frac{f}{f(1)} = \frac{\chi_\rho}{\chi_\rho(1)}$, for a unique $\rho \in \mathcal{R}(G)$.

Proof. We extend the $\#$ operator defined earlier on functions to representations. For any finite dimensional continuous unitary representation, ρ, let $\rho^\#(x) = \int_G \rho(gxg^{-1})dg$, giving an operator valued function on G. For $y \in G$ using invariance of dg we get

$$\rho(y)\rho^\#(x)\rho(y)^{-1} = \int_G \rho(y)\rho(gxg^{-1})\rho(y)^{-1}dg = \int_G \rho((gy)x(gy)^{-1})dg$$

$$= \int_G \rho(gxg^{-1})dg = \rho^\#(x).$$

Thus $\rho^\#(x)$ is an intertwining operator. If ρ is irreducible $\rho^\#(x) = \lambda(x)I$ and taking traces shows $\lambda(x) = \frac{\operatorname{tr}(\rho^\#(x))}{d_\rho}$. On the other hand,

$$\operatorname{tr}(\rho^\#(x)) = \int_G \operatorname{tr}(\rho(gxg^{-1}))dg = \chi_\rho(x),$$

so that for all $x \in G$,

$$\int_G \rho(gxg^{-1})dg = \frac{\chi_\rho(x)}{d_\rho}I.$$

Hence
$$\int_G \rho(gxg^{-1}y)dg = \frac{\chi_\rho(x)}{d_\rho}\rho(y).$$

Taking traces yields the functional equation

$$\int_G \chi_\rho(gxg^{-1}y)dg = \frac{\chi_\rho(x)\chi_\rho(y)}{d_\rho}.$$

Conversely, let f be an arbitrary continuous function satisfying the functional equation. From it we see $f(1) \neq 0$ for otherwise $f \equiv 0$. Let $y = 1$ in the equation. Then $f(x)f(1) = f(1)\int_G f(gxg^{-1})dg = f(1)f^\#(x)$. Since $f(1) \neq 0$ $f(x) = f^\#(x)$ so f is central. We will show that for every $\rho \in \mathcal{R}(G)$ and every $x \in G$,

$$\frac{f(x)}{f(1)}\langle f, \chi_\rho \rangle = \frac{\chi_\rho(x)}{\chi_\rho(1)}\langle f, \chi_\rho \rangle. \tag{5.2}$$

Having done so we complete the proof by choosing a $\rho \in \mathcal{R}(G)$ such that $\langle f, \chi_\rho \rangle \neq 0$. For then we can cancel and conclude from (5.2) that

$$\frac{f(x)}{f(1)} = \frac{\chi_\rho(x)}{\chi_\rho(1)}. \tag{5.3}$$

Since f is central such a ρ must exist by Theorem 5.5.9. The ρ satisfying (5.3) is unique because the characters of distinct representations are linearly independent. It remains only to prove (5.2). To do so consider $\int_G \int_G f(gxg^{-1}y)\overline{\chi_\rho(y)}dgdy$. By hypothesis this is

$$\int_G \int_G \overline{\chi_\rho(y)}\frac{f(x)f(y)}{f(1)}dgdy = \frac{f(x)}{f(1)}\int_G f(y)\overline{\chi_\rho(y)}dy = \frac{f(x)}{f(1)}\langle f, \chi_\rho \rangle.$$

On the other hand by Fubini's theorem, left translating

$$\int_G \int_G f(gxg^{-1}y)\overline{\chi_\rho(y)}dgdy = \int_G \left(\int_G f(gxg^{-1}y)\overline{\chi_\rho(y)}dy\right)dg$$
$$= \int_G \left(\int_G f(y)\overline{\chi_\rho(gx^{-1}g^{-1}y)}dy\right)dg.$$

Using $\chi_\rho(t^{-1}) = \overline{\chi_\rho(t)}$, the latter is

$$\int_G \int_G f(y)\chi_\rho(y^{-1}gxg^{-1})dydg = \int_G f(y)(\int_G \chi_\rho(y^{-1}gxg^{-1})dg)dy$$
$$= \int_G f(y)(\int_G \chi_\rho(gxg^{-1}y^{-1})dg)dy.$$

By the part of the theorem already proved this is just

$$\int_G f(y)\frac{\chi_\rho(x)\chi_\rho(y^{-1})}{\chi_\rho(1)}dy = \frac{\chi_\rho(x)}{\chi_\rho(1)}\int_G f(y)\overline{\chi_\rho(y)}dy$$

or

$$\frac{\chi_\rho(x)}{\chi_\rho(1)}\langle f, \chi_\rho\rangle.$$

$\square$

5.6   Induced Representations

We now study induced representations of a compact group, G. Recall (see Theorem 2.3.5) that if H is a closed subgroup of G and dg and dh are the respective normalized Haar measures, then there is a (finite) G-invariant measure μ on the homogeneous space G/H satisfying

$$\int_G f(g)dg = \int_{G/H} \int_H f(gh)dhd(\mu).$$

Now suppose we have a finite dimensional representation σ of H on V_σ. We now define the induced representation of σ to G. This representation, written $\text{Ind}(H \uparrow G, \sigma)$, will be infinite dimensional unless H has finite index in G. We consider only finite dimensional, σ, to avoid technical difficulties and because most of the applications we are interested in are in this situation.

Consider the vector space $\mathcal{W}$ consisting of functions $F : G \to V_\sigma$ satisfying

(1) F is measurable,

(2) $F(gh) = \sigma(h)^{-1}F(g)$, for $h \in H$ and $g \in G$,

(3) $\int_{G/H} ||F(g)||^2_{V_\sigma} d(\mu)(\bar{g}) < \infty$.

Such functions clearly form a complex vector space under pointwise operations. If $\langle \cdot, \cdot \rangle_{V_\sigma}$ denotes the Hermitian inner product on V_σ we can use this to define an inner product on this space as follows. For F_1 and F_2 here, the function $g \mapsto \langle F_1(g), F_2(g) \rangle_{V_\sigma}$ is a continuous function on G, which by condition 2 descends to a function on G/H. In particular, $g \mapsto ||F(g)||^2_{V_\sigma}$ is a non-negative measurable function on G/H.

Now $\mathcal{W}$ is actually a Hilbert space whose inner product is given by $\langle F_1, F_2 \rangle = \int_{G/H} \langle F_1(g), F_2(g) \rangle_{V_\sigma} d(\mu)(\bar{g})$. This inner product converges by the Schwarz inequality which comes built in.

$$
\int_{G/H} \langle F_1(g), F_2(g) \rangle_{V_\sigma} d(\mu)(\bar{g})
$$

$$
\leq \int_{G/H} ||F_1(g)||^2_{V_\sigma} d(\mu)(\bar{g}) \int_{G/H} ||F_2(g)||^2_{V_\sigma} d(\mu)(\bar{g}).
$$

Now let G act on $\mathcal{W}$ by left translation $(x \cdot F)(g) = F(x^{-1}g)$, where $F \in \mathcal{W}$ and $x, g \in G$.

Proposition 5.6.1. Ind$(H \uparrow G, \sigma)$ *is a unitary representation of G on* $\mathcal{W}$.

Exercise 5.6.2. The proof of this is routine and is left to the reader. We also leave to the reader to check that the left regular representation, L acting on L_2, is an induced representation. Here σ is the trivial 1-dimensional representation of $H = \{1\}$. (This is a good example of an induced representation to keep in mind.)

We now show $\mathcal{W}$ contains a certain dense set of functions to be described below. Let $f \in C(G, V_\sigma)$, the continuous vector valued functions on G and define for $x \in G$, $F_f(x) = \int_H \sigma(h) f(xh) dh$. Since the integrand is a V_σ valued continuous function on H which is compact the integral exists and is a V_σ valued function $F_f : G \to V_\sigma$ on G.

Lemma 5.6.3. *The F_f are continuous and in* $\mathcal{W}$.

Proof. We prove 2). After we show F_f is continuous, 1) and 3) follow automatically.

$$F_f(gh_1) = \int_H \sigma(h)f(gH_1h)dh = \int_H \sigma(h_1^{-1}h)f(gh)dh$$

$$= \int_H \sigma(h_1^{-1})\sigma(h)f(gh)dh = \sigma(h_1^{-1})\int_H \sigma(h)f(gh)dh.$$

Thus $F_f(gh_1) = \sigma(h_1^{-1})F_f(g)$ proving 2).

Now since f is uniformly continuous given $\epsilon > 0$ there is a neighborhood U of 1 in G so that $\|f(xh) - f(yh)\|_{V_\sigma} < \epsilon$, whenever $h \in H$ and $xy^{-1} \in U$. Therefore

$$\|F_f(x) - F_f(y)\|_{V_\sigma} \le \int_H \|\sigma(h)\|\|f(xh) - f(yh)\|_{V_\sigma}dh.$$

Since σ is unitary we see if $xy^{-1} \in U$, then $\|F_f(x) - F_f(y)\|_{V_\sigma} < \epsilon$. $\square$

Lemma 5.6.4. *The F_f are dense in $\mathcal{W}$.*

Proof. Clearly the continuous functions in $\mathcal{W}$ form a dense subspace of $\mathcal{W}$. We will actually show that the F_f are not only dense, but actually comprise all continuous functions in $\mathcal{W}$. Let F_1 be any continuous function satisfying 2). We want to find an $f \in C(G, V_\sigma)$ so that

$$\|F_1 - F_f\|^2 = \int_{G/H} \langle F_f - F_1, F_f - F_1 \rangle d\mu(\bar{g})$$

is small. Now $F_1(g) = \sigma(h)F_1(gh)$, so

$$F_1(g) = \int_H F_1(g)dh = \int_H \sigma(h)F_1(gh)dh.$$

On the other hand, $F_f(g) = \int_H \sigma(h)f(gh)dh$. Therefore since σ is unitary $\|F_1 - F_f\|^2 = \int_{G/H}\int_H \|f(gh) - F_1(gh)\|_{V_\sigma}^2 dh d\mu(\bar{g})$. Taking $f = F_1$, then this last integral is zero so $F_f = F_1$. Since they are both continuous they are identically equal on G. Therefore the F_f consist of all continuous functions in $\mathcal{W}$, and hence they are dense in $\mathcal{W}$. $\square$

Corollary 5.6.5. Ind$(H \uparrow G, \sigma)$ *is a strongly continuous unitary representation of G on $\mathcal{W}$.*

Proof. We will show that if $g_\nu \to g$, and $F \in \mathcal{W}$ is fixed, then $||L_{g_\nu}(F) - L_g(F)|| \to 0$. Since Ind$(H \uparrow G, \sigma)$ is a unitary representation we have $||L_{g_\nu}(F) - L_g(F)|| = ||L_{g^{-1}g_\nu}(F) - F||$ so we may as well assume $g = 1$. Also if we were to prove this for all F_f, then by density it would hold for all $F \in \mathcal{W}$. We leave this to be checked by the reader. (It is essentially the same argument as in the one for the regular representation in the third paragraph of section 4.) Thus we may assume F is continuous. Hence F is uniformly continuous and $||F(x^{-1}g) - F(g)||_{V_\sigma}^2 < \epsilon^2$, if $x \in U$ a neighborhood of 1 and $g \in G$. Then if $x \in U$,

$$||L_x(F) - F||^2 = \int_{G/H} ||L_x F(\bar{g}) - F(\bar{g})||_{V_\sigma}^2 < \epsilon^2 \mu(G/H).$$

So $||L_x(F) - F|| < \epsilon$. $\qquad\square$

Since by Theorem 5.4.19 any continuous unitary representation of a compact group on a Hilbert space is the direct sum of finite dimensional continuous irreducible unitary representations. In particular,

Corollary 5.6.6. Ind$(H \uparrow G, \sigma)$ *is a direct sum of finite dimensional, continuous, irreducible unitary representations of G on $\mathcal{W}$.*

Exercise 5.6.7. The following is a consequence of the fact that a continuous function $F : G \to V_\sigma$ which satisfies condition 2) is determined by its values on coset representatives of H in G and its proof is left to the reader.

Corollary 5.6.8. *(1)* $\dim_{\mathbb{C}} \mathcal{W}$ *is infinite unless* $[G : H]$ *is finite.*
(2) If $[G : H]$ *is finite, then* $\dim_{\mathbb{C}} \mathcal{W} = [G : H] \dim_{\mathbb{C}} V_\sigma$.

The next result includes the possibility that γ could be induced, or finite dimensional.

Proposition 5.6.9. *Let G be a compact group and γ a strongly continuous representation of G on a Hilbert space, V and $\rho \in \mathcal{R}(G)$. Then $[\gamma : \rho] = \dim_{\mathbb{C}} \mathrm{Hom}_G(V_\rho, V)$.*

Proof. By 5.4.19 we can write V as the orthogonal direct sum of finite dimensional irreducible continuous unitary subrepresentations, $(V_i, \gamma|_{V_i})$, where $i \in I$ and $V = \bigoplus V_i$. Let π_i be the orthogonal projection of V onto V_i. Partition $I = I_1 \cup I_2$, where I_1 contains those representations equivalent to ρ, while I_2 contains those representations which are not equivalent to ρ. For $T \in \mathrm{Hom}_G(V_\rho, V)$, $\pi_i \circ T \in \mathrm{Hom}_G(V_\rho, V_i)$. So if $i \in I_2$, Schur's lemma tells us $\pi_i \circ T = 0$, while if $i \in I_1$, $\pi_i \circ T$ is a scalar multiple of the identity. Thus the former components have dimension 0 while the latter have dimension 1. Let W be the closure of the sum of the V_i for $i \in I_1$. Hence the dimension of $\mathrm{Hom}_G(V_\rho, V)$ is the same as that of $\mathrm{Hom}_G(V_\rho, W)$ which is the cardinality of I_1. $\square$

We now come to the Frobenius reciprocity theorem a particular case of which states that each irreducible representation ρ of G is contained in the induced from σ with the same multiplicity that its restriction contains the irreducible σ of H. In particular the multiplicity of ρ in the induced is always finite.

Theorem 5.6.10. *Let G be a compact group, H be a closed subgroup, σ be any finite dimensional continuous unitary representation of H and ρ a finite dimensional continuous unitary representation of G. Then*

$$\mathrm{Hom}_G(\rho, \mathrm{Ind}(H \uparrow G, \sigma)) \simeq \mathrm{Hom}_H(\rho|_H, \sigma). \qquad (5.4)$$

In particular, these have the same dimension. Hence by Proposition 5.6.9 if ρ and σ are each irreducible, then $[\mathrm{Ind}(H \uparrow G, \sigma) : \rho] = [\rho|_H : \sigma]$.

Proof. Our proof of this result is functorial, In this way it does not really depend on compactness of G at all. For example it also works for any (not necessarily unitary, but) finite dimensional representations if $[G : H] < \infty$. Nor does it depend on irreducibility!

We will prove (5.4) by constructing a vector space isomorphism between them. Let T be a G-linear map $T : V_\rho \to \mathcal{W}$. Then for each $v_\rho \in V_\rho$ we know $T(v_\rho) \in \mathcal{W}$ and so $T(v_\rho)(1) \in V_\sigma$. This gives us a linear map T^* from V_ρ to V_σ. So $T^* \in \mathrm{Hom}_{\mathbb{C}}(V_\rho, V_\sigma)$. Moreover, $T \mapsto T^*$ is itself a $\mathbb{C}$-linear map.

Now let us consider how the action of H fits into this picture. Let $h \in H$. Then $T^*(\rho_h(v_\rho)) = T(\rho_h(v_\rho))(1)$ and since by 2) $F(h^{-1}) = \sigma(h)F(1)$ we get

$$T(\rho_h(v_\rho))(1) = L_h T(v_\rho)(1) = T(v_\rho)(h^{-1}) = \sigma(h)T(v_\rho)(1) = \sigma(h)T^*(v_\rho).$$

This says $T^*\rho_h = \sigma_h T^*$, for all $h \in H$. Thus $T^* \in \mathrm{Hom}_H(\rho|_H, \sigma)$. We now construct the inverse of this map.

Let $S \in \mathrm{Hom}_H(\rho|_H, \sigma)$ and define $S_* : V_\rho \to \mathcal{W}$ by $S_*(v_\rho)(g) = S(\rho_g^{-1}(v_\rho)) \in V_\sigma$. Since $S_*(v_\rho)$ is a mapping from G to V_σ, it has a chance of being in $\mathcal{W}$. Because ρ is continuous as is S, we see $S_*(v_\rho)$ is continuous and so measurable.

It also satisfies 2).

$$\begin{aligned}
S_*(v_\rho)(gh) &= S(\rho(gh)^{-1}(v_\rho)) = S(\rho(h)^{-1}\rho(g)^{-1}(v_\rho)) \\
&= \sigma(h)^{-1}S(\rho(g)^{-1}(v_\rho)) = \sigma(h)^{-1}S_*(v_\rho)(g).
\end{aligned}$$

Since G/H is compact and this function is continuous, it has finite square integrable norm. Thus $S_*(v_\rho) \in \mathcal{W}$ and so we have a linear map $S_* : V_\rho \to \mathcal{W}$.

For $g \in G$, one checks easily that $S \mapsto S_*$ is linear and $L_g S_* = S_* \rho_g$. Hence $S_* \in \mathrm{Hom}_G(V_\rho, \mathcal{W})$. It remains only to see that these maps invert one another.

Now

$$\begin{aligned}
T^*(v_\rho)_*(g) &= T^*(\rho(g)^{-1})(v_\rho) = T\rho(g)^{-1}(v_\rho)(1) \\
&= L_{g^{-1}}T(v_\rho)(1) = T(v_\rho)(g).
\end{aligned}$$

Since this holds for all $g \in G$ and $v_\rho \in V_\rho$ we conclude $T_*^* = T$. Also, $S_*(v_\rho)(1) = S(v_\rho)$. Hence $(S_*)^*(v_\rho) = S(v_\rho)$ so $(S_*)^* = S$. $\square$

5.7   Some Consequences of Frobenius Reciprocity

Let $SO(3, \mathbb{R})$ act on S^2 with isotropy group $SO(2, \mathbb{R})$. Then $SO(3, \mathbb{R})$ also acts on $C(S^2)$ and therefore on $L_2(S^2)$. Hence (see 5.1.3) we can decompose this representation into irreducible components, called *spherical harmonics*. An interesting question is then which spherical harmonics occur and with what multiplicities?

It is easy to verify directly that $SU(2, \mathbb{C})$ is a compact real form of $SL(2, \mathbb{C})$, that is the complexification of $\mathfrak{su}(2, \mathbb{C})$ is $\mathfrak{sl}(2, \mathbb{C})$. The general fact follows from Corollary 7.4.10. Since $SU(2, \mathbb{C})$ is simply connected, its finite dimensional irreducible representations are the same as those of $\mathfrak{su}(2, \mathbb{C})$ by Corollary 1.4.15. Hence the finite dimensional irreducible representations of $SU(2, \mathbb{C})$ are in bijective correspondence with those of $\mathfrak{sl}(2, \mathbb{C})$, that is to say the positive integers by the degree of the representation (see Section 3.1.5). Since $SU(2, \mathbb{C})$ is two-sheeted covering of $SO(3, \mathbb{R})$, its irreducibles are those of odd degree.

Exercise 5.7.1. Show that the irreducibles of $SU(2, \mathbb{C})$ which are trivial on $\pm id$ are those of odd degree.

Theorem 5.7.2. *In the action of* $SO(3, \mathbb{R})$ *on* $L_2(S^2)$ *each irreducible representation of* $SO(3, \mathbb{R})$ *occurs and with multiplicity* 1.

Proof. We know $S^2 = SO(3, \mathbb{R}) / SO(2, \mathbb{R})$. Consider the trivial irreducible representation σ of $SO(2, \mathbb{R})$. Then the representation of $SO(3, \mathbb{R})$ on $L_2(S^2)$ is $\mathrm{Ind}(SO(2, \mathbb{R}) \uparrow SO(3), \sigma)$. If ρ is an irreducible representation of $SO(3)$, then $[\mathrm{Ind}(SO(2) \uparrow SO(3), \sigma) : \rho]$ is the same as $[\rho|_{SO(2,\mathbb{R})} : 1_{SO(2,\mathbb{R})}]$. But we know what the irreducibles of $SO(3)$ are; these are just the irreducibles of $SU(2, \mathbb{C})$ of odd degree, or what is the same thing, the complex Lie algebra irreducibles of $\mathfrak{sl}(2, \mathbb{C})$ of odd degree. So the question is: given an irreducible representation of $\mathfrak{sl}(2, \mathbb{C})$ of odd degree, how many times does its restriction to $\mathfrak{h}$ (the line through H) contain the 0 representation of $\mathfrak{h}$? Our study of these representations tells us the answer is 1. $\square$

We now study the relationship between the representations of G and those of a proper subgroup H.

Proposition 5.7.3. *Let* G *be a compact group and* H *a proper closed subgroup. Then there exists* $\rho \in \mathcal{R}(G) \setminus \{1\}$ *whose restriction to* H *contains the 1-dimensional trivial representation. That is, there exists*

$v_0 \in V_\rho$ with $\rho_h(v_0) = v_0$ for all $h \in H$.

Proof. If $\rho \in \mathcal{R}(G) \setminus \{1\}$ the orthogonality relations show $\int_G \rho_{ij}(x)\overline{1(x)}dx = 0$ for all $i, j = 1, \ldots, d_\rho$. Hence $\int_G r(x)dx = 0$ for all $r \in R(\rho)$ and all such ρ. For each such ρ, $\rho|_H$ is a direct sum of irreducibles, $\sigma^1, \ldots, \sigma^m$ of H. If the statement of the proposition were false none of the σ^i would be 1_H. If $r \in R(\rho)$, then $r|_H$ is a linear combination of the coefficients of the σ^i. By the orthogonality relations on H, $\int_H \sigma^i_{kl}(h)\overline{1(h)}dh = 0$ for all i. Hence $\int_H r(h)dh = 0$. Thus

$$\int_G r(x)dx = \int_H r(h)dh \text{ for all } r \in R(\rho), \rho \in \mathcal{R}(G) \setminus \{1\}. \qquad (5.5)$$

On the other hand consider $\rho = 1_G \in \mathbb{R}(G)$. Here $\rho|_H = 1$ and the representative functions associated with this are $r(x) = \lambda \cdot 1 = \lambda$ so $\int_G r(x)dx = \lambda = \int_H r(h)dh$. Hence (5.5) holds for all $\rho \in \mathcal{R}(G)$. Now let $f \in C(G)$ and $\epsilon > 0$. Choose $r \in R(G)$ so that $\|f - r\|_G < \epsilon$. Then

$$|\int_H f(h)dh - \int_G f(g)dg| \leq |\int_H f(h)dh - \int_H r(h)dh| +$$
$$|\int_H r(h)dh - \int_G r(g)dg| +$$
$$|\int_G r(g)dg - \int_G f(g)dg|$$
$$\leq 2\epsilon$$

and since ϵ is arbitrary, $\int_G f(g)dg = \int_H f(h)dh$ for all $f \in C(G)$. Now $H \neq G$ so there must be another coset, $x_0 H$. Choose a neighborhood U of $x_0 H$ which is disjoint from H and a continuous non-negative real valued function, f, which is $\equiv 1$ on $x_0 H$ and $\equiv 0$ on H. Then $\int_G f(g)dg > 0$ and $\int_H f(h)dh = 0$, a contradiction.   □

Because of Frobenius reciprocity and the fact that $[\rho|_H : 1] \geq 1$ this proposition has the following corollary.

Corollary 5.7.4. *Let G be a compact group and H a proper closed subgroup. Then there exists $\rho \in \mathcal{R}(G) \setminus \{1\}$ for which $[\mathrm{Ind}(H \uparrow G, 1) : \rho] \geq 1$.*

We will prove the following

Theorem 5.7.5. *Let H be a closed subgroup of the compact group G. For each $\sigma \in \mathcal{R}(H)$ there is some $\rho \in \mathcal{R}(G)$ with $[\rho|_H : \sigma] \geq 1$.*

Hence by Frobenius reciprocity we would get

Corollary 5.7.6. *Let H be a closed subgroup of the compact Lie group G. For each $\sigma \in \mathcal{R}(H)$ there is some $\rho \in \mathcal{R}(G)$ with $[\mathrm{Ind}(H \uparrow G, \sigma) : \rho] \geq 1$.*

Proof of Theorem 5.7.5. Let ρ_0 be faithful representation of G (and also H). Earlier 5.4.18 we proved that σ is a subrepresentation of $\rho_0^{(n)} \otimes \rho_0^{-(m)}|_H$ for some choice of n and m. Therefore σ is a subrepresentation of some irreducible component of ρ since ρ is an irreducible component of some $\rho_0^{(n)} \otimes \rho_0^{-(m)}$. Therefore some irreducible component ρ of $\rho_0^{(n)} \otimes \rho_0^{-(m)}$ must be restricted to σ.

Exercise 5.7.7. Let G be a compact group and H be a closed subgroup.

(1) Each $\sigma \in \mathcal{R}(H)$ is an irreducible component of the restriction $\rho|_H$ of some $\rho \in \mathcal{R}(G)$.

(2) If H happens to be a Lie group, then there is a finite dimensional continuous representation ρ of G whose restriction to H is faithful.

(3) The restriction map $R(G) \to R(H)$ is surjective.

We conclude this chapter with the following result connected with equivariant imbeddings of compact G-spaces.

Theorem 5.7.8. *Let H be a closed subgroup of the compact Lie group G. Then there exists a finite dimensional continuous unitary representation ρ of G on V_ρ and a nonzero vector $v_0 \in V_\rho$ so that $H = \mathrm{Stab}_G(v_0)$.*

We first need a lemma which tells us that the dimension together with the number of components determines the size of a compact Lie group.

Lemma 5.7.9. *Let G be a compact Lie group and $G \supseteq G_1 \supseteq G_2 \dots$ be a chain closed subgroups. Then this chain must eventually stabilize.*

Proof. Since $\dim G_i \geq \dim G_{i+1} \dots$ and all these dimensions are finite, then for $i \geq n_0$ all the dimensions must be constant. Hence for $i \geq n_0$ each G_{i+1} is open in G_i which is itself open in G_{n_0}, the number of components of G_{i+1} is $\leq$ the number of components of G_i which is $\leq$ the number of components of G_{n_0} Since G_{n_0} is closed in G, it is compact and therefore has a finite number of components. It follows that eventually these must also stabilize and hence the conclusion. $\square$

Proof of Theorem 5.7.8. We may assume $H < G$ since if $H = G$ we may take ρ to be the trivial 1-dimensional representation and v_0 any nonzero vector. We will now prove

(**) If $g_0 \in G - H$, then there exists a representation ρ of G on V and $v_0 \neq 0 \in V_\rho$ such that $\rho_{g_0}(v_0) \neq 0$ and $\rho_h(v_0) = 0$ for all $h \in H$.

Suppose we can do this. Then $G \supseteq \mathrm{Stab}_G(v_0) \supseteq H$ and g_0 is not in $\mathrm{Stab}_G(v_0)$. Replacing G by the closed and therefore compact group $\mathrm{Stab}_G(v_0)$ we can apply (**) again to this subgroup. In this way we get a descending chain of closed subgroups terminating in H which must terminate by Lemma 5.7.9. Therefore they must terminate in H. This would prove the Theorem.

Proof of (**). Since H and Hg_0^{-1} are disjoint compact sets we can find a continuous function f on G for which $f|_H = \alpha$ and $f|_{Hg_0^{-1}} = \beta$, where $\alpha < \beta$. Approximate f by $r \in R(G)$ to within $\epsilon = \frac{\beta - \alpha}{2}$. Let $F(g) = \int_H r(hg)dh$. Then F is continuous and therefore in $L_2(G)$. Since $r \in R(G)$, $F \in R(G)$ also. For $h_1 \in H$,

$$F(h_1) = \int_H r(hh_1)dh = \int_H r(h)dh \leq \epsilon + \alpha.$$

So $F|_H \leq \epsilon + \alpha$. On the other hand,

$$F(h_1 g_0^{-1}) = \int_H r(hh_1 g_0^{-1}) dh = \int_H r(hg_0^{-1}) dh \geq \beta - \epsilon.$$

So $F|_{Hg_0^{-1}} \geq \beta - \epsilon$. In particular, $F(1) \leq \epsilon + \alpha$ and $F(g_0^{-1}) \geq \beta - \epsilon$ so $F(1) \neq F(g_0^{-1})$. Now apply L, the left regular representation of G on L_2. Hence because $L_{g_0} F(1) = F(g_0^{-1})$ we see $L_{g_0} F \neq F$. On the other hand,

$$L_{h_1}(F)(g) = F(h_1^{-1} g) = \int_H r(hh_1^{-1} g) dh = \int_H r(hg) dh = F(g).$$

Thus $L_{h_1}(F) = F$ for all $h_1 \in H$. Since $F \in R(G)$ it lies in a finite dimensional L-invariant subspace V_ρ of $C(G)$. So there is a finite dimensional continuous unitary representation ρ of G and a nonzero vector F in it with $\rho_h(F) = F$ for all $h \in H$ and $\rho_{g_0}(F) \neq F$, proving (**).

Chapter 6

Symmetric Spaces of Non-compact Type

6.1 Introduction

In this chapter we shall give an introduction to symmetric spaces of non-compact type. This subject, largely the creation of Élie Cartan (1869-1951), is of fundamental importance both to geometry and Lie theory. Indeed, one of the great achievements of the mathematics of the first half of the twentieth century was E. Cartan's discovery of the fact that these two categories correspond exactly. Namely, given a connected, centerless, real semisimple Lie group G without compact factors there is associated to it a unique symmetric space of non-compact type. This is G/K, where K is a maximal compact subgroup of G and G/K takes the Riemannian metric induced from the Killing form of G. Conversely, if one starts with an arbitrary symmetric space, X, none of whose irreducible constituents is either compact or $\mathbb{R}^n$, then $X = G/K$, where G is the identity component of the isometry group of X. Here G is a centerless, real semisimple Lie group without compact factors. Thus, we have a bijective correspondence between the two categories and this fact underlies an important reason why differential geometry and Lie theory are so closely bound. As one might expect, this close relationship between the two will show up in some of the proofs. For the details of all this, see [32] and [61]. Also, [32] has a particularly convenient and useful chapter on differential geometry. Concerning this correspondence, the same may be said of Euclidean space and its group of isometries, or of compact semisimple groups and symmetric spaces of compact type, which were also studied by E. Cartan. However, we shall not deal with these here. Taken as a whole, Cartan's work on symmetric spaces can be considered as the completion of the well-known "Erlanger Program" first formulated by F. Klein in 1872. In particular, it ties together Euclidean, elliptic and hyperbolic geometry in any dimension.

Before turning to our subject proper it might be helpful to consider a most important example, namely that of $G = \text{SL}(2, \mathbb{R})$ and X the hyperbolic plane, which we view here as the Poincaré upper half plane, H^+, consisting of all complex numbers $z = x + iy$, where $y > 0$. We let G act on H^+ by fractional linear transformations, $g \cdot z = \frac{az+b}{cz+d}$,

$$g = \begin{pmatrix} a & b \\ c & d \end{pmatrix}$$

where a, b, c and d are real and $\det g = 1$. Since $\Im(\frac{az+b}{cz+d}) = \frac{\Im(z)}{|cz+d|^2} > 0$, we see that $g \cdot z \in H^+$. It is easy to verify that this is an action. Now this action is transitive. Let $c = 0$, then $a \neq 0$ and $d = \frac{1}{a}$. Then $g \cdot i = a^2 i + ab$. Evidently, by varying $a > 0$ and $b \in \mathbb{R}$ this gives all of H^+. A moment's reflection tells us that the isotropy group, $\text{Stab}_G(i)$, is given by $a = d$ and $c = -b$. Since $\det g = a^2 + b^2 = 1$, we see

$$\text{Stab}_G(i) = \{g : g = \begin{pmatrix} \cos t & \sin t \\ -\sin t & \cos t \end{pmatrix} : t \in \mathbb{R}\}.$$

On H^+ we place the Riemannian metric $ds^2 = \frac{dx^2 + dy^2}{y^2}$ (meaning the hyperbolic metric $ds = \frac{ds_{Euc}}{\Im(z)}$) and check that G acts by isometries on H^+ (for this see, for example, p. 118 of [55]). Since G is connected, its image, $\text{PSL}(2, \mathbb{R})$, is contained in $\text{Isom}_0(H^+)$. Actually it is $\text{Isom}_0(H^+)$ but that will not matter. From the point of view of the symmetric space it does not even matter whether we take $\text{SL}(2, \mathbb{R})$ or $\text{PSL}(2, \mathbb{R})$. However, we note that $\text{PSL}(2, \mathbb{R})$, the group that is really acting, is the centerless version.

Another model for this symmetric space is the unit disk, $D \subseteq \mathbb{C}$, called the disk model. It takes the metric $ds^2 = 4\frac{dx^2+dy^2}{(1-r^2)^2}$ and has the advantage of radial symmetry about the origin, 0. Here r is the usual radial distance from 0. The quantity 4, as we shall see, makes D isometric with H^+, or put another way, it normalizes the curvature on D to be -1. Now the Cayley transform $c(z) = \frac{z-i}{z+i}$ maps H^+ diffeomorphically onto D. Its derivative is $c'(z) = \frac{2i}{(z+i)^2}$. A direct calculation shows that for $z \in H^+$

$$\frac{2|c'(z)|}{1 - |c(z)|^2} = \frac{1}{\Im(z)}.$$

Using this we see that if $w = c(z)$, then $|dw| = |c'(z)||dz|$ and so

$$\frac{2|dw|}{1 - |w|^2} = \frac{2|c'(z)|}{1 - |c(z)|^2}|dz| = \frac{|dz|}{|\Im(z)|}.$$

Thus c is an isometry. Of course in the form of the disk, the group of isometries and its connected component will superficially look different.

Example 6.1.1. The action of $SL(2, \mathbb{R})$ on the upper half plane can be generalized in two different ways. One is $SO(n, 1)$ acting on hyperbolic n-space which will be discussed in detail in Section 6.4. The other is the *Siegel generalized upper half space* consisting of $z = x + iy$, where $x \in$ Symm$(n, \mathbb{R})$, $n \times n$ the real symmetric matrices, and $y \in$ Symm$(n, \mathbb{R})^+$, the positive definite real symmetric matrices. The action of $Sp(n, \mathbb{R})$ on Z is given by $g \cdot z = \frac{az+b}{cz+d}$.
Notice when $n = 1$, just as $Sp(1, \mathbb{R}) = SL(2, \mathbb{R})$, Z is the usual upper half plane and the action is also the usual one. In general, since as we shall see Symm$(n, \mathbb{R})$ and Symm$(n, \mathbb{R})^+$ are diffeomorphic and each has dimension $\frac{n(n+1)}{2}$, so Z has dimension $n(n + 1)$. Finally, because a maximal compact subgroup of $Sp(n, \mathbb{R})$ is $U(n, \mathbb{C})$, $Sp(n, \mathbb{R})/U(n, \mathbb{C}) = Z$.

Exercise 6.1.2. Prove:

(1) $g \cdot z = \frac{az+b}{cz+d}$ defines a transitive action of $Sp(n, \mathbb{R})$ on Z.
(2) The isotropy group of $i1_{n \times n}$ is $U(n, \mathbb{C})$.
(3) $U(n, \mathbb{C})$ is a maximal compact subgroup of $Sp(n, \mathbb{R})$.

6.2   The Polar Decomposition

We shall begin by studying the exponential map on certain specific manifolds.

As usual $n \times n$ complex matrices will be denoted by $\mathfrak{gl}(n, \mathbb{C})$ and the real ones by $\mathfrak{gl}(n, \mathbb{R})$. Denote by $\mathcal{H}$ the set of all Hermitian matrices in $\mathfrak{gl}(n, \mathbb{C})$ and by H the positive definite ones. It is easy to see that $\mathcal{H}$ is a real (but not a complex!) vector space of $\dim n^2$. Similarly, we denote by $\mathcal{P}$ the symmetric matrices in $\mathfrak{gl}(n, \mathbb{R})$ and by P those that are positive definite. $\mathcal{P}$ is a real vector space of $\dim \frac{n(n+1)}{2}$. As we shall see, H and P and some of their subspaces will actually comprise all symmetric spaces of non-compact type.

Proposition 6.2.1. *P and H are open in $\mathcal{P}$ and $\mathcal{H}$, respectively. As open sets in a real vector space each is, in a natural way, a smooth manifold of the appropriate dimension.*

Proof. Let $p(z) = \sum_i p_i z^i$ and $q(z) = \sum_i q_i z^i$ be polynomials of degree n with complex coefficients, let $z_1 \dots z_n$ and $w_1 \dots w_n$ denote their respective roots counted according to multiplicity and let $\epsilon > 0$. It follows from Rouché's theorem (see [55]) that there exists a sufficiently small $\delta > 0$ so that if for all $i = 0, \dots, n$, $|p_i - q_i| < \delta$, then after a possible reordering of the w_i's, $|z_i - w_i| < \epsilon$ for all i. Suppose H were not open in $\mathcal{H}$. Then there would be an $h \in H$ and a sequence $x_j \in \mathcal{H} - H$ converging to h in $\mathfrak{gl}(n, \mathbb{C})$. Since h is positive definite, all its eigenvalues are positive. Choose ϵ so small that the union of the ϵ balls about the eigenvalues of h lies in the right half plane. Since the coefficients of the characteristic polynomial of an operator are polynomials and therefore continuous functions of the matrix coefficients and x_j converges to h, for j sufficiently large, the coefficients of the characteristic polynomial of x_j are in a δ-neighborhood of the corresponding coefficient of the characteristic polynomial of h. Hence all the eigenvalues of such an x_j are positive. This contradicts the fact that none of the x_j are in H, proving H is open in $\mathcal{H}$. Intersecting everything in sight with $\mathfrak{gl}(n, \mathbb{R})$ shows that P is also open in $\mathcal{P}$. $\qquad\square$

Proposition 6.2.2. *Upon restriction, the exponential map of* $\mathfrak{gl}(n,\mathbb{C})$ *is a diffeomorphism between* $\mathcal{H}$ *and* H. *Its inverse is given by*

$$\operatorname{Log} h = \log(\operatorname{tr} h)I - \sum_{i=1}^{\infty}(I - \frac{h}{\operatorname{tr} h})^i/i,$$

which is a smooth function on H.

As a consequence we see that the restriction of Exp to any real subspace of $\mathcal{H}$ gives a diffeomorphism of the subspace with its image. In particular, Exp is a diffeomorphism between $\mathcal{P}$ and P. In particular, in all these cases Exp is a bijection.

Proof. We shall do this for $\mathcal{H}$, the real case being completely analogous. Suppose $h \in H$ is diagonal with eigenvalues $h_i > 0$. Then $\operatorname{tr}(h) > 0$ and $0 < \frac{h_i}{\operatorname{tr}(h)}$ so $\log(\operatorname{tr}(h))$ is well-defined and $\log(\frac{h_i}{\operatorname{tr}(h)})$ is defined for all i. But since $0 < \frac{h_i}{\operatorname{tr}(h)} < 1$, we see that $0 < (1 - \frac{h_i}{\operatorname{tr}(h)}) < 1$ for all positive integers k. Hence $\operatorname{Log}(\frac{h}{\operatorname{tr}(h)})$ is given by an absolutely convergent power series $-\sum_{i=1}^{\infty}(I - \frac{h}{\operatorname{tr} h})^i/i$. If u is a unitary operator so that uhu^{-1} is diagonal, then $\operatorname{tr}(uhu^{-1}) = \operatorname{tr}(h)$ and since conjugation by u commutes with any convergent power series, this series actually converges for all $h \in H$ and is a smooth function Log on H. Because on the diagonal part of H this function inverts Exp, and both Exp and this power series commute with conjugation, it inverts Exp everywhere on H. Finally, $\log(\operatorname{tr}(h))I$ and $\operatorname{Log}(\frac{h}{\operatorname{tr} h})$ commute and Exp of a sum of commuting matrices is the product of the Exp's. Since Log inverts Exp on the diagonal part of H it follows that

$$\operatorname{Log}(h) = \log(\operatorname{tr}(h))I + \operatorname{Log}(\frac{h}{\operatorname{tr} h}) = \log(\operatorname{tr}(h))I - \sum_{i=1}^{\infty}(I - \frac{h}{\operatorname{tr} h})^i/i.$$

$\square$

We shall need the following elementary fact whose proof is left to the reader.

Lemma 6.2.3. *For any* $g \in \operatorname{GL}(n,\mathbb{C})$, $g^*g \in H$.

It follows that for all $g \in \mathrm{GL}(n, \mathbb{C})$, $\mathrm{Log}(g^*g) \in \mathcal{H}$ and since this is a real linear space also $\frac{1}{2}\mathrm{Log}(g^*g) \in \mathcal{H}$. This means we can apply Exp and conclude the following:

Corollary 6.2.4. $h(g) = \mathrm{Exp}(\frac{1}{2}\mathrm{Log}(g^*g)) \in H$ *is a smooth map from* $\mathrm{GL}(n, \mathbb{C}) \to H$.

Hence $h(g)^n = \mathrm{Exp}(\frac{n}{2}\mathrm{Log}(g^*g)) \in H$ for every $n \in \mathbb{Z}$. In particular, $h(g)^{-2} = \mathrm{Exp}(\frac{2}{2}\mathrm{Log}(g^*g)) = g^*g$. So that

$$gh(g)^{-1}(gh(g)^{-1})^* = gh(g)^{-1}h(g)^{-1*}g^* = gh(g)^{-2}g^*$$

and, since $h(g)^{-1} \in H$, $g(g^*g)^{-1}g^* = I$. Thus, $gh(g)^{-1} = u(g)$ is unitary for each $g \in \mathrm{GL}(n, \mathbb{C})$. Since group multiplication and inversion are smooth, $g \mapsto u(g)$ is also a smooth function on $\mathrm{GL}(n, \mathbb{C})$ (as is $h(g)$). Now this decomposition $g = uh$, where $u \in \mathrm{U}(n, \mathbb{C})$ and $h \in H$ is actually unique. To see this, suppose $u_1h_1 = g = u_2h_2$. Then $u_2^{-1}u_1 = h_2h_1^{-1}$ so that $h_2h_1^{-1}$ is unitary. This means $(h_2h_1^{-1})^* = (h_2h_1^{-1})^{-1}$ and hence $h_1^2 = h_2^2$. But since h_1 and $h_2 \in H$, each is an exponential of something in $\mathcal{H}$; $h_i = \exp x_i$. But then $h_i^2 = \exp 2x_i$ and since exp is $1:1$ on $\mathcal{H}$, we get $2x_1 = 2x_2$ so $x_1 = x_2$ and therefore $h_1 = h_2$ and $u_1 = u_2$. The upshot of all this is that we have a smooth map $\mathrm{GL}(n, \mathbb{C}) \to \mathrm{U}(n, \mathbb{C}) \times H$ given by $g \mapsto (u(g), h(g))$. Since $g = u(g)h(g)$ for every g (multiplication in the Lie group $\mathrm{GL}(n, \mathbb{C})$), this map is surjective and has a smooth inverse. We summarize these facts as follows:

Theorem 6.2.5. *(polar decomposition) The map* $g \mapsto (u(g), h(g))$ *gives a real analytic diffeomorphism* $\mathrm{GL}(n, \mathbb{C}) \to \mathrm{U}(n, \mathbb{C}) \times H$. *Identical reasoning also shows that as a smooth manifold* $\mathrm{GL}(n, \mathbb{R})$ *is diffeomorphic to* $\mathrm{O}(n, \mathbb{R}) \times P$.

From this it follows that, since H and P are each diffeomorphic with a Euclidean space, and therefore are topologically trivial, in each case the topology of the non-compact group is completely determined by that of the compact one. In this situation, one calls the compact group a *deformation retract* of the non-compact group. Since P and H are diffeomorphic images under Exp of some Euclidean space, one calls them *exponential submanifolds.* For example, connectedness, the number of components, simple connectedness and the fundamental group of the non-compact group are each the same as that of the compact one. Thus for all $n \geq 1$, $\mathrm{GL}(n, \mathbb{C})$ is connected and its fundamental group is $\mathbb{Z}$, while for all n, $\mathrm{GL}(n, \mathbb{R})$ has 2 components and the fundamental group of its identity component is $\mathbb{Z}_2$ for $n \geq 3$ and $\mathbb{Z}$ for $n = 2$ (see Section 1.5).

As a final application of the polar decomposition theorem we have the following inequality which is a variant of one in Margulis [41] p. 169.

Corollary 6.2.6. *Let T be a linear transformation on a finite dimensional real or complex vector space V of dimension n and $|| \cdot ||$ be the Hilbert-Schmidt norm on $\mathrm{End}(V)$. Then $|\det T| \leq ||T||^n$. See Section 6.5 for the definition of the Hilbert-Schmidt norm.*

Proof. Clearly we may assume T is invertible, since otherwise $|\det T| = 0$. For $T \in \mathrm{GL}(V)$ write the polar decomposition $T = kp$. Then since $|\det k| = 1$, $|\det T| = |\det p|$ and

$$||T||^2 = \mathrm{tr}(kp(kp)^*) = \mathrm{tr}(kpp^*k^{-1}) = \mathrm{tr}(pp^*) = ||p||^2.$$

Thus we may assume T is positive definite symmetric, or Hermitian. As such it is diagonalizable $T = kDk^{-1}$. Thus $|\det T| = |\det D|$ and $||T||^2 = ||D||^2$ so we may actually assume T is diagonal with positive eigenvalues, $d_1, \ldots, d_n$. We have to show $(d_1 \ldots d_n)^{\frac{1}{n}} \leq \sqrt{\sum_{i=1}^{n} d_i^2}$. Now the geometric mean is less than or equal to the arithmetic mean $(d_1 \ldots d_n)^{\frac{1}{n}} \leq \frac{1}{n} \sum_{i=1}^{n} d_i$, so we show $\sum_{i=1}^{n} d_i \leq n\sqrt{\sum_{i=1}^{n} d_i^2}$, or $(\sum_{i=1}^{n} d_i)^2 \leq n^2 \sum_{i=1}^{n} d_i^2$. By the Schwarz inequality $(\sum_{i=1}^{n} d_i)^2 \leq n \sum_{i=1}^{n} d_i^2$. Thus, the question is just $n \sum_{i=1}^{n} d_i^2 \leq n^2 \sum_{i=1}^{n} d_i^2$ which is true since $\sum_{i=1}^{n} d_i^2 > 0$ and $n \geq 1$. $\qquad\square$

6.3   The Cartan Decomposition

We now turn to more general groups G and also streamline our notation. Instead of $\mathcal{H}$, we shall consider certain real subspaces of $\mathcal{H}$ denoted by $\mathfrak{p}$ whose exponential image will be P and make the following definition.

Definition 6.3.1. Let G be a Lie subgroup of $\mathrm{GL}(n, \mathbb{R})$ with Lie algebra $\mathfrak{g}$. We denote by $K = \mathrm{O}(n, \mathbb{R}) \cap G$, by P the positive definite symmetric matrices in G, by $\mathfrak{p}$ the symmetric matrices of $\mathfrak{g}$, and by $\mathfrak{k}$ the skew symmetric matrices in $\mathfrak{g}$. In the case that G is a Lie subgroup of $\mathrm{GL}(n, \mathbb{C})$ we again denote its Lie algebra by $\mathfrak{g}$, but now $K = \mathrm{U}(n, \mathbb{C}) \cap G$, P is the positive definite Hermitian matrices of G, $\mathfrak{p}$ is Hermitian matrices of $\mathfrak{g}$ and $\mathfrak{k}$ the skew Hermitian matrices in $\mathfrak{g}$.

Lemma 6.3.2. Let $q(t) = \sum_{j=1}^{n} c_j \exp(b_j t)$ be a trigonometric polynomial, where $c_j \in \mathbb{C}$, and b_j and $t \in \mathbb{R}$. If q vanishes for an unbounded set of real t's, then $q \equiv 0$.

An immediate consequence is that for a polynomial $p \in \mathbb{C}[z_1, \ldots, z_n]$ in n complex variables with complex coefficients and $(x_1, \ldots, x_n) \in \mathbb{R}^n$, if $p(\exp(tx_1), \ldots, \exp(tx_n))$ vanishes for an unbounded set of real t's, then it vanishes identically in t.

Proof. First we can assume that the t's for which q vanishes tend to $+\infty$. Otherwise, they would have to tend to $-\infty$ and in this case we just let $p(t) = q(-t)$. Then p is also a trigonometric polynomial and if $p = 0$, then so is q. Reorder the b_j's, if necessary, so that they are strictly increasing by combining terms by adding the corresponding c_j's. Of course, we can now assume that all the c_j's are nonzero. Let t_k be a sequence tending to $+\infty$ on which q vanishes. Suppose there are two or more b_j's. Since

$$\frac{q(t)}{c_n \exp(b_n t)} = \sum_{j=1}^{n-1} \frac{c_j}{c_n} \exp((b_j - b_n)t) + 1,$$

it follows that $\frac{q(t)}{c_n \exp(b_n t)} \to 1$ as $k \to \infty$. But since q is identically 0 in k so is this quotient, a contradiction. This means that all the b_j's are equal and so $q(t) = c \exp(bt)$ for some $c \in \mathbb{C}$ and $b \in \mathbb{R}$. This function cannot have an infinite number of zeros unless $c = 0$, that is $q = 0$.

□

Proposition 6.3.3. *Suppose M is an algebraic subgroup of* $GL(n, \mathbb{C})$ *and G be a Lie subgroup of* $GL(n, \mathbb{R})$ *(or* $GL(n, \mathbb{C})$*) with Lie algebra* $\mathfrak{g}$. *Let G have finite index in $M_{\mathbb{R}}$ (respectively M). If $X \in \mathcal{H}$ and* $\exp X \in G$, *then* $\exp tX \in P$ *for all real t. In particular, $X \in \mathfrak{g}$ and hence $X \in \mathfrak{p}$.*

Proof. To avoid circumlocutions we shall prove the complex case, the real case being completely analogous. Choose $u \in U(n, \mathbb{C})$ so that uXu^{-1} is diagonal with real eigenvalues λ_j. Replace G by uGu^{-1}, a Lie subgroup of $GL(n, \mathbb{C})$ which is contained in uMu^{-1} with finite index. Now uMu^{-1} is an algebraic subgroup of $GL(n, \mathbb{C})$ (and in the real case $uM_{\mathbb{R}}u^{-1} = (uMu^{-1})_{\mathbb{R}}$). Hence we can assume X is diagonal. Let $p(z_{ij})$ be one of the complex polynomials defining M. Since $\exp X \in G$ and G is a group, $\exp kX \in G \subseteq M$ for all $k \in \mathbb{Z}$. But $\exp kX$ is diagonal with diagonal entries $\exp(k\lambda_j)$. Applying p to $\exp kX$, we get $p(\exp kX) = 0$ for all k. By the corollary, $p(\exp tX) = 0$ for all t. Because p was an arbitrary polynomial defining M, it follows that $\exp tX \in M$ for all real t. Since G has finite index in M and the 1-parameter group $\exp tX$ is connected, it must lie entirely in G and therefore in P. Hence $X \in \mathfrak{g}$. $\qquad\square$

Definition 6.3.4. A subgroup G of $GL(n, \mathbb{R})$ (or $GL(n, \mathbb{C})$) is called *self-adjoint* if it is stable under taking transpose (respectively $*$). Here transpose and $*$ refer to any linear involution (respectively conjugate linear involution) on $\mathbb{R}^n$ (respectively $\mathbb{C}^n$).

For example, $SL(n, \mathbb{R})$ and $SL(n, \mathbb{C})$ are self-adjoint since $\det g^t = \det g$ ($\det g^* = \overline{\det(g)}$). The routine calculations showing $O(n, C)$, $SO(n, \mathbb{C})$, $O(p, q)$ and $SO(p, q)$ are also self-adjoint are left to the reader. In fact, the reader can check that any classical non-compact simple group in E. Cartan's list (see [32]) is self-adjoint. Clearly by their very definition these groups are either algebraic or have finite index in the real points of an algebraic group (essentially algebraic). Now it is an important insight of Mostow [57] that any linear real semisimple Lie group is self-adjoint under an appropriate involution. Moreover, by the root space decomposition, Section 7.3, the adjoint group of any semisimple group without compact factors is algebraic (actually over $\mathbb{Q}$). Thus here we are really talking about all the semisimple groups without compact factors and, of course, this means our construction actually gives all symmetric spaces of non-compact type. *But even if we did not know this, since any classical non-compact simple group is easily seen to be self-adjoint as well as essentially algebraic, we already get a plethora of symmetric spaces from them.*

Particular cases of Theorem 6.3.5 below are the following. We shall leave their routine verification to the reader. $\mathrm{SL}(n, \mathbb{R})$ is diffeomorphic with $\mathrm{SO}(n) \times P_1$, where P_1 is the positive definite symmetric matrices of det 1, which in turn is diffeomorphic under exp with the linear space of real symmetric matrices of trace 0. Similarly, $\mathrm{SL}(n, \mathbb{C})$ is diffeomorphic with $\mathrm{SU}(n) \times H_1$, where H_1 is the positive definite Hermitian matrices of det 1, which in turn is diffeomorphic with the linear space of Hermitian matrices trace 0. As deformation retracts, similar conclusions can be drawn about the topology of these, as well as the other groups mentioned earlier.

The following result is a special case of the general Iwasawa decomposition theorem which holds for an arbitrary Lie group with a finite number of components, but with a somewhat more elaborate formulation (see G.P. Hochschild [33]). Here, we content ourselves with the matter at hand, namely self-adjoint algebraic groups, or their real points. In this context, it is called the *Cartan decomposition*. By a *maximal compact subgroup* of G we mean one not properly contained in a larger compact subgroup of G. Our next result is the Cartan decomposition.

Theorem 6.3.5. *Let G be a self-adjoint subgroup of $\mathrm{GL}(n, \mathbb{C})$ (or $\mathrm{GL}(n, \mathbb{R})$) with Lie algebra $\mathfrak{g}$. Suppose that G has finite index in an algebraic subgroup M of $\mathrm{GL}(n, \mathbb{C})$ (G has finite index in $M_\mathbb{R}$, the real points of M). Then*

(1) $G = K \times P$ as smooth manifolds.

(2) $\mathfrak{g} = \mathfrak{k} \oplus \mathfrak{p}$ as a direct sum of $\mathbb{R}$-vector spaces.

(3) $\exp : \mathfrak{p} \to P$ is a diffeomorphism whose inverse is given by the global power series of Proposition 6.2.2.

(4) K is a maximal subgroup of G. In particular, P is simply con-

nected and K is a deformation retract of G.

Proof. Here again we deal with the complex case, the real case being similar. First we show each $g \in G$ can be written uniquely as $g = u \exp X$, where $u \in K$ and $X \in \mathfrak{p}$. By Theorem 6.2.5, $g = up$, where $u \in \mathrm{U}(n, \mathbb{C})$ and $p \in \mathcal{H}$. Now $g^* = (up)^* = p^*u^* = pu^{-1}$, so $g^*g = pu^{-1}up = p^2$. Since G is self-adjoint, $p^2 \in G$. Now $p = \exp X$ for some Hermitian X, then $p^2 = \exp 2X$ where $2X$ is also Hermitian. By Proposition 6.3.3, $\exp t2X \in P$ for all real t, in particular for $t = \frac{1}{2}$ for which we get $\exp X = p \in P \subseteq G$ and $X \in \mathfrak{p}$. But then $gp^{-1} = u \in G$, therefore $u \in K$. Also, since $\exp tX \in P$ for all real t, $X \in \mathfrak{p}$. Thus $g = up$, where $u \in K$ and $p \in P$. Thus we have a map $g \mapsto (u, p)$ from G to $K \times P$. As above, if we can show uniqueness of the representation $g = up$, then the map is onto. But since $K \subseteq \mathrm{U}(n, \mathbb{C})$ and $P \subseteq$ the positive definite Hermitian matrices, this follows from the uniqueness result proven earlier. Since multiplication inverts this map it is one-to-one and has a smooth inverse. The formula, $p(g) = \exp(\frac{1}{2} \log(g^*g)) \in P$ derived in the case of $\mathrm{GL}(n, \mathbb{C})$ is still valid, if suitably interpreted, and gives a smooth map $G \to P$. Arguing exactly as in the case of $\mathrm{GL}(n, \mathbb{C})$ we see that part 1 is true. Part 3 follows immediately from the case of $\mathrm{GL}(n, \mathbb{C})$ treated earlier.

For part 2, write $X = \frac{X - X^*}{2} + \frac{X + X^*}{2}$. Since the first term is skew Hermitian, the second is Hermitian and each is an $\mathbb{R}$-linear function of $X \in \mathfrak{gl}(n, \mathbb{C})$, this proves part 2 for the case $\mathfrak{gl}(n, \mathbb{C})$. To prove it in general we need only show that $\frac{X - X^*}{2} \in \mathfrak{k}$ and $\frac{X + X^*}{2} \in \mathfrak{p}$ and for this it suffices to show that $\mathfrak{g}$ is stable under map $X \mapsto X^*$. Note that for $X \in \mathfrak{g}$, $\exp tX \in G$ for all t. Since G is self-adjoint and $(\exp tX)^* = \exp t(X^*)$, it follows that $X^* \in \mathfrak{g}$.

To prove part 4, we first consider the basic cases, $\mathrm{GL}(n, \mathbb{R})$ and $\mathrm{GL}(n, \mathbb{C})$.

Proposition 6.3.6. *Let L be a compact subgroup of $\mathrm{GL}(n, \mathbb{C})$ (or $\mathrm{GL}(n, \mathbb{R})$). Then some conjugate gLg^{-1}, $g \in \mathrm{GL}(n, \mathbb{C})$ (respectively in $\mathrm{GL}(n, \mathbb{R})$) is contained in $\mathrm{U}(n, \mathbb{C})$ (respectively $\mathrm{O}(n, \mathbb{R})$). In particular, $\mathrm{U}(n, \mathbb{C})$ is a maximal compact subgroup of $\mathrm{GL}(n, \mathbb{C})$ and $\mathrm{O}(n, \mathbb{R})$ a maximal compact subgroup of $\mathrm{GL}(n, \mathbb{R})$. In $\mathrm{GL}(n, \mathbb{C})$ and $\mathrm{GL}(n, \mathbb{R})$ any* *two maximal compact subgroups are conjugate.*

Proof. We deal with the complex case, the other being completely analogous. If $(\cdot,\cdot)$ is a Hermitian inner product on $\mathbb{C}^n$, using (finite) Haar measure dl on L we can form an L-invariant Hermitian inner product on $\mathbb{C}^n$ given by $\langle v, w \rangle = \int_L (lv, lw) dl$. Thus for some $g \in \mathrm{GL}(n, \mathbb{C})$, gLg^{-1} is contained in $\mathrm{U}(n, \mathbb{C})$.

If $L \supset \mathrm{U}(n, \mathbb{C})$, where L is a compact subgroup of $\mathrm{GL}(n, \mathbb{C})$, then by the previous discussion $gLg^{-1} \subset \mathrm{U}(n, \mathbb{C})$ for some g, so both have the same dimension. Therefore $\mathrm{U}(n, \mathbb{C})$ is an open subgroup of L. Since $\mathrm{U}(n, \mathbb{C})$ is connected, we conclude that $\mathrm{U}(n, \mathbb{C}) = L_0$, the identity component of L. On the other hand $gL_0 g^{-1} \subset gLg^{-1} \subset \mathrm{U}(n, \mathbb{C}) = L_0$, since L_0 is connected, therefore they are all equal, in particular L is connected and $L = L_0 = \mathrm{U}(n, \mathbb{C})$.

In the real case we just work with the compact connected group $\mathrm{SO}(n, \mathbb{R})$ instead of $\mathrm{U}(n, \mathbb{C})$. Thus $\mathrm{U}(n, \mathbb{C})$ and $\mathrm{O}(n, \mathbb{R})$ are maximal compact subgroups of $\mathrm{GL}(n, \mathbb{C})$ and $\mathrm{GL}(n, \mathbb{R})$, respectively. That any other maximal compact subgroup is conjugate to one of these now follows from the first statement of the proposition. $\square$

In particular, if L is any compact subgroup of $\mathrm{GL}(n, \mathbb{C})$, all its elements have their eigenvalues on the unit circle. From this we see that if an element $l \in L$ has all its eigenvalues equal to 1, then $l = I$. This is because gLg^{-1} is unitary for some g. Hence for some u we know $uglg^{-1}u^{-1}$ is diagonal and also has all eigenvalues equal to 1. Thus $uglg^{-1}u^{-1} = I$ and hence l itself equals I.

Finally, we turn to the proof of part 4 itself. First suppose L is any compact subgroup of G. Then $L \cap P = \{1\}$. To see this just observe that, by the previous result, since L is compact, all its elements have all their eigenvalues on the unit circle. But the eigenvalues of elements of P are all positive. Hence all the elements of $L \cap P$ have all their eigenvalues equal to 1 as above, so $L \cap P = \{1\}$. Now we prove that K is a maximal compact subgroup. Suppose that $L \supseteq K$, then each $l \in L$ can be written $l = up$, where $u \in K \subset L$ and $p \in P$. But since $u \in L$, so is p. Hence by the above $p = I$ and $l = k$. Hence $L \subseteq K$, so that actually $L = K$. $\square$

We have essentially used the conjugacy of maximal compact subgroups in $\mathrm{GL}(n, \mathbb{C})$ and $\mathrm{GL}(n, \mathbb{R})$ to show that K is a maximal compact subgroup of G, in general. However to prove, in general, that any two maximal compact subgroups of G are conjugate will require something more. For this we will rely on the important differential geometric fact, called *Cartan's fixed point theorem*, that a compact group of isometries acting on a complete simply connected Riemannian manifold of nonpositive sectional curvature at every point (Hadamard manifold) always has a unique fixed point and, for the reader's convenience, we will prove Cartan's result as well in the next section. However, we will only prove it for symmetric spaces of non-compact type. This will also establish the fact that for each $p \in P$, $\mathrm{Stab}_G(p)$ is a maximal compact subgroup of G.

We note that the Cartan involution of $\mathfrak{g}$ is given by $k + p \mapsto k - p$. It is an automorphism of $\mathfrak{g}$ whose fixed point set is $\mathfrak{k}$. We also mention the *Cartan relations*, which were also proved earlier. If the Cartan decomposition of $\mathfrak{g}$ is $\mathfrak{g} = \mathfrak{k} \oplus \mathfrak{p}$, since $\mathfrak{k}$ is a subalgebra and $[x^*, y^*] = -[x, y]^*$ and $[x^t, y^t] = -[x, y]^t$ it follows that

(1) $[\mathfrak{k}, \mathfrak{k}] \subseteq \mathfrak{k}$,

(2) $[\mathfrak{k}, \mathfrak{p}] \subseteq \mathfrak{p}$,

(3) $[\mathfrak{p}, \mathfrak{p}] \subseteq \mathfrak{k}$.

We conclude this section by observing that for all the Lie group G considered in this section, there is a natural smooth action of G on P given by $(g, p) \mapsto g^t p g$. Now this action is transitive. To see this, consider the G orbit of $I \in P$, $\mathcal{O}_G(I) = \{g^t g : g \in G\}$. As we saw earlier, this is $\{p^2 ; p \in P\}$. But since everything in P is exp of a unique element X of $\mathfrak{p}$, it follows that everything in P has a unique square root in P, namely $\exp \frac{1}{2} X$. This means the action is transitive. What is the isotropy group $\mathrm{Stab}_G(I)$ of I? This is $\{g \in G : g^t g = I\} = G \cap \mathrm{O}(n, \mathbb{R}) = K$. Hence, by general principles, $P \simeq \mathcal{O}_G(I)$ is G-equivariantly diffeomorphic with G/K, endowed with the action G by right translation. As we shall see, this transitive action will be of great importance in what follows.

Observe that this action does not have the two-point homogeneity property. That is, given p, q and p', q', all in P, there may not be a $g \in G$ so that $g(p) = p'$ and $g(q) = q'$, even when $\dim P = 1$. Note also that $g^t(\exp X)g$ is not equal to $\exp(g^t X g)$, so this is not G-equivalent with the $\mathbb{R}$-linear representation of G acting on $\mathfrak{p}$ by $(g, X) \mapsto g^t X g$, $X \in \mathfrak{p}$. Concomitantly, the latter is not a transitive action because it is linear, so 0 is a single orbit. In fact, here the orbit space can be parameterized by the number of positive, negative and zero eigenvalues of a representative.

Corollary 6.3.7. *For all $n \geq 1$, $\mathrm{SL}(n, \mathbb{C})$ is simply connected.*

Proof. This follows from the Cartan decomposition that the homotopy type of $\mathrm{SL}(n, \mathbb{C})$ is that of its maximal compact subgroup $\mathrm{SU}(n, \mathbb{C})$, which is simply connected (see Corollary 1.5.2).   □

Exercise 6.3.8. Find the Cartan decompositions of $\mathrm{Sp}(n, \mathbb{R})$ and $\mathfrak{sp}(n, \mathbb{R})$.

6.4 The Case of Hyperbolic Space and the Lorentz Group

We now make explicit the Cartan decomposition in an important special case and give the Lorentz model for hyperbolic n space, H^n. We consider $O(n, 1)$ the subgroup of $\mathrm{GL}(n+1, \mathbb{R})$ leaving invariant the nondegenerate quadratic form $q(v, t) = v_1^2 + \ldots + v_n^2 - t^2$, where $v \in \mathbb{R}^n$ and $t \in \mathbb{R}$. Equivalently, by polarization, this means leaving invariant the nondegenerate symmetric bilinear form $\langle (v, t), (w, s) \rangle = (v, w) - ts$, where (v, w) is the usual (positive definite) inner product in $\mathbb{R}^n$. Thus G is defined by the condition $g^{-1} = g^t$ (transpose with respect to $\langle \cdot, \cdot \rangle$). It is easy to check that G is the set of $\mathbb{R}$-points of a self-adjoint algebraic group and, in particular, is a Lie group. Now G is not compact. For example, $\mathrm{SO}(1, 1) \subseteq O(1, 1)$, which sits inside $O(n, 1)$, consists of all matrices

$$g = \begin{pmatrix} a & b \\ c & d \end{pmatrix}$$

with $a^2 - c^2 = 1$, $ab - cd = 0$ and $b^2 - d^2 = -1$. In particular, taking an arbitrary a and $c = (a^2 - 1)^{\frac{1}{2}}$, where $a^2 - 1 = c^2 > 0$ and letting b and d be determined by the remaining two equations we see that $b = (a^2 - 1)^{\frac{1}{2}} = c$ and $d = a$. Now consider the identity component $SO(1,1)_0$. Since the locus $a^2 - c^2 = 1$ has two connected components, if $g \in SO(1,1)_0$, then $a > 0$ and so there is a unique $t \in \mathbb{R}$ for which $a = \cosh t$ and $b = \sinh t$. Thus

$$g(t) = \begin{pmatrix} \cosh t & \sinh t \\ \sinh t & \cosh t \end{pmatrix}.$$

Because these hyperbolic functions are unbounded, we see that even $SO(1,1)_0$ is not compact. The identities satisfied by the hyperbolic functions show that this is an abelian subgroup. However, we shall see this without these identities; in fact, we will derive the identities. Let

$$X = \begin{pmatrix} 0 & 1 \\ 1 & 0 \end{pmatrix}.$$

A direct calculation using the fact that $X^2 = I$ shows that $\exp tX = I \cosh t + X \sinh t = g(t)$, from which it follows that $g(s+t) = g(s)g(t)$. This equation gives all the identities satisfied by the hyperbolic functions $\sinh$ and $\cosh$ and g is a smooth isomorphism of $SO(1,1)_0$ with $\mathbb{R}$. The geometric importance of such 1-parameter subgroups will be seen in a moment.

By Theorem 6.3.5 a maximal compact subgroup of G is given by $O(n+1,\mathbb{R}) \cap O(n,1)$. Because subgroups of $GL(n,\mathbb{R})$ can be regarded as subgroups of $GL(n+1,\mathbb{R})$ via the imbedding $g \mapsto \operatorname{diag}(g,1)$, we may think of $O(n,\mathbb{R})$ as a subgroup of $GL(n+1,\mathbb{R})$ and, in fact, of $O(n,1)$. Thus $O(n,\mathbb{R}) \subseteq O(n+1,\mathbb{R}) \cap O(n,1)$. Clearly these are equal. Since $O(n,\mathbb{R})$ has two components, so does $O(n,1)$ which equals $O(n,\mathbb{R}) \times P$, where P an exponential submanifold. Therefore, $O(n,1)_0 = SO(n,\mathbb{R}) \times P$.

Note that for $g \in O(n,1)$ we have $gg^t = I$, so $(\det g)^2 = 1$, thus $\det g = \pm 1$, a discrete set. It follows that $SO(n,1)$ is open in $O(n,1)$ and hence has the same P. The same is true of $SO(n,1)_0$ because we are dealing with Lie groups. Thus $SO(n,1)_0 = SO(n,\mathbb{R}) \times P$ and we now work with this connected group $G = \mathrm{SO}(n,\mathbb{R}) \times P^1$. The Lie algebra $\mathfrak{g}$ of $G = \mathrm{SO}(n,1)_0$ is

$$\{X \in \mathfrak{gl}(n+1,\mathbb{R}) : X^t = -X\}$$

which has dimension $\frac{(n+1)n}{2}$. Now consider the subspace of $\mathfrak{gl}(n+1,\mathbb{R})$ consisting of

$$\begin{pmatrix} X & v \\ v & 0 \end{pmatrix},$$

where $X \in \mathfrak{so}(n,\mathbb{R})$ the Lie algebra of $\mathrm{SO}(n,\mathbb{R})$ and $v \in \mathbb{R}^n$. It is clearly a subspace and has dimension $\frac{(n-1)n}{2} + n = \frac{(n+1)n}{2}$ and it consists of skew symmetric matrices with respect to $\langle \cdot, \cdot \rangle$. Hence it must coincide with $\mathfrak{g}$. Here the Cartan decomposition is perfectly clear. The $\mathfrak{k}$ part is

$$\begin{pmatrix} X & 0 \\ 0 & 0 \end{pmatrix},$$

for $X \in \mathfrak{so}(n,\mathbb{R})$, while the $\mathfrak{p}$ part is

$$\begin{pmatrix} 0 & v \\ v & 0 \end{pmatrix},$$

for $v \in \mathbb{R}^n$. Consider the locus of points,

$$H = \{(v,t) \in \mathbb{R}^{n+1} : q(v,t) = -1\}.$$

For $g \in \mathrm{O}(n,1)$, $q(g(v,t)) = q(v,t)$. In particular, if $q(v,t) = -1$, then $q(g(v,t)) = -1$. Thus H is invariant under $\mathrm{O}(n,1)$. Now H is a hyperboloid of two sheets: $1 + \|v\|^2 = t^2$. So $t = \pm(1 + \|v\|^2)^{\frac{1}{2}}$. Write $H = H^+ \cup H^-$, a disjoint union of the upper and lower sheets. Both sheets are open subsets of H since they are the intersection of H with a half space. Each is diffeomorphic with $\mathbb{R}^n$. In particular, each is connected and simply connected. We show that $G = \mathrm{SO}(n,1)_0$ leaves both H^+ and H^- invariant. Note that

$$g(H^+) = (g(H^+) \cap H^+) \cup (g(H^+) \cap H^-),$$

---

[1]Actually, $\mathrm{SO}(n,1)$ is connected if n is even, and has two components if n is odd.

and $g(H^+)$ is connected. Therefore $g(H^+) \subseteq H^+$ or $g(H^+) \subseteq H^-$. Since g is a diffeomorphism of H, $g(H^+) = H^+$ or $g(H^+) = H^-$. We show that the former must hold. Since G is arcwise connected, there must be a smooth path g_t in G joining $g = g_1$ to $I = g_0$. Consider the disjoint sets $T^+ = \{t \in [0,1] : g_t(H^+) = H^+\}$ and $T^- = \{t \in [0,1] : g_t(H^+) = H^-\}$. Note that $[0,1] = T^+ \cup T^-$ and $T^+ \neq \emptyset$ as $t = 0 \in T^+$. We prove that T^+ and T^- are closed. For if $t_k \to t$ and say $g_{t_k}(H^+) = H^+$, for all k, but $g_t(H^+) = H^-$, then for $x \in H^+$, $g_{t_k}(x) \to g_t(x)$. This is impossible as the distance between H^+ and H^- is 2. Therefore $[0,1] = T^+$ and $g(H^+) = g_1(H^+) = H^+$.

We now know G operates on H^+ which we shall call H^n, the *Lorentz model* of hyperbolic n-space. Consider the lowest point, $p_0 = (0, \ldots, 0, 1) \in H^n$. What is $\mathrm{Stab}_G(p_0)$? This is clearly a subgroup which does not change the t coordinate and is arbitrary in the other coordinates since it is linear and so always fixes 0. Hence, $\mathrm{Stab}_G(p_0) = \mathrm{SO}(n, \mathbb{R})$, a maximal compact subgroup of G. Next we look at the G-orbit $\mathcal{O}(p_0)$ and show that G acts transitively on H^n. Let $p = (v, t)$, where $t = (1 + \|v\|^2)^{\frac{1}{2}}$, be any point in H^n and apply $\mathrm{SO}(n, \mathbb{R})$ on the first n coordinates to bring p to $(\|v\|, 0, \ldots, 0, t)$. Now the problem is reduced to a two-dimensional situation, let us consider $(x, y) = (\|v\|, t)$, where $y^2 - x^2 = 1$. We want to transform $(0, 1)$ to (x, y) by something on the 1-parameter group

$$g(s) = \begin{pmatrix} \cosh s & \sinh s \\ \sinh s & \cosh s \end{pmatrix}.$$

But this is just the fundamental property of the right hand branch of the hyperbola mentioned earlier. Therefore, G acting transitively on H^n is equivariantly equivalent to the action by left translation on $\mathrm{SO}_0(n, 1) / \mathrm{SO}(n, \mathbb{R})$.

Now consider the hyperplane $t = 1$ in $\mathbb{R}^{n+1}$. This is the tangent space T_{p_0} to H^n at p_0. Consider $(\cdot, \cdot)$ the standard Euclidean metric on T_{p_0}. If p is another point of H^n, choose $g \in G$ such that $g(p) = p_0$. Then its derivative $d_p g$ at p maps T_p to T_{p_0} bijectively. Use this to transfer the inner product from T_{p_0} to T_p. Now if $h(p)$ also equals p_0, then $gh^{-1} \in \mathrm{Stab}_G(p_0) = \mathrm{SO}(n, \mathbb{R})$. Therefore $d_{p_0}(gh^{-1}) = d_p g \, d_{p_0} h^{-1}$

is a linear isometry of T_{p_0}. This shows that the inner product on T_p is independent of g and is well defined. Hence we get a Riemannian metric on H^n because G is a Lie group acting smoothly on H^n. Evidently, G acts by isometries, the action is transitive and H^n can be identified with $G/\operatorname{Stab}_G(p_0) = \operatorname{SO}(n,1)_0/\operatorname{SO}(n,\mathbb{R})$.

Notice that $\operatorname{SO}(n,\mathbb{R}) = \operatorname{Stab}_G(p_0)$ acts transitively on k-dimensional subspaces for all $1 \le k \le n$. In particular, this is so for 2-planes in $\mathbb{R}^n = T_{p_0}(H^n)$. Since it acts by isometries, this means the sectional curvature is constant as both the point and the plane section vary.

6.5   The G-invariant Metric Geometry of P

Here we introduce a Riemannian metric on any P and study its most basic differential geometric properties. From now on we will write exp and log instead of Exp and Log. Much if this section is an elaboration of results in [61].

Lemma 6.5.1. *If A and B are $n \times n$ complex matrices, then $\operatorname{tr}(AB) = \operatorname{tr}(BA)$. Also $\operatorname{tr}(B^*B) \ge 0$ and equals 0 if and only if $B = 0$. Evidently, $\operatorname{tr}(B)^- = \operatorname{tr}(B^*)$.*

Proof. Suppose $A = (a_{ij})$ and $B = (b_{kl})$. Then $(AB)_{il} = \sum_j a_{ij} b_{jl}$. Therefore $\operatorname{tr}(AB) = \sum_{i,j} a_{ij} b_{ji}$. But then $\operatorname{tr}(BA) = \sum_{i,j} b_{ij} a_{ji} = \sum_{i,j} a_{ji} b_{ij} = \sum_{j,i} a_{ij} b_{ji} = \operatorname{tr}(AB)$. Taking B^* for A we get $\operatorname{tr}(B^*B) = \sum_{i,j} \bar{b}_{ji} b_{ji} \ge 0$ and equals 0 if and only if $B = 0$. □

This enables us to put a Hermitian inner product on $\mathfrak{gl}(n,\mathbb{C})$ called the *Hilbert-Schmidt inner product* and a symmetric inner product on $\mathfrak{gl}(n,\mathbb{R})$ by defining

$$\langle Y, X \rangle = \operatorname{tr}(Y^*X).$$

For X Hermitian (or symmetric), we now study the linear operator ad X on $\mathfrak{gl}(n,\mathbb{C})$ (respectively $\mathfrak{gl}(n,\mathbb{R})$). As we saw from the Cartan relations for $T \in \mathfrak{gl}(n,\mathbb{C})$ and X Hermitian, $[X,T]^* = [T^*,X] = -[X,T^*]$.

Lemma 6.5.2. *If X is Hermitian, $\langle \operatorname{ad} X(T), S \rangle = \langle T, \operatorname{ad} X(S) \rangle$ for all S and T; that is, $\operatorname{ad} X$ is self-adjoint. In particular, the eigenvalues of such an $\operatorname{ad} X$ are all real.*

Proof. We calculate $\text{tr}([X,T]^*S) = \text{tr}(-[X,T^*]S) = -\text{tr}((XT^* - T^*X)S) = \text{tr}(T^*XS) - \text{tr}(XT^*S)$. On the other hand, $\text{tr}(T^*[X,S]) = \text{tr}(T^*XS) - \text{tr}(T^*SX)$. Thus we must show that $\text{tr}(XT^*S) = \text{tr}(T^*SX)$. But this follows from the lemma above. □

A formal calculation, which we leave to the reader, proves the following:

Lemma 6.5.3. *For each* $U \in \mathfrak{gl}(n,\mathbb{C})$, $L_{\exp(U)} = \exp(L_U)$ *and* $R_{\exp(U)} = \exp(R_U)$.

Definition 6.5.4. For X and $Y \in \mathfrak{gl}(n,\mathbb{C})$ let

$$D_X(Y) = \frac{d}{dt}\exp(-X/2)\exp(X+tY)\exp(-X/2)|_{t=0}.$$

Proposition 6.5.5. *For* $X \in \mathfrak{p}$, *the operator* d_X *is self-adjoint on* $\mathfrak{gl}(n,\mathbb{C})$. *Using functional calculus, this operator is given by the formula*

$$D_X = \sinh(\frac{\text{ad } X}{2})/(\frac{\text{ad } X}{2}).$$

Proof. Let $t \in \mathbb{R}$, X, $Y \in \mathfrak{gl}(n,\mathbb{C})$ and $X(t) = X + tY$. Then

$$D_X(Y) = \exp(-X/2)\frac{d}{dt}\exp(X(t))|_{t=0}\exp(-X/2).$$

Now for all t,

$$X(t) \cdot \exp(X(t)) = \exp(X(t)) \cdot X(t).$$

Differentiating we get

$$X'(t) \cdot \exp(X(t)) + X(t) \cdot \frac{d}{dt}\exp(X(t)) = \frac{d}{dt}\exp(X(t)) \cdot X(t)$$
$$+ \exp(X(t)) \cdot X'(t).$$

Evaluating at $t = 0$ and subtracting gives $X \cdot \frac{d}{dt}\exp(X(t))|_{t=0} - \frac{d}{dt}\exp(X(t))|_{t=0} \cdot X = \exp(X)Y - Y\exp(X)$. Multiplying on both the left and right by $\exp(-X/2)$ and taking into account the fact that $\exp(-X/2)$ and X commute, we get

$$X \cdot \exp(-X/2)\frac{d}{dt}\exp(X(t))|_{t=0}\exp(-X/2)$$
$$- \exp(-X/2)\frac{d}{dt}\exp(X(t))|_{t=0}\exp(-X/2)X$$
$$= \exp(X/2)Y\exp(-X/2) - \exp(-X/2)Y\exp(X/2).$$

Substituting for $D_X(Y)$, the left hand side becomes

$$XD_X(Y) - D_X(Y)X = \operatorname{ad}XD_X(Y),$$

while the right hand side is

$$L_{\exp(X/2)}R_{\exp(-X/2)}(Y) - L_{\exp(-X/2)}R_{\exp(X/2)}(Y).$$

But by the lemma above

$$L_{\exp(U)} = \exp(L_U) \text{ and } R_{\exp(U)} = \exp(R_U).$$

Substituting we get

$$\operatorname{ad}XD_X(Y) = \exp(L_{X/2})\exp(R_{-X/2})(Y) - \exp(L_{-X/2})\exp(R_{X/2})(Y).$$

Since L_U and $R_{U'}$ commute for all U and U', we see that

$$\exp(L_{X/2})\exp(R_{-X/2}) = \exp(L_{X/2} + R_{-X/2}) = \exp(L_{X/2} - R_{X/2})$$
$$= \exp(\operatorname{ad}X/2).$$

Similarly,

$$\exp(L_{-X/2})\exp(R_{X/2}) = \exp(L_{-X/2} + R_{X/2}) = \exp(-\operatorname{ad}X/2).$$

So for all Y,

$$\operatorname{ad}X \cdot D_X(Y) = (\exp(\operatorname{ad}X/2) - \exp(-\operatorname{ad}X/2))\,(Y).$$

Now let

$$f(z) = e^{z/2} - e^{-z/2} = z + 2(z/2)^3/3! + 2(z/2)^5/5! + \cdots.$$

Then f is an entire function and $f(0) = 0$. In terms of f, the equation above says

$$\operatorname{ad} X D_X = f(\operatorname{ad} X).$$

This means if we let

$$g(z) = f(z)/z = 1 + (z/2)^2/3! + (z/2)^4/5! + \cdots,$$

with $g(0) = 1$, then g is also entire and $D_X = g(\operatorname{ad} X)$. Now $\sinh z = z + z^3/3! + z^5/5! + \cdots$ so $g(z) = \sinh(z/2)/(z/2)$ and hence the conclusion. Finally, because $D_X = g(\operatorname{ad} X)$, $\operatorname{ad} X$ is self-adjoint and the Taylor coefficients of g are real, D_X is also self-adjoint. $\square$

Exercise 6.5.6. Using the same method in the proof of Proposition 6.5.5, prove that

$$d_X \exp(Y) = \phi(-\operatorname{ad} X)(Y)$$

where $\phi(z) = \sum_{n=0}^{\infty} \frac{z^n}{(n+1)!}$.

Corollary 6.5.7. *For $X \in \mathfrak{p}$, $\operatorname{Spec}(\frac{\sinh(\operatorname{ad} X)}{\operatorname{ad} X})$ consists of real numbers greater than or equal to 1. The same is true for the operator D_X.*

Proof. Since for $t \in \mathbb{R}$, $\frac{\sinh t}{t} = 1 + t^2/3! + t^4/5! + \cdots$, we see that $\frac{\sinh t}{t} > 1$ unless $t = 0$. Now by Exercise 0.5.13, $\operatorname{Spec}(\operatorname{ad} X) \subseteq \{\lambda_i - \lambda_j | \lambda_i, \lambda_j \in \operatorname{Spec} X\}$, therefore

$$\operatorname{Spec}(\frac{\sinh(\operatorname{ad} X)}{\operatorname{ad} X}) = \{\frac{\sinh(\lambda)}{\lambda} : \lambda \in \operatorname{Spec} \operatorname{ad} X\}$$

$$\subseteq \{\frac{\sinh(\lambda_i - \lambda_j)}{\lambda_i - \lambda_j} : \lambda_i, \lambda_j \in \operatorname{Spec} X\}.$$

If $\lambda = \lambda_i - \lambda_j$ for distinct eigenvalues of X, then $\frac{\sinh(\lambda)}{\lambda} > 1$. If λ_i and λ_j are equal, then $\lambda = 0$ and $\frac{\sinh(\lambda)}{\lambda} = 1$. $\square$

We now work exclusively over $\mathbb{R}$. The same type of arguments also work just as well over $\mathbb{C}$.

Corollary 6.5.8. *For $X \in \mathfrak{p}$ and $Y \in \mathfrak{gl}(n, \mathbb{R})$, $\operatorname{tr}(Y^2) \leq \operatorname{tr}((D_X(Y))^2)$. Equality occurs if and only if $[X, Y] = 0$.*

Proof. Because ad X is self-adjoint, we can choose an orthonormal basis of real eigenvectors of ad X, $Y_1, \ldots, Y_j \in \mathfrak{gl}(n, \mathbb{R})$ which, since $D_X = g(\mathrm{ad}\, X)$ are also eigenvectors for D_X with corresponding real eigenvalues $\mu_1, \ldots, \mu_j$. Then $D_X(Y_k) = \mu_k Y_k$ for all k. If $Y = \sum_k a_k(Y) Y_k$, then $D_X(Y) = \sum_k a_k(Y) D_X(Y_k) = \sum_k a_k(Y) \mu_k Y_k$. Since the Y_k form an orthonormal basis, we see that $\mathrm{tr}(D_X(Y)^2) = \sum_k a_k(Y)^2 \mu_k^2$, while $\mathrm{tr}(Y^2) = \sum_k a_k(Y)^2$. Thus we are asking whether $\sum_k a_k(Y)^2 \leq \sum_k a_k(Y)^2 \mu_k^2$. Since each $\mu_k \geq 1$, this is clearly so and equality occurs only if $\mu_k = 1$ whenever $a_k(Y) \neq 0$. Rearrange the eigenvectors so that the $\mu_k = 1$ come first and for $k \geq k_0$, $\mu_k > 1$. Hence $\mathfrak{gl}(n, \mathbb{R}) = W_1 \oplus W_\infty$ is the orthogonal direct sum of two ad X-invariant subspaces. Here W_1 is the 1-eigenspace, and W_∞ the sum of all the others. But since $a_k(Y) = 0$ for $k \geq k_0$, $Y \in W_1$. But on W_1 all eigenvalues of $g(\mathrm{ad}\, X) = D_X$ are 1, and the eigenvalues of ad X are 0 so ad $X = 0$ on W_1 and hence $[X, Y] = 0$.

Conversely, if $[X, Y] = 0$, then $\mathrm{ad}\, X(Y) = 0$. Therefore $D_X = g(\mathrm{ad}\, X) = I$. $\qquad\square$

Theorem 6.5.9. *Along any smooth path $p(t)$ in P we have*

$$\mathrm{tr}[(\frac{d}{dt} \log p(t))^2] \leq \mathrm{tr}[(p^{-1}(t) p'(t))^2]$$

with equality if and only if $p(t)$ and $p'(t)$ commute for that t.

Proof. For each t, it is easy to see that

$$p^{1/2}(p^{-1} p')^2 p^{-1/2} = (p^{-1/2} p' p^{-1/2})^2.$$

It follows that $\mathrm{tr}[(p^{-1} p')^2] = \mathrm{tr}[(p^{-1/2} p' p^{-1/2})^2]$. Set $X(t) = \log p(t)$. Then $X(t)$ is a smooth path in $\mathfrak{p}$ and $p(t)^{-\frac{1}{2}} = \exp(-X(t)/2)$. Let t be fixed and $Y = X'(t)$. Since $D_X(Y) = \exp(-X/2) \frac{d}{ds} \exp(X + sY)|_{s=0} \exp(-X/2)$, this is $p^{-\frac{1}{2}} p' p^{-\frac{1}{2}}$, where $p' = \frac{d}{ds} \exp(X + sY)|_{s=0}$ (the tangent vector to curve $p(t)$ at $p = \exp X$). Hence $\mathrm{tr}[(p^{-\frac{1}{2}} p' p^{-\frac{1}{2}})^2] = \mathrm{tr}[(D_X(X'))^2]$. Also $\mathrm{tr}[(\frac{d}{dt} \log p(t))^2] = \mathrm{tr}[X'(t)^2]$. Now by the corollary, for each t,

$$\mathrm{tr}[X'(t)^2] \leq \mathrm{tr}[(d_{X(t)}(X'(t)))^2]$$

with equality if and only if $X(t)$ and $X'(t)$ commute for that t.

Finally we show that for all fixed t, $X(t)$ and $X'(t)$ commute if and only if $p(t)$ and $p'(t)$ commute. For by the chain rule and the formula Exercise 6.5.6,

$$p'(t) = d_{X(t)} \exp(X'(t)) = \phi(-\operatorname{ad} X(t))X'(t),$$

where ϕ is the entire function given by $\phi(z) = \sum_{n=0}^{\infty} \frac{z^n}{(n+1)!}$. If X' commutes with X for fixed t, then since $\phi(0) = 1$, we see that $\phi(-\operatorname{ad} X)X' = X'$ so that $p' = X'$. In particular, p' commutes with X and therefore with $\exp X = p$. On the other hand, if $\phi(-\operatorname{ad} X)X'$ commutes with $\exp X = p$, then since $\log : P \to \mathfrak{p}$ is given by a convergent power series in p (see Theorem 6.3.5, part 3), it must also commute with $\log p = X$. Looking at the specific form of the function ϕ, it follows that

$$[X, X' - \operatorname{ad} X(X')/2! + \operatorname{ad}_X^2(X')/3! + \cdots] = 0.$$

That is, $\operatorname{ad} X(X') - \operatorname{ad}_X^2(X')/2! + \operatorname{ad}_X^3(X')/3! + \cdots = 0$. Hence $\exp(-\operatorname{ad} X(X')) = X'$. Taking $\exp(\operatorname{ad} X)$ on both sides tells us $\exp(\operatorname{ad} X)(X') = X'$. Therefore $\operatorname{Ad}(\exp X)(X') = X'$ so X' commutes with $\exp X$. But then, reasoning as above, X' must commute with $\log(\exp X) = X$. $\square$

Since what is inside the square root is real and positive, we make the following definition.

Definition 6.5.10. Let $p(t)$ be a smooth path in P, where $a \leq t \leq b$. Then its length $l(p)$ equals $\int_a^b [\operatorname{tr}((p^{-1}p'(t))^2)]^{\frac{1}{2}} dt$. The Riemannian metric is given by $ds^2 = \operatorname{tr}((p^{-1}p')^2)dt^2$. We call this metric d.

Proposition 6.5.11. *G acts isometrically on P.*

Proof. We calculate that

$$(g^t p g)^{-1}(g^t p g)' = g^{-1}p^{-1}(g^t)^{-1}g^t p' g = g^{-1}p^{-1}p'g.$$

Hence $((g^t p g)^{-1}(g^t p g)')^2 = g^{-1}(p^{-1}p')^2 g$. Taking traces we get

$$\operatorname{tr}[((g^t p g)^{-1}(g^t p g)')^2] = \operatorname{tr}[(p^{-1}p')^2].$$

$\square$

On $\mathfrak{p}$ we place the metric given infinitesimally by $ds^2 =$ $\text{tr}[(\frac{d}{dt}\log p(t))^2]dt^2$, that is, if $X(t)$ is a smooth path in $\mathfrak{p}$, then $ds^2 = \text{tr}(X'(t)^2)dt^2$. We call this metric $d_{\mathfrak{p}}$. Earlier we defined an inner product on $\mathfrak{gl}(n,\mathbb{R})$ by $\langle Y, X\rangle = \text{tr}(Y^t X)$. Hence the linear subspace $\mathfrak{p}$ has an inner product on it by restriction, namely $\langle Y, X\rangle = \text{tr}(YX)$. The associated norm is $\|Y\|^2 = \text{tr}(Y^2)$. This, together with the formula above, shows $d_{\mathfrak{p}}$ is the Euclidean metric. If we transfer $d_{\mathfrak{p}}$ to P, then $d_{\mathfrak{p}}(p, q) = \|\log p - \log q\|$. This will give us the opportunity to compare $d_{\mathfrak{p}}$ and d on P. Since by Theorem 6.5.9 along any smooth path $p(t)$ in P we have,

$$\text{tr}[(\frac{d}{dt}\log p(t))^2] \le \text{tr}[(p(t)^{-1}p'(t))^2],$$

we see that infinitesimally and hence globally $d_{\mathfrak{p}} \le d$.

Now for $X \in \mathfrak{p}$, $D_X = \frac{\sinh(\text{ad } X/2)}{\text{ad } X/2}$. By Corollary 6.5.7 we have,

$$\text{Spec } D_X = \{\frac{\sinh(\lambda/2)}{\lambda/2} : \lambda \in \text{Spec ad } X\}.$$

As $\frac{\sinh t}{t}$ is analytic, by continuity $\frac{\sinh t}{t} \to 1$ as $t \to 0$. This tells us that from the formulas for $\text{tr}[(D_X(Y))^2]$ and $\text{tr}(Y^2)$, if $X \to 0$, then independently of Y, $\text{tr}[(D_X(Y))^2]$ can be made as near as we want to $\text{tr}(Y^2)$. This last statement implies that for p and q in a sufficiently small neighborhood of a point p_0, which by transitivity of G we may assume to be I, the nonpositively curved symmetric space and Euclidean distances approach one another.

$$\lim_{p,q\to p_0} \frac{d(p,q)}{d_{\mathfrak{p}}(\log p, \log q)} = 1.$$

This has the interesting philosophical consequence that in the nearby part of the universe that man inhabits, because of experimental error in making measurements, nonpositively curved symmetric space distances and Euclidean ones are (locally) indistinguishable. As we shall show below, angles at I are in any case identical. This means no experiment can tell us if we "really" live in a hyperbolic or Euclidean world.

Corollary 6.5.12. *If $p = \log X \in P$, the 1-parameter subgroup $\exp tX$ is the unique geodesic in (P, d) joining I with p. Moreover, any two points of P can be joined by a unique geodesic.*

Proof. Consider a path $p(t)$ in P which happens to be a 1-parameter subgroup. Since $p(t) = \exp tX$, $\log p(t) = tX$ and its derivative is X. Thus for each t, $\log p(t)$ and its derivative commute. Hence, as we showed, $p(t)$ and $p'(t)$ also commute. This tells us that all along $p(t)$, $d_{\mathfrak{p}}$ and d coincide. But the 1-dimensional subspaces of $\mathfrak{p}$ are geodesics for $d_{\mathfrak{p}}$. Hence if $p = \log X \in P$, the 1-parameter subgroup $\exp tX$ is the unique geodesic in (P, d) joining I with p. Let p and q be distinct points of P. Since G acts transitively on P, we can choose g so that $g(q) = I$. Connect I with $g(p)$ by its unique geodesic γ. Since G acts isometrically, $g^{-1}(I) = q$, $g^{-1}(g(p)) = p$ and $g^{-1}(\gamma)$ is the unique geodesic joining them.   □

This corollary also follows from more general facts in differential geometry. This is because as a 1-parameter subgroup every geodesic emanating from I has infinite length. Since G acts transitively by isometries, this is true at every point. Hence by the Hopf-Rinow theorem (see [23]) P is complete. In particular, any two points can be joined by a shortest geodesic (also Hopf-Rinow). Being diffeomorphic to Euclidean space, P is simply connected. If P had nonpositive sectional curvature in every section and at every point, then this geodesic would be unique. This last fact is actually valid for any Hadamard manifold and is called the Cartan-Hadamard theorem. We will give a direct proof of completeness of P shortly.

Corollary 6.5.13. *A curve $p(t)$ in P is a geodesic through $p_0 \in P$ if and only if $p(t) = g(\exp tX)g^t$, where $X \in \mathfrak{p}$ and $g \in G$.*

Proof. Since G acts transitively by isometries on P, choose $g \in G$ so that $gIg^t = p_0$. The result follows from the above since the 1-parameter subgroup $\exp tX$ is the unique geodesic in (P, d) beginning at I in the direction X.   □

Corollary 6.5.14. *At I the angles in the two metrics coincide.*

Proof. Let X and Y be two vectors in $\mathfrak{p}$ and $p(t)$ and $q(t)$ be curves in P passing through I with tangent vectors X and Y, respectively, and let $p_0(t) = \exp tX$ and $q_0(t) = \exp tY$ be two 1-parameter groups in P. Then since X and Y are also the tangent vectors of p_0 and q_0, respectively, the angle between p and q equals that between p_0 and q_0. We may therefore replace p and q by p_0 and q_0. Now $p^{-1}p'q^{-1}q'(0)$ is just XY so that $\operatorname{tr}(p^{-1}p'q^{-1}q'(0)) = \operatorname{tr}(XY)$.   □

Corollary 6.5.15. *For $X \in \mathfrak{p}$, $d(I, \exp X) = [\operatorname{tr}(X^2)]^{\frac{1}{2}}$.*

Proof. The 1-parameter group $\exp tX$ is a geodesic in P passing through I at $t = 0$. Hence, infinitesimally along this curve, $d = d_{\mathfrak{p}}$. This implies the same is true globally along it. Put another way, at each point of $\exp tX$, for $0 \le t \le 1$, the theorem tells us the metric is $\operatorname{tr}[(\frac{d}{dt}(tX))^2] = \operatorname{tr}(X^2)$. Since this is independent of t, integrating from 0 to 1 gives $\operatorname{tr}(X^2)$.   □

Corollary 6.5.16. *For X and $Y \in \mathfrak{p}$,*

$$d(\exp X, \exp Y) \ge (\operatorname{tr}[(X - Y)^2])^{\frac{1}{2}}.$$

Corollary 6.5.17. *P is complete.*

Proof. Let p_k be a Cauchy sequence in (P, d). By the inequality above, $X_k = \log p_k$ is a Cauchy sequence in $(\mathfrak{p}, d_{\mathfrak{p}})$ which must converge to X since Euclidean space is complete. By continuity, p_k converges to $\exp X = p$.   □

Corollary 6.5.18. *(Law of Cosines) Let a, b and c be the lengths of the sides of a geodesic triangle in P and A, B and C be the corresponding vertices. Then*

$$c^2 \ge a^2 + b^2 - 2ab \cos C$$

and the sum of the angles $A + B + C \le \pi$. Moreover, if the vertex C is at I then the equality holds if and only if

(1) The triangle lies in a connected abelian subgroup of P, or equivalently,

(2) $A + B + C = \pi$.

Proof. Put C at the identity via an isometry from G. Then the Euclidean angle at C equals the angle in the metric d. Also, $l_{\mathfrak{p}}(c) \leq c$ and $l_{\mathfrak{p}}(a) = a$ and $l_{\mathfrak{p}}(b) = b$. The inequality now follows from the Euclidean Law of Cosines.

The equality holds if and only if $l_{\mathfrak{p}}(c) = c$. This occurs if and only if log takes the side c to a geodesic in $\mathfrak{p}$ (*i.e.* a straight line) of the same length. This is also equivalent to $\mathrm{tr}[(\frac{d}{dt} \log p(t))^2] = \mathrm{tr}[(p^{-1}p'(t))^2]$, for all t, where $p(t)$ denotes the geodesic side of length c. This occurs if and only if $p(t)$ satisfies the condition that $p(t)$ and $p'(t)$ commute for all t which, as we showed, is equivalent to $[X, Y] = 0$, where X and Y are the infinitesimal generators of the sides a and b. Thus the equality in the Law of Cosines holds if and only if the Euclidean triangle lies in a 2-dimensional abelian subalgebra of $\mathfrak{g}$ contained in $\mathfrak{p}$. Equivalently, the geodesic triangle lies in a 2-dimensional abelian subgroup of G contained in P.

Next we show that in general the sum of the angles is at most π. Since d is a metric and $c = d(A, B)$, *etc.*, it follows that each length a, b, or c is less than the sum of the other two. Therefore there is an ordinary plane triangle with sides a, b and c. Denote its angles by A', B' and C'. Then $A \leq A'$, $B \leq B'$ and $C \leq C'$. For by the Law of Cosines $c^2 \geq a^2 + b^2 - 2ab\cos C$ and $c^2 = a^2 + b^2 - 2ab\cos C'$. This means $\cos C' \leq \cos C$. But then because C and C' are between 0 and π and cos is monotone decreasing there, we see $C \leq C'$. Similarly, this holds for the others. Since $A' + B' + C' = \pi$, it follows that $A + B + C \leq \pi$.

If $c^2 > a^2 + b^2 - 2ab\cos C$, then, as above, construct an ordinary plane triangle with sides a, b and c and angles A', B' and C'. Then since here we have a strict inequality, it follows as above that $C < C'$. But it is always the case that $A \leq A'$ and $B \leq B'$. Hence $A + B + C < A' + B' + C' = \pi$. Conversely, if $A + B + C = \pi$, then $c^2 = a^2 + b^2 - 2ab\cos C$ and $[X, Y] = 0$. Therefore X and Y generate an abelian subalgebra, and the triangle lies in a flat. $\qquad\square$

Our next result is of fundamental importance. Nonpositive and positive sectional curvature distinguish the symmetric spaces of non-compact type from those of compact type.

Corollary 6.5.19. *The sectional curvature of P is nonpositive and strictly negative off the flats. In particular, P is a Hadamard manifold.*

Proof. Each geodesic triangle lies in a plane section. We have just shown that each geodesic triangle in each such section has the sum of the angles $\leq \pi$ and the sum of the angles $< \pi$ if we are off a flat. It is a standard result of 2-dimensional Riemannian geometry (Gauss-Bonnet theorem) that these conditions are equivalent to $K \leq 0$ and $K < 0$, respectively, where K denotes the Gaussian curvature of the section, that is, the sectional curvature. $\square$

Remark 6.5.20. We remark that when $X, Y \in \mathfrak{p}$ and are orthonormal with respect to the Killing form, one actually has $K(X,Y) = -\|[X,Y]\|^2$. See [14] for more details.

Definition 6.5.21. A submanifold N of a Riemannian manifold M is called *totally geodesic* if given any two points of N and a geodesic γ in M joining them, γ lies entirely in N.

Corollary 6.5.22. *P is a totally geodesic submanifold in the set of all positive definite symmetric matrices.*

Proof. Let p and $q \in P$ be two arbitrary points. Since $p^{\frac{1}{2}}$ and $p^{-\frac{1}{2}}$ are self-adjoint, $p^{-\frac{1}{2}} q p^{-\frac{1}{2}}$ is positive definite and symmetric. But as we showed earlier, $p^{-\frac{1}{2}} \in G$. Hence $p^{-\frac{1}{2}} q p^{-\frac{1}{2}} \in G$. Because $p^{-\frac{1}{2}}$ is self-adjoint we see that $p^{-\frac{1}{2}} q p^{-\frac{1}{2}} \in P$. Let $X \in \mathfrak{p}$ be its log. Then $\exp tX$ lies in P, for all real t. Therefore $\gamma(t) = p^{\frac{1}{2}} (\exp tX) p^{\frac{1}{2}}$ is a geodesic in P. Clearly, $\gamma(0) = p$ and $\gamma(1) = q$. Therefore there is a unique geodesic in P joining p and q, which means P is a totally geodesic submanifold. $\square$

We conclude this section with the standard definition of a *symmetric space*.

Definition 6.5.23. A Riemannian manifold M is called a symmetric space if for each point $p \in M$ there is an isometry σ_p of M satisfying the following conditions.

(1) $\sigma_p^2 = I$, but $\sigma_p \neq I$,

(2) σ_p has only isolated fixed points among which is p,

(3) $d_p\sigma = -id_{T_pM}$.

Thus the main feature of the definition is that for each point p there is an isometry which leaves p fixed and reverses geodesics through p.

Corollary 6.5.24. *P is a symmetric space.*

Proof. Since G acts transitively and by isometries, we may restrict ourselves to the case $p = I$. Take $\sigma_I = \sigma(p) = p^{-1}$, for each $p \in P$. This map is clearly of order 2. If p is σ fixed, then $p^2 = I$. Since $p \in P$, $p^t = p = p^{-1}$, hence $p \in K \cap P$ which is trivial. Thus I is the only fixed point. Let $p = \exp X$, then $\sigma(p) = \exp(-X)$ so that $d(\sigma_p)_{\mathfrak{p}} = -I$, where here we identify $T_I(P)$ with $\mathfrak{p}$.

It remains to see that σ is an isometry. For a curve $p(t)$ in P, since $p(t)p(t)^{-1} = I$, differentiating tells us

$$p(t)\frac{d}{dt}(p(t)^{-1}) + \frac{dp}{dt}p(t)^{-1} = 0,$$

hence,

$$p(t)\frac{d}{dt}(p(t)^{-1}) = -\frac{dp}{dt}p(t)^{-1}.$$

By taking the trace we obtain

$$\mathrm{tr}[(p(t)\frac{d}{dt}(p(t)^{-1}))^2] = \mathrm{tr}[(\frac{dp}{dt}p(t)^{-1})^2] = \mathrm{tr}[\frac{dp}{dt}p(t)^{-1}\frac{dp}{dt}p(t)^{-1}]$$

$$= \mathrm{tr}[(p(t)^{-1}\frac{dp}{dt})^2].$$

Hence σ is an isometry of P and the latter is a symmetric space. $\square$

6.6   The Conjugacy of Maximal Compact Subgroups

The theorem on the conjugacy of maximal compact subgroups of G in the present context is due to E. Cartan. Actually, the result is true for an arbitrary connected Lie group, and due to K. Iwasawa, and in the case of a Lie group with finitely many components, to G.D. Mostow. In this more general context see [33]. We shall deal with this problem in the present context by means of Cartan's fixed point theorem which states that a compact group of isometries acting on a complete, simply connected Riemannian manifold of nonpositive sectional curvature (Hadamard manifold) has a unique fixed point. However, here we will prove the fixed point theorem where we need it, namely, in the special case when the manifold is a symmetric space of non-compact type.

Theorem 6.6.1. *Let $f : C \to (P, d)$ be a continuous map where d denotes the distance on a symmetric space P of non-compact type and C is a compact space with a positive finite regular measure, μ. Then the functional*

$$J(p) = \int_C d^2(p, f(c)) d\mu(c), \quad p \in P$$

attains its minimum value at a unique point of P called the center of gravity of $f(C)$ with respect to μ.

Proof. Fix a point $p_0 \in P$. Since C is compact, there is a ball $B_r(p_0)$ centered at p_0 such that if $p \notin B_r(p_0)$ then $J(p) > J(p_0)$. As the closure of $B_r(p_0)$ is compact, J takes its minimum at some point $q_0 \in B_r(p_0)$. To prove that q_0 is unique, it suffices to show that

$$J(q) > J(q_0), \text{ if } q \neq q_0.$$

Let $q(t)$ be the geodesic joining q and q_0, $q(0) = q_0$ and $q(1) = q$. By Lemma 6.6.2 below

$$\frac{d}{dt} d^2(q(t), f(c)) = \begin{cases} \|q'(t)\| d(q(t), f(c)) \cos \alpha_t(c) & \text{if } f(c) \neq q(t), \\ 0 & \text{otherwise,} \end{cases}$$

where $\alpha_t(c)$ is the angle between the unique geodesic $f(c)q(t)$ and $q(t)q$. One can prove that the map $(t, c) \mapsto \frac{d}{dt} d^2(q(t), f(c))$ is continuous, we leave this as an exercise to the reader. So $t \mapsto J(q(t))$ is differentiable and since $t = 0$ is a minimal point for $J(q_t)$, by differentiating we obtain

$$\|q(0)'\| \int_C d(q_0, f(c)) \cos \alpha_0(c) d\mu(c) = 0,$$

which implies

$$\int_C d(q_0, f(c)) \cos \alpha_0(c) d\mu(c) = 0. \tag{6.1}$$

Since the curvature is non-positive, by cosine inequality, if $f(c) \neq q_0$ then

$$d^2(q, f(c)) \geq d^2(q_0, f(c)) + d^2(q_0, q) - 2d(q_0, q)d(q_0, f(c)) \cos(\pi - \alpha_0(c)).$$

A similar inequality trivially holds if $f(c) = q_0$. After integrating both sides and using (6.1) we get,

$$J(q) \geq J(q_0) + d^2(q_0, q),$$

which proves that $J(q) > J(q_0)$. $\qquad\qquad\qquad\qquad\qquad\qquad\qquad\square$

Lemma 6.6.2. *Let $q(t)$ be a curve not passing through $p \in P$. Then*

$$\frac{d}{dt} d(q(t), p)|_{t=0} = \|q'(0)\| \cos \alpha$$

where α is the angle between the geodesic $pq(0)$ and $q(0)q(1)$.

Proof. Let $Q(t)$ be the curve in the tangent space T_pP such that $\exp_p(Q(t)) = q(t)$, where $\exp_p : T_pP \to P$ is the exponential map at the p. We also think of p as the origin of T_pP. Then

$$
\begin{aligned}
\frac{d}{dt} d(q(t), p)|_{t=0} &= \lim_{t \to 0} \frac{1}{t}(d(q(t), p) - d(q(0), p)) \\
&= \lim_{t \to 0} \frac{1}{2d(q(0), p)t}(d^2(q(t), p) - d^2(q(0), p)) \\
&= \lim_{t \to 0} \frac{1}{2d_p(Q(0), p)t}(d_p^2(Q(t), p) - d_p^2(Q(0), p)),
\end{aligned}
$$

$$\tag{6.2}$$

where d_p is the metric of the tangent space T_pP. By the cosine law in the Euclidean space T_pP we have

$$
\begin{aligned}
d_p(Q(t), p)^2 - d_p(Q(0), p)^2 &= d_p(Q(t), Q(0))^2 \\
&\quad + 2d_p(Q(0), p)d_p(Q(0), Q(t)) \cos \beta(t)
\end{aligned}
$$

where $\beta(t)$ is the angle between the lines $pQ(0)$ and $Q(0)Q(t)$. Let L_t be the arc length from $Q(0)$ to $Q(t)$. Then we have

$$\lim_{t \to 0} \frac{d_p(Q(t), Q(0))}{L_t} = 1$$

and

$$\lim_{t \to 0} \frac{L_t}{t} = \|Q'(0)\|.$$

Combining these we get

$$\lim_{t \to 0} \frac{d_p(Q(t), Q(0))^2}{t} = 0.$$

Continuing with (6.2), we have

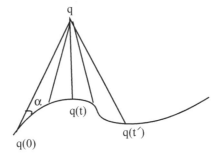

$$\frac{d}{dt} d(q(t), p)|_{t=0} = \frac{d_p(Q(0), Q(t))}{t} \cos \beta(t) = \|Q'(0)\| \cos \beta(0) \qquad (6.3)$$

Writing $Q'(0) = X_1 + X_2$ where X_1 is in the direction of the line pQ_0 and Y_1 is perpendicular to it. This orthogonal decomposition is preserved under the map $d_{Q_0} \exp$ and $\|d_{Q_0} \exp(X_1)\| = \|X_1\|$, therefore,

$$\|Q'(0)\| \cos \beta(0) = \|X_1\| = \|d_{Q_0} \exp_p(X_1)\| = \|d_{Q_0} \exp_p(Q'(0))\| \cos \alpha$$
$$= \|q'(0)\| \cos \alpha.$$

$\square$

As usual, G is a self-adjoint essentially algebraic subgroup of $\mathrm{GL}(n, \mathbb{R})$, or $\mathrm{GL}(n, \mathbb{C})$ acting on P by $(g, p) \mapsto g^t p g$. The following is Cartan's fixed point theorem for symmetric spaces of non-compact type.

Corollary 6.6.3. *If C is a compact subgroup of G, then C has a simultaneous fixed point acting on P.*

Proof. Let $\mu = dc$ be the normalized Haar measure on C, p_0 a point of P and $f : C \to P$ be the continuous function given by $f(c) = c \cdot p_0$. Then $J(p) = \int_C d^2(p, c \cdot p_0) dc$. Now for $c' \in C$, $J(c'p) = \int_C d^2(c'p, c \cdot p_0) dc$. Since C acts by isometries this is $\int_C d^2(p, (c')^{-1} c \cdot p_0) dc$. By left invariance of dc we get $\int_C d^2(p, c \cdot p_0) dc$. Thus $J(p) = J(c \cdot p)$ for all $c \in C$ and $p \in P$. But by Theorem 6.6.1, J has a unique minimum value at some $p \in P$. This means $c.p = p$ for all $c \in C$ since $J(p) = J(c \cdot p)$. Therefore p is a simultaneous fixed point.   □

We now prove the conjugacy theorem for maximal compact subgroups of G. The proof in [33] is similar to the one given here, but rather than involving differential geometry itself, it uses a convexity argument and a function which mimics the metric.

Theorem 6.6.4. *Let G be a self-adjoint essentially algebraic subgroup of $\mathrm{GL}(n, \mathbb{R})$, or $\mathrm{GL}(n, \mathbb{C})$. Then all maximal compact subgroups of G are conjugate. Any compact subgroup of G is contained in a maximal one.*

Proof. Let C be a compact subgroup of G. By Corollary 6.6.3 there is a point $p_0 \in P$ fixed under the action of C. Thus $C \subseteq \mathrm{Stab}_G(p_0)$. Since G acts transitively so $\mathrm{Stab}_G(p_0) = gKg^{-1}$ for some $g \in G$. Since K is a maximal compact subgroup by Theorem 6.3.5, so is the conjugate gKg^{-1}. This proves the second statement. If C is itself maximal then $C = gKg^{-1}$.   □

6.7   The Rank and Two-Point Homogeneous Spaces

Let $\mathfrak{g}$ be the Lie algebra of G a self-adjoint algebraic subgroup of $GL(n, \mathbb{R})$ or $GL(n, \mathbb{C})$, as discussed earlier in this chapter. Let $\mathfrak{g} = \mathfrak{k} \oplus \mathfrak{p}$ be a Cartan decomposition. By abuse of notation we shall call a subalgebra of $\mathfrak{g}$ contained in $\mathfrak{p}$ a subalgebra of $\mathfrak{p}$. Such subalgebras are abelian since $[\mathfrak{p}, \mathfrak{p}] \subset \mathfrak{k}$ and they will play an important role in what follows. By finite dimensionality, maximal abelian subalgebras of $\mathfrak{p}$ clearly exist. In fact, any abelian subset of $\mathfrak{p}$ is contained in a maximal abelian subalgebra of $\mathfrak{p}$.

Consider the adjoint representation of K on $\mathfrak{g}$. Then the subspace $\mathfrak{p}$ is invariant under this action. Since $\operatorname{Ad} k(\mathfrak{p}) \subseteq \mathfrak{g}$ for $k \in K$, to see this we only need to check that $\operatorname{Ad} k(\mathfrak{p})$ is symmetric (Hermitian). We shall always deal with the symmetric case except when the Hermitian one is harder. So for $X \in \mathfrak{p}$ and $k \in K$ we have $\operatorname{Ad} k(X) = kXk^{-1} = kXk^t$. Hence the transpose is $(kXk^t)^t = kXk^t = \operatorname{Ad} k(X)$.

Theorem 6.7.1. *In $\mathfrak{g}$ any two maximal abelian subalgebras $\mathfrak{a}$ and $\mathfrak{a}'$ of $\mathfrak{p}$ are conjugate by some element of K. In particular, their common dimension is an invariant of $\mathfrak{g}$ called $r = \operatorname{rank}(\mathfrak{g})$.*

This theorem was originally proved by E. Cartan. Here we adapt the argument of Theorem 4.3.1.

Proof. Let $\langle \cdot, \cdot \rangle$ be the Killing form on $\mathfrak{g}$. This is positive definite on $\mathfrak{p}$ and negative definite on $\mathfrak{k}$. Since K is compact and acts on $\mathfrak{g}$, by averaging with respect to Haar measure on K we can, in addition, assume this form to be K-invariant. That is, each $\operatorname{Ad} k$ preserves the form. Let $A \in \mathfrak{a}$ and $A' \in \mathfrak{a}'$ and consider the smooth numerical function on K given by $f(k) = \langle \operatorname{Ad} k(A), A' \rangle$. By compactness of K, this continuous function has a minimum value at k_0 and by calculus, at this point the derivative is zero. Thus for each $X \in \mathfrak{k}$, $\frac{d}{dt} \langle \operatorname{Ad}(\exp tX \cdot k_0)(A), A' \rangle_{|t=0} = 0$. But $\langle \operatorname{Ad}(\exp tX \cdot k_0)(A), A' \rangle = \langle \operatorname{Ad}(\exp tX) \operatorname{Ad} k_0(A), A' \rangle = \langle \operatorname{Exp}(t \operatorname{ad} X) \operatorname{Ad} k_0(A), A' \rangle$. Hence differentiating with respect to t at $t = 0$ gives $\langle \operatorname{ad} X \operatorname{Ad} k_0 A, A' \rangle = 0$

for all $X \in \mathfrak{k}$. A calculation similar to the one just given shows that the K-invariance of the form on $\mathfrak{k}$ has an infinitesimal version, $\langle [X, Y], Z \rangle + \langle Y, [X, Z] \rangle = 0$, valid for all $X \in \mathfrak{k}$ and $Y, Z \in \mathfrak{p}$. Hence, also for all $X \in \mathfrak{k}$, we get $\langle x, [\operatorname{Ad} k_0(A), A'] \rangle = 0$. Now $\operatorname{Ad} k_0(A)$ and $A' \in \mathfrak{p}$ and $[\mathfrak{p}, \mathfrak{p}] \subseteq \mathfrak{k}$. Hence $[\operatorname{Ad} k_0(A), A'] \in \mathfrak{k}$ and because $\langle X, [\operatorname{Ad} k_0(A), A'] \rangle = 0$ for all $X \in \mathfrak{k}$ and $\langle \cdot, \cdot \rangle$ is nondegenerate on $\mathfrak{k}$, it follows that $[\operatorname{Ad} k_0(A), A'] = 0$. Now hold $A \in \mathfrak{a}$ fixed. Because $[\operatorname{Ad} k_0(A), \mathfrak{a}'] = 0$ we see by maximality of $\mathfrak{a}'$ that $\operatorname{Ad} k_0(A) \in \mathfrak{a}'$ and since A is arbitrary $\operatorname{Ad} k_0 \mathfrak{a} \subseteq \mathfrak{a}'$. Thus $\mathfrak{a} \subseteq \operatorname{Ad} k_0^{-1}(\mathfrak{a}')$. The latter is an abelian subalgebra of $\mathfrak{p}$ and by maximality of $\mathfrak{a}$ they coincide. Thus $\operatorname{Ad} k_0(\mathfrak{a}) = \mathfrak{a}'$.   □

It might be helpful to mention the significance of this theorem in the most elementary situation, namely, when $G = \operatorname{GL}(n, \mathbb{R})$, or $\operatorname{GL}(n, \mathbb{C})$. As usual, we restrict our remarks to the real case. Here $\mathfrak{p}$ is the set of all symmetric matrices of order n. Let $\mathfrak{d}$ denote the diagonal matrices. These evidently form an abelian subalgebra of $\mathfrak{p}$. Now $\mathfrak{d}$ is actually maximal abelian. To see this, suppose there were a possibly larger abelian subalgebra $\mathfrak{a}$. Each element of $\mathfrak{a}$ is diagonalizable being symmetric. Since all these elements commute they are simultaneously diagonalizable. This means, in effect, that $\mathfrak{a} = \mathfrak{d}$. Thus $\mathfrak{d}$ is a maximal abelian subalgebra of $\mathfrak{p}$. Similarly, over $\mathbb{C}$ it says any commuting family of Hermitian matrices is simultaneously conjugate by a unitary matrix to the diagonal matrices. This is exactly the content of the theorem in these two cases. Thus Theorem 6.7.1 is a generalization of the classic result on simultaneous diagonalization of commuting families of quadratic or Hermitian forms.

We also note that the statement of Theorem 6.7.1 without the stipulation that the subalgebras are in $\mathfrak{p}$ is false. That is, in general, maximal abelian subalgebras of $\mathfrak{g}$ are not conjugate. For example, in $\mathfrak{g} = \mathfrak{sl}(2, \mathbb{R})$, the diagonal elements, the skew symmetric elements and the unitriangular elements are each maximal abelian subalgebras of $\mathfrak{g}$, but no two of them are conjugate (by an element of K or anything else, see Section 4.5).

Corollary 6.7.2. *In $\mathfrak{g}$ let $\mathfrak{a}$ be a maximal abelian subalgebra of $\mathfrak{p}$. Then the conjugates of $\mathfrak{a}$ by K fill out $\mathfrak{p}$, that is, $\bigcup_{k \in K} \operatorname{Ad} k(\mathfrak{a}) = \mathfrak{p}$. Of course, exponentiating and taking into account that $\exp$ commutes with conjugation, this translates on the group level to $P = \bigcup_{k \in K} kAk^{-1}$, where A is the connected abelian subgroup of G with Lie algebra $\mathfrak{a}$.*

This corollary is the analogue for non-compact groups of Corollary 4.3.9.

Proof. Let $X \in \mathfrak{p}$ and choose a maximal abelian subalgebra $\mathfrak{a}'$ containing it. By our theorem there is some $k \in K$ conjugating $\mathfrak{a}'$ to $\mathfrak{a}$. In particular, $\operatorname{Ad} k(X) \in \mathfrak{a}$ for some $k \in K$ and so $X \in \operatorname{Ad} k^{-1}(\mathfrak{a})$.   □

Our next corollary, also called the Cartan decomposition, follows from this last fact together with the usual Cartan decomposition, Theorem 6.3.5.

Corollary 6.7.3. *Under the same hypothesis $G = KAK$.*

Proof. $G = KP \subseteq KKAK = KAK \subseteq G$.   □

Remark 6.7.4. An important use of this form of the Cartan decomposition is that it reduces the study of the asymptotic at ∞ on G to A. That is, suppose g_i is a sequence in G tending to ∞. Now $g_i = k_i a_i l_i$, where k_i and $l_i \in K$ and $a_i \in A$. Since both k_i and l_i have convergent subsequences, again denoted by k_i and l_i, which converge to k and l, respectively, the sequence a_i must also tend to ∞. Thus in certain situations we can assume the original sequence started out in A.

We now make explicit the notions of a *homogeneous space* and *two-fold transitivity* from differential geometry mentioned earlier. If X is a connected Riemannian manifold, we shall say X is a homogeneous space if the isometry group $\operatorname{Isom}(X)$ acts transitively on X. Now even when the action may not be transitive it is a theorem of Myers and Steenrod (see [32]) that $\operatorname{Isom}(X)$ is a Lie group and the stabilizer K_p of any point p is a compact subgroup. In the case of a transitive action it follows from general facts about actions that X is equivariantly equivalent as a Riemannian manifold to $\operatorname{Isom}(X)/K_p$ with the quotient structure. Of course, if some subgroup of the isometry group acted transitively then these same conclusions could be drawn replacing the isometry group by the subgroup. Clearly, by its very construction, every symmetric space of non-compact type is a homogeneous space.

Now suppose in our symmetric space P we are given points p and q and p' and q' of P with $d(p,q) = d(p',q')$. We shall say that a subgroup of the isometry group acts *two-fold transitively* if there is always an isometry g in the subgroup taking p to p' and q to q' for any choices of such points. When this occurs we shall say P is a *two-point homogeneous space*. Clearly, every two-point homogeneous space is a homogeneous space. As we shall see the converse is not true and we will learn which of our symmetric spaces is actually a two-point homogeneous space. Before doing so, we make a simple observation which follows immediately from transitivity.

Proposition 6.7.5. *Let G be as above and K be a maximal compact subgroup. Then $G/K = P$ is a two-point homogeneous space if and only if K acts transitively on the unit geodesic sphere U of P.*

For example, when $G = SO(n,1)_0$ and $K = SO(n)$, then $G/K = H^n$, hyperbolic n-space. Here K acts transitively on U. Hence $SO(n,1)_0$ acts two-fold transitively on H^n. As we shall see in Theorem 6.7.6, this fact is a special case of a more general result. We also remark that this definition can be given for any connected Riemannian manifold and indeed such a manifold is of necessity, a symmetric space (see [32]).

Our last result tells us the significance of the rank in this connection. Before proving it we observe that for all semisimple or reductive groups under consideration $\dim \mathfrak{p} \geq 2$. The lowest dimension arising, is the case of the upper half plane introduced in Section 6.4. Indeed, suppose $\dim \mathfrak{p} = 1$. Then since $\mathfrak{p}$ is abelian and $\exp$ is a global diffeomorphism $\mathfrak{p} \to P$, it follows easily from $\exp(X + Y) = \exp X \exp Y$, where $X, Y \in \mathfrak{p}$, that P is a connected 1-dimensional abelian Lie group. Now since K acts on P by conjugation, and in this case these form a connected group of automorphisms of P we see that this action is trivial because $\operatorname{Aut}(P)_0 = \{1\}$. Thus K centralizes P and we have a direct product of groups. Such a group is not semisimple. It is clearly also not $GL(n, \mathbb{R})$

or $\mathrm{GL}(n, \mathbb{C})$ for $n \geq 2$.

We now characterize two-point homogeneous symmetric spaces.

Theorem 6.7.6. *Let G be as above, $\mathfrak{g}$ be its Lie algebra and K be a maximal compact subgroup. Then G/K is a two-point homogeneous space if and only if $\mathrm{rank}(\mathfrak{g}) = 1$.*

Proof. We first assume $\mathrm{rank}(\mathfrak{g}) = 1$. By Proposition 6.7.5, to see that G/K is a two-point homogeneous space, it is sufficient to show K acts transitively on geodesic spheres of P. Of course, we know $\mathrm{Ad}\, K$ acts linearly and isometrically on $\mathfrak{p}$. Now by Corollary 6.7.2 $\bigcup_{k \in K} \mathrm{Ad}\, k(\mathfrak{a}) = \mathfrak{p}$. Hence each point $p \in U$ is a conjugate by something in K to a point on the unit sphere of $\mathfrak{a}$. Since the dimension of this sphere is zero, it consists of two points, $\pm a_0$. Hence $U = \mathrm{Ad}\, K(a_0) \cup \mathrm{Ad}\, K(-a_0)$. In any case, U is a union of a finite number of orbits all of which are compact and therefore closed since K itself is compact. Since these are closed, so is the union of all but one of them. Hence U is the disjoint union of two nonempty closed sets. This is impossible since U is connected because $\dim \mathfrak{p} \geq 2$. Thus there is only one orbit and therefore K acts transitively on U.

Before proving the converse, the following generic example will be instructive. Let $G = \mathrm{SL}(n, \mathbb{R})$, $n \geq 2$. We shall see $\mathrm{SL}(n, \mathbb{R})/\mathrm{SO}(n)$ is a two-point homogeneous space if and only if $n = 2$. This suggests that unless the rank $= 1$, one can never have a two-point homogeneous symmetric space.

To see this, observe that since $G/K = P$ is the set of positive definite $n \times n$ symmetric matrices of $\det 1$, it follows that $\dim P = \frac{n(n+1)}{2} - 1$. Also $\dim K = \frac{n(n-1)}{2}$. Hence if U denotes the geodesic unit sphere in P, its dimension is $\frac{n(n+1)}{2} - 2$. Let K act on P and U by $(k, p) \mapsto kpk^{-1} = kpk^t$. For $p \in U$ the dimension of $\mathcal{O}_K(p)$, the K-orbit of p, is

$$\dim \mathcal{O}_K(p) = \frac{n(n-1)}{2} - \frac{(n-1)(n-2)}{2} = n - 1.$$

Now if K were to act transitively on U, then $\dim \mathcal{O}_K(p) = \dim U$. That is, $n - 1 = \frac{n(n+1)}{2} - 2$. Alternatively, $(n-2)(n+1) = 0$. Since $n \geq 2$, this holds if and only if $n = 2$.

We conclude by proving the converse. Suppose (P, G) is a two-point homogeneous space and hence K acts transitively (by conjugation) on the unit geodesic sphere U in $\mathfrak{p}$. Then $U = \mathcal{O}_K(A_0)$, where $A_0 \in \mathfrak{p}$ and $\|A_0\| = 1$. Since A_0 is conjugate to something in $\mathfrak{a}$, we may assume $A_0 \in \mathfrak{a}$. In particular, everything in $U \cap \mathfrak{a}$ is K-conjugate to everything else. Because these matrices commute, they can be simultaneously diagonalized by some u_0 (which may not be in K). By replacing these A_0's by their u_0 conjugates we may assume they are all diagonal. Being conjugate under K these matrices have the same spectrum S. Since S is finite and K is connected, K cannot permute this finite set. Thus the action of K leaves each of these matrices fixed. But K acts transitively on $U \cap \mathfrak{a}$ so $U \cap \mathfrak{a}$ must be a point. Hence it has dim 0 and dim $\mathfrak{a} = 1$.   □

Exercise 6.7.7. Show the rank of $\mathrm{Sp}(n, \mathbb{R})$ is n.

6.8   The Disk Model for Spaces of Rank 1

In this section we focus on the *classical* simple groups of rank 1 and their associated irreducible symmetric spaces which we view in the disk model. We will then use the geometry of the latter to indicate that the exponential map of the corresponding centerless simple group is surjective.

In this section, whose material is mostly taken from [19], we unify the study of the three infinite families of classical simple rank 1 groups or classical non-compact irreducible rank 1 symmetric spaces, $H_n(F)$ by considering a field F, where $F = \mathbb{R}$, $\mathbb{C}$, or $\mathbb{H}$, the quaternions and define $G = \mathrm{U}(n, 1, F)$ as follows: Let F^{n+1} be the right vector space over F consisting of $(n + 1)$-tuples of points from F. For such $(n + 1)$-tuples $x = (x_0, \ldots, x_n)$ and $y = (y_0, \ldots, y_n) \in F^{n+1}$, consider $\langle \cdot, \cdot \rangle$ defined

$$\langle x, y \rangle = x_0 \bar{y}_0 - \sum_{i=1}^{n} x_i \bar{y}_i.$$

This is a nondegenerate form over F which is linear in x and conjugate linear in y, where the conjugation is the natural one coming from F. (In the case $F = \mathbb{R}$ conjugation is the identity.) $G = \mathrm{U}(n, 1, F)$ is then defined to be those $g \in \mathrm{GL}(n+1, F)$ which preserve this form. G is evidently a Lie group, we denote its Lie algebra by $\mathfrak{u}(n, 1, F)$. Just as in Section 6.3 above, G is actually a self-adjoint algebraic subgroup of $\mathrm{GL}(n+1, F)$.

When $F = \mathbb{R}$ taking the identity component we get $\mathrm{SO}(n, 1)_0$ which is centerless and simple. When $F = \mathbb{C}$, $\mathrm{U}(n, 1, \mathbb{C})$ is merely reductive and non-semisimple. But, of course, $\mathrm{Ad}(\mathrm{U}(n, 1, \mathbb{C})) = \mathrm{PSU}(n, 1)$ is centerless and simple as is $\mathrm{Ad}(\mathrm{U}(n, 1, \mathbb{H})) = \mathrm{Ad}(\mathrm{Sp}(n, 1))$. A direct calculation tells us that respectively $K = \mathrm{SO}(n)$, $\mathrm{U}(n)$, and $\mathrm{Sp}(n)$, operating on $\mathbb{R}^n$, $\mathbb{C}^n$, $\mathbb{H}^n$ in the usual manner. The unit sphere, U, being S^{n-1}, S^{2n-1}, S^{4n-1}, respectively. Thus in all three cases K operates transitively on U. From this it follows that $\mathrm{Ad}\, G$ operates 2-fold transitively on G/K and hence $\mathrm{Ad}\, G$ always has rank 1. Using classification, these are the classical irreducible non-compact rank 1 symmetric spaces and except for one irreducible rank 1 symmetric space of non-compact type (or real simple non-compact Lie group) this accounts for all non-compact, centerless, simple groups of rank 1 (see [32]). The missing one called the exceptional group is related to the Cayley numbers which we will not deal with here. However all the properties of the disk model of the exceptional rank 1 symmetric space are actually the same as those of the classical ones.

Let $P(F^{n+1})$ be the projective space corresponding to F^{n+1} and $\pi : F^{n+1} \setminus \{0\} \to P(F^{n+1})$ the canonical map taking $x \mapsto [x]$. Now $\mathrm{GL}(n+1, F)$ operates of $P(F^{n+1})$ through F^{n+1}. Thus if $g \in \mathrm{GL}(n+1, F)$ and $x \in F^{n+1} - (0)$ we take $g[x] = [g(x)]$. This map is well-defined and gives an action of $\mathrm{GL}(n+1, F)$ on $P(F^{n+1})$ and upon restriction the same is true of any subgroup of $\mathrm{GL}(n+1, F)$. Thus $\mathrm{U}(n, 1, F) = G$ acts on $P(F^{n+1})$. Let

$$\Omega = \pi\{x \in F^{n+1} \setminus \{0\} : \langle x, x \rangle > 0\},$$

be the projective image of the interior of the *light cone*. Since G preserves $\langle \cdot, \cdot \rangle$, it leaves Ω invariant. Now let $x = (x_0, \ldots, x_n)$ be a vector with $[x] \in \Omega$. Then $|x_0|^2 > \sum_{i=1}^{n} |x_i|^2$. Next we show that G operates transitively on Ω. Let $y = (y_0, \ldots, y_n) \in F^{n+1}$ and assume $|y_0|^2 - \sum_{i=1}^{n} |y_i|^2 = 1$. Chose $t \geq 0$ so that $|y_0| = \cosh t$ and

$\sqrt{\sum_{i=1}^{n} |y_i|^2} = \sinh t$. Choose $u \in U(1, F)$ and $v \in U(n, F)$ so that $y_0 = u \cosh t$ and $(y_1, \ldots, y_n) = v(0, \ldots, 0, \sinh t)$. This is possible since $U(n, F)$ operates transitively on spheres in F^n for all n. Let

$$k = \begin{pmatrix} u & 0 \\ 0 & v \end{pmatrix}.$$

Then $k \in K$. If x_0 is the point in $P(F^{n+1})$ represented by $(1, 0, \ldots, 0)$, then $x_0 \in \Omega$. We show that $y = k a_t x_0$, where $a_t = \operatorname{Exp} t X_0$ and the matrix X_0 of order $n + 1$ given by

$$X_0 = \begin{pmatrix} 0 & v_0 & 1 \\ w_0 & O & w_0 \\ 1 & v_0 & 0 \end{pmatrix},$$

where $v_0 = (0, \ldots, 0)$ of order $n - 1$, $w_0 = v_0^t$ and O is the zero matrix of order $n - 1$. Hence $X_0 \in \mathcal{P}$ as in Section 6.2. Since the rank of G is 1 because it operates two fold transitively, therefore $\{a_t : t \in \mathbb{R}\}$ is a maximal abelian subgroup of P. A direct calculation shows that

$$a_t = \begin{pmatrix} \cosh t & v_0 & \sinh t \\ w_0 & I & w_0 \\ \sinh t & v_0 & \cosh t \end{pmatrix},$$

where $v_0 = (0, \ldots, 0)$ of order $n - 1$, $w_0 = v_0^t$ and I is the identity matrix of order $n - 1$. Hence $a_t(1, 0, \ldots, 0) = (\cosh t, 0, \ldots 0, \sinh t)$ and therefore

$$k a_t(1, 0, \ldots, 0) = k(\cosh t, 0, \ldots 0, \sinh t) = (u \cosh t, v(0, \ldots, 0, \sinh t))$$
$$= (y_0, y_1, \ldots, y_n).$$

Since K and a_t are connected we actually get G_0 acts transitively on Ω. Here we take the upper component of the hyperboloid $y_0^2 = 1 + \sum_{i=1}^{n} |y_i|^2$, $y_0 > 0$, projectivizing this, gives everything.

We now calculate the isotropy group of $[(1, 0, \ldots, 0)]$ within G_0. Since this group is connected we can calculate the isotropy group within the Lie algebra and exponentiate. Here we take

$$X = \begin{pmatrix} A & B \\ B^* & C \end{pmatrix},$$

where $\bar{A} = -A \in F$, $B \in F^n$, B^* is B conjugate transposed and C is an $n \times n$ matrix from F with $C^* = -C$. Since $[(1, 0, \ldots, 0)] = [(\lambda, 0, \ldots, 0)]$ for any $\lambda \neq 0 \in F$, if $X \in \mathrm{Stab}_\mathfrak{g}[(1, 0, \ldots, 0)]$, then $X(\lambda, 0, \ldots, 0) = 0$ so $B^* = 0$. Hence also $B = 0$. Therefore $A \in \mathfrak{u}(1, F)$ and $C \in \mathfrak{u}(n, F)$, and $\mathrm{Stab}_{G_0}[(1, 0, \ldots, 0)] = K$ (see Corollary 1.4.15). Thus $\Omega = G/K$, where G is the connected component of the identity in G and K is a maximal compact subgroup of G_0.

Finally, we come to the ball model of $H_n(F)$. We will denote this by

$$B(F^n) = \{(x_1, \ldots, x_n) \in F^n : \sum_{i=1}^n |x_i|^2 < 1\}.$$

Define $\phi : \Omega \to B(F^n)$ as follows. If $x \in \Omega$, then $|x_0|^2 > 0$. Since $x_0 \neq 0$ we can form $x_i x_0^{-1}$ for each i. Let ϕ be defined by $(x_0, \ldots, x_n) \mapsto [(x_1 x_0^{-1}, \ldots, x_n x_0^{-1})]$. Then for $x \in \Omega$, $|x_1 x_0^{-1}|^2 + \ldots + |x_n x_0^{-1}|^2 < 1$. So $\phi(x) \in B(F^n)$. Conversely, let $(y_1, \ldots, y_n) \in B(F^n)$. Then $\sum_{i=1}^n |y_i|^2 < 1$. Therefore, $(1, y_1, \ldots, y_n) \in \Omega$ and $\phi(1, y_1, \ldots, y_n) = (y_1, \ldots, y_n)$. Evidently, $\phi^{-1}(y_1, \ldots, y_n) = \{x_0, y_1 x_0, \ldots, y_n x_0 : x_0 \neq 0\}$. Thus ϕ maps $H^n(F)$ bijectively to $B(F^n)$. How does G operate on $B(F^n)$? Let $g \in G$ and $y \in B(F^n)$. Then

$$(gy)_i = (g_{i0} + \sum_{j=1}^n g_{ij} y_j)(g_{00} + \sum_{j=1}^n g_{0j} y_j)^{-1}.$$

Thus,

$$1 - \|gy\|^2 = (1 - \|y\|^2)(g_{00} + \sum_{j=1}^n g_{0j} y_j)^{-2}.$$

It follows that g operates by the same formula on the boundary, $\partial(B(F^n)) = \{y \in F^n : \|y\|^2 = 1\}$, as well. We conclude this section with introduction the *Lorentz model* and its connection of the ball or disk model.

We show $[x] = [y]$, for some $y = (y_0, \cdots, y_n)$ such that $|y_0|^2 - \sum_{i=1}^n |y_i|^2 = 1$. That is, we can change coordinates and sharpen the inequality to an equality. We can assume that $y_0 \geq 0$, otherwise we replace $(y_0, \ldots, y_n)$ by $(-y_0, \ldots, -y_n)$. Therefore Ω is diffeomorphic to

$$\mathcal{L} = \{(y_0, \ldots y_n) \in F^{n+1} : y_0 = \sqrt{1 + \sum_{i=1}^{n} |y_i|^2}, y_0 > 0\}$$

which is called the *hyperboloid* or *Lorentz model.*

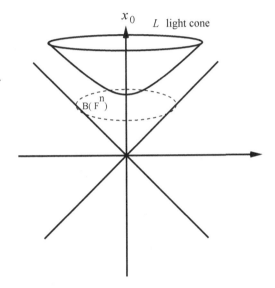

Figure 6.1: Lorentz model and disk model

To see this just choose $\lambda \neq 0 \in F$. Then $x\lambda = (x_0, \lambda, \ldots, x_n\lambda) = y$. Writing down the equation we want for y we see that $(|x_0|^2 - \sum_{i=1}^{n} |x_i|^2)|\lambda|^2 = 1$. Since $|x_0|^2 > \sum_{i=1}^{n} |x_i|^2$ the first term is nonzero forcing

$$\lambda = \sqrt{\frac{1}{|x_0|^2 - \sum_{i=1}^{n} |x_i|^2}} \in \mathbb{R} \subseteq F.$$

Now the projection from the hyperboloid with respect to *vertex of the light cone* $(1, 0 \ldots, 0)$ to the unit disk in $\{0\} \times \mathbb{R}^n \subset \mathbb{R}^{n+1}$ identifies these two models.

6.9   Exponentiality of Certain Rank 1 Groups

As above we call a connected Lie group *exponential* if exp is surjective, or alternatively if every point lies on a 1-parameter subgroup. For example, compact connected groups are exponential (see Theorem 4.3.8), but often non-compact simple groups are not. The purpose of this section is to indicate that the groups $SO(n, 1)_0$, $n \geq 2$, $PSU(n, 1)$, $n \geq 1$ and $Ad Sp(n, 1)$, $n \geq 1$ are all exponential [52]. We remark that the exceptional non-compact, centerless, rank 1 simple group, $Ad F_{(4, -20)}$, was proved to be non-exponential by D. Djokovic and N. Thang in [17] and we will see where our line of argument breaks down in this case.

Theorem 6.9.1. *The exponential map is surjective for all classical, connected, non-compact, centerless, rank 1 simple Lie groups.*

This will be proven by means of geometry; that is, studying the action of G as the connected component of the isometry group of the symmetric space $X = G/K$. A self evident principle which we shall employ here is the fact that if a Lie group is a union of exponential Lie subgroups, it must itself be exponential.

Now G/K is diffeomorphic to the interior of the closed unit ball B^n of dimension n, where K is a maximal compact subgroup of G and $n = \dim(G/K)$. As we noted above each isometry g of G extends continuously to the boundary of G/K. Thus G acts on B^n and since G is arcwise connected each isometry g is homotopic to the identity. By the Brouwer fixed point theorem each g has a fixed point in B^n. If this is in the interior of the ball, that is in G/K, then since G acts transitively with K as the isotropy group of 0, g lies in some conjugate of K. But because K is compact and connected, g lies on a 1-parameter subgroup of this conjugate of K and therefore of G. The other possibility is that there is a g-fixed point p on the boundary. Let K_p be the compact Lie subgroup of K consisting of the isometries leaving p fixed. We shall first prove

Proposition 6.9.2. K_p *is connected.*

Proof. By Proposition 6.7.5 and Theorem 6.7.6, K acts transitively on all geodesic spheres centered at 0. Now each point on the boundary of B^n lies on a unique geodesic emanating from 0. Let q and r be boundary points and $\gamma_q(t)$ and $\gamma_r(t)$, $t \in \mathbb{R}$ be the corresponding geodesics. Then for $n \in \mathbb{Z}^+$, $\gamma_q(n) = q_n$ and $\gamma_r(n) = r_n$ converge respectively to q and r and for each such n there is a $k_n \in K$ such that $k_n(q_n) = r_n$. By compactness some subsequence, which we again call $k_n \to k \in K$. Since also the corresponding subsequences $q_n \to q$ and $r_n \to r$, and isometries extend continuously to the boundary we see that $k(q) = r$. Thus K also acts transitively on the boundary of B^n and therefore K/K_p is (K-equivariantly) homeomorphic with S^{n-1}. But the latter is simply connected for $n > 2$ and K is connected so we conclude from the long exact sequence for homotopy of a fibration that K_p is also connected. This leaves only the case $n = 2$, *i.e.* the hyperbolic disk. Here, by direct calculation, one sees easily that $K_p = \{1\}$. □

Continuing the proof of Theorem 6.9.1, let G_p denote the subgroup of all elements of G fixing p. Then according to p. 154 of [61]

$$G_p = K_p A U_p, \tag{6.4}$$

where U_p is the unipotent radical of G_p. In particular, U_p is normal in G_p and connected and simply connected. This means, in particular, that $H_p = K_p U_p$ is a subgroup of G_p and that U_p contains no nontrivial compact subgroups. Thus $U_p \cap K_p = \{1\}$ and $H_p = K_p \times_\eta U_p$ (semidirect product).

Now suppose that there is another g-fixed point q on the boundary. In this case p and q can be joined by a *unique* geodesic γ of G/K. Since g is an isometry, $g(\gamma)$ is also a geodesic joining p and q and therefore it must coincide with γ so that γ is stabilized by g. By conjugation by an isometry in G taking a point on the interior of γ to 0 we can assume that p and q are opposite to one another. Since as we saw earlier K acts transitively on the boundary we can also assume that q is any particular boundary point. Passing to the upper half space model we can therefore take $q = \infty$, $p = 0$ then γ is a vertical half line. In this model if $g(\infty) = \infty$ then using (6.4), $g = \theta \circ \lambda \circ u$ where $\lambda > 0$, $\theta \in K_p$

and $u \in U_p$. Since u is unipotent,

$$g(x, z) = (\lambda \theta(x) + b(u), \lambda z) \qquad (6.5)$$

for $x \in \mathbb{R}^n$ and $z > 0$ where $b(u) \in \mathbb{R}^n$. Now if in addition $g(0) = 0$ then $b(u) = 0$ and

$$g(x, z) = (\lambda \theta(x), \lambda z) \qquad (6.6)$$

and so on γ our map is $z \mapsto \lambda z$. Now compose with the isometry $h : (x, z) \mapsto \lambda^{-1}(x, z)$. By (6.6), h fixes 0 and ∞ and the composition leaves γ pointwise fixed and has many fixed points in the interior, so $\in K$. In fact the composition is in K_p and hence $g \in \mathbb{R}_+^\times \times K_p$ (direct product). Thus the original g lies in H, a conjugate by something in G of this direct product subgroup. (Notice that since A commutes with K_p and $hg = \theta$ leaves γ fixed it must leave the orthogonal hyperplane in the disk model *i.e.* the boundary in the upper half space invariant. Hence by (6.6) so does g. In any case, since K_p is connected, such a group is clearly of exponential type (this will also be included in the result below), so $g = \exp Y$, where $Y \in$ the Lie algebra of H.

If g has more than two fixed points say p, q and r on the boundary, reasoning as above we would then have, in the upper half space model $q = \infty$, $p = 0$, γ is a vertical half line and r is a boundary point $\neq p$. Hence, by (6.6), $\theta = 1$. But then, after applying g, the distance $d(p, r)$ is multiplied by λ. Since both points are g-fixed this means that $\lambda = 1$ and $g = I$ and so every interior point is also g-fixed.

The only other possibility is that p is the unique g-fixed point on the boundary. Passing again to the upper half space model we can take $p = \infty$. Since G_p is a group, g^{-1} also leaves p fixed. Clearly $1/\lambda(g) = \lambda(g^{-1})$ and so if $\lambda(g) \neq 1$ then we may assume that each of these has norm $\neq 1$. But then applying (6.5) to a boundary point $(x, 0)$ we have $g(x, 0) = (\lambda \theta(x) + b(u), 0)$. This means that x is g-fixed if and only if $x = \lambda \theta(x) + b(u)$ or alternatively $(\theta - \lambda^{-1}I)x = -b(u)/\lambda$. Since all the eigenvalues of θ are of absolute value 1 and $|\lambda^{-1}| \neq 1$ it follows that $(\theta - \lambda^{-1}I)$ is invertible and this equation can be solved giving a g-fixed point on the (finite) boundary. Hence $x \neq \infty$ and we are back in the case of a screw motion ($g \in \mathbb{R}_+^\times \times K_p e$) . Thus we may assume that $\lambda = 1$ and g has no A part and this means g lies in $H_p = K_p U_p$ (semidirect product)[2].

We will complete the proof of Theorem 6.9.1 by showing that in all three cases, $H_p = H$ is exponential. Now for hyperbolic space since N is abelian, this can be done directly as follows.

Theorem 6.9.3. *The identity component,* $\mathrm{SO}(n) \times_\eta \mathbb{R}^n = H$, *of the Euclidean motion group is exponential.*

That is, we will show the exponential map is surjective for the connected group of isometries of a (simply connected) space form of zero curvature,

Proof. Take faithful matrix representations of H and its Lie algebra $\mathfrak{h}$ of order $n + 1$ as follows.

$$H = \{\begin{pmatrix} \alpha & v \\ 0 & 1 \end{pmatrix} : \alpha \in \mathrm{SO}(n), v \in \mathbb{R}^n\}$$

while

$$\mathfrak{h} = \{\begin{pmatrix} X & w \\ 0 & 0 \end{pmatrix} : X \in M_n(\mathbb{R}), X^t = -X, w \in \mathbb{R}^n\}.$$

If we denote the elements of $\mathfrak{h}$ by (X, w), then by a direct calculation

$$\exp(X, w) = \begin{pmatrix} \exp X & \frac{\exp X - I}{X}(w) \\ 0 & 1 \end{pmatrix}$$

where, by $\frac{\exp X - I}{X}$ we understand the functional calculus, *i.e.* this matrix valued function of a matrix argument, has a removable singularity at 0 with value I.

---

[2]It should be remarked that another way of looking at the three possibilities which arise in the proof below is by means of the following classification scheme for isometries of spaces of negative curvature (X, d), due to M. Gromov (see p. 77 of [3]), which, in our situation, gives the following trichotomy for an isometry g. If $\inf_{x \in X} d(x, gx)$ is 0 and is assumed, or is 0 and is not assumed, or is positive (when it must be assumed), then g is *elliptic*, *parabolic*, or *hyperbolic*, respectively.

Let $(\alpha, v) \in H$. Since $SO(n)$ is a compact connected Lie group we can choose $X \in \text{End}(V)$, so that $\exp X = \alpha$. Then $\exp(X, w) = (\alpha, v)$ for some w if and only if $\frac{\exp X - I}{X}(w) = v$. Since v is arbitrary this amounts to knowing that the linear transformation $\frac{\exp X - I}{X}$ is onto, *i.e.* all its eigenvalues are $\neq 0$. But by functional calculus the eigenvalues of this operator are either 1, or are of the form $\frac{e^{\lambda} - 1}{\lambda}$, where λ is a nonzero eigenvalue of X. Clearly such an eigenvalue is 0 if and only if $\lambda = 2\pi i m$ for some integer m. Hence this operator is invertible if and only if X has no eigenvalues of the form $2\pi i m$ for some $m \neq 0 \in \mathbb{Z}$. Now by an orthonormal change of basis (which affects nothing since exp is constant on conjugacy classes), α and X are respectively given by

$$\alpha = \text{diag}(R(t_1), \dots, R(t_j), 1, \dots 1),$$

and

$$X = \text{diag}(S(t_1), \dots, S(t_j), 0, \dots 0),$$

where $R(t_k)$ and $S(t_k)$ are the planar rotation and infinitesimal rotation determined by t_k:

$$R(t_k) = \begin{pmatrix} \cos t_k & \sin t_k \\ -\sin t_k & \cos t_k \end{pmatrix}$$

$$S(t_k) = \begin{pmatrix} 0 & t_k \\ -t_k & 0 \end{pmatrix}.$$

We may assume none of the t_k is an integral multiple of 2π. For if there were such a t_k it would just produce additional 1's in the block diagonalization of α, above and we exponentiate onto I by O. Hence we may assume that each $t_k \neq 0$ and satisfies $-\pi \leq t_k \leq \pi$. But then, X has no eigenvalues of the form $2\pi i m$, where $m \neq 0 \in \mathbb{Z}$. □

Turning to the other two cases, in general since K_p and U_p are both connected so is H_p. Denote U_p by U and K_p by C. If U were trivial, then H would be compact and connected and so would be of exponential type. Otherwise we shall show that $Z_C(U) = \{v\}$. In the disk model let q be the point opposite p on the boundary, $k_0 \neq 1 \in C$ and suppose that $k_0 u = u k_0$ for all $u \in U$. Then $k_0 u(q) = u k_0(q)$. Since $k_0 \in C$ and

$k_0(p) = p$ it follows that $k_0(q)$ also equals q and therefore each $\mathrm{U}(q)$ is k_0 fixed. On the other hand, each u stabilizes each horosphere and in particular the boundary of B^n. Since the only points on B^n which k_0 leaves fixed are p and q, either $\mathrm{U}(q) = p$ or q. But if $\mathrm{U}(q) = p$ for some u then $u^{-1}(u(q)) = q = u^{-1}(p) = p$, a contradiction. Thus for each u, $\mathrm{U}(q) = q$ and $\mathrm{U}(p) = p$. Since p and q are also C fixed and U and C generate H, we see that H itself leaves p and q fixed so H is contained in a group isomorphic with the direct product of a compact connected group with $\mathbb{R}^{\times}_{+}$ and therefore is also exponential. The remaining case is that only the identity of C centralizes U so C acts faithfully as a group of automorphisms of U.

In order to complete the proof of Theorem 6.9.1 in the remaining two cases we must now investigate the exponentiality of certain semidirect products where N is non-abelian. In these cases although H is not solvable, the solvable methods of [56] can be applied. Namely, we consider a compact connected group, L, of automorphisms of a simply connected nilpotent group, N with Lie algebra, $\mathfrak{n}$, and $H = L \times_{\eta} N$, the natural semi-direct product. Since N is simply connected and nilpotent we can identify $\mathrm{Aut}(N)_0$ with $\mathrm{Aut}(\mathfrak{n})_0$.

In [56] the following result is proved. Here $T(X)$ is the subgroup of T fixing $X \in \mathfrak{n}$.

Theorem 6.9.4. *Let N be a connected nilpotent Lie group and $G = T \times_{\eta} N$ be a semi-direct product of N with a torus. Then G is exponential if and only if $T(X)$ is connected for each $X \in \mathfrak{n}$.*

Using Theorem 6.9.4 we can get at the exponentiality of H as follows.

Corollary 6.9.5. *Suppose a maximal torus of T of L is the set of diagonal matrices in L whose coefficients vary independently. Then $H = L \times_{\eta} N$ is exponential.*

Proof. Let $X = (x_1, \ldots, x_n) \in \mathfrak{n}$ and $t = (t_1, \ldots, t_n) \in T$, we have

$$T(X) = \{X \in \mathfrak{n} : t \cdot X = (t_1 x_1, \ldots, t_n x_n) = X, t \in T\}.$$

If $x_i = 0$, then there is no condition on t_i and $T(X)_i = \mathbb{T}^1$. Whereas if $x_i \neq 0$, then $t_i x_i = x_i$ if and only if $t_i = 1$ so $T(X)_i = \{1\}$. In any case, for each i, $T(X)_i$ is connected. As $T(X)$ is a direct product of its components and these are all connected so is $T(X)$. Since this is true for every $X \in \mathfrak{n}$, it follows from Theorem 3 of [56] that $T \cdot N$ is exponential. Now N is L-invariant, and the conjugates of T fill out L so

$$\bigcup_{l \in L} l(T \cdot N)l^{-1} = \bigcup_{l \in L} l(T)l^{-1} \cdot N = L \cdot N.$$

Thus the conjugates of $T \cdot N$ fill out H. Since exp is constant on conjugacy classes of H, the latter is also exponential. $\qquad\square$

Now consider $\mathfrak{n}$ the Heisenberg Lie algebra of dimension $2n + 1$, viewed as $\mathbb{C}^n \oplus i\mathbb{R}$ and N is the Heisenberg group. Here the bracketing relations are $[v, w] = \mathfrak{Im}\langle v, w \rangle$, where v and $w \in \mathbb{C}^n$, $\langle -, - \rangle$ is the standard Hermitian form on $\mathbb{C}^n$ and all other brackets are zero. Then the natural action of $U(n)$ on $\mathbb{C}^n$, leaving the center, $i\mathbb{R}$, pointwise fixed is evidently by Lie algebra automorphisms. To identify a maximal compact subgroup, L, of $\mathrm{Aut}(\mathfrak{n})_0$, we proceed as follows: By a calculation (see [47]) one sees that the identity component of the group of measure preserving automorphisms is $Sp(n, \mathbb{R})$. Since L must be contained in this group, L is a maximal compact subgroup of $Sp(n, \mathbb{R})$. Thus by Theorem 6.3.5 L is conjugate to $Sp(n, R) \cap SO(2n, R)$. Its Lie algebra is therefore conjugate to

$$\mathfrak{l} = \{ \begin{pmatrix} A & B \\ -B & A \end{pmatrix} : A^t = -A, B^t = B \},$$

and so $\dim \mathfrak{l} = n^2$. Since L is connected and $U(n)$ is a compact connected subgroup of $\mathrm{Aut}(\mathfrak{n})_0$ of the same dimension, by the conjugacy of maximal compact subgroups L and $U(n)$ are conjugate. Hence $U(n)$ is a maximal compact subgroup of $\mathrm{Aut}(\mathfrak{n})_0$. Now a maximal torus of T of $U(n)$ is the set of diagonal matrices in $U(n)$. Since the action of $U(n)$ on $\mathfrak{n}$ is linear on $\mathbb{C}^n$ and leaves $\mathfrak{z}(\mathfrak{n})$ fixed, it follows that a maximal torus is $\mathbb{T}^n$ of $U(n)$ consisting of the diagonal matrices in $U(n)$ together with a 1 in the $\mathfrak{z}(\mathfrak{n})$ component. By Corollary 6.9.5 we see that $U(n) \times_\eta N$ is exponential.

Our last case is $\mathfrak{n}$, the Lie algebra of dimension $4n + 3$, defined as follows. $\mathfrak{n} = H^n \oplus \mathfrak{Im}(H)$, where H is the quaternions and $\mathfrak{Im}(H)$ is the 3-dimensional subspace of pure quaternions. Here we take for bracketing relations $[v, w] = \mathfrak{Im}\langle v, w \rangle$, where v and $w \in H^n$, $\langle \cdot, \cdot \rangle$ is the standard H-Hermitian form on H^n, all other brackets are zero, and $\mathfrak{Im}$ is the projection from H onto $\mathfrak{Im}(H)$. This construction gives us a 2-step nilpotent Lie algebra with 3-dimensional center. Let N be the corresponding simply connected group. Calculations of the automorphism group of $\mathfrak{n}$ in the dissertation of P. Barbano, [4] show that the natural action of $\mathrm{Sp}(n)$ on H^n together with a 3-dimensional surjective representation, ρ, acting on the center, $\mathfrak{Im}(H)$ is a maximal compact subgroup of $\mathrm{Aut}(\mathfrak{n})_0$. We shall restrict our attention to $\mathrm{Sp}(n)$. A maximal torus of T of $\mathrm{Sp}(n)$ is the set of block diagonal matrices (D, D^-), where $D \in \mathrm{U}(n)$ and $^-$ is complex conjugation. As above, at each H^n component either we get a circle, if that component is zero, or a point, if it is nonzero (complex conjugation leaving this situation unchanged). It follows that $T(X)$ is connected for all $X \in \mathfrak{n}$. Hence, by Corollary 6.9.5, $\mathrm{Sp}(n) \times_\eta N$ is also exponential. Thus H is exponential in the remaining two cases and this completes the proof of Theorem 6.9.1.

Theorem 6.9.6. *Let L be a maximal compact subgroup of* $\mathrm{Aut}(N)_0$, *where N is Heisenberg group, or let L be $\mathrm{Sp}(n)$, where N is the simply connected group of Heisenberg type, based on the quaternions given above. Then $H = L \times_\eta N$ is exponential.*

We remark that this type of argument also works for the Euclidean motion group, $\mathrm{SO}(n) \times_\eta \mathbb{R}^n$. However, if one attempts to apply the reasoning above to the exceptional rank 1 group, one must look at the analogous 15-dimensional Heisenberg type Lie algebra, $\mathfrak{n}$, based on the Cayley numbers, with 7 dimensional center, $\mathfrak{z}$. In this case $K = \mathrm{Spin}(9)$, $K_p = \mathrm{Spin}(7)$, and the action on $\mathfrak{z}$ is by the full rotation group, $\mathrm{SO}(7) = \mathrm{Ad}(\mathrm{Spin}(7))$. Since a maximal torus, $\mathbb{T}^3$, of $\mathrm{Spin}(7)$ is a two-fold covering of a maximal torus of $\mathrm{SO}(7)$ one sees, for suitable $X \in \mathfrak{n}$, that $\mathbb{T}^3(X)$ can be finite and nontrivial. This means $K_p \times_\eta N$ is not exponential. Indeed, it could not be, for if it were then $\mathrm{Ad}\, F_{(4,-20)}$ would also be exponential.

Chapter 7

Semisimple Lie Algebras and Lie Groups

7.1   Root and Weight Space Decompositions

Here we will discuss the root space decomposition and the existence and fundamental properties of Cartan subalgebras, particularly when the complex Lie algebra is semisimple. Let $\mathfrak{h}$ be a (finite dimensional) complex Lie algebra and $\rho : \mathfrak{h} \to \mathfrak{gl}(V)$ a finite dimensional complex Lie algebra representation of $\mathfrak{h}$ on V. For $\lambda \in \mathfrak{h}^*$, its dual, we consider

$$V_\lambda = \{ v \in V : (\rho_H - \lambda(H)I)^k(v) = 0, \text{ for some } k \},$$

for all $H \in \mathfrak{h}$ and some integer k. Here k could, in principle, depend on H and v. However, by the Jordan canonical form, for fixed H and $v \neq 0 \in V$, $(\rho_H - \lambda(H)I)^k(v)$ only takes value 0 for an eigenvalue. Hence k can always be taken to be $\dim V$.

We call V_λ a *weight space*, its vectors *weight vectors*, and λ is called a *weight*. Of course there are at most $\dim V$ nonzero weight spaces.

For a fixed $H \in \mathfrak{h}$ it is also convenient to write

$$V_{\lambda,H} = \{ v \in V : (\rho_H - \lambda(H)I)^k(v) = 0 \},$$

so that

$$V_\lambda = \cap_{H \in \mathfrak{h}} V_{\lambda,H}.$$

313

The following result shows that complex representations of nilpotent algebras are rather special.

Theorem 7.1.1. *Suppose $\mathfrak{h}$ is nilpotent. Then*

(1) *V is the direct sum of the nonzero weight spaces.*

(2) *Each weight space is ρ-invariant.*

Proof. We first show that each V_λ is invariant under $\mathfrak{h}$. To do so it is sufficient to show each $V_{\lambda,H}$ is invariant. Since $\mathfrak{h}$ is nilpotent we know by Engel's theorem $\operatorname{ad} H$ is nilpotent for all H. Let $H \neq 0 \in \mathfrak{h}$ be fixed and define

$$\mathfrak{h}_k = \{Y \in \mathfrak{h} : \operatorname{ad} H^k(Y) = 0\}.$$

Then $\mathfrak{h}_k$'s form an increasing sequence of sets whose union is $\mathfrak{h}$. We first show by induction on k that for $Y \in \mathfrak{h}_k$ we have $\rho(Y)V_{\lambda,H} \subseteq V_{\lambda,H}$. If $k = 0$, then $\mathfrak{h}_0 = \{Y \in \mathfrak{h} : \operatorname{ad} H^0(Y) = Y = 0\}$ so $\rho(0) = 0$ which leaves everything invariant. Now suppose $Y \in \mathfrak{h}_k$. Then $[H, Y] \in \mathfrak{h}_{k-1}$ and since ρ is a representation,

$$(\rho(H) - \lambda(H)I)\rho(Y) = \rho([H,Y]) + \rho(Y)\rho(H) - \lambda(H)\rho(Y)$$
$$= \rho(Y)(\rho(H) - \lambda(H)I) + \rho([H,Y]).$$

Iterating this several times we get for each integer k,

$$(\rho(H) - \lambda(H)I)^k \rho(Y) = \rho(Y)(\rho(H) - \lambda(H)I)^k$$
$$+ \sum_{j=0}^{k-1} (\rho(H) - \lambda(H)I)^{k-1-j}\rho([H,Y])$$
$$\times (\rho(H) - \lambda(H)I)^j. \tag{7.1}$$

Now let $v \in V_{\lambda,H}$ and $(\rho(H) - \lambda(H)I)^n(v) = 0$. Take $k \geq 2n$. Then if $j \geq n$ the right side of (7.1) gives zero. so we can assume $j < n$. But then $k - 1 - j \geq 2n - 1 - j \geq n$. Note that $(\rho(H) - \lambda(H)I)^j(v) \in V_{\lambda,H}$, and since $[H, Y] \in \mathfrak{h}_{k-1}$ we know by induction that $\rho([H,Y])$ preserves $V_{\lambda,H}$. This means for large enough k

$$(\rho(H) - \lambda(H)I)^{k-1-j}\rho([H,Y])(\rho(H) - \lambda(H)I)^j(v) = 0$$

and by (7.1), $(\rho(H) - \lambda(H)I)^k \rho(Y)(v) = 0$. Thus $\rho(Y)$ stabilizes $V_{\lambda,H}$. This completes the induction and shows each V_λ is invariant under $\mathfrak{h}$.

We now turn to the direct sum decomposition. Let $H_1, \ldots, H_n$ be a basis of $\mathfrak{h}$. By the Jordan canonical form applied to $\rho(H_1)$ we get $V = \bigoplus V_{c_i,H_1}$ where the c_i are the eigenvalues of ρ_{H_1}. Here, since we have a single operator, c_i can be regarded as $\lambda_i(H_1)$ so that V_{λ_i,H_1} is the generalized eigenspace of a single operator defined earlier, which is stable under $\rho(\mathfrak{h})$. Hence each of these spaces can be further decomposed under $\rho(H_2)$ so that $V = \bigoplus_{i,j} V_{\lambda_i,H_1} \cap V_{\lambda_j,H_2}$ etc. until we finally get

$$V = \bigoplus_{\lambda(H_1),\ldots,\lambda(H_n)} \cap_{i=1}^n V_{\lambda(H_i),H_i}$$

with each of these spaces stable under $\rho(\mathfrak{h})$. But since $\mathfrak{h}$ is nilpotent and therefore solvable, Lie's theorem 3.2.18 tells us that the whole Lie algebra $\mathfrak{h}$ and in particular each of the $\rho(H_i)$, $i = 1, \ldots, n$ acts as simultaneous triangular operators on $\cap_{i=1}^n V_{\lambda(H_i),H_i}$ with diagonal entries the $\lambda(H_i)$. Hence the linear span, $\rho(\mathfrak{h})$ must also do this. Taking for our linear functional $\lambda(\sum z_i H_i) = \sum z_i \lambda(H_i)$ we get our result.   □

Now let $\mathfrak{g}$ be a finite dimensional Lie algebra, $\mathfrak{h}$ a nilpotent subalgebra and ρ be the adjoint representation of $\mathfrak{g}$ restricted to $\mathfrak{h}$. Then we write $\mathfrak{g}_\lambda$ instead of V_λ and call this a *root space*, its elements *root vectors* and λ a *root* if $\lambda \neq 0$. Similarly $\mathfrak{g}_{\lambda,X}$ replaces $V_{\lambda,X}$ for a fixed element $X \in \mathfrak{g}$.

Before turning to our next result we need the following lemma which, just as the binomial theorem, can easily be proved by induction on k. We leave it to the reader as an exercise.

Lemma 7.1.2. *Let $\mathfrak{g}$ be a complex Lie algebra, D be a derivation, Y and $Z \in \mathfrak{g}$ and a and $b \in \mathbb{C}$. Then for each positive integer, k,*

$$(D - (a+b)I)^k [Y, Z] = \sum_{r=0}^k \binom{k}{r} [(D - aI)^r (Y), (D - bI)^{k-r} (Z)].$$

In particular, for $X \in \mathfrak{g}$,

$$(\operatorname{ad} X - (a+b)I)^k [Y, Z] = \sum_{r=0}^{k} \binom{k}{r} [(\operatorname{ad} X - aI)^r (Y), (\operatorname{ad} X - bI)^{k-r}(Z)].$$

Corollary 7.1.3. *Suppose $\mathfrak{h}$ is nilpotent. Then $\mathfrak{g}$ is the direct sum of the nonzero root spaces each of which is* ad-*invariant. Moreover, $\mathfrak{h} \subseteq \mathfrak{g}_0$ and $[\mathfrak{g}_\lambda, \mathfrak{g}_\mu] \subseteq \mathfrak{g}_{\lambda+\mu}$, where the latter is understood to be zero if $\lambda + \mu$ is not a root. Finally, $\mathfrak{g}_0$ is a subalgebra of $\mathfrak{g}$.*

Proof. Now $\mathfrak{g}_0 = \{H \in \mathfrak{h} : \operatorname{ad} H^k = 0\}$, for some k. As we already saw this includes $\operatorname{ad} \mathfrak{h}$ since this is nilpotent. $[\mathfrak{g}_\lambda, \mathfrak{g}_\mu] \subseteq \mathfrak{g}_{\lambda+\mu}$ follows immediately from Lemma 7.1.2. Finally, the last statement follows from this relation by taking λ and μ both zero. $\qquad\square$

7.2   Cartan Subalgebras

Definition 7.2.1. A nilpotent subalgebra, $\mathfrak{h}$ is called a *Cartan subalgebra* of $\mathfrak{g}$ if $\mathfrak{h} = \mathfrak{g}_0$.

We now come to Chevalley's characterization of Cartan subalgebras.

Proposition 7.2.2. *A nilpotent subalgebra, $\mathfrak{h}$ is a Cartan subalgebra if and only if it equals its own normalizer. That is $\mathfrak{h} = \mathfrak{n}_\mathfrak{g}(\mathfrak{h})$.*

Proof. Now, in general, $\mathfrak{h} \subseteq \mathfrak{n}_\mathfrak{g}(\mathfrak{h})$ since it is a subalgebra. Moreover, if $[H, X] \in \mathfrak{h}$ for every $H \in \mathfrak{h}$, then since $\operatorname{ad} H^k(X) = \operatorname{ad} H^{k-1}[H, X]$ and $\operatorname{ad} H$ is nilpotent we see $\mathfrak{n}_\mathfrak{g}(\mathfrak{h}) \subseteq \mathfrak{g}_0$. If $\mathfrak{h}$ were a Cartan subalgebra then it would equal $\mathfrak{g}_0$ and hence all three subalgebras would coincide. Thus $\mathfrak{h}$ would equal its normalizer. Suppose $\mathfrak{h}$ were not a Cartan, *i.e.* were strictly smaller than $\mathfrak{g}_0$. Then $\operatorname{ad} \mathfrak{h}$ acts as a Lie algebra of linear operators on the nonzero vector space $\mathfrak{g}_0/\mathfrak{h}$. Since $\operatorname{ad} \mathfrak{h}$ is nilpotent and therefore solvable, Proposition 3.2.14 gives us a vector $X + \mathfrak{h} \in \mathfrak{g}_0/\mathfrak{h}$, where X is not in $\mathfrak{h}$ satisfying $\operatorname{ad} H(X) - \lambda(H)X \in \mathfrak{h}$ for all H. But since, as we know from Engel's theorem, $\operatorname{ad} \mathfrak{h}$ consists of nilpotent operators, $\lambda(H) = 0$. Hence X normalizes $\mathfrak{h}$. This means $\mathfrak{h}$ is strictly smaller than its normalizer. $\qquad\square$

Lemma 7.2.3. *A Cartan subalgebra $\mathfrak{h}$ of a Lie algebra $\mathfrak{g}$ is a maximal nilpotent subalgebra.*

Proof. Suppose $\mathfrak{n}$ is a nilpotent subalgebra of $\mathfrak{g}$ strictly containing $\mathfrak{h}$. Consider the representation $\operatorname{ad}\mathfrak{h} : \mathfrak{h} \to \mathfrak{gl}(\mathfrak{n}/\mathfrak{h})$ induced by the adjoint representation of $\mathfrak{n}$. Since $\mathfrak{h}$ is nilpotent, there is a nontrivial class $[U] \in \mathfrak{n}/\mathfrak{h}$ such that $\operatorname{ad}|_{\mathfrak{h}}(U) = [0]$. This means $\operatorname{ad}\mathfrak{h}(U) \in \mathfrak{h}$. Hence U normalizes of $\mathfrak{h}$. Because the latter is a Cartan subalgebra $U \in \mathfrak{h}$, a contradiction. $\square$

Notice that we do not yet know whether Cartan subalgebras exist. As before let ρ be a representation of $\mathfrak{g}$ on V. For $X \in \mathfrak{g}$ we consider the algebraic eigenspace, $V_{0,X}$, of $\rho(X)$ with eigenvalue 0. Define $\min(\rho, \mathfrak{g}, V)$ to be the smallest $\dim V_{0,X}$, as X varies over $\mathfrak{g}$ and $\operatorname{reg}(\rho, \mathfrak{g}, V)$ to be those $X \in \mathfrak{g}$ which achieve this minimal dimension. Let $n = \dim V$ and consider the characteristic polynomial

$$\det(\lambda I - \rho(X)) = \lambda^n + \sum_{j=0}^{n-1} d_j(X)\lambda^j$$

of X. Here the $d_j(X)$ are polynomial functions in the coefficients of $\rho(X)$. For instance $d_{n-1}(X) = -\operatorname{tr}(\rho(X))$ and $d_0(X) = \pm \det(\rho(X))$. In fact, each $d_j(X)$ is a homogeneous polynomial of degree $n-j$. Since ρ is linear these are polynomial functions on $\mathfrak{g}$. Let X be fixed and consider the smallest j where $d_j(X) \neq 0$. This j must be $\dim V_{0,X}$ because the multiplicity of 0 in the characteristic polynomial as an algebraic eigenvalue is $\dim V_{0,X}$. Hence $\min(\rho, \mathfrak{g}, V)$ is the smallest j for which d_j is not identically zero and $\operatorname{reg}(\rho, \mathfrak{g}, V)$ consists of those $X \in \mathfrak{g}$ where $d_{\min}(X) \neq 0$. We call such elements *regular elements* for $(\rho, \mathfrak{g}, V)$. The minimum value of j is called the *rank*, which we denote by r. Clearly, regular elements always exist. When ρ is the adjoint representation we just say regular elements.

We will need the following lemma which shows that $(\rho, \mathfrak{g}, V)$ is an open, dense and connected subset of $\mathfrak{g}$. Compare Lemma 7.4.12.

Lemma 7.2.4. *Let p be a non-identically zero polynomial function defined on $\mathbb{C}^n$. If p vanishes on a non-empty open set in U in $\mathbb{C}^n$, then*

*it vanishes everywhere. The set where $p \neq 0$ is Euclidean dense in $\mathbb{C}^n$
and is also connected.*

Proof. The first statement follows from the identity theorem since p is
entire. Let $V = \{x \in \mathbb{C}^n : p(x) \neq 0\}$. If V were not dense there would
be a disk $D \subseteq \mathbb{C}^n$ with positive radius on which p vanishes. Hence $p \equiv 0$,
a contradiction. To see that V is connected, let x and y be two distinct
elements of V. These can be joined by a (complex) line segment, L.
Since L is compact and has dim $= 2$ over $\mathbb{R}$ it can only hit the zero set
in at most a finite number of points. For otherwise we would have a
limit point and again by the identity theorem $p \equiv 0$. Since removal of a
finite number of points cannot disconnect L we see that L is connected.
Hence so is V.   □

Theorem 7.2.5. *Every complex Lie algebra has a Cartan subalgebra.
This is because $\mathfrak{g}_{0,X}$ is a Cartan subalgebra for any regular element,
$X \in \mathfrak{g}$. Its dimension is r, the rank.*

Proof. To prove this we must show $\mathfrak{g}_{0,X}$ is nilpotent and coincides with
its own normalizer. To see that $\mathfrak{g}_{0,X}$ is nilpotent it suffices by Engel's
theorem to show for each $Y \in \mathfrak{g}_{0,X}$ that $\operatorname{ad} Y$ restricted to $\mathfrak{g}_{0,X}$ is a
nilpotent operator. Call this restriction $\operatorname{ad} Y^1$ and denote by $\operatorname{ad} Y^2$ the
induced endomorphism on $\mathfrak{g}/\mathfrak{g}_{0,X}$. Let $d = \dim(\mathfrak{g}_{0,X}) = r$ which is the
rank as X is regular.

   Let $U = \{Y \in \mathfrak{g}_{0,X} : (\operatorname{ad} Y^1)^d \neq 0\}$. In other words, $U = \{Y \in \mathfrak{g}_{0,X} :
\operatorname{ad} Y^1$ is not nilpotent$\}$. Let $V = \{Y \in \mathfrak{g}_{0,X} : \operatorname{ad} Y^2$ is invertible$\}$. Both
U and V are open in $\mathfrak{g}_{0,X}$. Then V is non-empty as it contains X. This
is because $(\operatorname{ad} X^2)(U + \mathfrak{g}_{0,X}) = [X, U] + \mathfrak{g}_{0,X}$. If this operator were
singular there would be $U \in \mathfrak{g} \setminus \mathfrak{g}_{0,X}$ with $[X, U] \in \mathfrak{g}_{0,X}$. But then this
would force $U \in \mathfrak{g}_{0,X}$, a contradiction. Since V is the complement of the
zero set of a polynomial, namely det, we see by the lemma just above
that V is dense in $\mathfrak{g}_{0,X}$. Suppose U were non-empty. Since U is open
it would have to intersect V. Let $Y \in U \cap V$. Because $Y \in U$, $\operatorname{ad} Y^1$
has 0 as an eigenvalue with multiplicity strictly smaller than d. On the
other hand since $Y \in V$, 0 is not an eigenvalue of $\operatorname{ad} Y^2$. Therefore
the multiplicity of the eigenvalue 0 of $\operatorname{ad} Y$ on $\mathfrak{g}$ is strictly $< r$. This

contradicts the definition of r. We conclude that U must be empty and therefore $\operatorname{ad} Y^1$ is nilpotent for every $Y \in \mathfrak{g}_{0,X}$.

Now we show $\mathfrak{g}_{0,X}$ is its own normalizer. Suppose $\operatorname{ad} Y$ preserves $\mathfrak{g}_{0,X}$. Then $[Y, X] \in \mathfrak{g}_{0,X}$. Hence by definition there is some integer k so that $(\operatorname{ad} X)^k[X, Y] = 0$. Hence $(\operatorname{ad} X)^{k+1} Y = 0$ so indeed $Y \in \mathfrak{g}_{0,X}$.   □

In the next section we shall see that the Cartan subalgebras of a complex semisimple Lie algebra are all conjugate and therefore this method of constructing Cartan subalgebras gives them all.

Proposition 7.2.6. *If $\mathfrak{g}$ is a complex semisimple Lie algebra then any Cartan subalgebra, $\mathfrak{h}$, is abelian.*

Proof. We know $\mathfrak{h}$ is nilpotent. Hence so is $\operatorname{ad} \mathfrak{h}$. Therefore this algebra is solvable, so Lie's theorem 3.2.13 tells us that these operators on $\mathfrak{g}$ are in simultaneous triangular form. Let H_1 and $H_2 \in \mathfrak{h}$, $X \in \mathfrak{g}$ and B denote the Killing form. We want to calculate

$$B([H_1, H_2], X) = \operatorname{tr}(\operatorname{ad}[H_1, H_2] \operatorname{ad} X)$$
$$= \operatorname{tr}(\operatorname{ad} H_1 \operatorname{ad} H_2 \operatorname{ad} X) - \operatorname{tr}(\operatorname{ad} H_2 \operatorname{ad} H_1 \operatorname{ad} X).$$

When $X \in \mathfrak{h}$, and therefore all three matrices are triangular we see $B([H_1, H_2], X) = 0$. Now let $X \in \mathfrak{g} = \sum \mathfrak{g}_\lambda$ as in Corollary 7.1.3. In fact suppose $X \in \mathfrak{g}_\lambda$, for a root, λ and let $H \in \mathfrak{h}$. We know by Corollary 7.1.3, $\operatorname{ad} H \operatorname{ad} X(\mathfrak{g}_\mu) \subseteq \mathfrak{g}_{\lambda+\mu}$. Since $\mathfrak{g}$ is a direct sum of these root spaces and tr is linear, $\operatorname{tr}(\operatorname{ad} H \operatorname{ad} X) = 0$. Now take $H = [H_1, H_2]$ in the calculation above. Since $B([H_1, H_2], X) = 0$ for all X and B is nondegenerate we conclude $[H_1, H_2] = 0$.   □

Corollary 7.2.7. *Suppose $\mathfrak{g}$ is a complex semisimple Lie algebra. Let $X \in \mathfrak{g}$ and $\operatorname{ad} X = S + N$ be its Jordan decomposition. Then there exist X_s and $X_n \in \mathfrak{g}$ so that $\operatorname{ad} X_s = S$ and $\operatorname{ad} X_n = N$.*

Proof. We first note that on the root space, $\mathfrak{g}_\lambda$, $S = \lambda I$. This is because S and $\operatorname{ad} X$ share the same invariant subspaces and on such an invariant subspace they share the same eigenvalues (see Jordan decomposition, Theorem 3.3.2). But on $\mathfrak{g}_\lambda$, $\operatorname{ad} X$ has only λ as an eigenvalue. Since S is semisimple on $\mathfrak{g}$ and hence also on $\mathfrak{g}_\lambda$ we see $S = \lambda I$.

On the other hand because $[\mathfrak{g}_\lambda, \mathfrak{g}_\mu] \subseteq \mathfrak{g}_{\lambda+\mu}$, for $X \in \mathfrak{g}_\lambda$ and $Y \in \mathfrak{g}_\mu$ we get $S[X, Y] = (\lambda + \mu)[X, Y]$, while $S(X) = \lambda X$ and $S(Y) = \mu Y$. This means $S[X, Y] = [S(X), Y] + [X, S(Y)]$ so S is a derivation. By Corollary 3.3.23, it must be inner; $S = \operatorname{ad} X_s$. Hence $N = \operatorname{ad} X - \operatorname{ad} X_s$ so $N = \operatorname{ad} X_n$, where $X_n = X - X_s$. $\square$

In the case of a semisimple algebra we get another useful characterization of a Cartan subalgebra.

Proposition 7.2.8. *Let $\mathfrak{g}$ be a complex semisimple Lie algebra. Then a subalgebra, $\mathfrak{h}$, is a Cartan subalgebra if (and only if) it is a maximal abelian diagonalizable subalgebra.*

We will complete the proof of this by dealing with "only if" part in Proposition 7.3.3.

Proof. Suppose $\mathfrak{h}$ is abelian, $\operatorname{ad} \mathfrak{h}$ is simultaneously[1] diagonalizable and there is no larger such subalgebra of $\mathfrak{g}$. We will show that $\mathfrak{h}$ is a Cartan by proving it coincides with its normalizer.

Since $\mathfrak{h}$ is abelian and therefore nilpotent we can apply Corollary 7.1.3 to get $\mathfrak{g} = \mathfrak{g}_0 \oplus_{\lambda \neq 0} \mathfrak{g}_\lambda$, where these weight spaces are invariant under $\operatorname{ad} \mathfrak{h}$. Since the latter is simultaneously diagonalizable and $\mathfrak{h}$ is a subspace of $\mathfrak{g}_0$ (because $\mathfrak{h}$ is abelian) which is clearly $\operatorname{ad} \mathfrak{h}$-invariant, we see $\mathfrak{g}_0 = \mathfrak{h} \oplus \mathfrak{l}$ with $[\mathfrak{h}, \mathfrak{l}] = 0$. We know $\mathfrak{h} \subseteq \mathfrak{n}_\mathfrak{g}(\mathfrak{h}) \subseteq \mathfrak{g}_0$ so all we need to do is to show $\mathfrak{l} = 0$. Suppose $X \neq 0 \in \mathfrak{l}$. Then $\mathfrak{h} + \mathbb{C}X$ is a properly larger abelian subalgebra of $\mathfrak{g}$ so $\operatorname{ad} X$ cannot be diagonalizable. But we can still decompose $\mathfrak{g}$ into weight spaces according to this abelian subalgebra and get a (perhaps) more refined decomposition.

By Corollary 7.2.7 there exist X_s and $X_n \in \mathfrak{g}$ so that $\operatorname{ad} X_s = S$. Since S is a polynomial in $\operatorname{ad} X$ without constant term and X centralizes $\mathfrak{h}$ so does X_s. Hence by maximality $X_s \in \mathfrak{h}$. But then $X_n = X - X_s$ must be in $\mathfrak{h} + \mathbb{C}X$. Hence we may assume $\operatorname{ad} X$ is actually nilpotent on $\mathfrak{g}$. Because $\mathfrak{g}_0$ is a subalgebra and X is in it, $\mathfrak{g}_0$ is $\operatorname{ad} X$ invariant and acts as a nilpotent operator on it. As X was taken arbitrarily from $\mathfrak{l}$ we

---

[1]As we know, an abelian family of operators which is individually diagonalizable is simultaneously so.

find $\mathfrak{l}$ and therefore all of $\mathfrak{l} + \mathfrak{h} = \mathfrak{g}_0$ acts nilpotently on $\mathfrak{g}_0$. By Engel's theorem $\mathfrak{g}_0$ is nilpotent. Again using the weight space decomposition, this time under all of $\mathfrak{g}_0$, we find that the zero weight space cannot possibly be any bigger than what we got under $\mathfrak{h}$ itself. This means $\mathfrak{g}_0$ is a Cartan. So we have

$$\mathfrak{g} = \mathfrak{g}_0 \oplus_{\lambda \neq 0} \mathfrak{g}_\lambda, \tag{7.2}$$

where the λ's are weights of $\mathfrak{g}_0$. Let $X_0 \in \mathfrak{g}_0$ and $X_\lambda \in \mathfrak{g}_\lambda$, $\lambda \neq 0$. For our original X we calculate $B(X, X_0)$ and $B(X, X_\lambda)$ where B is the Killing form. By bilinearity of B we get

$$B(X, X_0) = \sum_\lambda (\dim \mathfrak{g}_\lambda) \lambda(X) \lambda(X_0).$$

Since $\operatorname{ad} X$ is nilpotent on all of $\mathfrak{g}_0$, $\lambda(X) = 0$ for every λ. Hence $B(X, X_0) = 0$. As for $B(X, X_\lambda)$, this is also zero since we are just shifting these weight spaces. By (7.2) it follows that $B(X, Y) = 0$ for all $Y \in \mathfrak{g}$. Since B is nondegenerate $X = 0$ and this contradiction completes the proof. $\square$

Let $\operatorname{Aut}(\mathfrak{g})$ denote the full group of Lie algebra automorphisms, while $\operatorname{Inn}(\mathfrak{g})$ be the inner automorphisms; that is the subgroup of $\operatorname{Aut}(\mathfrak{g})$ generated by $\operatorname{Exp}(\operatorname{ad} X)$, for $X \in \mathfrak{g}$.

Before turning to the proof of the conjugacy of Cartan subalgebras of a complex semisimple Lie algebra we need two lemmas.

Lemma 7.2.9. *For a Cartan subalgebra $\mathfrak{h}$ of $\mathfrak{g}$, $\operatorname{reg}(\operatorname{ad}|_\mathfrak{h}, \mathfrak{h}, \mathfrak{g})$ consists of those $X \in \mathfrak{h}$ for which $\lambda(X) = 0$ for all $\lambda \neq 0$. Alternatively, it consists of all $X \in \mathfrak{h}$ for which $\mathfrak{g}_{0,X} = \mathfrak{h}$.*

Proof. For a Cartan subalgebra $\mathfrak{h}$ of $\mathfrak{g}$ we know $\operatorname{reg}(\operatorname{ad}|_\mathfrak{h}, \mathfrak{h}, \mathfrak{g})$ consists of those $X \in \mathfrak{h}$ where $d_{\min}(X) \neq 0$, or alternatively, where $\dim \mathfrak{g}_{0,X}$ is minimal. Now $\mathfrak{g} = \mathfrak{h} \oplus_{\lambda \neq 0} \mathfrak{g}_\lambda$. Let $X \in \mathfrak{h}$ then $\mathfrak{g}_{0,X}$ consists of those $Y \in \mathfrak{g}$ such that $\operatorname{ad} X^{\dim(\mathfrak{g})}(Y) = 0$. Hence $\mathfrak{g}_{0,X} = \mathfrak{h} \oplus_{\lambda \neq 0, \lambda(X) = 0} \mathfrak{g}_\lambda$. Now we have a finite number of nonzero linear functionals λ on $\mathfrak{h}$ each of which has for its zero set a hyperplane in $\mathfrak{h}$. The union of finitely many (or by countable subadditivity of Lebesgue measure even countably many) of such hyperplanes cannot exhaust $\mathfrak{h}$. Hence $\mathfrak{g}_{0,X}$ is smallest when it is $\mathfrak{h}$. $\square$

Consider the natural action $\phi : \mathrm{Inn}(\mathfrak{g}) \times \mathrm{reg}(\mathrm{ad}\,|_{\mathfrak{h}}, \mathfrak{h}, \mathfrak{g}) \to \mathfrak{g}$ given by $(\alpha, X) \mapsto \alpha(X)$.

Lemma 7.2.10. *ϕ is an open map.   Its image is contained in* $\mathrm{reg}(\mathrm{ad}\,\mathfrak{g}, \mathfrak{h}, \mathfrak{g})$.

Proof. Since we have a transitive group action it is sufficient to prove each $Y \in \mathrm{reg}(\mathrm{ad}\,|_{\mathfrak{h}}, \mathfrak{h}, \mathfrak{g})$ has a neighborhood contained in the orbit of the action ϕ. To this end we linearize and calculate the derivative of ϕ at (I, Y). As $\mathrm{reg}(\mathrm{ad}\,|_{\mathfrak{h}}, \mathfrak{h}, \mathfrak{g})$ is open in $\mathfrak{h}$, the latter is the tangent space at any point. Similarly, because $\mathrm{reg}(\mathrm{ad}\,\mathfrak{g}, \mathfrak{h}, \mathfrak{g})$ is open in $\mathfrak{g}$ the tangent space at $\phi(I, Y)$ can be regarded as $\mathfrak{g}$ itself. The tangent space to I in $\mathrm{Inn}(\mathfrak{g})$ is of course its Lie algebra, $\mathrm{ad}\,\mathfrak{g}$ (see Theorem 1.4.29). Thus $d_{(I,Y)}\phi : \mathrm{ad}\,\mathfrak{g} \times \mathfrak{h} \to \mathfrak{g}$. Evidently, $d_{(I,Y)}\phi(\mathrm{ad}\,X, 0) = [X, Y]$ and $d_{(I,Y)}\phi(0, H) = H$ so because the derivative is linear we get,

$$d_{(I,Y)}\phi(\mathrm{ad}\,X, H) = d_{(I,Y)}\phi(\mathrm{ad}\,X, 0) + d_{(I,Y)}\phi(0, H) = H + [X, Y].$$

For $X \in \mathrm{reg}(\mathrm{ad}\,|_{\mathfrak{h}}, \mathfrak{h}, \mathfrak{g})$, $\mathrm{ad}\,X$ is non singular when restricted to the sum of the nonzero weight spaces. This is because it acts on each $\mathfrak{g}_\lambda$ with eigenvalue $\lambda(X)$, which is nonzero by definition. Hence $d_{(I,Y)}\phi$ maps onto $\mathfrak{g}$ and so has maximal rank. Hence the image of ϕ itself contains a neighborhood in $\mathfrak{g}$ by the implicit function theorem. This proves the first statement.

Since $\mathrm{reg}(\mathrm{ad}\,\mathfrak{g}, \mathfrak{g}, \mathfrak{g})$ is dense in $\mathfrak{g}$ it must meet this neighborhood in say Z. Hence there is some $\alpha \in \mathrm{Inn}(\mathfrak{g})$ (and recall $Y \in \mathfrak{h}$) with $\alpha(Y) = Z$. Since α is an automorphism of $\mathfrak{g}$ we have $\alpha(\mathfrak{g}_{0,Y}) = \mathfrak{g}_{0,Z}$. In particular these have the same dimension and since $\dim \mathfrak{g}_{0,Z} = \min \dim \mathfrak{g}_{0,X}$ for $X \in \mathfrak{g}$ and $\dim \mathfrak{g}_{0,Y} = \min \dim \mathfrak{g}_{0,H}$ for $H \in \mathfrak{h}$, we get $\mathrm{reg}(\mathrm{ad}\,|_{\mathfrak{h}}, \mathfrak{h}, \mathfrak{g}) \subseteq \mathrm{reg}(\mathrm{ad}\,\mathfrak{g}, \mathfrak{g}, \mathfrak{g})$. Because $\mathrm{reg}(\mathrm{ad}\,|_{\mathfrak{h}}, \mathfrak{h}, \mathfrak{g})$ is stable under every automorphism of $\mathfrak{g}$, $\phi(\mathrm{Inn}(\mathfrak{g}) \times \mathrm{reg}(\mathrm{ad}\,\mathfrak{g}, \mathfrak{h}, \mathfrak{g}))$ is contained in $\mathrm{reg}(\mathrm{ad}\,\mathfrak{g}, \mathfrak{h}, \mathfrak{g})$. $\square$

We can now prove the Cartan subalgebras of $\mathfrak{g}$ are conjugate. This gives another proof that the rank is an invariant.

Theorem 7.2.11. *Let $\mathfrak{g}$ be a complex semisimple Lie algebra. Then any two Cartan subalgebras are conjugate.*

Proof. We shall prove this theorem by contradiction by assuming there were two non-conjugate Cartan subalgebra $\mathfrak{h}_1$ and $\mathfrak{h}_2$.

We first show their images under ϕ must be disjoint. For suppose $\phi(X_1) = \phi(X_2)$, where each $X_i \in \mathrm{reg}(\mathrm{ad}\,|_{\mathfrak{h}_i}, \mathfrak{h}_i, \mathfrak{g})$. Then $\alpha(X_1) = \beta(X_2)$ so $\beta^{-1}\alpha(X_1) = X_2$. Hence, we see as above there is some $\gamma \in \mathrm{Inn}(\mathfrak{g})$ with $\gamma(\mathfrak{g}_{0,X_1}) = \mathfrak{g}_{0,X_2}$. By Lemma 7.2.9, $\mathfrak{g}_{0,X_i} = \mathfrak{h}_i$. So $\gamma(\mathfrak{h}_1) = \mathfrak{h}_2$, a contradiction.

Thus by Lemma 7.2.10, $\mathrm{reg}(\mathrm{ad}\,\mathfrak{g}, \mathfrak{g}, \mathfrak{g})$ is a nontrivial disjoint union of open sets. On the other hand by Lemma 7.2.4, we also know it is connected. This contradiction completes the proof.   □

7.3   Roots of Complex Semisimple Lie Algebras

Throughout this section $\mathfrak{g}$ will stand for a complex semisimple Lie algebra, $\mathfrak{h}$ a fixed Cartan subalgebra and $\mathfrak{g} = \mathfrak{h} \oplus_{\lambda \neq 0} \mathfrak{g}_\lambda$ the corresponding root space decomposition and Λ will denote the set of nonzero roots and B the Killing form of $\mathfrak{g}$.

Proposition 7.3.1.   *(1) If λ and $\mu \in \Lambda \cup \{0\}$ and $\lambda + \mu \neq 0$, then $\mathfrak{g}_\lambda$ and $\mathfrak{g}_\mu$ are orthogonal under B.*

(2) For $\lambda \in \Lambda \cup \{0\}$, B induces a nondegenerate form on $\mathfrak{g}_\lambda \times \mathfrak{g}_{-\lambda}$.

(3) If $\lambda \in \Lambda$, then $-\lambda \in \Lambda$.

(4) B restricted to $\mathfrak{h} \times \mathfrak{h}$ is nondegenerate.

(5) For each $\lambda \in \Lambda$ there exists a unique $H_\lambda \in \mathfrak{h}$ so that $\lambda(H) = B(H, H_\lambda)$.

(6) Λ spans $\mathfrak{h}^*$, the dual space.

Proof. 1. Choose X_λ and X_μ in each of the corresponding root spaces. Then $\mathrm{ad}\,X\,\mathrm{ad}\,Y(\mathfrak{g}_\nu) \subseteq \mathfrak{g}_{\lambda+\mu+\nu}$. Since $\lambda + \mu \neq 0$, $\mathfrak{g}_\nu \cap \mathfrak{g}_{\lambda+\mu+\nu} = \{0\}$.

Taking a basis of each root space including $\mathfrak{h}$ one sees that the matrix of $\operatorname{ad} X \operatorname{ad} Y$ has trace 0 because all diagonal entries are zero.

2. Let $X_\lambda \in \mathfrak{g}_\lambda$ and suppose $B(X_\lambda, \mathfrak{g}_{-\lambda}) = 0$. Then since $\mathfrak{g}_\lambda$ is orthogonal to everything else by 1, and therefore by the root space decomposition to everything in $\mathfrak{g}$, it follows that $X_\lambda = 0$ since B is nondegenerate.

3. Suppose to the contrary that $\mathfrak{g}_{-\lambda} = \{0\}$. Then $B(\mathfrak{g}_\lambda, X) = 0$ for all $X \in \mathfrak{g}_{-\lambda}$ which is impossible by 2. Hence $-\lambda$ is also a root.

4. Follows from 2 by taking $\lambda = 0$.

5. Follows from 4.

6. We first show that if $H \in \mathfrak{h}$ and $\lambda(H) = 0$ for all $\lambda \in \Lambda$, then $H = 0$. This is because the root space decomposition Corollary 7.1.3 then forces $[H, X] = 0$ for all $X \in \sum_{\lambda \in \Lambda} \mathfrak{g}_\lambda$ and since $\mathfrak{h}$ is abelian, by Corollary 7.2.6, it follows that $[H, X] = 0$ for all $X \in \mathfrak{g}$. But then $H \in \mathfrak{z}(\mathfrak{g}) = \{0\}$, again by semisimplicity. Therefore Λ separates the points of $\mathfrak{h}$ and so spans the dual.   □

For each root λ, consider the adjoint representation of $\mathfrak{h}$ on $\mathfrak{g}_\lambda$ and apply Corollary 7.1.3 to choose a fixed nonzero $E_\lambda \in \mathfrak{g}_\lambda$ satisfying $[H, E_\lambda] = \lambda(H)E_\lambda$ on $\mathfrak{h}$.

Proposition 7.3.2.   *(1) If λ is a root and $X \in \mathfrak{g}_{-\lambda}$, then $[E_\lambda, X] = B(E_\lambda, X)H_\lambda$.*

(2) If λ and $\mu \in \Lambda$, then $\mu(H_\lambda)$ is a rational multiple of $\lambda(H_\lambda)$.

(3) For $\lambda \in \Lambda$, $\lambda(H_\lambda) \neq 0$.

Proof. 1. By Corollary 7.1.3, $[\mathfrak{g}_\lambda, \mathfrak{g}_{-\lambda}] \subseteq \mathfrak{g}_0 = \mathfrak{h}$. Hence $[E_\lambda, X] \in \mathfrak{h}$. Let $H \in \mathfrak{h}$. Then, by invariance and skew symmetry,

$$B([E_\lambda, X], H) = -B(X, [E_\lambda, H]) = B(X, [H, E_\lambda]) = \lambda(H)B(X, E_\lambda).$$

This in turn is

$$\lambda(H)B(X, E_\lambda) = B(H_\lambda, H)B(E_\lambda, X) = B(B(E_\lambda, X)H_\lambda, H).$$

Since this holds for all H and B is nondegenerate on $\mathfrak{h}$, we conclude $B(E_\lambda, X)H_\lambda = [E_\lambda, X]$.

2. By the above choose $X_{-\lambda} \in \mathfrak{g}_{-\lambda}$ so that $B(E_\lambda, X_{-\lambda}) = 1$. By 1 we see

$$H_\lambda = [E_\lambda, X_{-\lambda}]. \tag{7.3}$$

Let $\mu \in \Lambda$ be fixed and consider $W = \oplus_{n \in \mathbb{Z}} \mathfrak{g}_{\mu + n\lambda}$. Then W is a subspace of $\mathfrak{g}$ which is invariant under $\operatorname{ad} H_\lambda$ because $\operatorname{ad} H_\lambda$ leaves $\mathfrak{g}_{\mu + n\lambda}$ invariant and acts on it by a single algebraic eigenvalue, $\mu + n\lambda$. Since the sum is direct we see the trace of $\operatorname{ad} H_\lambda$ on W is

$$\sum_{n \in \mathbb{Z}} (\mu(H_\lambda) + n\lambda(H_\lambda)) \dim \mathfrak{g}_{\mu + n\lambda}.$$

On the other hand W is invariant under $\operatorname{ad} E_\lambda$ and $\operatorname{ad} X_{-\lambda}$ and therefore also their bracket. Since ad is a representation this is $\operatorname{ad} H_\lambda$ so by (7.3) this trace is zero and hence the conclusion.

3. If $\lambda(H_\lambda) = 0$, then $\mu(H_\lambda)$ being a multiple of it would also be zero for every μ. Since Λ spans $\mathfrak{h}^*$ this forces $H_\lambda = 0$. But then by 5. of the previous proposition we would then have λ itself zero, a contradiction. $\qquad\square$

Proposition 7.3.3. *(1) For $\lambda \in \Lambda$, $\dim \mathfrak{g}_\lambda = 1$.*

(2) If $\lambda \in \Lambda$, the only multiples of it which are also in Λ are $\pm\lambda$.

(3) $\operatorname{ad} \mathfrak{h}$ acts simultaneously diagonally on $\mathfrak{g}$.

(4) On $\mathfrak{h} \times \mathfrak{h}$ the Killing form is given by $B(H, H') = \sum_{\lambda \in \Lambda} \lambda(H)\lambda(H')$.

(5) The pair $E_\lambda, E_{-\lambda}$ can be normalized so $B(E_\lambda, E_{-\lambda}) = 1$.

Proof. As above, choose $X_{-\lambda} \in \mathfrak{g}_{-\lambda}$ so that $B(E_\lambda, X_{-\lambda}) = 1$ so that $[E_\lambda, X_{-\lambda}] = H_\lambda$. Now consider the (complex) subspace U of $\mathfrak{g}$ spanned by E_λ, H_λ together with $\mathfrak{g}_{n\lambda}$, where $n < 0$. Because $[\mathfrak{g}_\lambda, \mathfrak{g}_\mu] \subseteq \mathfrak{g}_{\lambda+\mu}$ we see U is invariant under bracketing by E_λ and H_λ. By part (1) of Proposition 7.3.2 it is also invariant under bracketing by $X_{-\lambda}$. Since $[E_\lambda, X_{-\lambda}] = H_\lambda$ we know $\operatorname{ad} H_\lambda$ has trace zero on U. Since this operator acts with a single algebraic eigenvalue on each summand we conclude

$$\lambda(H_\lambda) + 0 + \sum_{n < 0} n\lambda(H_\lambda) \dim(\mathfrak{g}_{n\lambda}) = 0.$$

Since $\lambda(H_\lambda) \neq 0$ this tells us

$$\sum_{n>0} n \dim(\mathfrak{g}_{n\lambda}) = 1.$$

Hence $\dim(\mathfrak{g}_{-\lambda}) = 1$ and $\dim(\mathfrak{g}_{-n\lambda}) = 0$ for $n \geq 2$. Since we already know that Λ is symmetric under taking negatives this proves (1) and (2).

(3) Using the root space decomposition and the fact that $\mathfrak{h}$ is abelian and so acts trivially on itself we see (3) follows immediately from (1).

(4) Follows from (3).

(5) Follows from the fact that B is nondegenerate on $\mathfrak{g}_\lambda \times \mathfrak{g}_{-\lambda}$ and these spaces are 1-dimensional.   □

Remark 7.3.4. We see here that for each $\lambda \in \Lambda$ we get a 3-dimensional subalgebra of $\mathfrak{g}$. This is because we now know each $\mathfrak{g}_\lambda$ is 1-dimensional. Hence $\mathfrak{g}_\lambda$ is generated by E_λ and $\mathfrak{g}_{-\lambda}$ by $E_{-\lambda}$. Now $[H_\lambda, E_\lambda] = \lambda(H_\lambda)E_\lambda$ and $[H_\lambda, E_{-\lambda}] = -\lambda(H_\lambda)E_{-\lambda}$. Since $B(E_\lambda, E_{-\lambda}) = 1$ we also have $[E_\lambda, E_{-\lambda}] = H_\lambda$. Thus E_λ, H_λ and $E_{-\lambda}$ generate a 3-dimensional subalgebra. But what subalgebra is this?

Since $\lambda(H_\lambda) \neq 0$ we can further normalize the basis (and therefore the structure constants) to get $H'_\lambda = \frac{2}{\lambda(H_\lambda)} H_\lambda$, $E'_\lambda = \frac{2}{\lambda(E_\lambda)} E_\lambda$ and $E'_{-\lambda} = E_{-\lambda}$. Then

$$[H'_\lambda, E'_\lambda] = 2E'_\lambda,$$

$$[H'_\lambda, E'_{-\lambda}] = -2E'_{-\lambda},$$

and

$$[E'_\lambda, E'_{-\lambda}] = H'_\lambda.$$

So this is just $\mathfrak{sl}(2, \mathbb{C})$.

We now come to the definition of a *root string*.

Definition 7.3.5. Let $\lambda \in \Lambda$ and $\mu \in \Lambda \cup \{0\}$. We shall call all elements of $\Lambda \cup \{0\}$ of the form $\mu + n\lambda$, where $n \in \mathbb{Z}$, the λ *root string* through μ.

Before stating our next result it will be convenient to transfer $B|_{\mathfrak{h} \times \mathfrak{h}}$ to $\mathfrak{h}^* \times \mathfrak{h}^*$ as follows: For ϕ and $\psi \in \mathfrak{h}^*$ we take $\langle \phi, \psi \rangle = B(H_\phi, H_\psi)$, where the isomorphism between $\mathfrak{h}$ and $\mathfrak{h}^*$ is given Proposition 7.3.1, part 5. This gives us a nondegenerate symmetric form on $\mathfrak{h}^* \times \mathfrak{h}^*$. Notice that because of the way we identify $\mathfrak{h}$ and $\mathfrak{h}^*$, $B(H_\phi, H_\psi) = \phi(H_\psi)$, or $\psi(H_\phi)$.

We now turn to a proposition which will play an important role in finding compact real forms of complex semisimple algebras. As we shall see, our previous work on the representation theory of $\mathfrak{sl}(2, \mathbb{C})$ will play an important role here.

Proposition 7.3.6. *Let* $\lambda \in \Lambda$ *and* $\mu \in \Lambda \cup \{0\}$. *Then*

(1) *The* λ *string through* μ *has the form* $\mu + n\lambda$ *(*$-p \le n \le q$*) i.e. the string is uninterrupted. Here* p *and* $q \ge 0$.

(2) $p - q = \frac{2\langle \mu, \lambda \rangle}{\langle \lambda, \lambda \rangle}$. *In particular, the latter quantity is in* $\mathbb{Z}$.

(3) *If* $\mu + n\lambda$ *is never zero, then the adjoint action of* $\mathfrak{sl}(2, \mathbb{C})$ *on* $W = \oplus_{n \in \mathbb{Z}} \mathfrak{g}_{\mu + n\lambda}$ *is irreducible.*

Proof. If $\mu + n\lambda = 0$ for some n then by Proposition 7.3.3, $\mu = 0$, or $\mu = \pm\lambda$. In either case there are no gaps and $p - q = \frac{2\langle \mu, \lambda \rangle}{\langle \lambda, \lambda \rangle}$ while the third conclusion has no content since its hypothesis is false.

We may therefore assume $\mu + n\lambda$ is never 0 and will prove all conclusions simultaneously.

By Proposition 7.3.3, ad H'_λ is diagonal on W with distinct eigenvalues. The eigenvalues are $(\mu + n\lambda)(H'_\lambda) = \frac{2}{\langle \lambda, \lambda \rangle}(\mu + n\lambda)(H_\lambda)$. But this is $\frac{2\langle \mu, \lambda \rangle + n\langle \lambda, \lambda \rangle}{\langle \lambda, \lambda \rangle}$ so

$$(\mu + n\lambda)(H'_\lambda) = \frac{2\langle \mu, \lambda \rangle}{\langle \lambda, \lambda \rangle} + 2n. \tag{7.4}$$

This tells us that an ad H'_λ-invariant subspace of W is a sum of certain of the $\mathfrak{g}_{\mu + n\lambda}$. In particular this must also be true of any $\mathfrak{sl}(2, \mathbb{C})$ invariant subspace of W. To show $\mathfrak{sl}(2, \mathbb{C})$ acts irreducibly on W, suppose V is any $\mathfrak{sl}(2, \mathbb{C})$ invariant (and therefore by Weyl's theorem 3.4.3, we may assume to be an irreducible) subspace and let $-p$ and q be the smallest and largest integers n appearing in this representation. Our

328 Chapter 7 Semisimple Lie Algebras and Lie Groups

study of irreducible representations $\mathfrak{sl}(2,\mathbb{C})$ tells us that the eigenvalues of $\operatorname{ad} h = \operatorname{ad} H'_\lambda$ on V are $N - 2i$ that is they drop by 2, where here $N = \dim(V) - 1$. This tells us that all n between $-p$ and q come up. Furthermore by (7.4), we get $N = \frac{2\langle \mu, \lambda \rangle}{\langle \lambda, \lambda \rangle} + 2q$, and $-N = \frac{2\langle \mu, \lambda \rangle}{\langle \lambda, \lambda \rangle} - 2p$ so adding gives

$$p - q = \frac{2\langle \mu, \lambda \rangle}{\langle \lambda, \lambda \rangle}. \tag{7.5}$$

But the theorem on representations of $\mathfrak{sl}(2,\mathbb{C})$ tell us that W is the direct sum of its irreducible components. If W_0 is another such irreducible subspace and $-p_0$ and q_0 the corresponding smallest and largest integers appearing, then (7.5) tells us $p_0 - q_0 = \frac{2\langle \mu, \lambda \rangle}{\langle \lambda, \lambda \rangle}$. Hence $p_0 - q_0 = p - q$. But all the n's between $-p$ and q are accounted for by W_0 so either $q_0 < -p$ or $-p_0 > q$. Using symmetry we may assume the former. Hence $p_0 < -q$. But then $q_0 \geq -p_0 > q \geq -p$. So we find $p_0 - q_0 < p - q$, a contradiction. $\qquad\square$

Definition 7.3.7. Let V be a complex vector space and U be a real subspace of $V_\mathbb{R}$ (this means considering V as a real vector space). If $U_\mathbb{C} = V$, then we shall call U a *real form* of V.

Put more simply, U is a real form of V if, $U \oplus iU = V$.

Corollary 7.3.8. *(1) Let λ and $\mu \in \Lambda \cup \{0\}$ with $\lambda + \mu \neq 0$. Then*
$$[\mathfrak{g}_\lambda, \mathfrak{g}_\mu] = \mathfrak{g}_{\lambda+\mu}.$$

(2) Let $X_\lambda \in \mathfrak{g}_\lambda$, $X_{-\lambda} \in \mathfrak{g}_{-\lambda}$ and $X_\mu \in \mathfrak{g}_\mu$, where $\mu + n\lambda$ is never zero for any $n \in \mathbb{Z}$. Then

$$[X_\lambda, [X_{-\lambda}, X_\mu]] = \frac{p(1+q)}{2} \langle \lambda, \lambda \rangle B(X_\lambda, X_{-\lambda}) X_\mu.$$

(3) Let $\lambda \in \Lambda$ and $\mu \in \Lambda \cup \{0\}$. Then $\mu(H_\lambda)$ is rational.

(4) Let V be the $\mathbb{R}$-linear span of Λ in $\mathfrak{h}^*$. Then V is a real form of $\mathfrak{h}^*$ and the restriction of our symmetric bilinear form to $V \times V$ is positive definite.

(5) Let $\mathfrak{h}_0$ be the $\mathbb{R}$-linear span of $\{H_\lambda : \lambda \in \Lambda\}$. Then $\mathfrak{h}_0$ is a real form of $\mathfrak{h}$ and V consists of those linear functionals on $\mathfrak{h}$ which take real values on $\mathfrak{h}_0$.

Proof. 1. Let λ and $\mu \in \Lambda \cup \{0\}$ with $\lambda + \mu \neq 0$. We may evidently assume $\lambda \neq 0$. Also, we know by Corollary 7.1.3 that $[\mathfrak{g}_\lambda, \mathfrak{g}_\mu] \subseteq \mathfrak{g}_{\lambda+\mu}$. If μ is an integral multiple of λ, then we know $\mu = 0$, or $\pm\lambda$. If $\mu = 0$, then $[\mathfrak{h}, \mathfrak{g}_\lambda] = \mathfrak{g}_\lambda$ because ad $\mathfrak{h}$ acts diagonally with a nonzero eigenvalue. Since our hypothesis rules out $\mu = -\lambda$, we are left with $\mu = \lambda$. But then $\mathfrak{g}_{\lambda+\mu} = 0$ so the statement is trivially true.

Thus we can assume that μ is not an integral multiple of λ. Hence by the previous proposition a copy of $\mathfrak{sl}(2, \mathbb{C})$ acts irreducibly on the invariant subspace $W = \oplus_{n \in \mathbb{Z}} \mathfrak{g}_{\mu+n\lambda}$. Now we have classified the finite dimensional irreducible representations of $\mathfrak{sl}(2, \mathbb{C})$ in Section 3.1.5. When we match this up with X^+, H and X^- in Section 3.1.5, ignoring constant factors, the vectors $E_{\mu+n\lambda}$ correspond to the v_i where the only i for which e maps v_i to zero is $i = 0$ and v_0 corresponds to $E_{\mu+q\lambda}$. Hence if $[\mathfrak{g}_\lambda, \mathfrak{g}_\mu] = \{0\}$ we must have $q = 0$ which means $\mu + \lambda$ is not a root. Hence $\mathfrak{g}_{\lambda+\mu}$ is also zero.

2. Suppose $X_\lambda \in \mathfrak{g}_\lambda$, $X_{-\lambda} \in \mathfrak{g}_{-\lambda}$ and $X_\mu \in \mathfrak{g}_\mu$, where $\mu + n\lambda$ is never 0. We want to prove $[X_\lambda, [X_{-\lambda}, X_\mu]] = \frac{p(1+q)}{2} \langle \lambda, \lambda \rangle B(X_\lambda, X_{-\lambda}) X_\mu$, where p and q are the positive integers defined earlier. Now both sides are linear in X_λ and $X_{-\lambda}$. We can therefore normalize them as in Remark 7.3.4 so that $B(X_\lambda, X_{-\lambda}) = 1$. Therefore we can make the identification of the linear span of X_λ, H_λ and $X_{-\lambda}$ with X^+, H and X^- in $\mathfrak{sl}(2, \mathbb{C})$ as in Remark 7.3.4. Using the fact that

$$B(X_\lambda, X_{-\lambda}) = \frac{2}{\langle \lambda, \lambda \rangle}$$

the formula we wish to prove in these terms yields $[X_\lambda, [X_{-\lambda}, X_\mu]] = p(1 + q)X_\mu$. So the question is, is this true? By Proposition 7.3.6, $\mathfrak{sl}(2, \mathbb{C})$ acts irreducibly on $W = \sum_{n \in \mathbb{Z}} \mathfrak{g}_{\mu+n\lambda}$. In these identifications $X_{\mu+q\lambda}$ corresponds to a multiple of v_0 in Section 3.1.5. Because X_μ is a multiple of ad $X_{-\lambda}{}^q(X_{\mu+q\lambda})$, it follows from our work on the irreducible representations of $\mathfrak{sl}(2, \mathbb{C})$ that

$$[X_\lambda, [X_{-\lambda}, X_\mu]] = \text{ad}\, X_\lambda \, \text{ad}\, X_{-\lambda} \, \text{ad}\, X_{-\lambda}{}^q(X_{\mu+q\lambda})$$
$$= (q + 1)(N - q - 1 + 1)X_\mu,$$

where $N = \dim W - 1$; that is $N = q+p+1-1$. Hence $(q+1)(N-q) = p(q+1)$.

3. For $\phi, \psi \in \mathfrak{h}^*$ we have $\langle \phi, \psi \rangle = B(H_\phi, H_\psi)$. Hence by Proposition 7.3.3, part 4, we get $\sum_{\lambda \in \Lambda} \lambda(H_\phi)\lambda(H_\psi)$, that is

$$\langle \phi, \psi \rangle = \sum_{\lambda \in \Lambda} \langle \lambda, \phi \rangle \langle \lambda, \psi \rangle. \tag{7.6}$$

Let $p_{\lambda,\mu}$ and $q_{\lambda,\mu}$ be the integers associated with the λ root string containing μ. Setting $\phi = \lambda = \psi$ we get $\langle \lambda, \lambda \rangle = \sum_{\mu \in \Lambda} \langle \mu, \lambda \rangle^2$ which by Proposition 7.3.6, part 2, is $\sum_{\mu \in \Lambda}[(p_{\lambda,\mu} - q_{\lambda,\mu})\frac{\langle \lambda, \lambda \rangle}{2}]^2$. Since as we saw in Proposition 7.3.2, part 3, $\lambda(H_\lambda) = \langle \lambda, \lambda \rangle \neq 0$, dividing by it and solving for $\langle \lambda, \lambda \rangle$ we get

$$|\lambda|^2 = \langle \lambda, \lambda \rangle = \frac{4}{\sum_{\mu \in \Lambda}(p_{\lambda,\mu} - q_{\lambda,\mu})^2}. \tag{7.7}$$

This shows that $\langle \lambda, \lambda \rangle$ is positive rational. By Proposition 7.3.2, part 2, $\mu(H_\lambda)$ is always rational.

4. Suppose $\dim_{\mathbb{C}} \mathfrak{h} = r$. Since by Proposition 7.3.1, part 6, Λ spans $\mathfrak{h}^*$ we can choose roots $\lambda_1, \cdots, \lambda_r$ so that $H_{\lambda_1}, \cdots, H_{\lambda_r}$ is a basis for $\mathfrak{h}$. Let $\phi_1, \cdots \phi_r$ be the dual basis of $\mathfrak{h}^*$. Thus $\phi_i(H_{\lambda_j}) = \delta_{ij}$. Let V be the $\mathbb{R}$-subspace of all functionals in $\mathfrak{h}^*$ which take real values on all H_{λ_j}. Then V is the direct sum of the $\mathbb{R}$-lines through the ϕ_i. Hence V is a real form of $\mathfrak{h}^*$. Since $\lambda_1, \cdots, \lambda_r$ are linearly independent over $\mathbb{C}$ and hence also over $\mathbb{R}$ we see that V is the $\mathbb{R}$ span of Λ. Finally, for $\phi \in V$ since $\phi(H_\mu)$ is real for each $\mu \in \Lambda$ so by (7.6) we have $\langle \phi, \phi \rangle = \sum_{\mu \in \Lambda} \langle \mu, \phi \rangle^2 = \sum_{\mu \in \Lambda} \phi(H_\mu)^2$. As a sum of squares of real numbers we see $\langle \cdot, \cdot \rangle$ is positive definite on $V \times V$.

5. $\phi \mapsto H_\phi$ is an isomorphism of V with $\mathfrak{h}_0$. Since V has real dimension r and is the linear span of Λ, hence the real span of the H_λ must also have real dimension r. But they are independent over $\mathbb{C}$ and so also over $\mathbb{R}$. Thus they form an $\mathbb{R}$-basis of $\mathfrak{h}_0$. This means V is the set of functionals in $\mathfrak{h}^*$ that are real on $\mathfrak{h}_0$ and the restriction of the isomorphism above gives an isomorphism of V onto $\mathfrak{h}_0$. $\square$

We now define new structure constants and study their relations. If $\lambda + \mu \in \Lambda$, define $[X_\lambda, X_\mu] = C_{\lambda,\mu} X_{\lambda+\mu}$. If $\lambda + \mu$ is not in Λ, omit $C_{\lambda,\mu} = 0$.

Lemma 7.3.9. *(1)* $C_{\lambda,\mu} = -C_{\mu,\lambda}$.

(2) If λ, μ and $\nu \in \Lambda$ and $\lambda + \mu + \nu = 0$, then $C_{\lambda,\mu} = C_{\mu,\nu} = C_{\nu,\lambda}$.

(3) If λ, μ, ν and $\xi \in \Lambda$, $\lambda + \mu + \nu + \xi = 0$, and ξ is not one of $-\lambda$, $-\mu$ and $-\nu$, then

$$C_{\lambda,\mu} C_{\nu,\xi} + C_{\mu,\nu} C_{\lambda,\xi} + C_{\nu,\lambda} C_{\mu,\xi} = 0.$$

Proof. The first relation follows immediately from the skew symmetry of the bracket. As to the second, by the Jacobi identity

$$[[X_\lambda, X_\mu], X_\nu] + [[X_\mu, X_\nu], X_\lambda] + [[X_\nu, X_\lambda], X_\mu] = 0.$$

Hence $C_{\lambda,\mu} H_\nu + C_{\mu,\nu} H_\lambda + C_{\nu,\lambda} H_\mu = 0$. Note that $H_\nu = -H_\lambda - H_\mu$. The linear independence of H_λ and H_μ yields the conclusion.

The third relation follows from the Jacobi identity once we prove that

$$C_{\lambda,\mu} C_{\nu,\xi} = B([[X_\lambda, X_\mu] X_\nu], X_\xi).$$

Since B is the Killing form,

$$B([[X_\lambda, X_\mu] X_\nu], X_\xi) = B([X_\lambda, X_\mu], [X_\nu, X_\xi]).$$

First suppose that $\lambda + \mu \in \Lambda$. Then since $\lambda + \mu = -(\nu + \xi) \neq 0$ we have

$$
\begin{aligned}
B([X_\lambda, X_\mu], [X_\nu, X_\xi]) &= C_{\lambda,\mu} C_{\nu,\xi} B(X_{\lambda+\mu}, X_{\nu+\xi}) \\
&= C_{\lambda,\mu} C_{\nu,\xi} B(X_{\lambda+\mu}, X_{-(\lambda+\mu)}) = C_{\lambda,\mu} C_{\nu,\xi}.
\end{aligned}
\tag{7.8}
$$

If $\lambda + \mu \notin \Lambda$ then the identity holds as $C_{\lambda,\mu} = 0$ and $[X_\lambda, X_\mu] = 0$.

$\square$

Lemma 7.3.10. *Let λ, μ and $\lambda + \mu \in \Lambda$ and let $\mu + n\lambda$, where $-p \leq n \leq q$ be the λ string containing μ. Then $C_{\lambda,\mu} C_{-\lambda,-\mu} = -\frac{p(1+q)}{2} |\lambda|^2$.*

Proof. By Corollary 7.3.8, taking into account the fact that $\langle \cdot, \cdot \rangle$ is positive definite on a real form of $\mathfrak{h}$, we know

$$[X_{-\lambda}, [X_\lambda, X_\mu]] = \frac{p(1+q)}{2} |\lambda|^2 B(X_\lambda, X_{-\lambda}) X_\mu.$$

Taking our normalizations into account as well gives,

$$[X_{-\lambda}, [X_\lambda, X_\mu]] = \frac{p(1+q)}{2} |\lambda|^2 X_\mu.$$

Since the left side of this equation is $C_{-\lambda, \lambda+\mu} C_{\lambda, \mu} X_\mu$ we see

$$C_{-\lambda, \lambda+\mu} C_{\lambda, \mu} = \frac{p(1+q)}{2} |\lambda|^2. \tag{7.9}$$

But because $-\lambda + (\lambda + \mu) + -\mu = 0$ the two lemmas just above show $C_{-\lambda, \lambda+\mu} = C_{-\mu, -\lambda} = -C_{-\lambda, -\mu}$. Substituting into (7.9) completes the proof. □

We conclude this section with the Chevalley normalization theorem whose proof requires the Serre's isomorphism theorem. This shows that a complex semi-simple Lie algebra is completely determined by a Cartan subalgebra and the root pattern.

Theorem 7.3.11. *Let $\mathfrak{g}$ and $\mathfrak{g}'$ be two complex semisimple Lie algebras with Cartan subalgebras $\mathfrak{h}$ and $\mathfrak{h}'$. Suppose that $\mu \mapsto \mu'$ is a bijection between the roots of $\mathfrak{g}$ and $\mathfrak{g}'$ such that $\mu' + \lambda'$ is a root for $\mathfrak{g}'$ if and only if $\mu + \lambda$ is a root for $\mathfrak{g}$ and that $\mu' + \lambda' = (\mu + \lambda)'$. Then there is a Lie algebra isomorphism $f : \mathfrak{g} \to \mathfrak{g}'$ such that $f(\mathfrak{h}) = \mathfrak{h}'$ and for every root μ of $\mathfrak{g}$ we have*

$$\mu' \circ f|_{\mathfrak{h}} = \mu.$$

Proof. It follows from the assumption that for any two roots μ and λ of $\mathfrak{g}$, the length of the root string of μ through λ and μ' through λ' are identical. Therefore by (7.7),

$$\lambda(H_\lambda) = \lambda'(H_{\lambda'}) \tag{7.10}$$

and by Proposition 7.3.6, part 2,

$$\lambda(H_\mu) = \lambda'(H_{\mu'})$$

for all μ and λ. Let $\mu_1, \ldots, \mu_n$ be a maximal set of linearly independent roots of $\mathfrak{g}$ with the corresponding $H_{\mu_i} \in \mathfrak{h}$, $1 \le i \le n$. Since $\mu_1, \ldots, \mu_n$ is a basis for $\mathfrak{h}^*$, then $H_{\mu_1}, \ldots, H_{\mu_n}$ is a basis for $\mathfrak{h}$ and the determinant $\det((\mu_i(H_{\mu_j}))_{i,j})$ is nonzero.

Now consider $\mu'_1, \ldots, \mu'_n$ with their corresponding $H_{\mu'_i} \in \mathfrak{h}'$, $1 \le i \le n$. Since $\mu_i(h_{\mu_j}) = \mu'_i(h'_{\mu_j})$, μ'_i are also linearly independent. Interchanging the role of $\mathfrak{g}$ and $\mathfrak{g}'$, we conclude that $\mu'_1, \ldots, \mu'_n$ is a maximal linearly independent set.

Define the isomorphism $f_H : \mathfrak{h} \to \mathfrak{h}'$ by setting $f_H(H_{\mu_i}) = H_{\mu'_i}$, then by (7.10) we get

$$\mu' \circ f_H = \mu \tag{7.11}$$

for all i. So to prove the theorem we must show that f_H extends to a Lie algebra isomorphism from $f : \mathfrak{g} \to \mathfrak{g}'$. For each root μ_i of $\mathfrak{g}$ consider vectors X_μ and $X_{-\mu} \in \mathfrak{g}$ such that $B(X_\mu, X_{-\mu}) = 1$ and $[X_\mu, X_{-\mu}] = H_\mu$ (see Remark 7.3.4). We make a similar choice of $X_{\mu'}$ and $X_{-\mu'}$ for $\mathfrak{g}'$. Then f is defined once we have determined $f(X_\mu)$ and $f(X_{-\mu})$. These are defined by $f(X_\mu) = c_\mu X_{\mu'}$ and $f(X_{-\mu}) = c_{-\mu} X_{-\mu'}$ for a suitable choice of c_μ and $c_{-\mu}$. The identity $f([X, Y]) = [f(X), f(Y)]$ requires that

(1) $c_\mu c_{-\mu} = 1$, obtained for $X = X_\mu$ and $Y = X_{-\mu}$,
(2) $C_{\mu,\lambda} c_{\mu+\lambda} = C_{\mu',\lambda'} c_\mu c_\lambda$ if $\mu + \lambda$ is also a root, obtained for $X = X_\mu$ and $Y = X_\lambda$.

Here $C_{\mu,\lambda}$'s and $C_{\mu',\lambda'}$'s are the structure constants relative to the basis X_μ and $X_{\mu'}$ for $\mathfrak{g}$ and $\mathfrak{g}'$ respectively.

In fact these conditions are sufficient because $f([X_\mu, X_\lambda]) = [f(X_\mu), f(X_\lambda)]$ is basically conditions 1 and 2,

$$f([H_\mu, X_\lambda]) = f(\lambda(H_\mu)X_\lambda) = \lambda(H_\mu)f(X_\lambda) = \lambda'(H_{\mu'})c_\lambda X_{\lambda'}$$
$$= [H_{\mu'}, c_\lambda X_{\lambda'}] = [f(H_\mu), f(X_\lambda)]$$

follows from (7.11), and $f([X, Y]) = [f(X), f(Y)] = 0$ if $X, Y \in \mathfrak{h}$.

Now we construct c_μ's, this is done inductively. Note that the rational linear combinations of all μ_i give all roots of $\mu \in \mathfrak{g}$, because if $\mu = \sum_{i=1}^n a_i \mu_i$ then

$$\mu(H_{\mu_j}) = \sum_{i=1}^n a_i \mu_i(H_{\mu_j}),$$

since all $\mu_i(H_{\mu_j})$'s and $\mu(H_{\mu_j})$'s are rational (see Proposition 7.3.2) therefore all a_i are rational. We order lexicographically all rational linear combinations of μ_i's. That is $\sum_i a_i \mu_i > \sum_i b_i \mu_i$ if and only if the first nonzero $a_i - b_i$ is positive. Once we have defined c_λ for $\lambda > 0$, we set $c_{-\lambda} = c_\lambda^{-1}$ to assure condition 1. Let λ be a root of $\mathfrak{g}$ and suppose that c_μ has been defined for all $-\lambda < \mu < \lambda$. We show that we can define c_λ while maintaining condition 2. If we cannot write $\lambda = \mu + \nu$ where $-\lambda < \mu, \nu < \lambda$, we can simply define $c_\lambda = c_{-\lambda} = 1$. If $\lambda = \mu + \nu$ where $-\lambda < \mu, \nu < \lambda$, guided by condition 2 we define $c_\lambda = C_{\mu,\nu}^{-1} C_{\mu',\nu'} c_\mu c_\nu$, this is well-defined since Lemma 7.3.10 guarantees that $C_{\mu,\nu}$ is nonzero. Then c_λ is nonzero as $C_{\mu',\nu'}$ is nonzero again by Lemma 7.3.10. We have to check that condition 2 also holds for the pair $(-\mu, -\nu)$. By Lemma 7.3.10 we have

$$C_{\mu,\nu} C_{-\mu,-\nu} = C_{\mu',\nu'} C_{-\mu',-\nu'},$$

therefore

$$
\begin{aligned}
C_{-\mu,-\nu} c_{-\mu-\nu} &= C_{-\mu,-\nu} c_{\mu+\nu}^{-1} = C_{-\mu,-\nu} C_{\mu,\nu} C_{\mu',\nu'}^{-1} c_\mu^{-1} c_\nu^{-1} \\
&= C_{\mu',\nu'} C_{-\mu',-\nu'} C_{\mu',\nu'}^{-1} c_\mu^{-1} c_\nu^{-1} = C_{-\mu',-\nu'} c_\mu^{-1} c_\nu^{-1} \\
&= C_{-\mu',-\nu'} c_{-\mu} c_{-\nu}.
\end{aligned}
$$

To complete the proof we prove that condition 2 holds for any other pair (μ_1, ν_1) such that $\lambda = \mu_1 + \nu_1$ and $-\lambda < \mu_1, \nu_1 < \lambda$. Since ν_1 is different from μ, ν and $-\mu_1$ by Lemma 7.3.9 applied to the quadruple $-\mu, -\nu, \mu_1$ and ν_1 we

$$C_{-\mu,-\nu} C_{\mu_1,\nu_1} + C_{-\nu,\mu_1} C_{-\mu,\nu_1} + C_{\mu_1,-\mu} C_{-\nu,\nu_1} = 0. \qquad (7.12)$$

Because of the orderings we have that μ, ν, μ_1 and ν_1 are all positive therefore the difference of any two of them is between $-\lambda$ and λ. By condition 2 we have

$$C_{-\nu,\mu_1}c_{-\nu+\mu_1} = C_{-\nu',\mu_1'}c_{-\nu}c_{\mu_1}, \quad C_{-\mu,\nu_1}c_{-\mu+\nu_1} = C_{-\mu',\nu_1'}c_{-\mu}c_{\nu_1},$$

$$C_{\mu_1,-\mu}c_{\mu_1-\mu} = C_{\mu_1',-\mu'}c_{\mu_1}c_{-\mu}, \quad C_{-\nu,-\nu_1}c_{-\nu+\nu_1} = C_{-\nu',-\nu_1'}c_{-\nu}c_{\nu_1}.$$

These relations are true even if $-\nu + \mu_1$ etc. is not a root, with convention that $c_\rho = 1$ if ρ is not a root. These last four relations imply that

$$C_{-\nu,\mu_1}C_{-\mu,\nu_1} = C_{-\nu',\mu_1'}C_{-\mu',\nu_1'}c_{-\nu}c_{\mu_1}c_{-\mu}c_{\nu_1} \tag{7.13}$$

$$C_{\mu_1,-\mu}C_{-\nu,-\nu_1} = C_{\mu_1',-\mu'}C_{-\nu',\nu_1'}c_{\mu_1}c_{-\mu}c_{-\nu}c_{\nu_1} \tag{7.14}$$

because $c_{-\nu+\mu_1}c_{-\mu+\nu_1} = 1 = c_{\mu_1-\mu}c_{-\nu+\nu_1}$ as $-\nu + \mu_1 = -(-\mu + \nu_1)$ and $\mu_1 - \mu = -(-\nu + \nu_1)$.

By inserting the left-hand side of (7.13) and (7.14) in (7.12) and multiplying by $c_\nu c_{-\mu_1}c_\mu c_{-\nu_1} = (c_{-\nu}c_{\mu_1}c_{-\mu}c_{\nu_1})^{-1}$ we obtain,

$$C_{-\mu,-\nu}C_{\mu_1,\nu_1}c_\nu c_{-\mu_1}c_\mu c_{-\nu_1} + C_{-\nu',\mu_1'}C_{-\mu',\nu_1'} + C_{\mu_1',-\mu'}C_{-\nu',\nu_1'} = 0. \tag{7.15}$$

Since (7.12) hold for $\mu', \nu', \mu_1', \nu_1'$ in the place of μ, ν, μ_1, ν_1 and comparing with (7.15) we get

$$C_{-\mu',-\nu'}C_{\mu_1',\nu_1'} = C_{-\mu,-\nu}C_{\mu_1,\nu_1}c_\nu c_{-\mu_1}c_\mu c_{-\nu_1}. \tag{7.16}$$

Condition 2 holds for the pair $(-\mu, -\nu)$, that is

$$C_{-\mu',-\nu'} = C_{-\mu,-\nu}c_{-\nu}c_{-\mu}c_{-\nu}^{-1}c_{-\mu}^{-1} = C_{-\mu,-\nu}c_\nu c_\mu c_{\nu+\mu}^{-1}$$
$$= C_{-\mu,-\nu}c_\nu c_\mu c_{\nu_1+\mu_1}^{-1},$$

which together with (7.16) imply that

$$C_{\mu_1',\nu_1'} = C_{\mu_1,\nu_1}c_{\nu_1+\mu_1}c_{-\mu_1}c_{-\nu_1} = C_{\mu_1,\nu_1}c_{\nu_1+\mu_1}c_{\mu_1}^{-1}c_{\nu_1}^{-1}$$

or

$$C_{\mu_1',\nu_1'}c_{\mu_1}c_{\nu_1} = C_{\mu_1,\nu_1}c_{\nu_1+\mu_1}.$$

So condition 2 holds for the pair (μ_1, ν_1) as well. $\qquad\square$

Theorem 7.3.12. *Let $\mathfrak{g}$ be a complex semisimple Lie algebra, $\mathfrak{h}$ be a Cartan subalgebra and Λ be the corresponding roots. For each $\lambda \in \Lambda$ we can choose $X_\lambda \in \mathfrak{g}_\lambda$ so that*

(1) $[X_\lambda, X_{-\lambda}] = H_\lambda$,

(2) $[X_\lambda, X_\mu] = N_{\lambda,\mu} X_{\lambda+\mu}$, if $\lambda + \mu \in \Lambda$ and

(3) $[X_\lambda, X_\mu] = 0$, if $\lambda + \mu \neq 0$ and is not in Λ.

Moreover, these constants satisfy $N_{\lambda,\mu} = -N_{-\lambda,-\mu}$ and $N_{\lambda,\mu}^2 = \frac{p(1+q)}{2}|\lambda|^2$, where p and q are the integers associated with the λ string containing μ. In particular since $N_{\lambda,\mu}^2 \geq 0$, $N_{\lambda,\mu}$ is real.

Proof. Consider the linear map $-I : \mathfrak{h} \to \mathfrak{h}$ which is a linear isomorphism. Its transpose (also $-I$) carries Λ bijectively onto itself and hence by Theorem 7.3.11, extends to an automorphism α of $\mathfrak{g}$. Now by our choice of α, $\alpha(X_\lambda) \in \mathfrak{g}_{-\lambda}$. Hence there is some constant $c_{-\lambda}$ so that $\alpha(X_\lambda) = c_{-\lambda} X_{-\lambda}$. As we shall see below in Lemma 7.5.2, the Killing form is preserved by all automorphisms of $\mathfrak{g}$. Hence $B(\alpha(X), \alpha(Y)) = B(X, Y)$ for all $X, Y \in \mathfrak{g}$. Taking $X = X_\lambda$ and $Y = X_{-\lambda}$ we get

$$c_{-\lambda} c_\lambda = c_{-\lambda} c_\lambda B(X_\lambda, X_{-\lambda}) = B(\alpha(X_\lambda), \alpha(X_{-\lambda})) = B(X_\lambda, X_{-\lambda}) = 1.$$

Thus $c_{-\lambda} = \frac{1}{c_\lambda}$. Now for each $\lambda \in \Lambda$ choose $z_\lambda \in \mathbb{C}$ so that $z_{-\lambda} = \frac{1}{z_\lambda}$ and $z_\lambda^2 = -c_\lambda$. This can be done because $c_{-\lambda} c_\lambda = 1$. For instance, if $c_\lambda = re^{i\theta}$ and $c_{-\lambda} = r^{-1}e^{-i\theta}$ then let $z_\lambda = ir^{\frac{1}{2}} e^{\frac{1}{2}i\theta}$ and $z_{-\lambda} = ir^{-\frac{1}{2}} e^{-\frac{1}{2}i\theta}$. This consistent choice of the z_λ and $z_{-\lambda}$ gives us multiples $Z_\lambda = z_\lambda X_\lambda \in \mathfrak{g}_\lambda$ which satisfy

$$[Z_\lambda, Z_{-\lambda}] = [z_\lambda X_\lambda, z_\lambda X_{-\lambda}] = z_\lambda z_{-\lambda}[X_\lambda, X_{-\lambda}] = [X_\lambda, X_{-\lambda}] = H_\lambda.$$

Also,

$$\alpha(Z_\lambda) = \alpha(z_\lambda X_\lambda) = z_\lambda \alpha(X_\lambda) = z_\lambda c_{-\lambda} X_{-\lambda} = -z_\lambda^{-1} X_{-\lambda} = -z_{-\lambda} X_{-\lambda}.$$

But this last term is $-z_{-\lambda} X_{-\lambda} = -Z_{-\lambda}$. Hence

$$\alpha(Z_\lambda) = -Z_{-\lambda}. \tag{7.17}$$

Now define the constants $N_{\lambda,\mu}$ just the way we defined $C_{\lambda,\mu}$ earlier. Then

$$-N_{\lambda,\mu}Z_{-\lambda-\mu} = \alpha(N_{\lambda,\mu}Z_{\lambda+\mu}) = [\alpha(Z_\lambda), \alpha(Z_\mu)]$$
$$= [-Z_{-\lambda}, -Z_{-\mu}] = N_{-\lambda,-\mu}Z_{-\lambda-\mu}.$$

Therefore $-N_{\lambda,\mu} = N_{-\lambda,-\mu}$. The relation $N_{\lambda,\mu}^2 = \frac{p(1+q)}{2}|\lambda|^2$ follows immediately from Lemma 7.3.10. $\qquad\square$

7.4 Real Forms of Complex Semisimple Lie Algebras

The main purpose of this section is to prove E. Cartan's theorem on the existence of a compact real form for any complex semisimple Lie algebra.

We now extend our definition of a real form from vector spaces to Lie algebras.

Definition 7.4.1. Let $\mathfrak{g}$ be a complex Lie algebra and $\mathfrak{h}$ be a real subalgebra of $\mathfrak{g}_\mathbb{R}$ (this means considering $\mathfrak{g}$ as a real Lie algebra). If $\mathfrak{h}_\mathbb{C} = \mathfrak{g}$, then we shall call $\mathfrak{h}$ a *real form* of $\mathfrak{g}$.

Just as before $\mathfrak{h}$ is a real form of $\mathfrak{g}$ if, $\mathfrak{h} \oplus i\mathfrak{h} = \mathfrak{g}$.

Exercise 7.4.2. For example, $\mathfrak{sl}(n, \mathbb{R})$ is a real form of $\mathfrak{sl}(n, \mathbb{C})$. The reader should check that $\mathfrak{su}(n, \mathbb{C})$ is also a real form of $\mathfrak{sl}(n, \mathbb{C})$.

Definition 7.4.3. If in addition $\mathfrak{h}$ is of compact type (see Definition 3.9.2), we shall call it a *compact real form* of $\mathfrak{g}$.

We leave the following observation to the reader.

Lemma 7.4.4. *Let $\mathfrak{h}$ be a real form of $\mathfrak{g}$. For X and $Y \in \mathfrak{h}$ we define $\alpha(X + iY) = X - iY$. Then α is an automorphism of the real Lie algebra, $\mathfrak{g}$. Moreover, for any $X \in \mathfrak{g}$ and $c \in \mathbb{C}$, $\alpha(cX) = \bar{c}\alpha(X)$. Also, $\alpha^2 = I$.*

Definition 7.4.5. Whenever we have a complex Lie algebra, $\mathfrak{g}$, and an automorphism α of it as a real Lie algebra satisfying

  (1) $\alpha(cX) = \bar{c}\alpha(X)$, $X \in \mathfrak{g}$, $c \in \mathbb{C}$
  (2) $\alpha^2 = I$.

We call α a *conjugation*. The conjugation given in Lemma 7.4.4 is called the *conjugation relative to the real form* $\mathfrak{h}$.

Lemma 7.4.6. *Let $\mathfrak{g}$ be a complex Lie algebra and α be a conjugation. Then the α fixed points, $\mathfrak{g}_\alpha$, of $\mathfrak{g}$ is a real form of $\mathfrak{g}$ and the conjugation relative to this real form is α.*

Proof. Let X and $Y \in \mathfrak{g}_\alpha$. Then $\alpha(X) = X$ and $\alpha(Y) = Y$. Hence $\alpha[X,Y] = [\alpha(X),\alpha(Y)] = [X,Y]$ and similarly for the linear combinations. Thus $\mathfrak{g}_\alpha$ is a real subalgebra of $\mathfrak{g}$. Notice that if $\alpha(Z) = -Z$, then $Z \in i\mathfrak{g}_\alpha$. That is $-iZ \in \mathfrak{g}_\alpha$. To see that it is a real form, let $X \in \mathfrak{g}$ and write $X = \frac{1}{2}(X+\alpha(X))+\frac{1}{2}(X-\alpha(X))$. Now $\alpha(X+\alpha(X)) = \alpha(X)+X$ and $\alpha(X-\alpha(X)) = \alpha(X)-X = -(X-\alpha(X))$. Hence $\frac{1}{2}(X+\alpha(X)) \in \mathfrak{g}_\alpha$ and $\frac{1}{2}(X-\alpha(X)) \in i\mathfrak{g}_\alpha$. This shows $\mathfrak{g} = \mathfrak{g}_\alpha+i\mathfrak{g}_\alpha$. Clearly, $\mathfrak{g}_\alpha \cap i\mathfrak{g}_\alpha = \{0\}$ so $\mathfrak{g}_\alpha$ is a real form. The last statement is left to the reader. $\square$

Theorem 7.4.7. *Any complex semisimple Lie algebra has a compact real form.*

Before turning to the proof of this important fact we mention that for classical groups and their Lie algebras one can actually verify the result by inspection.

Example 7.4.8. In the case of classical Lie algebras we can verify Cartan's theorem by hand. We have the following examples.

$$\mathfrak{gl}(n,\mathbb{C}) = \mathfrak{u}(n,\mathbb{C})_\mathbb{C}, \tag{7.18}$$
$$\mathfrak{sl}(n,\mathbb{C}) = \mathfrak{su}(n,\mathbb{C})_\mathbb{C}, \tag{7.19}$$
$$\mathfrak{sp}(n,\mathbb{C}) = \mathfrak{sp}(n)_\mathbb{C}, \tag{7.20}$$
$$\mathfrak{so}(n,\mathbb{C}) = \mathfrak{so}(n,\mathbb{R})_\mathbb{C}. \tag{7.21}$$

To see that the real Lie algebras $\mathfrak{u}(n,\mathbb{C}), \mathfrak{su}(n,\mathbb{C}), \mathfrak{sp}(n)$ and $\mathfrak{so}(n,\mathbb{R})$ are of compact type it is sufficient to show that they are respectively the Lie algebras of appropriate compact Lie groups. Indeed,

$$U(n,\mathbb{C})^{\bullet} = \mathfrak{u}(n,\mathbb{C})$$
$$SU(n,\mathbb{C})^{\bullet} = \mathfrak{su}(n,\mathbb{C})$$
$$Sp(n)^{\bullet} = \mathfrak{sp}(n)$$
$$SO(n,\mathbb{R})^{\bullet} = \mathfrak{so}(n,\mathbb{R}),$$

where the $\bullet$ signifies the associated Lie algebra.

As a result, since all the Lie groups above are compact we conclude by Corollary 3.9.7 that the corresponding Lie algebras are all of compact type. On the other hand the dimension of the real Lie algebras are respectively $n^2, n^2 - 1, 2n^2 + n$ and $\frac{n(n-1)}{2}$ which equal the complex dimension of the corresponding complex Lie algebra on the left hand side of (7.18).

We now show this is true in general.

Proof of Theorem 7.4.7: Let $\mathfrak{h}$ be a Cartan subalgebra. Define root vectors as in Theorem 7.3.12 and $\mathfrak{g}_k$ as follows:

$$\mathfrak{g}_k = \sum_{\lambda \in \Lambda} \mathbb{R}(iH_\lambda) + \sum_{\lambda \in \Lambda} \mathbb{R}(X_\lambda - X_{-\lambda}) + \sum_{\lambda \in \Lambda} \mathbb{R}i(X_\lambda + X_{-\lambda}).$$

Evidently $\mathfrak{g}_k$ is a real subspace of $\mathfrak{g}_{\mathbb{R}}$. Moreover, $\mathfrak{g}_k + i\mathfrak{g}_k$ contains the $\mathbb{C}$ span of the H_λ, the $X_\lambda - X_{-\lambda}$ and the $X_\lambda + X_{-\lambda}$. Therefore it contains the $\mathbb{C}$ span of the H_λ, the X_λ and $X_{-\lambda}$ and by the root space decomposition this is $\mathfrak{g}$.

To show $\mathfrak{g}_k$ is a Lie algebra, we write it as $\mathfrak{g}_k = I + II + III$. We first consider the case when we are in the same root space.

Since $[iH_\lambda, (X_\lambda - X_{-\lambda})] = i|\lambda|^2 (X_\lambda + X_{-\lambda})$ we see that $[I_\lambda, II_\lambda] \subseteq III_\lambda$. On the other hand $[iH_\lambda, i(X_\lambda + X_{-\lambda})] = -|\lambda|^2 (X_\lambda - X_{-\lambda})$ so we get $[I_\lambda, III_\lambda] \subseteq II_\lambda$, and

$$[(X_\lambda - X_{-\lambda}), i(X_\lambda + X_{-\lambda})] = 2iH_\lambda$$

says that $[III_\lambda, II_\lambda] \subseteq I_\lambda$.

Now suppose $\lambda \neq \pm\mu$. Then

$$
\begin{aligned}
[(X_\lambda - X_{-\lambda}), (X_\mu - X_{-\mu})] &= N_{\lambda,\mu} X_{\lambda+\mu} + N_{-\lambda,-\mu} X_{-\lambda-\mu} \\
&\quad - N_{-\lambda,\mu} X_{-\lambda+\mu} - N_{\lambda,-\mu} X_{\lambda-\mu} \\
&= N_{\lambda,\mu}(X_{\lambda+\mu} - X_{-(\lambda+\mu)}) \\
&\quad - N_{-\lambda,\mu}(X_{-\lambda+\mu} - X_{-(-\lambda+\mu)}).
\end{aligned}
$$

Using the relations $N_{\lambda,\mu} = -N_{-\lambda,-\mu}$ we see $[II, II] \subseteq II$. Similarly we get, $[II, III] \subseteq III$ and of course $[I, I] = 0$ since $\mathfrak{h}$ is abelian. Thus $\mathfrak{g}_k$ is closed under bracketing and so is a Lie subalgebra of $\mathfrak{g}_\mathbb{R}$. Hence $\mathfrak{g}_k$ is a real form of $\mathfrak{g}$.

Finally we will show the Killing form of $\mathfrak{g}_k$, which is the restriction of B to $\mathfrak{g}_k \times \mathfrak{g}_k$ (see Lemma 3.1.62) is negative definite. Hence by Theorem 3.9.4, $\mathfrak{g}_k$ is of compact type. Now by Proposition 7.3.1, part 1, $B(I, II + III) = 0$. Also, B is positive definite on $\sum_{\lambda \in \Lambda} \mathbb{R}(H_\lambda)$ by Corollary 7.3.8, part 4. Hence B is negative definite on I. Now if $\lambda \neq \pm\mu$, Proposition 7.3.1 again shows

$$
B(X_\lambda - X_{-\lambda}, X_\mu - X_{-\mu}) = 0,
$$

$$
B(X_\lambda - X_{-\lambda}, i(X_\mu + X_{-\mu})) = 0,
$$

and

$$
B(i(X_\lambda + X_{-\lambda}), i(X_\mu + X_{-\mu})) = 0.
$$

Finally,

$$
B(X_\lambda - X_{-\lambda}, X_\lambda - X_{-\lambda}) = -2B(X_\lambda, X_{-\lambda}) = -2,
$$

and

$$
B(i(X_\lambda + X_{-\lambda}), i(X_\lambda + X_{-\lambda})) = -2B(X_\lambda, X_{-\lambda}) = -2,
$$

showing B is negative on $\mathfrak{g}_k$.

Exercise 7.4.9. Let $\mathfrak{g} = \mathfrak{sl}(2, \mathbb{C})$ with the usual basis $\{X^+, H, X^-\}$ (which gives roots). Show Theorem 7.4.7 gives $\mathfrak{g}_k = \mathfrak{su}(2, \mathbb{C})$.

The examples given just above also suggest the following

Corollary 7.4.10. *Let G be a connected complex semisimple Lie group, $\mathfrak{k}$ be a compact real form of $\mathfrak{g}$, and K be the connected Lie subgroup with Lie algebra $\mathfrak{k}$. Then K is a maximal compact subgroup of G.*

Proof. We know K is compact by Theorem 3.9.4 since its Killing form is negative definite. Let L be a maximal connected compact subgroup of G containing K. If $x \in Z(L)$ since x then commutes with all of K, $\operatorname{Ad} G(x)$ leaves $\mathfrak{k}$ pointwise fixed. As a complex linear automorphism it must then leave $\mathfrak{g}$ pointwise fixed. Therefore $\operatorname{Ad} G(x) = I$ and $x \in Z(G)$. Thus $Z(L) \subseteq Z(G)$. In particular $Z(L)$ is discrete. It follows from Theorem 3.9.4 that L is semisimple. Since $\mathfrak{g} = \mathfrak{k} \oplus i\mathfrak{k}$ we know $\mathfrak{l} = \mathfrak{k} \oplus i\mathfrak{s}$, where $\mathfrak{s}$ is a vector subspace of $\mathfrak{k}$. But

$$i[\mathfrak{s}, \mathfrak{k}] = [i\mathfrak{s}, \mathfrak{k}] \subseteq \mathfrak{l} \cap i\mathfrak{k} = i\mathfrak{s}.$$

Hence $[\mathfrak{s}, \mathfrak{k}] \subseteq \mathfrak{s}$ so $\mathfrak{s}$ is an ideal in $\mathfrak{k}$. This means $\mathfrak{s} + i\mathfrak{s}$ is an ideal in $\mathfrak{g}$. In particular $\mathfrak{s} + i\mathfrak{s}$ is itself semisimple (see Corollary 3.3.21). Let P be the corresponding complex semisimple subgroup of G. On the other hand $\mathfrak{s} + i\mathfrak{s}$ is also a real subalgebra of $\mathfrak{l}$. Therefore P is compact semisimple group. But a complex connected Lie group which is compact must be abelian by Proposition 1.1.10. Therefore P is abelian. But it is also semisimple therefore $P = \{1\}$. This means $\mathfrak{s} = \{0\}$ and hence $\mathfrak{l} = \mathfrak{k}$ so $L = K$.   □

We will now give an alternative proof of the theorem of Hermann Weyl concerning complete reducibility of representations of a complex semisimple Lie group using the so called unitarian trick. This was actually the first proof of this result. Of course, Theorem 3.4.3 is more general that Theorem 7.4.11, but the present one retains great appeal to the authors. Here is its statement. Because of Theorem 7.4.7 it applies to any complex semisimple (or reductive) Lie group.

Theorem 7.4.11. *Let G be a complex connected Lie group whose Lie algebra has a compact real form, $\mathfrak{k}$. Then every finite dimensional holomorphic representation is completely reducible.*

Of course just as in Theorem 3.4.3 once we know the complete reducibility for holomorphic representations of complex groups we also get complete reducibility for real groups.

The proof of Theorem 7.4.11 below requires the following simple lemma.

Lemma 7.4.12. *Let $\phi : \mathbb{C}^n \to \mathbb{C}$ be an entire function which vanishes identically on $\mathbb{R}^n$. Then $\phi \equiv 0$.*

Proof. $\phi(z_1, \ldots, z_n)$ vanishes when all z_i are real, so consider $\phi(z_1, x_2, \ldots, x_n)$ where the x_i are real. This is an entire function of z_1 and vanishes on the real axis. By the identity theorem, it vanishes identically. Let z_1 be fixed, but arbitrary and consider $\phi(z_1, z_2, x_3, \ldots, x_n)$ where the x_i are real. This is an entire function of z_2 which vanishes when z_2 is real and, therefore, identically in z_2. Continuing by induction, we see that $\phi(z_1, \ldots, z_n) \equiv 0$. □

It also requires the following results:

(1) Let ρ be a representation of a connected Lie group H on V and ρ' its differential representation on the Lie algebra $\mathfrak{h}$. Then a subspace W of V is H-invariant if and only if it is $\mathfrak{h}$-invariant.

(2) The Lie algebra $\mathfrak{g}$ of a complex semisimple Lie group G has a compact real form $\mathfrak{k}$ and the Lie subgroup K of G with Lie algebra $\mathfrak{k}$ is compact. (In fact its a maximal compact subgroup of G.)

Proof. Let ρ be a holomorphic representation of G on V. We will show that if W is a K-invariant subspace of V, then it is actually G-invariant. This would imply

(1) If ρ is irreducible, then so is ρ_K.

(2) ρ is completely reducible since ρ_K is.

Proof of the first statement. Suppose not, then since K is compact, $\rho_K = \sum \rho_i$, a direct sum of irreducibles. Each of the corresponding subspaces V_i is K and, therefore, G-invariant. Hence ρ is reducible, a contradiction.

Proof of the second statement. Let W be a G-invariant subspace of V. Then W is K-invariant. Since K is compact, there is a complementary K-invariant subspace W' which would be therefore G-invariant. Thus ρ would be completely reducible.

Therefore, it only remains to show that if W is a $\mathfrak{k}$-stable, $\mathbb{C}$-subspace of V, then it is $\mathfrak{g}$-stable. Let $\lambda \in (V/W)^*$, the $\mathbb{C}$-dual of V/W, and $w \in W$. Then for $k \in \mathfrak{k}$, we know $\rho_k(w) \in W$ and hence $\lambda(\rho_k(w)) = 0$. For $X \in \mathfrak{g}$ let $\phi(X) = \lambda(\rho_X(w))$, where $w \in W$ and $\lambda \in (V/W)^*$. Then $\phi : \mathfrak{g} \to \mathbb{C}$ is an entire function which vanishes on $\mathfrak{k}$ and hence, by Lemma 7.4.12, it vanishes on all of $\mathfrak{g}$. Since this is true for all $w \in W$ and all $\lambda \in (V/W)^*$, it follows that W is $\mathfrak{g}$-stable. $\square$

7.5 The Iwasawa Decomposition

Definition 7.5.1. Let $\mathfrak{g}$ be a real semisimple Lie algebra. An automorphism, α, of $\mathfrak{g}$ is called an *involution* if $\alpha^2 = I$. Now let B be the Killing form of $\mathfrak{g}$ and θ be an involution. We call θ a *Cartan involution* if the bilinear form $B_\theta(X, Y) := -B(X, \theta Y)$ on $\mathfrak{g}$ is symmetric and positive definite.

Actually, by the following Lemma B_θ is always symmetric.

Lemma 7.5.2. *(1) Let α be an automorphism of any Lie algebra $\mathfrak{g}$. Then for each $X \in \mathfrak{g}$ we have $\operatorname{ad} \alpha(X) = \alpha(\operatorname{ad} X)\alpha^{-1}$.*

(2) The Killing form B of $\mathfrak{g}$ is preserved by $\operatorname{Aut}(\mathfrak{g})$. That is, $B(\alpha(X), \alpha(Y)) = B(X, Y)$ for all $\alpha \in \operatorname{Aut}(\mathfrak{g})$.

(3) If α is an involution of $\mathfrak{g}$, then $B_\alpha(X, Y) = -B(\alpha(X), Y)$ is always symmetric.

Proof. (1) For X and $Y \in \mathfrak{g}$,

$$\operatorname{ad} \alpha(X)(Y) = [\alpha(X), Y] = \alpha[X, \alpha^{-1}(Y)] = \alpha(\operatorname{ad} X)\alpha^{-1}(Y).$$

(2) $B(\alpha(X), \alpha(Y)) = \operatorname{tr}(\operatorname{ad} \alpha(X) \operatorname{ad} \alpha(Y))$. But by 1) this is,

$$\operatorname{tr}(\alpha(\operatorname{ad} X)\alpha^{-1}\alpha(\operatorname{ad} Y)\alpha^{-1}) = \operatorname{tr}(\alpha(\operatorname{ad} X)(\operatorname{ad} Y)\alpha^{-1})$$
$$= \operatorname{tr}(\operatorname{ad} X \operatorname{ad} Y) = B(X, Y).$$

(3) By 2) $B(\alpha(X),Y) = B(\alpha^2(X),\alpha(Y)) = B(X,\alpha(Y))$. But because
   B is symmetric $B(X,\alpha(Y)) = B(\alpha(Y),X)$. Hence

$$B_\alpha(X,Y) = -B(X,\alpha(Y)) = -B(\alpha(Y),X) = B_\alpha(Y,X).$$

$\square$

Example 7.5.3. Suppose $\mathfrak{g}$ is a linear semisimple Lie algebra which is
stable under taking transpose. Let $\theta(X) = -X^t$. Then θ is a Cartan
involution. Evidently, θ is a linear operator and $\theta(\theta(X)) = -((-X^t))^t =
X$. Also,

$$\theta[X,Y] = -[X,Y]^t = -[Y^t,X^t] = [-X^t,-Y^t] = [\theta(X),\theta(Y)],$$

so θ is an involution. To see that it is a Cartan involution since symmetry is automatic we show $B_\theta(X,Y) = -B(X,\theta Y)$ is positive definite,
i.e. $-\operatorname{tr}(\operatorname{ad} X \operatorname{ad} \theta(X)) \geq 0$ and positive unless $X = 0$. But this is
$-\operatorname{tr}(\operatorname{ad} X \operatorname{ad} -X^t) = \operatorname{tr}(\operatorname{ad} X \operatorname{ad} X^t) \geq 0$. If it were zero then $\operatorname{ad} X$
would be zero, since this is the Hilbert-Schmidt norm. By semisimplic-
ity $X = 0$.

Remark 7.5.4. Notice that in these examples (see Chapter 6) the
Cartan decomposition $\mathfrak{g} = \mathfrak{k} \oplus \mathfrak{p}$ is given by $X = \frac{X-X^t}{2} + \frac{X+X^t}{2}$,
where the first factor is in $\mathfrak{k}$ and the second in $\mathfrak{p}$. Then we have
$\mathfrak{k} = \{X \in \mathfrak{g} : \theta(X) = X\}$ and $\mathfrak{p} = \{X \in \mathfrak{g} : \theta(X) = -X\}$. Thus $\theta|_{\mathfrak{k}} = I$
and $\theta|_{\mathfrak{p}} = -I$. Since $\mathfrak{g} = \mathfrak{k} \oplus \mathfrak{p}$ this means that θ is diagonalizable with
eigenvalues ± 1. $\mathfrak{k}$ is the 1-eigenspace and $\mathfrak{p}$ the -1-eigenspace.

   One further observation is that for X and $Y \in \mathfrak{k}$, $B_\theta(X,Y) =
-B(X,\theta(Y)) = -B(X,Y)$ so that $B_\theta(X,X) = -B(X,X)$. Similarly
for X and $Y \in \mathfrak{p}$ we have $B_\theta(X,X) = B(X,X)$. Since (see Chapter
6) $B_{\mathfrak{k} \times \mathfrak{k}}$ is negative definite and $B_{\mathfrak{p} \times \mathfrak{p}}$ is positive definite, we see that
B_θ is positive definite on $\mathfrak{k}$ and negative definite on $\mathfrak{p}$. All these are
characteristic properties of a Cartan involution.

   We do not yet know that Cartan involutions exist. To this end we
need the following.

Proposition 7.5.5. *Let $\mathfrak{g}_o$ be a real Lie algebra, $\mathfrak{g}$ its complexification, and $\mathfrak{g}_\mathbb{R}$ be $\mathfrak{g}$ regarded as a real Lie algebra. Then the Killing forms are related as follows: $B_0(X,Y) = B(X,Y)$, X and $Y \in \mathfrak{g}_o$ and $B_\mathbb{R}(X,Y) = 2\Re(B(X,Y))$, X and $Y \in \mathfrak{g}$. If any of these algebras is semisimple so are all the others.*

Proof. We first show if $\mathfrak{g}_o$ is semisimple then so are the other two. Suppose $\mathfrak{g}_o$ is semisimple. Then B_0 is nondegenerate. Let $B(Z_1, Z_2) = 0$ for all $Z_2 \in \mathfrak{g}$. Then for $j = 1, 2$, $Z_j = X_j + iY_j$ and $B(Z_1, Z_2) = B(X_1, X_2) + iB(X_1, Y_2) + iB(Y_1, X_2) - B(Y_1, Y_2) = 0$. Hence $B_0(X_1, X_2) = B_0(Y_1, Y_2)$ and $B_0(X_1, Y_2) = -B_0(Y_1, X_2)$. Take $Z_2 = X_2$ (and $Y_2 = 0$) and get X_1 and $Y_1 = 0$ i.e. $Z_1 = 0$. Hence $\mathfrak{g}$ is semisimple.

Since $\mathfrak{g}$ is semisimple, B is nondegenerate. Suppose $B_\mathbb{R}(Z_1, Z_2) = 0$ for all Z_2. Then $B_0(X_1, X_2) = B_0(Y_1, Y_2)$ for all X_2 and Y_2. Taking each of these in turn to be zero shows X_1 and Y_1 are both zero. Hence $Z_1 = 0$.

Now to the computation of the Killing forms. The first of these is obvious since $\mathfrak{g}_o$ is a real subalgebra of $\mathfrak{g}_\mathbb{R}$. If $A + iB$ is the matrix of $\mathrm{ad}\, X\, \mathrm{ad}\, Y$ with respect to a basis $X_1 + iY_1, \ldots, X_n + iY_n$ of $\mathfrak{g}$ (here A and B are real), then the matrix of this same operator on $\mathfrak{g}_\mathbb{R}$ is the $2n \times 2n$ matrix $\begin{pmatrix} A & B \\ -B & A \end{pmatrix}$, so the second relation follows. $\square$

Corollary 7.5.6. *Let $\mathfrak{g}$ be a complex semisimple Lie algebra and $\mathfrak{g}_\mathbb{R}$ be $\mathfrak{g}$ considered as a real (semisimple) Lie algebra. Let $\mathfrak{u}$ be a compact real form of $\mathfrak{g}$ and τ be the associated conjugation. Then τ is a Cartan involution of $\mathfrak{g}_\mathbb{R}$.*

Proof. τ is evidently an involution. To see that τ is a Cartan involution of $\mathfrak{g}_\mathbb{R}$ we must show $B_{\mathfrak{g}_{\mathbb{R}\tau}}$ is positive definite. But $B_{\mathfrak{g}_\mathbb{R}}(Z_1, Z_2) = 2\Re B_\mathfrak{g}(Z_1, Z_2)$. Writing $Z \in \mathfrak{g}$ as $X + iY$, where X and $Y \in \mathfrak{u}$ we get $B_{\mathfrak{g}_\mathbb{R}}(Z, \tau(Z)) = B_{\mathfrak{g}_\mathbb{R}}(X + iY, X - iY) = B_\mathfrak{g}(X, X) + B_\mathfrak{g}(Y, Y) = B_\mathfrak{u}(X, X) + B_\mathfrak{u}(Y, Y)$, which is ≥ 0 and > 0 unless $X = 0 = Y$ i.e. $Z = 0$. $\square$

Proposition 7.5.7. *Let $\mathfrak{g}$ be a real semisimple Lie algebra and θ be a Cartan involution. For any involution σ of $\mathfrak{g}$ there exists $\alpha \in \operatorname{Inn} \mathfrak{g}$ so that $\alpha\theta\alpha^{-1}$ commutes with σ.*

Proof. Let B_θ be the associated positive definite inner product on $\mathfrak{g}$. Then $\eta = \sigma\theta \in \operatorname{Aut}(\mathfrak{g})$. Now because $\theta^2 = I$ we see that $\eta\theta = \sigma\theta^2 = \sigma$. Taking inverses we get $\theta^{-1}\eta^{-1} = \sigma^{-1}$. But because σ and θ are each of order 2 we see that $\theta\eta^{-1} = \sigma$. Hence $\eta\theta = \theta\eta^{-1}$. Alternatively, $\eta^{-1}\theta = \theta\eta$.

Now Lemma 7.5.2 tells us that $\operatorname{Aut}(\mathfrak{g})$ leaves B invariant. Hence $B(\eta(X), \theta(Y)) = B(X, \eta^{-1}\theta(Y)) = B(X, \eta^{-1}\theta(Y)) = B(X, \theta\eta(Y))$. Thus $B_\theta(\eta(X), Y) = B_\theta(X, \eta(Y))$ so η is self adjoint. Taking $X = \eta(Y)$ we get $B_\theta(\eta^2(X), X) = B_\theta(\eta(X), \eta(X))$ which tells us that since η is an automorphism, η^2 is positive definite and hence is diagonalizable with positive real eigenvalues. Now we make use of the fact that Exp is a diffeomorphism between $\mathcal{P}$ and P in Chapter 6. In particular, $\eta^2 = \operatorname{Exp}(W)$ for some self-adjoint operator W with respect to B_θ. Here W is in the Lie algebra of $\operatorname{Aut}(\mathfrak{g})$. Hence $\operatorname{Exp}(tW) \in \operatorname{Aut}(\mathfrak{g})_0 = \operatorname{Inn}(\mathfrak{g})$ for all real t. The latter because $\mathfrak{g}$ is semisimple and so all derivations are inner. Because W is diagonal so is $\operatorname{Exp}(tW)$ for all real t. Hence each of the $\operatorname{Exp}(tW)$ commutes with η. The relation $\eta^{-1}\theta = \theta\eta$ then propagates to the whole 1-parameter group $\operatorname{Exp}(-tW)\theta = \theta\operatorname{Exp}(tW)$. Hence

$$\operatorname{Exp}(\tfrac{1}{4}W)\theta\operatorname{Exp}(-\tfrac{1}{4}W)\sigma = \operatorname{Exp}(\tfrac{1}{2}W)\theta\sigma = \operatorname{Exp}(\tfrac{1}{2}W)\eta^{-1}$$

$$= \eta\operatorname{Exp}(-\tfrac{1}{2}W) = \sigma\theta\operatorname{Exp}(-\tfrac{1}{2}W)$$

$$= \sigma\operatorname{Exp}(\tfrac{1}{4}W)\theta\operatorname{Exp}(-\tfrac{1}{4}W).$$

Taking $\alpha = \operatorname{Exp}(\tfrac{1}{4}W)$, we see σ commutes with $\alpha\theta\alpha^{-1}$ for some $\alpha \in \operatorname{Inn}(\mathfrak{g})$. $\square$

Corollary 7.5.8. *Any real semisimple Lie algebra, $\mathfrak{g}$, has a Cartan involution.*

Proof. Let $\mathfrak{g}_{\mathbb{C}}$ be the complexification of $\mathfrak{g}$. Then by Proposition 7.5.5, $\mathfrak{g}_{\mathbb{C}}$ is a complex semisimple Lie algebra and hence by Theorem 7.4.7 has a compact real form $\mathfrak{u}$. Let σ and τ be conjugations of $\mathfrak{g}_{\mathbb{C}}$ with respect to $\mathfrak{g}$ and $\mathfrak{u}$, respectively (see Lemma 7.4.4). Then they are each involutions of $\mathfrak{g}_{\mathbb{C}}$ regarded as a real Lie algebra. Here we write $\mathfrak{l}$ for $\mathfrak{u}_{\mathbb{C}} = \mathfrak{g}_{\mathbb{C}}$. By Proposition 7.5.5 $\mathfrak{l}$ is semisimple. Hence by Corollary 7.5.6, τ is a Cartan involution of $\mathfrak{l}$. By Proposition 7.5.7, we can find $\alpha \in \mathrm{Inn}(\mathfrak{l})$ so that $\alpha\tau\alpha^{-1}$ commutes with σ. Now $\alpha\tau\alpha^{-1}$ is the conjugation of $\mathfrak{l}$ with respect to $\alpha(\mathfrak{u})$, which is also a compact real form of $\mathfrak{g}$. Hence, $B_{\alpha\tau\alpha^{-1}}(Z_1, Z_2) = -2\Re B_{\mathfrak{g}_{\mathbb{C}}}(Z_1, \alpha\tau\alpha^{-1}Z_2)$ is positive definite on $\mathfrak{l}$.

Now $\mathfrak{g}$ is precisely the fixed set under σ. But if $\sigma(X) = X$, then $\sigma\alpha\tau\alpha^{-1}(X) = \alpha\tau\alpha^{-1}\sigma(X) = \alpha\tau\alpha^{-1}(X)$ so that $\alpha\tau\alpha^{-1}$ restricts to an involution, θ, of $\mathfrak{g}$ and $B_\theta(X, Y) = -B_{\mathfrak{g}}(X, \theta(Y)) = -B_{\mathfrak{g}}(X, \alpha\tau\alpha^{-1}(Y)) = \frac{1}{2}B_{\alpha\tau\alpha^{-1}}(X, Y)$ so that B_θ is positive definite and θ is a Cartan involution.   □

Before turning to the Iwasawa decomposition we need the following lemma. We denote by $(\cdot, \cdot)$ the inner product on $\mathfrak{g}_{\mathbb{C}}$ associated with a Cartan involution (by Corollary 7.5.8).

Lemma 7.5.9. *Let $\mathfrak{g}$ be a real semisimple Lie algebra and θ be a Cartan involution. Then for each $X \in \mathfrak{g}$ as an operator on $\mathfrak{g}_{\mathbb{C}}$, $\mathrm{ad}\,X^* = -\mathrm{ad}\,\theta(X)$. In particular $\mathrm{ad}\,\mathfrak{k}$ acts on $\mathfrak{g}_{\mathbb{C}}$ by skew Hermitian operators while $\mathrm{ad}\,\mathfrak{p}$ acts by Hermitian operators. Hence each of these is diagonalizable with purely imaginary eigenvalues, or real eigenvalues, respectively.*

Proof. Let Y and $Z \in \mathfrak{g}_{\mathbb{C}}$. Then $(\mathrm{ad}\,XY, Z) = -B(\mathrm{ad}\,X(Y), \theta(Z))$. By the invariance of the Killing form (see 3.1.60), this is $B(Y, \mathrm{ad}\,X(\theta(Z)))$. Because θ is an involution this last term is just $B(Y, \theta[\theta(X), Z]) = -B_\theta(X, \mathrm{ad}\,\theta(X)Z)$. Thus $\mathrm{ad}\,X^* = -\mathrm{ad}\,\theta(X)$.   □

We first formulate the Iwasawa decomposition for a (non-compact) real semisimple Lie algebra. Let $\mathfrak{g} = \mathfrak{k} \oplus \mathfrak{p}$ be the Cartan decomposition of $\mathfrak{g}$ (see Chapter 6). Let $\mathfrak{a}$ be a maximal abelian subspace of $\mathfrak{p}$. Then by Lemma 7.5.9 just above, the elements of $\mathrm{ad}\,\mathfrak{a}$ are simultaneously diagonalizable with real eigenvalues. Let θ be the corresponding Cartan involution of $\mathfrak{g}$.

This leads to the following

Definition 7.5.10. For $\lambda \in \mathfrak{a}^*$, the real dual space of $\mathfrak{a}$, $\lambda \neq 0$ we form the *restricted root space*,

$$\mathfrak{g}_\lambda = \{X \in \mathfrak{g} : \operatorname{ad} HX = \lambda(H)X \text{ for all } H \in \mathfrak{a}\}.$$

We write Λ for the set of restricted roots with $\mathfrak{g}_\lambda \neq 0$. We also write $\mathfrak{m} = \mathfrak{z}_{\mathfrak{k}}(\mathfrak{a})$, the centralizer of $\mathfrak{a}$ in $\mathfrak{k}$.

Proposition 7.5.11. $\mathfrak{g}$ *is a direct sum of subspaces*, $\mathfrak{a} \oplus \mathfrak{m} \oplus \sum_{\lambda \in \Lambda} \mathfrak{g}_\lambda$.

Proof. Let $\mathfrak{g}_0 = \{X \in \mathfrak{g} : \operatorname{ad} HX = 0 \text{ for all } H \in \mathfrak{a}\}$. Then $\mathfrak{g} = \mathfrak{g}_0 \oplus \sum_{\lambda \in \Lambda} \mathfrak{g}_\lambda$. Now θ is an automorphism of $\mathfrak{g}$ which sends each element of $\mathfrak{p}$ and hence of $\mathfrak{a}$ to its negative. So if $[H, X] = 0$ for all $H \in \mathfrak{a}$, then $\theta[H, X] = [-H, \theta(X)] = 0$. So $[\theta(X), H]$ is also 0 for all $H \in \mathfrak{a}$. Hence $\mathfrak{g}_0$ is θ-stable. Applying this to the Cartan decomposition and taking into account that θ also preserves $\mathfrak{k}$ and $\mathfrak{p}$ tells us $\mathfrak{g}_0 = \mathfrak{g}_0 \cap \mathfrak{k} \oplus \mathfrak{g}_0 \cap \mathfrak{p}$. But $\mathfrak{g}_0 \cap \mathfrak{k} = \mathfrak{m}$ and by maximality of $\mathfrak{a}$, $\mathfrak{g}_0 \cap \mathfrak{p} = \mathfrak{a}$. $\square$

Now let $H_1, \ldots, H_r$ be a basis of $\mathfrak{a}$. Order $\mathfrak{a}^*$ lexicographically relative to this ordered basis. If Λ^+ is the positive roots and Λ^- the negative roots, then Λ is the disjoint union of Λ^+ and Λ^-. Also, if λ and $\mu \in \Lambda^+$ and $\lambda + \mu \in \Lambda$, then $\lambda + \mu \in \Lambda^+$ and finally $-\Lambda^+ = \Lambda^-$.

Let $\mathfrak{n}^+ = \sum_{\lambda \in \Lambda^+} \mathfrak{g}_\lambda$. Then $\mathfrak{n}^+$ is a subalgebra of $\mathfrak{g}$. Since Λ^+ is finite and if λ and $\mu \in \Lambda^+$, then $\lambda + \mu$ is larger than either of them we see that $\mathfrak{n}^+$ is nilpotent. Similarly let $\mathfrak{n}^- = \sum_{\lambda \in \Lambda^-} \mathfrak{g}_\lambda$, *i.e.* $\mathfrak{n}^- = \theta(\mathfrak{n}^+)$. Then we get another nilpotent subalgebra and $\mathfrak{g} = \mathfrak{n}^- \oplus \mathfrak{m} \oplus \mathfrak{a} \oplus \mathfrak{n}^+$.

We now come to the Iwasawa decomposition of a real semisimple Lie algebra.

Theorem 7.5.12. $\mathfrak{g} = \mathfrak{k} \oplus \mathfrak{a} \oplus \mathfrak{n}^+$ *(direct sum of subspaces)*.

Proof. Let $N^- \in \mathfrak{n}^-$, then $N^- = N^- + \theta(N^-) - \theta(N^-) \in \mathfrak{k} + \mathfrak{n}^+$ as $\theta(N^-) \in \mathfrak{n}^+$. Hence $\mathfrak{n}^- \subset \mathfrak{n}^+ + \mathfrak{k}$. Since $\mathfrak{m} \subset \mathfrak{k}$ and $\mathfrak{g} = \mathfrak{a} \oplus \mathfrak{m} \oplus \mathfrak{n}^- \oplus \mathfrak{n}^+ \subset \mathfrak{a} + \mathfrak{n}^+ + \mathfrak{k}$, $\mathfrak{g} = \mathfrak{a} + \mathfrak{n}^+ + \mathfrak{k}$.

Let $X \in \mathfrak{k}$, $H \in \mathfrak{a}$ and $N^+ \in \mathfrak{n}^+$ and assume $X + H + N^+ = 0$. Then applying θ we get $0 = X - H + \theta(N^+)$. Subtracting we find, $2H + N^+ - \theta(N^+) = 0$, where $\theta(N^+) \in \mathfrak{n}^-$. But $\mathfrak{a} \cap (\mathfrak{n}^+ + \mathfrak{n}^-) = \{0\}$ so that $H = 0$ and $N^+ = N^-$. Since $\mathfrak{n}^+ \cap \mathfrak{n}^-$ is also $\{0\}$, $N^+ = 0 = N^-$. Therefore $X = 0$ and the sum is direct.

$\square$

Before turning to the Iwasawa decomposition of a real semisimple Lie group G, we first deal with a few preliminaries.

Proposition 7.5.13. *Let G be a real semisimple Lie group. Then*

(1) $\operatorname{Ad} G$ *is closed in* $\operatorname{GL}(\mathfrak{g})$.

(2) If G is linear, then $Z(G)$ is finite and $Z(G) \subseteq K$.

Proof. 1. This is because $\operatorname{Aut}(\mathfrak{g})$, as the real points of an algebraic group (see Proposition 1.4.26), is definitely closed in $\operatorname{GL}(\mathfrak{g})$, as is its identity component. Alternatively see Corollary 3.4.5. Since every derivation of $\mathfrak{g}$ is inner and $\operatorname{Ad} G$ is connected we see $(\operatorname{Aut}(\mathfrak{g}))_0 = \operatorname{Ad} G$.

2. Suppose $G \subseteq \operatorname{GL}(n, \mathbb{C}) = \operatorname{GL}(V)$. Since G is semisimple, by Theorem 3.4.3 V is the direct sum of invariant irreducible subspaces V_i. For each i, the map $g \mapsto g|_{V_i}$ is a smooth homomorphism. Hence $G|_{V_i}$ is also a semisimple group. By Schur's lemma for each i, $Z(G)$ acts by scalars λ_i on V_i. Since $g \mapsto (g|_{V_1}, \dots, g|_{V_r})$ is injective it suffices to prove each $Z(G)|_{V_i}$ is finite. Thus we have replaced G by $G|_{V_i}$. In other words we may assume $Z(G)$ acts by scalars on V. Since G is a semisimple group it has no characters. In particular, $\det g \equiv 1$. Restricting to the center we see $\lambda^n \equiv 1$ so $Z(G)$ is finite. Also, K is compact (Chapter 6). Therefore, $Z(G)K$ is a compact subgroup of G containing K. But since K is actually a maximal compact subgroup of G, $Z(G)K = K$ so $Z(G) \subseteq K$.

$\square$

Exercise 7.5.14. Under the hypothesis of 2) show that the order of $Z(G)$ can be estimated by $|Z(G)| \leq n$.

We now set notation and some preparatory ideas for the Iwasawa decomposition.

In $GL(n, \mathbb{R})$ we let D_n and N_n stand for the diagonal and strictly upper triangular matrices. If $\mathfrak{g}$ is a real semisimple Lie algebra, let $(X, Y) = -B(X, \theta Y)$, where B is the Killing form. Then $(\cdot, \cdot)$ is a positive definite inner product on $\mathfrak{g}$. Let Λ be the restricted roots relative to $\mathfrak{a}$. Choose a linear ordering on $\mathfrak{a}^*$ as before. Then we know by Proposition 7.5.11 that $\mathfrak{g} = \mathfrak{a} \oplus \mathfrak{m} \oplus \mathfrak{n}^- \oplus \mathfrak{n}^+$. Let $\Lambda^+ = \{\lambda_1 < \ldots < \lambda_r\}$. Suppose the dimension of $\mathfrak{g}_{\lambda_i}$ is p_i, $i = 1, \ldots, r$. Choose an orthonormal basis for each $\mathfrak{g}_{\lambda_i}$ putting them together in reverse order. Set $q = p_1 + \ldots + p_r$, the dimension of $\mathfrak{n}^+$. Let $\{X_{q+1}, \ldots, X_{q+m}\}$ be an orthonormal basis of $\mathfrak{a} \oplus \mathfrak{m}$. Finally let $X_{q+m+j} = \theta(X_{q-j+1})$ giving an orthonormal basis of $\mathfrak{n}^-$. Then $X_1, \ldots, X_n$ is an orthonormal basis of $\mathfrak{g}$, where $n = 2q + m$. Relative to this basis ad $\mathfrak{k}$ is skew symmetric ad $\mathfrak{a}$ is diagonal and ad $\mathfrak{n}^+$ is strictly upper triangular.

In our formulation of the Iwasawa decomposition, just below, one must assume G is linear (or at least has a finite center) in order to be sure that K is compact. Examples of when this difficulty can arise are provided by the universal covering group of $SL(2, \mathbb{R})$, or more generally by the universal covering group of $Sp(n, \mathbb{R})$.

Theorem 7.5.15. *Let G be a linear real semisimple Lie group with Lie algebra $\mathfrak{g} = \mathfrak{k} \oplus \mathfrak{a} \oplus \mathfrak{n}^+$ and let K, A and N be the corresponding connected Lie subgroups (see Section 1.6). Then*

(1) $\exp : \mathfrak{a} \to A$ *is a Lie isomorphism. So A is a simply connected abelian group.*

(2) $\exp : \mathfrak{n}^+ \to N$ *is a (surjective) diffeomorphism. N is a simply connected nilpotent group.*

(3) The multiplication map $(k, a, n) \mapsto kan$ is a diffeomorphism from $K \times A \times N \to G$.

Proof. We first prove 1) and 2) in general.

Now ad $: \mathfrak{g} \to \text{End}(\mathfrak{g})$. Here ad is injective since $\mathfrak{g}$ is semisimple. Hence for $X \in \mathfrak{a}$ or $\mathfrak{n}^+$, respectively, we can regard X as a diagonal, respectively upper nil-triangular operator on $\mathfrak{g}$. (However, in doing so we are identifying Exp with exp which is not strictly correct since $\text{Ad} \exp X = \text{Exp}(\text{ad} X)$. Thus we must keep in mind that Exp actually takes us into $\text{Ad} G$.) When $X \in \mathfrak{a}$, respectively, $\mathfrak{n}^+$ then $\exp(X)$ is diagonal with positive entries, respectively upper unitriangular with real entries and in each case is a diffeomorphism onto A, or N respectively. In the former case this is essential because $\exp : \mathbb{R} \to \mathbb{R}_+^\times$ is a global diffeomorphism and since $\mathfrak{a}$ is abelian $\exp$ is also a homomorphism, proving 1).

Now in the latter case as we saw in Proposition 7.5.13 $\mathrm{Ad}\, G$ is closed in $\mathrm{GL}(\mathfrak{g})$ and as a linear subspace $\mathrm{ad}\, \mathfrak{n}^+$ is closed in the nil-triangular matrices of $\mathrm{End}(\mathfrak{g})$. Therefore $\mathrm{Exp}(\mathfrak{n}^+)$ is closed in $\mathrm{GL}(\mathfrak{g})$ and therefore also in $\mathrm{Ad}\, G$. Because on $\mathfrak{n}^+$ we know $\exp$ and $\log$ are inverses of one another (see Proposition 6.2.2) this proves 2). In order to prove 3) we first consider the case when G is the adjoint group, *i.e.* when $Z(G)$ is trivial. Now $\mathrm{Ad}(A) \subseteq D_n$, $\mathrm{Ad}(N) \subseteq N_n$ and each is closed in the respective linear group. By Proposition 1.6.2, $\mathrm{GL}(n, \mathbb{R})$ the multiplication map there is a global diffeomorphism. It follows that $\mathrm{Ad}(A)\,\mathrm{Ad}(N)$ is closed in $D_n N_n$ and hence in $\mathrm{GL}(n, \mathbb{R})$. But $\mathrm{Ad}(K)$ is compact hence $\mathrm{Ad}(K)\,\mathrm{Ad}(A)\,\mathrm{Ad}(N) = \mathrm{Ad}(KAN)$ is closed in $\mathrm{GL}(n, \mathbb{R})$ and hence in $\mathrm{Ad}\, G$.

By the $\mathrm{GL}(n, \mathbb{R})$ case (see Proposition 1.6.2) the multiplication map, ϕ, taking $\mathrm{Ad}(k)\,\mathrm{Ad}(a)\,\mathrm{Ad}(n) \mapsto \mathrm{Ad}(kan)$ is injective. We calculate its derivative at a general point kan. Let $X \in \mathrm{ad}\, \mathfrak{k}$, $Y \in \mathrm{ad}\, \mathfrak{a}$ and $Z \in \mathrm{ad}\, \mathfrak{n}^+$. Then we will show $d_{kan}\phi(X, Y, Z) = \mathrm{Ad}(an)^{-1}X + \mathrm{Ad}(n)^{-1}Y + Z$. The reader will notice how similar this calculation is to that in Lemma 7.2.10. We have

$$d_{kan}\phi(X, 0, 0)f = \frac{d}{dt}f(k\exp(tX)an))|_{t=0}$$
$$= \frac{d}{dt}f(kan\exp(\mathrm{Ad}(an)^{-1}tX))|_{t=0}$$
$$= \mathrm{Ad}(an)^{-1}Xf,$$

$$d_{kan}\phi(0, Y, 0)f = \frac{d}{dt}f(ka\exp(tY)n)|_{t=0}$$
$$= \frac{d}{dt}f(kan\exp(t\,\mathrm{Ad}(n)^{-1}Y))|_{t=0} = \mathrm{Ad}(n)^{-1}Yf,$$

and

$$d_{kan}\phi(0,0,Z)f = Zf.$$

Hence using the linearity of $d_0\phi$ we see

$$d_{kan}\phi(X,Y,Z) = \text{Ad}(an)^{-1}X + \text{Ad}(n)^{-1}Y + Z.$$

Hence if $d_{kan}\phi = 0$ we get $\text{Ad}(an)^{-1}X + \text{Ad}(n)^{-1}Y + Z = 0$ so that $X = -\text{Ad}(a)Y - \text{Ad}(an)Z$. Since $Y \in \text{ad }\mathfrak{a}$, $Z \in \text{ad }\mathfrak{n}^+$ and N is normal in AN (*i.e.* N is normalized by A), the same is true of $-\text{Ad}(a)Y$ and $-\text{Ad}(an)Z$ respectively. By Theorem 7.5.12, we see $X = 0$ and hence $\text{Ad}(a)Y + \text{Ad}(an)Z = 0$. But then $\text{Ad}(a)Y$ and $\text{Ad}(an)Z$ are each zero. Finally, this yields $Y = 0 = Z$ so that $d_{kan}\phi$ is non singular at every point. By the inverse function theorem ϕ is a global diffeomorphism. In particular the image is open in $\text{Ad }G$. But by connectedness this has no open subgroups so ϕ is also surjective. This proves 3) when G is the adjoint group.

Now in general for $g \in G$, $\text{Ad }G = \text{Ad}(k)\,\text{Ad}(a)\,\text{Ad}(n)$. Therefore $g = zkan$, where $z \in Z(G)$. But $Z(G)$ is itself in K so any $g \in G$ can be written as $g = kan$ and the multiplication map, ϕ, is surjective here as well. If $k'a'n' = kan$, then taking Ad of everything in sight and applying what we already know tells us that $\text{Ad}(k') = \text{Ad}(k)$, $\text{Ad}(a') = \text{Ad}(a)$ and $\text{Ad}(n') = \text{Ad}(n)$. But then by 1) and 2) of the theorem which we have already proved we get $a' = a$ and $n' = n$. Hence also $k' = k$ so ϕ is injective here as well. Thus ϕ is a bijection. Using this ϕ in the derivative calculations just above shows it too is a diffeomorphism at every point proving 3) and with it the theorem. $\square$

We conclude this chapter with the following important global result which shows, for example, that the Iwasawa decomposition theorem applies to all complex semisimple groups. It is important for other reasons as well. Our proof will use a theorem of algebraic groups due to Chevalley. The original proof due to Goto was different. Chevalley's theorem had not yet been discovered. However, before turning to this theorem we need the following lemma.

Lemma 7.5.16. *Let G be any complex Lie group with a faithful representation $\rho : G \to \mathrm{GL}(n, \mathbb{C})$ and F be a finite central subgroup. Then G/F also has a faithful representation.*

Proof. The finite group $\rho(F)$ is Zariski closed in $\mathrm{GL}(V)$ as is its normalizer N in $\mathrm{GL}(V)$. Therefore by the theorem of Chevalley [9] $N/\rho(F)$ is an algebraic group (in say $\mathrm{GL}(W)$) and the projection, $\pi : N \to N/\rho(F)$, is a rational morphism. In particular π is holomorphic. Now consider $\pi \circ \rho$ which is a holomorphic representation of G on W. Its kernel is exactly F.   □

Theorem 7.5.17. *A complex semisimple Lie group G always has a faithful holomorphic linear representation. In particular, by Proposition 7.5.13 a complex semisimple Lie group always has a finite center.*

Proof. Let $(\tilde{G}, \pi)$ be the universal covering of G. Then $\tilde{G}$ is also a complex semisimple group. If we can show it has a faithful representation then $Z(\tilde{G})$ must be finite by Proposition 7.5.13. Since $\mathrm{Ker}\,\pi = F$ is a (discrete) central subgroup of $\tilde{G}$, we may assume by Lemma 7.5.16 that G itself is simply connected.

Let $\mathfrak{k}$ be a compact real form of $\mathfrak{g}$. Then the corresponding group, K, is a maximal compact subgroup (see Corollary 7.4.10). By a corollary to the Peter-Weyl theorem, K has a faithful smooth representation on U. Its derivative gives a faithful representation of $\mathfrak{k}$ on U which extends canonically to a complex representation of $\mathfrak{g} = \mathfrak{k} \oplus i\mathfrak{k}$ on $U_{\mathbb{C}}$. By simple connectivity of G there is a holomorphic representation σ of G on $U_{\mathbb{C}}$ whose derivative, $d\sigma$ is this representation of $\mathfrak{g}$. Now $\mathrm{Ker}\,\mathrm{Ad} = Z(G) \subseteq K$ (Proposition 7.5.13). Thus $\sigma \oplus \mathrm{Ad}$ is a holomorphic representation of G on $U_{\mathbb{C}} \oplus \mathfrak{g}$. If $\mathrm{Ad}\,g = I$, then $g \in Z(G) \subseteq K$ so if in addition $\sigma(g) = I$ we get $g = 1$. Hence $\mathrm{Ker}(\sigma \oplus \mathrm{Ad})$ is trivial and $\sigma \oplus \mathrm{Ad}$ is faithful.   □

Theorem 7.5.17 is not true, in general, for real groups for a very simple reason. Namely, if such a group were linear it would have to have finite center. But we have seen many examples of real semisimple groups with an infinite center.

Exercise 7.5.18. Show the intermediate coverings of $SL(2, \mathbb{R})$ also have no faithful representations (in spite of the fact that they have finite centers).

Chapter 8

Lattices in Lie Groups

In this chapter we will consider a Lie group G and a lattice (or a uniform lattice) Γ and we will ask how much of G can be recovered, or is determined by Γ? Another perhaps even more fundamental question is when is there such a Γ, or how can one construct a Γ? A third might be to investigate the properties of such Γ's and to distinguish between lattices and uniform lattices in G. Finally we should ask, just how different is a lattice in a general Lie group in relation to that group in comparison to a lattice in Euclidean space as compared to $\mathbb{R}^n$? These are all aspects of a fundamental issue in mathematics. Namely, the comparison of the discrete to the continuous. We begin with the progenitor, namely Euclidean space.

8.1   Lattices in Euclidean Space

In this section we discuss some results concerning lattices in Euclidean space. These are fundamental to further developments and, as the reader will see, are of considerable interest in their own right. Here by a lattice we mean a discrete subgroup Γ of $\mathbb{R}^n$ with finite volume quotient; in other words a subgroup of $\mathbb{R}^n$ with n linearly independent generators.

Exercise 8.1.1. Show that in $\mathbb{R}^n$ a closed subgroup H has finite volume quotient if and only if the quotient is compact. Hint: Consult Exercise 0.1.7.

A typical example of a lattice is $\mathbb{Z}^n$. But of course if $g \in \mathrm{GL}(n,\mathbb{R})$ and Γ is a lattice then so is $g\Gamma$. In fact a moment's reflection tells us that we get all lattices in this way. Thus $\mathrm{GL}(n,\mathbb{R})$ acts transitively on the set $\mathcal{L}$ of lattices. Therefore we can choose any lattice as a base point for this orbit. Choosing the standard lattice, $\mathbb{Z}^n$, we see that the isotropy group is $\mathrm{GL}(n,\mathbb{Z})$. Thus $\mathcal{L}$ can be identified in a natural way with the homogeneous space $\mathrm{GL}(n,\mathbb{R})/\mathrm{GL}(n,\mathbb{Z})$. To topologize $\mathcal{L}$ we take that natural topology from this coset space. It does not depend on a choice of generators in the lattice and makes $\mathcal{L}$ into a locally compact, second countable, Hausdorff manifold. In this way the lattices in $\mathbb{R}^n$ and the homogeneous space $\mathrm{GL}(n,\mathbb{R})/\mathrm{GL}(n,\mathbb{Z})$ (as well as $\mathrm{SL}(n,\mathbb{R})/\mathrm{SL}(n,\mathbb{Z})$) are very closely related.

Let Γ be a lattice in $\mathbb{R}^n$, $d\mu$ be Lebesgue measure on $\mathbb{R}^n$ and $\pi\colon \mathbb{R}^n \to \mathbb{R}^n/\Gamma$ be the natural projection. Then (see Theorem 2.3.5) there is an invariant finite regular measure $d\bar{\mu}$ on the torus, $\mathbb{R}^n/\Gamma$ such that for a continuous function on $\mathbb{R}^n$ with compact support one has

$$\int_{\mathbb{R}^n} f(x)d\mu(x) = \int_{\mathbb{R}^n/\Gamma} \Big(\sum_{\gamma \in \Gamma} f(\gamma + x)\Big)d\bar{\mu}(\bar{x}).$$

Thus the three measures are related and normalizing any two of them (say Lebesgue measure and counting measure) determines the third, $d\bar{\mu}$. So for example if $\Gamma = g\mathbb{Z}^n$, then $\bar{\mu}(\mathbb{R}^n/\Gamma) = |\det g|$. (Notice that this statement is independent of the g doing this since if h leaves $\mathbb{Z}^n$ stable, then $|\det h| = 1$.)

Our study of lattices in $\mathbb{R}^n$ begins with Minkowski's theorem.

Theorem 8.1.2. *Let Γ be a lattice in $\mathbb{R}^n$ and Ω be an open convex set which is symmetric about the origin. If $\mathrm{vol}(\Omega) \geq 2^n \mathrm{vol}(\mathbb{R}^n/\Gamma)$, then $\overline{\Omega}$ meets Γ is a nontrivial lattice point.*

Proof. Let $\pi\colon \mathbb{R}^n \to \mathbb{R}^n/\Gamma$ be the natural projection. This map is either injective on $\frac{1}{2}\overline{\Omega}$, or it is not. In the latter case there must be a $\gamma \neq 0 \in \Gamma$ so that $\gamma + x \in \frac{1}{2}\overline{\Omega}$ and $x \in \frac{1}{2}\overline{\Omega}$. But then by symmetry and convexity of $\frac{1}{2}\overline{\Omega}$ we get $\frac{1}{2}(-x) + \frac{1}{2}(\gamma + x) = \frac{1}{2}\gamma \in \frac{1}{2}\overline{\Omega}$. Hence $\gamma \in \overline{\Omega}$ and we would be done.

We will show that the other alternative, namely that π is injective on $\frac{1}{2}\overline{\Omega}$ leads to a contradiction. Suppose π is injective on $\frac{1}{2}\overline{\Omega}$, then π would also be injective on $\frac{1}{2}\Omega$. Now $\text{vol}(\frac{1}{2}\Omega) = \text{vol}(\pi(\frac{1}{2}\Omega)) \leq \text{vol}(\mathbb{R}^n/\Gamma)$. But since $\text{vol}(\Omega) \geq 2^n \text{vol}(\mathbb{R}^n/\Gamma)$ we know $\text{vol}(\pi(\frac{1}{2}\Omega)) \geq \text{vol}(\mathbb{R}^n/\Gamma)$. It follows that π restricted to this set is surjective. For if the image were smaller, since $\mathbb{R}^n/\Gamma$ is of finite (regular) measure there would be an open set of positive measures left out, a contradiction. Because $\pi(\frac{1}{2}\Omega) = \mathbb{R}^n/\Gamma$ it follows $\mathbb{R}^n = \bigcup_{\gamma \in \Gamma} \gamma + \frac{1}{2}\Omega$, and since π is injective here this union is *disjoint*. Let $U = \frac{1}{2}\Omega$ and $V = \bigcup_{\gamma \neq 0 \in \Gamma}(\gamma + \frac{1}{2}\Omega)$. Then U and V are both closed (and open) and disjoint and $U \cup V = \mathbb{R}^n$ which contradicts the connectivity of $\mathbb{R}^n$.

$\square$

Applications of Minkowski's theorem:

Consider an $n \times n$ nonsingular real matrix (a_{ij}) and use it to define linear functionals λ_i where $i = 1, \ldots, n$ on $\mathbb{R}^n$ by

$$\lambda_i(x) = \sum_j a_{ij} x_j,$$

where $x = (x_1, \ldots, x_n)$. Then $A : \mathbb{R}^n \to \mathbb{R}^n$ defined by

$$A(x) = (\lambda_1(x), \ldots, \lambda_n(x))$$

is an invertible linear transformation whose determinant is $\det(a_{ij})$. Choose positive constants c_i, $i = 1, \ldots, n$ so that $c_1 \ldots c_n \geq |\det A|$ and let

$$\Omega = \{x \in \mathbb{R}^n : |\lambda_i(x)| \leq c_i \text{ for all } i\}.$$

Corollary 8.1.3. Ω *meets* $\mathbb{Z}^n$ *is a nontrivial lattice point.*

Proof. It is easy to see that Ω is a closed convex set which is symmetric about the origin. Now

$$\text{vol}(\Omega) = (\det A)^{-1} \prod (2c_i) \geq 2^n = 2^n \text{vol}(\mathbb{R}^n/\mathbb{Z}^n).$$

Hence by Minkowski's theorem Ω meets Γ nontrivially. $\square$

Minkowski's theorem can also be applied to positive definite, symmetric bilinear forms on $\mathbb{Z}^n$. Indeed this was its original purpose. We denote the ball in $\mathbb{R}^n$ centered at 0 of radius 1 by $B_1(0)$.

Exercise 8.1.4. Let $\beta(x, y) = (Bx, y)$, where $(\cdot, \cdot)$ is the usual inner product, $x, y \in \mathbb{Z}^n$ and B is a positive definite symmetric $n \times n$ matrix with integer coefficients. Prove that there exists a non zero $x \in \mathbb{Z}^n$ such that $\beta(x, x) \leq 4 \det^{\frac{1}{n}}(B) \cdot \operatorname{vol}(B_1(0))^{\frac{-2}{n}}$.

Suggestion: Orthogonally reduce β to diagonal form and use the fact that the volume of an ellipsoid is $\operatorname{vol} B_1(0)$ times the product of its various semiaxes.

Let $q(x) = \sum_{i,j=1}^{n} a_{ij} x_i x_j$ be a real quadratic form, where $x = (x_1, \ldots, x_n) \in \mathbb{R}^n$ and $(a_{ij}) = A$ is a positive definite real symmetric matrix. For $c > 0$ let

$$X_c = \{x \in \mathbb{R}^n : q(x) \leq c \det^{\frac{1}{n}}(a_{ij})\}.$$

Corollary 8.1.5. *Let* $q(x) = \sum_{i,j=1}^{n} a_{ij} x_i x_j$ *be a positive definite symmetric form. Given a lattice* Γ *for* c *sufficiently large* X_c *must meet* Γ.

Proof. We first prove that $\operatorname{vol}(X_c) = \operatorname{vol} B_1(0) c^n$. Since A is positive definite it can be diagonalized *i.e.* there is an orthonormal matrix B such that $B^t A B$ is diagonal with λ_i's as eigenvalues. Consider $x = By \in X_c$, we have $q(x) = \sum_i \lambda_i y_i^2$, therefore $\sum_i \lambda_i y_i^2 \leq c(\Pi \lambda_i)^{\frac{1}{n}}$. Hence for positive numbers $\mu_i = \frac{\lambda_i}{c(\Pi \lambda_i)^{\frac{1}{n}}}$ we see that $B^{-1}(X_c)$ is defined by $\sum \mu_i x_i^2 \leq 1$. Since this is an ellipsoid centered at 0 it is clearly closed and convex. Its volume is given by

$$\operatorname{vol} B_1(0) \Pi \mu_i = \operatorname{vol} B_1(0) \Pi \frac{\lambda_i}{c(\Pi \lambda_i)^{\frac{1}{n}}} = \operatorname{vol} B_1(0) c^n.$$

On the other hand $\operatorname{vol}(X_c) = \operatorname{vol}(B^{-1}(X_c)) = \operatorname{vol} B_1(0) c^n$ since B is orthonormal. Now X_c is a closed symmetric convex set centered at 0 in $\mathbb{R}^n$ satisfying $\operatorname{vol}(X_c) = \operatorname{vol} B_1(0) c^n$. In particular, this volume is independent of q, and for a lattice Γ, c can be chosen large enough so that $\operatorname{vol}(X_c) \geq 2^n \operatorname{vol}(\mathbb{R}^n/\Gamma)$. $\square$

An interesting application of Minkowski's theorem to number theory is the four square theorem which was first proved by Lagrange 100 years prior to Minkowski by other methods.

Corollary 8.1.6. *Any positive integer is the sum of at most 4 squares.*

Before turning to a sketch of the proof we mention that as it can be easily checked, for example, cannot be written as the sum of 3 or fewer squares.

Proof. If x and y are quaternions and $N(x) = x_1^2 + \cdots + x_4^2$ is its norm then it is well known that $N(x)N(y) = N(xy)$. Thus if one has a product of a sum of four squares by another sum of four squares the result is again a sum of four squares. It shows that it is sufficient to prove the 4 square theorem for primes, p and we may evidently assume p is odd. For such a prime there exist integers r and s such that $r^2 + s^2 + 1$ is divisible by p.

Proof of this: Let $S_+ = \{0^2, 1^2, \ldots, (\frac{p-1}{2})^2\}$. Then $|S_+| = \frac{p-1}{2} + 1 = \frac{p+1}{2}$. Similarly, $S_- = \{0^2 - 1, -1^2 - 1, \ldots, -(\frac{p-1}{2})^2 - 1\}$ and $|S_-| = \frac{p+1}{2}$. Now if $x^2 \equiv y^2 \bmod(p)$, p divides $x^2 - y^2$ so p divides $x - y$ or $x + y$. That is $x \equiv y \bmod(p)$ or $x \equiv -y \bmod(p)$. Therefore none of the elements of S_+ are congruent $\bmod(p)$ and similarly none of the elements of S_- are congruent $\bmod(p)$. But there are only p residue classes $\bmod(p)$ and $|S_+ \cup S_-| = |S_+| + |S_-| = p + 1$, hence there exist $r^2 \in S_+$ and $-s^2 - 1 \in S_-$ so that $r^2 \equiv -s^2 - 1 \bmod(p)$.

Now consider the matrix

$$T = \begin{pmatrix} p & 0 & r & s \\ 0 & p & s & -r \\ 0 & 0 & 1 & 0 \\ 0 & 0 & 0 & 1 \end{pmatrix}.$$

Then T is nonsingular so $\Gamma = T(\mathbb{Z}^4)$ is a lattice. In fact since $|\det T| = p^2$ we see that $\bar{\mu}(\mathbb{R}^4/\Gamma) \doteq p^2$. The volume of a ball $B^4(r)$ in $\mathbb{R}^4$ of radius $r > 0$ is $\frac{\pi^2}{2} r^4$ (see Proposition 2.1.10). Therefore the ball of radius $\sqrt{2p}$ which is a convex body symmetric about the origin has $\frac{\pi^2}{2} 4p^2 = 2\pi^2 p^2 > 2^4 p^2$. Therefore by Minkowski there is a nonzero lattice point γ in this ball and the sum of the squares of the γ_i is $< r^2 = 2p$. But a direct calculation using the matrix T, where $\gamma = T(x)$ shows that p divides $N(\gamma)$. So $N(\gamma) \geq p$. Hence $N(\gamma) = p$.   □

We shall return to questions concerning families of lattices in Euclidean space shortly.

8.2   GL$(n, \mathbb{R})/$GL$(n, \mathbb{Z})$ and SL$(n, \mathbb{R})/$SL$(n, \mathbb{Z})$

As we saw GL$(n, \mathbb{R})/$GL$(n, \mathbb{Z})$ can be identified with $\mathcal{L}$, the space of all lattices in $\mathbb{R}^n$ and in this way we can put a manifold structure on $\mathcal{L}$. This cuts both ways, we can also use our knowledge of $\mathcal{L}$ to learn something about the coset space. Given a lattice, Γ in Euclidean space, we take a basis $\{x_1, \ldots, x_n\}$ for it. Then consider the parallelepiped spanned by this basis. By abuse of notation we say vol$(\mathbb{R}^n/\Gamma)$ is the Euclidean volume of the parallelepiped. So vol$(g(\mathbb{Z}^n)) = |\det(g)|$ gives a well-defined map $\Delta : \mathcal{L} \to \mathbb{R}$. Now consider $\mathcal{L}_0$ the space of lattices whose parallelepiped has volume 1. Clearly SL$(n, \mathbb{R})$ operates transitively on $\mathcal{L}_0$ with isotropy group SL$(n, \mathbb{Z})$. Hence SL$(n, \mathbb{R})/$SL$(n, \mathbb{Z})$ can be identified with the space of lattices, $\mathcal{L}_0$. As we shall see this homogenous space has finite volume, but is non-compact. Whereas $\mathcal{L}$ itself does not even have finite volume.

To see that GL$(n, \mathbb{R})/$GL$(n, \mathbb{Z})$ cannot support a finite GL$(n, \mathbb{R})$-invariant measure. Suppose μ was such a measure, consider the det : GL$(n, \mathbb{R}) \to \mathbb{R}^\times$. It induces an onto map GL$(n, \mathbb{R})/$GL$(n, \mathbb{Z}) \to \mathbb{R}^\times/(\pm 1) \cong \mathbb{R}_+^\times$. Push μ forward (Proposition 2.3.6) with the latter map to get a finite invariant measure on $\mathbb{R}_+^\times$ which must be Haar measure. Hence this group would have to be compact, a contradiction.

We now define a Siegel domain for GL$(n, \mathbb{Z})$ within GL$(n, \mathbb{R})$. This was actually done by C.L. Siegel for all the classical non-compact simple groups ([8]). We shall only deal with GL$(n, \mathbb{R})$ and SL$(n, \mathbb{R})$.

Recall the Iwasawa decomposition for $G = $ GL$(n, \mathbb{R})$ which says that as a manifold $G = KAN$ where $K = $ O$(n, \mathbb{R})$, A consists of diagonal matrices with positive entries and N is all upper triangular matrices with eigenvalues 1. This is a direct product $K \times A \times N$ where the inverse map is given by group multiplication in G. For $t, u > 0$ we denote by

$$A_t = \{a \in A : \frac{a_{ii}}{a_{(i+1)(i+1)}} \leq t, \quad i = 1, \ldots, n-1\}$$

and by

$$N_u = \{n \in N : |n_{ij}| \leq u, 1 \leq i < j \leq n\}.$$

Since N is diffeomorphic to $\mathbb{R}^{\frac{n(n-1)}{2}}$ we know that N_u is compact. We define the Siegel domain $S_{t,u} = K A_t N_u$. Evidently Siegel domains are stable under left translation by K and by scalar multiples of the identity and are compact if and only if A_t is compact.

We will prove the following using a sequence of lemmas.

Theorem 8.2.1. $\mathrm{GL}(n, \mathbb{R}) = S_{\frac{2}{\sqrt{3}}, \frac{1}{2}} \mathrm{GL}(n, \mathbb{Z})$.

Before proving this we first deal with N.

Lemma 8.2.2. $N = N_{\frac{1}{2}} N_{\mathbb{Z}}$, where $N_{\mathbb{Z}} = N \cap \mathrm{GL}(n, \mathbb{Z})$.

Proof. Suppose that $u = (u_{ij}) \in N$. We shall find $z = (z_{ij}) \in N_{\mathbb{Z}}$ such that

$$|(u.z)_{ij}| \leq 1/2, \quad i < j. \tag{8.1}$$

As $u_{ik} = 0$ for $k < i$, $z_{jk} = 0$ for $k < j$ and $u_{ii} = z_{ii} = 1$ for all i, (8.1) reads,

$$|z_{ij} + u_{i,i+1} z_{i+1,j} + u_{i,i+2} z_{i+2,j} + \ldots + u_{ij}| \leq 1/2, \quad i < j. \tag{8.2}$$

We find z_{ij} recursively starting by $j = n$ and $i = n-1$. For these values (8.2) is

$$|z_{n-1,n} + u_{n-1,n}| \leq 1/2$$

where we can find $z_{n-1,n}$ such that the inequality (8.2) holds. Now by fixing $j = n$ and varying i we can recursively find all $z_{i,n}$ for all $i \leq n$ such that the inequality is satisfied. By the same process we can find all $z_{i,n-1}$ for $i \leq n-1$ and eventually $z_{i,j}$ for $i \leq j$, for all j's. $\square$

Let $e_1, e_2, ..., e_n$ be the standard basis for $\mathbb{R}^n$ and let $\Phi(g) = ||ge_1||$ for $g \in \mathrm{GL}(n, \mathbb{R})$. Then $\Phi : \mathrm{GL}(n, \mathbb{R}) \to \mathbb{R}^{\times}$ is a continuous map. $\Phi(g) = a_{11} = \Phi(a)$ where $g = kan$ is the Iwasawa decomposition of g.

Lemma 8.2.3. *Let $g \in \mathrm{GL}(n, \mathbb{R})$ be fixed and consider $\gamma \mapsto \Phi(g\gamma)$. Then this function has a positive minimum.*

Proof. $g\,\mathrm{GL}(n, \mathbb{Z})(e_1) \subseteq g(\mathbb{Z}^n \setminus \{0\})$ which is the nonzero elements of some lattice in $\mathbb{R}^n$, hence $||g\gamma(e_1)||$ has a positive minimum as γ varies over $\mathrm{GL}(n, \mathbb{Z})$. $\qquad\square$

Lemma 8.2.4. *Let $g = kan \in \mathrm{GL}(n, \mathbb{R})$ and suppose that $\Phi(g) \leq \Phi(g\gamma)$ for all $\gamma \in \mathrm{GL}(n, \mathbb{Z})$. Then $a_{11} \leq \frac{2}{\sqrt{3}} a_{22}$.*

Proof. Let $n_0 \in N$ then $gn_0 = kan\,n_0$. Since $n, n_0 \in N$ so $\Phi(gn_0) = a_{11} = \Phi(g)$. By Lemma 8.2.2 there is an $n_0 \in N_{\mathbb{Z}}$ so that $|(nn_0)_{ij}| \leq \frac{1}{2}$ for all i, j, so we can assume that $|n_{ij}| \leq 1/2$. Now we let $\gamma_0 \in \mathrm{GL}(n, \mathbb{Z})$ be the following element:

$$\begin{pmatrix} 0 & -1 & 0 \\ 1 & 0 & 0 \\ 0 & 0 & I_{n-2} \end{pmatrix}.$$

Then $\gamma_0(e_1) = -e_2, \gamma_0(e_2) = -e_1$ and $g\gamma_0(e_1) = ge_2 = kan(e_2) = ka(e_2 + e_{21}e_1) = k(a_{22}e_2 + a_{11}n_{12}e_1)$. So $||g\gamma_0(e_1)||^2 = a_{22}^2 + a_{11}^2 n_{12}^2 \leq a_{22}^2 + \frac{1}{4}a_{11}^2$. By the assumption $a_{11}^2 \leq a_{22}^2 + \frac{1}{4}a_{11}^2$ from which the conclusion follows. $\qquad\square$

Proof of Theorem 8.2.1: We prove Siegel's theorem by induction on n. When $n = 1$, $\mathrm{GL}(n, \mathbb{R}) = S_{t,u} = \mathbb{R}^{\times}$, therefore there is nothing to prove. Now let $g \in \mathrm{GL}(n, \mathbb{R})$ and $y \in g\,\mathrm{GL}(n, \mathbb{Z})$ so that $\Phi(y) \leq \Phi(g\gamma)$ for all $\gamma \in \mathrm{GL}(n, \mathbb{Z})$. Hence also $\Phi(y) \leq \Phi(y.\gamma)$ for all $\gamma \in \mathrm{GL}(n, \mathbb{Z})$. One can write

$$k_y^{-1} y = \begin{pmatrix} a_{11} & * \\ 0 & b \end{pmatrix}$$

where b is in $\mathrm{GL}(n - 1, \mathbb{R})$. So by inductive hypothesis there is $z' \in \mathrm{GL}(n-1, \mathbb{Z})$ such that $bz' \in S_{\frac{2}{\sqrt{3}}, \frac{1}{2}}$. Consider the Iwasawa decomposition of $bz' = k'a'n'$ and let

$$z = \begin{pmatrix} 1 & 0 \\ 0 & z' \end{pmatrix} \in \text{GL}(n, \mathbb{Z}).$$

Then

$$k_y^{-1} yz = \begin{pmatrix} a_{11} & * \\ 0 & k'a'n' \end{pmatrix},$$

and this has an Iwasawa decomposition $k''a''n''$, where

$$k'' = k_y \begin{pmatrix} 1 & 0 \\ 0 & k' \end{pmatrix} \in K, \quad a'' = \begin{pmatrix} a_{11} & 0 \\ 0 & a' \end{pmatrix} \in A, \quad n'' = \begin{pmatrix} 1 & 0 \\ 0 & n' \end{pmatrix} \in N.$$

By induction $a_{ii}'' \leq a_{(i+1)(i+1)}''$ for $2 \leq i$. Since z leaves e_1 fixed therefore $\Phi(yz) = \Phi(y)$ and consequently $\Phi(yz) = \Phi(y) \leq \Phi(yz\gamma)$ for all $\gamma \in$ GL$(n, \mathbb{Z})$. By Lemma 8.2.4 we have $a_{11}'' \leq \frac{2}{\sqrt{3}} a_{22}''$ therefore $yz \in KA_{\frac{2}{\sqrt{3}}} N$ and hence $g \in y \,\text{GL}(n, \mathbb{Z}) = yz\,\text{GL}(n, \mathbb{Z}) \subset KA_{\frac{2}{\sqrt{3}}} N\,\text{GL}(n, \mathbb{Z})$. By Lemma 8.2.2 $N = N_{1/2} N_{\mathbb{Z}}$ and therefore $KA_{\frac{2}{\sqrt{3}}} N = KA_{\frac{2}{\sqrt{3}}} N_{1/2} N_{\mathbb{Z}} \subset S_{\frac{2}{\sqrt{3}}, \frac{1}{2}} \text{GL}(n, \mathbb{Z})$.

We now turn to Mahler's compactness criterion. For a lattice Γ in $\mathbb{R}^n$ we know $\Gamma = g.\mathbb{Z}^n$ for some $g \in$ GL$(n, \mathbb{R})$ and g is uniquely determined up to an element of GL$(n, \mathbb{Z})$. Since $|\det| \equiv 1$ on GL$(n, \mathbb{Z})$ we get a well defined function $\Delta(\Gamma) = |\det g|$.

An important result concerning subsets $\mathcal{S} \subseteq \mathcal{L}$, the family of all lattices in $\mathbb{R}^n$ is Mahler's theorem first proved in 1946 in [38]. A very efficient proof of this result can be given by means of Siegel domains in GL$(n, \mathbb{R})$. Mahler's theorem, which bears a striking resemblance to the classical theorem of Ascoli, is the following:

Theorem 8.2.5. *A subset $\mathcal{S} \subseteq \mathcal{L}$ has compact closure if and only if:*

(1) Δ is bounded on $\mathcal{S}$.

(2) There exists a neighborhood U of 0 in $\mathbb{R}^n$ so that $\Gamma \cap U = \{0\}$ for all $\Gamma \in \mathcal{S}$.

The first condition is analogous to uniform boundedness while the second (often described as $\mathcal{S}$ being uniformly discrete) is analogous to equicontinuity in Ascoli's theorem.

Proof. Because of Theorem 8.2.1 we get all the lattices in $\mathbb{R}^n$ already from Siegel set. Hence the statement that is compact is equivalent to having a subset S of the Siegel set with compact closure and $S(\mathbb{Z}^n) = \mathcal{S}$. This is equivalent to having the A part of the Siegel set compact. That is to say that there should be α, β with

$$0 < \alpha \leq a_{ii} \leq \beta \tag{8.3}$$

where g varies over the Siegel set and $1 \leq i \leq n$. We will prove that (8.3) is equivalent to the following two statements:

   (a) $|\det|$ is bounded on $\mathcal{S}$.
   (b) There exists $c > 0$ such that $\|g(x)\| > c$ for each $x \in \mathbb{Z}^n \setminus 0$ and $g \in \mathcal{S}$.

These two conditions are exact reformulation of the two conditions in the theorem.

Suppose that (a) and (b) hold and (a_{ii}) is the A part of the Iwasawa decomposition of $g = kan$. By (b) $\|(g(e_1)\| = a_{11} \geq c > 0$ for every $g \in \mathcal{S}$. Since $\mathcal{S}$ is a subset of Siegel set we know that $a \in A_{\frac{2}{\sqrt{3}}}$ so we have $c \leq a_{11} \leq ta_{22}$ where $t = \frac{2}{\sqrt{3}}$. So $a_{22} \geq \frac{c}{t}$ and $a_{33} \geq \frac{c}{t^2}$ etc. By taking the minimum of this finite number of positive quantity we have $a_{ii} \geq \alpha > 0$ for all $1 \leq i \leq n$ and $g \in \mathcal{S}$. By (a), $|\det g| = \prod_{i=1}^{n} a_{ii} \leq M$ for some constant M. Let j be a fixed index and since $\alpha^{n-1} a_{jj} \leq a_{11} \ldots a_{jj} \ldots a_{nn} \leq M$ thus $a_{jj} \leq \frac{M}{\alpha^{n-1}} = \beta$.

Turning to the converse, suppose that (8.3) holds then $|\det g| \leq \beta^n$ for all $g \in \mathcal{S}$ proving (a). Let $x \in \mathbb{Z}^n \setminus 0$ and write $x = \sum m_i e_i$ where for some i, $m_i \neq 0$. Let k be the first of such i's, then $\|g(x)\| = \|an(x)\|$ and the kth coordinate of $an(x)$ is $a_{kk}m_k$. Hence $\|g(x)\| \geq a_{kk}|m_k| \geq a_{kk} \geq \alpha > 0$ for all $g \in \mathcal{S}$ proving the second condition. □

Since $\mathcal{L}_0$ is closed in $\mathcal{L}$ (prove!). A direct corollary of Mahler's theorem is then: A subset $\mathcal{S}$ of $\mathcal{L}_0$ has compact closure if and only if there exists some neighborhood of 0 in $\mathbb{R}^n$ so that $\Gamma \cap U = \{0\}$ for all $\Gamma \in \mathcal{S}$.

Corollary 8.2.6. *For $n \geq 2$, $\mathrm{SL}(n, \mathbb{R}) / \mathrm{SL}(n, \mathbb{Z})$ is non-compact.*

Proof. Now $\mathcal{L}_0$, the space of lattices with parallelepiped of volume 1 and is identified with $\mathrm{SL}(n, \mathbb{R})/\mathrm{SL}(n, \mathbb{Z})$. If the latter is compact so is $\mathcal{L}_0$. Since $\mathcal{L}_0$ has compact closure in $\mathcal{L}$, Mahler's criterion must be satisfied. However the second condition cannot be satisfied. Consider the matrix

$$g_k = \begin{pmatrix} 1/k & 0 & 0 \\ 0 & k & 0 \\ 0 & 0 & I_{n-2} \end{pmatrix}$$

which is an element of $\mathrm{SL}(n, \mathbb{R})$ and $\|g_k(e_1)\| = 1/k$. Since this tends to zero as $k \to \infty$ this violates the second condition.   □

We now make a brief digression to give the reader a longer view of the terrain. We first define the concept of the unipotent radical of a connected algebraic group.

Definition 8.2.7. Let G be a connected algebraic a group. Its unipotent radical, G_u, is the largest normal, connected, unipotent subgroup of G.

The basic examples just above illustrate an important result of Borel-Harish Chandra 8.2.8 whose proof is beyond the scope of this book [6].

Theorem 8.2.8. *If G is a connected algebraic group defined over $\mathbb{Q}$, then $G_{\mathbb{R}}/G_{\mathbb{Z}}$ has a finite invariant measure if and only if G has no non-trivial $\mathbb{Q}$-characters.*

Here $G_{\mathbb{R}}$ and $G_{\mathbb{Z}}$ are respectively the real and integer points of G. Moreover the results of both Mostow-Tamagawa [64] and Borel-Harish Chandra each tell us that under the same conditions $G_{\mathbb{R}}/G_{\mathbb{Z}}$ is compact if and only if G has no nontrivial $\mathbb{Q}$ characters and every unipotent element is in the unipotent radical of G.

To illustrate these results in a simple situation, let $G = \mathrm{GL}(1, \mathbb{C}) = \mathbb{C}^{\times}$. For every $n \in \mathbb{Z}$, $z \mapsto z^n$ is a $\mathbb{Q}$-character. $G_{\mathbb{R}} = \mathbb{R}^{\times}$ and $G_{\mathbb{Z}} = \mathbb{Z}_2$. So $G_{\mathbb{R}}/G_{\mathbb{Z}} = \mathbb{R}_+^{\times}$. Since this is non-compact and everything is abelian $G_{\mathbb{R}}/G_{\mathbb{Z}}$ does not have finite volume. On the other hand if G is the abelian subgroup $(= (\mathbb{C}, +))$ of unitriangular matrices in $\mathrm{GL}(2, \mathbb{C})$, then

$G_\mathbb{R}/G_\mathbb{Z}$ is compact and has finite volume. Concomitantly, as a unipotent $\mathbb{Q}$-group G has no nontrivial characters.

Since det is such a $\mathbb{Q}$ character for $GL(n, \mathbb{C})$, the Borel-Harish Chandra theorem gives an alternative proof that $GL(n, \mathbb{R})/GL(n, \mathbb{Z})$ cannot have a finite invariant measure. On the other hand, for a semisimple group G (such as $SL(n, \mathbb{C})$) there are no $\mathbb{Q}$ characters. Hence here $G_\mathbb{R}/G_\mathbb{Z}$ always has a finite invariant measure. In particular, for a semisimple group G (which also has no unipotent radical) the condition for compactness means G has no non-trivial unipotent elements at all.

It should be noted that a close look at the exposition of this compactness criterion (see [71]) shows that $SL(n, \mathbb{R})/SL(n, \mathbb{Z})$ is the crucial case of the compactness result. Later we shall deal with it directly.

In the case of semisimple Lie groups without compact factors and their lattices this has been further generalized by Kazdan and Margulis (see [71]) proving Selberg's conjecture (2. below).

Let G be connected linear semisimple Lie group without compact factors and μ be a fixed Haar measure on G. Then

(1) There is a constant $c(G) > 0$ such that for all lattices, Γ, in G the measure induced on $G/\Gamma \geq c(G)$.

(2) If Γ is a non-uniform lattice in G, then Γ has a non-trivial unipotent element.

(3) If Γ is a uniform lattice in G, then every element in it is Ad semisimple.

Another important general result is Mostow's rigidity theorem (or the Mostow-Margulis rigidity theorem).

Mostow's theorem is the following:

Theorem 8.2.9. *Let G and G' be connected semisimple linear groups without compact factors, or factors locally isomorphic with $SL(2, \mathbb{R})$ and let Γ and Γ' be uniform lattices in G and G', respectively. Then any isomorphism $\Gamma \to \Gamma'$ extends to a smooth isomorphism of $G \to G'$.*

This was proven by Mostow in stages starting with the group of hyperbolic motions $G = G' = SO(n,1)_0$, $n \geq 3$ and then extending it to the general case. When $G = G'$ the algebraic formulation can be replaced by a more geometric one. Consider the associated symmetric space of non-compact type $G/K = P$. Then $P/\Gamma = X$ and $P/\Gamma' = X'$ are compact manifolds of the same dimension. (Since any two maximal compact subgroups of G are conjugate, the dimension of G/K is an invariant of G called its *characteristic index.*) Since P is simply connected Γ and Γ' are the respective fundamental groups. Hence our hypothesis is that these two compact manifolds have isomorphic fundamental groups. On the other hand, G is essentially the connected isometry group of P so X and X' are isometric.

Why is $SL(2, \mathbb{R})$ excluded? Taking $G = G' = PSL(2, \mathbb{R})$. Here G/K is the Poincaré upper half plane, the universal covering surface of all compact oriented Riemann surfaces of genus, $g \geq 2$. Let X and X' be two such Riemann surfaces of the same genus $g \geq 2$. Then X and X' are homeomorphic and hence have isomorphic fundamental groups. But they need not be analytically equivalent because there are $6g - 6$ analytically inequivalent such surfaces [20]. Since for $PSL(2, \mathbb{R})$ analytic equivalence is the same as being isometric [20] this gives a counterexample.

For non-compact simple groups of real rank ≥ 2, Margulis has extended the rigidity theorem to non-uniform lattices. Closely connected with the Mostow-Margulis rigidity theorem is the following result of Prasad: Let G be a non-compact simple linear Lie group, not locally isomorphic with $SL(2, \mathbb{R})$ and Γ be a lattice in G. If G' is another such simple group and Γ' is a discrete subgroup isomorphic with Γ, then Γ' is a lattice in G' if and only if the characteristic index of G equals that of G'. So for example, Γ cannot be isomorphic to any of its subgroups of infinite index since such a subgroup cannot be a lattice. Also this recaptures the result of Furstenberg [21] that no lattice in $SL(n, \mathbb{R})$ can be isomorphic with a lattice in $SL(m, \mathbb{R})$, if $n \neq m$ and both are greater than 2. Since these matters are also beyond the scope of this book we will not pursue them further.

Returning to Siegel domains in $GL(n, \mathbb{R})$ another interesting consequence is Hermite's inequality. This tells us that given a lattice $\Gamma = g\mathbb{Z}^n$ there is a universal constant $c_n > 0$ such that $c_n |\det g|^{\frac{1}{n}}$ dominates the smallest nonzero length of all the lattice points in $\Gamma = g\mathbb{Z}^n$. In particular, if $\Gamma \in \mathcal{L} - 0$ then c_n itself dominates these lengths.

Corollary 8.2.10. *Let $g \in \mathrm{GL}(n, \mathbb{R})$. Then*

$$\min_{\gamma \in \mathbb{Z}^n \setminus \{0\}} \|g(\gamma)\| \leq (\frac{2}{\sqrt{3}})^{\frac{n-1}{2}} |\det g|^{\frac{1}{n}}.$$

Proof. Let $\Gamma = g(\mathbb{Z}^n)$ and choose $g' \in g\Gamma \cap S_{\frac{2}{\sqrt{3}}, \frac{1}{2}}$ so that $\Phi(g') \leq \Phi(g\gamma)$, for all $\gamma \in \mathrm{GL}(n, \mathbb{Z})$. Hence

$$\min_{\gamma \in \mathbb{Z}^n \setminus \{0\}} \|g(\gamma)\| \leq \min_{\gamma \in \mathrm{GL}(n, \mathbb{Z})} \|g\gamma(e_1)\| = \min_{\gamma \in \mathrm{GL}(n, \mathbb{Z})} \Phi(g(\gamma)) = \Phi(g') = a'_{11},$$

where a'_{11} is the first component of a', the a part of g'. But $g' \in S_{\frac{2}{\sqrt{3}}, \frac{1}{2}}$. Hence $a'_{11} \leq \frac{2}{\sqrt{3}} a'_{22}, \ldots, a'_{(n-1)(n-1)} \leq \frac{2}{\sqrt{3}} a'_{nn}$. Therefore

$$a'_{11}{}^n \leq (\frac{2}{\sqrt{3}})^{1+2+\ldots+n-1} a'_{11} \ldots a'_{nn}.$$

Thus

$$a'_{11}{}^n \leq (\frac{2}{\sqrt{3}})^{\frac{n(n-1)}{2}} |\det a'|.$$

But $|\det g'| = |\det a'|$ since $|\det k'| = 1 = |\det n'|$ and $|\det g'| = |\det g|$ since $g' = g\gamma$ and $|\det \gamma| = 1$. This means

$$(\min_{\gamma \in \mathbb{Z}^n \setminus \{0\}} \|g(\gamma)\|)^n \leq a'_{11}{}^n \leq (\frac{2}{\sqrt{3}})^{\frac{n(n-1)}{2}} |\det g|.$$

Taking nth roots proves the result. $\square$

We now apply some of the results above to the so-called "reduction theory" of quadratic forms. Let p be a positive definite symmetric matrix and $q(x) = (px, x)$ be the associated positive definite quadratic form on $\mathbb{R}^n$. As in Chapter 6 $G = \mathrm{GL}(n, \mathbb{R})$ operates transitively in a natural way on the space P of such forms via $p \mapsto gpg^t$, $g \in \mathrm{GL}(n, \mathbb{R})$ with isotropy group $\mathrm{Stab}_G(I) = \mathrm{O}(n, \mathbb{R})$. The projection, $\pi : G \to P$ given by $g \mapsto gg^t$ commutes with the action of G on itself by left translation. We also have the usual Siegel set $S_{(t,u)}$ and also $S'_{(t,u)} = $

$\{nan^t : a \in A_t, n \in N_u\}$. Since $(nak)(nak)^t = nakk^{-1}an^t = na^2n^t$ we see that if $g = nak \in S_{(t,u)}$, then $gg^t \in S'_{(t^2,u)}$. Hence $\pi(S_{(t,u)}) = S'_{(t^2,u)}$ and $\pi^{-1}(S'_{(t^2,u)}) = S_{(t,u)}$. Hence we have,

Corollary 8.2.11.   *(1) $P = S'_{(t,u)}(\text{GL}(n, \mathbb{Z}))$ whenever $t \geq \frac{4}{3}$ and $u \geq \frac{1}{2}$.*

(2) $\min_{x \in \mathbb{Z}^n \setminus \{0\}} q(x) \leq (\frac{4}{3})^{\frac{n-1}{2}} (\det p)^{\frac{1}{n}}$.

Similarly SL$(n, \mathbb{R})$ acts transitively on P^*, the positive definite symmetric matrices of determinant 1 with isotropy group SO$(n, \mathbb{R})$. Denoting the corresponding Siegel domains here by $S^*_{(t,u)}$ and $S^{*'}_{(t,u)}$, using the same actions restricted to SL$(n, \mathbb{R})$, we get

Corollary 8.2.12. *$P^* = S^{*'}_{(t,u)}(\text{SL}(n, \mathbb{Z}))$ whenever $t \geq \frac{4}{3}$ and $u \geq \frac{1}{2}$ and (by the result just below) since SO$(n, \mathbb{R})$ is compact $P^*/\text{SL}(n, \mathbb{Z})$ has finite volume.*

In the crucial case of SL$(n, \mathbb{R})/$ SL$(n, \mathbb{Z})$, finiteness of volume can be proved "by hand" using the method of Siegel as follows.

Theorem 8.2.13. SL$(n, \mathbb{R})/$ SL$(n, \mathbb{Z})$ *has finite volume.*

Proof. We intersect all the elements in Siegel's theorem, Theorem 8.2.1, with SL$(n, \mathbb{R})$. By Iwasawa decomposition SL$(n, \mathbb{R}) = \text{SO}(n, \mathbb{R}) \times A^* \times N$ where $A^* = \{a \in A | \det a = 1\}$. Hence here

$$S^*_{t,u} = \text{SO}(n, \mathbb{R}) \times A^*_t \times N_u.$$

Let μ be left Haar measure on SL$(n, \mathbb{R})$. We will prove that $\mu(S^*_{t,u})$ is finite. From this it will follow from Proposition 2.4.3 that SL$(n, \mathbb{R})/$ SL$(n, \mathbb{Z})$ has finite volume.

$d\mu = dk\, db^*$ where dk is left Haar measure on $K = \text{SO}(n, \mathbb{R})$ and db^* is left Haar measure on $B^* = A^* N$. Here $db^* = \rho(a^*) da^* dn$ where $\rho(a^*)$ is the distortion of the Euclidean volume of N by the automorphism $i_{a^*}|_N$ where i_{a^*} is the inner automorphism determined by a^* (see Section

2.3). $\rho(a^*) = \prod_{i<j} \frac{a_{ii}^*}{a_{jj}^*}$. So $d\mu = \prod_{i<j} \frac{a_{ii}^*}{a_{jj}^*} dk da^* dn$. Now we calculate $\mu(S_{t,u}^*)$.

$$\mu(S_{t,u}^*) = \int \int \int K A_t^* N_u \prod_{i<j} \frac{a_{ii}^*}{a_{jj}^*} dk da^* dn \overset{\text{Fubini}}{=} \int_K dk \int_{A_t^*} \prod_{i<j} \frac{a_{ii}^*}{a_{jj}^*} \int_{N_u} dn.$$

Since K and N_u are compact $\int_K dk$ and $\int_{N_u} dn$ are finite. It remains to show that $\int_{A_t^*} \prod_{i<j} \frac{a_{ii}^*}{a_{jj}^*}$ is finite. We use a change of coordinate to compute this integral. Our new coordinates are $b_i = \frac{a_{ii}^*}{a_{i+1,i+1}^*}, i = 1, \ldots, n$. On A_t^*, $b_i \leq t$ for all i. One sees directly that $\rho(a^*) = \prod_{i=1}^{n-1} b_i^{r_i}$ where r_i are certain positive integers combinatorially dependent on n. So

$$\int_{A_t^*} \rho(a^*) da^* \overset{Fubini}{=} \prod_{i=1}^{n-1} \int_{b_i \leq t} b_i^{r_i} db_i.$$

If $\mathfrak{a}^* = \{y | y \text{ real diagonal matrix of trace zero}\}$ and the exponential map defines a global diffeomorphism from $\mathfrak{a}^*$ to A^* and it takes the global measure to the Haar measure. Choose $y_i \in \mathfrak{a}^*, i = 1, \ldots, n-1$ so that $\exp y_i = b_i$ for all i. Hence $\exp r y_i = b_i^{r_i}$. Therefore

$$\int_{b_i \leq t} b_i^{r_i} db_i = \int_{-\infty}^{\log t} (\exp r_i y_i) dy_i.$$

For $\lambda > 0$,

$$\int_{-\infty}^{\log t} (\exp \lambda y) dy = \frac{e^{\lambda y}}{\lambda} \Big|_{-\infty}^{\log t} = \frac{t^\lambda}{\lambda}.$$

Thus $\int_{A_t^*} \rho(a^*) da^* = \prod_{i=1}^{n-1} \frac{t^{r_i}}{r_i} < \infty$. □

Remark 8.2.14. Here we again make contact with Riemann surface theory by showing that although we have constructed non-uniform lattices in, for example $\mathrm{SL}(2, \mathbb{R})$, we can construct an infinite family of examples of uniform lattices in $\mathrm{SL}(2, \mathbb{R})$ as well. Let S be a compact Riemann surface of genus $g \geq 2$. By uniformization theorem $S = \mathbb{H}_2^+ / \Gamma$

where $\mathbb{H}_2^+$ is the upper half plane and the universal cover of S and Γ is the fundamental group of S which is a discrete group of $\mathrm{SL}(2,\mathbb{R})$. Now $\mathbb{H}_2^+$ is $K \backslash \mathrm{SL}(2,\mathbb{R})$, the right cosets K in $\mathrm{SL}(2,\mathbb{R})$ where is $K = \mathrm{SO}(2,\mathbb{R})$ is a maximal compact subgroup of $\mathrm{SL}(2,\mathbb{R})$. Therefore $\mathrm{SL}(2,\mathbb{R})/\Gamma$ is compact. Another way of constructing uniform lattices in $\mathrm{SL}(2,\mathbb{R})$ is by means of quaternions [25].

8.3 Lattices in More General Groups

In this section we only sketch the results. It is now natural to turn from $\mathbb{R}^n$ to simply connected nilpotent Lie groups where the results are due to Malcev [71]. As with $\mathbb{R}^n$ these groups also have faithful linear (unipotent) representations. As mentioned earlier there is no distinction between lattices and uniform lattices. However, there are strict requirements for a discrete subgroup to be a lattice and in particular there are simply connected nilpotent Lie groups which have no lattices. In fact, G has a lattice if and only if its Lie algebra has a basis with respect to which all of the structure constants are *rational*. Hence by Proposition 3.1.69 there exist simply connected 2-step nilpotent groups which have no lattices at all. Further, an abstract group is isomorphic to a lattice in some simply connected nilpotent group if and only if it is finitely generated, nilpotent and torsion free. Thus, in this regard the situation here is similar to the abelian case.

If the simply connected nilpotent group is the full strictly triangular group *i.e.* the N of Proposition 8.2.2 then since $N = N_{\frac{1}{2}} N_{\mathbb{Z}}$ and $N_{\frac{1}{2}}$ is compact we see that $N_{\mathbb{Z}}$ is a uniform lattice in N. Actually, $N/N_{\mathbb{Z}}$ is compact for any unipotent (hence simply connected, nilpotent) group whose Lie algebra has rational structure constants. This follows from the Borel-Harish Chandra and Mostow-Tamagawa theorems since such a group has no nontrivial $\mathbb{Q}$ characters and, of course, every unipotent element lies in the unipotent radical. A lattice is provided by taking $N_{\mathbb{Z}}$, the matrices of N with integer coordinates.

Two other interesting things happen here. First the integer parameters for the set of all lattices are not arbitrary, but are governed by certain divisibility conditions. Secondly, in general the log of a lattice is not a lattice (or even a subgroup) of the additive group of $\mathfrak{g}$. When it is such a lattice it is called a *log lattice*. C. Moore has shown [43] that any lattice in a simply connected nilpotent group is always bracketed between two lattices both of which are log lattices.

Exercise 8.3.1. Find all lattices in the Heisenberg group G up to automorphisms, $\text{Aut}(G)$. See which ones have $\log(\Gamma)$ a lattice in $\mathfrak{g}$, the Lie algebra of G. Notice that here $\text{Aut}(G)$ does not act transitively on $\mathcal{L}(G)$.

If Γ is a lattice in a simply connected solvable group, G, a theorem of Mostow [61] $\Gamma \cap \text{Nil}(G)$ is a lattice in $\text{Nil}(G)$, the nilradical. Moreover as mentioned earlier, for a connected solvable Lie group cofinite volume and cocompactness of a closed subgroup are the same. This result is also due to Mostow [57].

Exercise 8.3.2. In connection with Mostow's theorem mentioned just above, the reader should construct an example of a lattice Γ in $\mathbb{R}^2$ and a closed connected subgroup H of $\mathbb{R}^2$ with the property that $H \cap \Gamma$ is not a lattice in H.

Let G be a connected Lie group with Levi decomposition $G = SR$, where S is a Levi factor and R is the radical. Since S is semisimple we can further decompose $S = S_0 C$, where S_0 is semisimple without compact factors (the product of all non-compact simple subgroups of S) and C is compact semisimple (the product of all compact simple subgroups of S). Then S_0 and C commute pointwise. For all this see Corollary 3.3.19. Since R is characteristic and therefore normalized by C, CR is a subgroup of G which is connected. It is closed since C is compact because of Weyl's theorem, Theorem 2.5.8, and R is closed because it is the radical. It is evidently also normal since R is and C commutes with S_0. Also CR while not solvable is almost as good; it is amenable and G/CR is semisimple without compact factors. Evidently no larger connected subgroup can be amenable so CR is itself a kind of radical. In this way we have separated the parts of G which are semisimple without compact factors from the rest.

Similarly to the result of Mostow mentioned above, a theorem of
H.C. Wang [79], Garland and Goto [24] states that if Γ is a lattice in a
Lie group G, then $\Gamma \cap CR$ is a uniform lattice in CR. This fact provides
a method for proving the folk theorem that lattices in a Lie group are
always finitely generated. (In particular lattices are always countable.)
For we have already proved, Proposition 2.5.2, that a uniform lattice
is finitely generated. Since the image of Γ mod CR is a lattice in a
semisimple group without compact factors one is reduced to showing
that a lattice in such a group is finitely generated. This is done by
different methods in the rank one and higher rank cases.

One final result along these lines (see [29]) is the following. Let G
be a connected Lie group containing a lattice Γ and $B(G)$ the *bounded
part* of the group G, namely the elements $g \in G$ whose conjugacy class
has compact closure. If $B(G) = Z(G)$ or even more generally if $B(\tilde{G}) =
Z(\tilde{G})$ (the universal coverings), then $\Gamma \cap Z(G)$ is a (uniform) lattice in
$Z(G)$.

For a connected Lie group G and a smooth automorphism α, by
taking its derivative $d_1\alpha$ at 1 we get a linear automorphism of its Lie
algebra $\mathfrak{g}$. This map preserves composition and is injective. If G is sim-
ply connected the map $\alpha \mapsto d_1\alpha$ is an isomorphism $\mathrm{Aut}(G) \to \mathrm{Aut}(\mathfrak{g})$.
In this way we can regard $\mathrm{Aut}(G)$ as the real points of a real linear
algebraic group. If $\mathfrak{g}$ has a basis whose structure constants are rational this linear algebraic group is defined over $\mathbb{Q}$. Its Lie algebra is the
derivations $\mathrm{Der}(\mathfrak{g})$ of $\mathfrak{g}$. If we consider the subgroup of Haar measure
preserving automorphisms, the Lie algebra of this subgroup consists of
$\mathrm{Der}_0(\mathfrak{g})$, the derivations of trace zero. Hence, if N is a simply connected
nilpotent group which has a lattice, then its automorphism group is an
algebraic $\mathbb{Q}$-group. It follows that the same is true of the group of
Haar measure preserving automorphisms $M(N)$, as well as its identity
component $M(N)_0$.

The following gives a new construction of both lattices and uniform
lattices. Of course, it depends on the theorems of Borel-Harish Chandra
[6] and Mostow-Tamagawa [64]. Let Γ be a lattice in N where N is
nilpotent part of the Iwasawa decomposition $G = KAN$ of a real rank 1
simple group G. Such N's always possess lattices. If $G = \mathrm{SO}_0(n,1)$ or

$\mathrm{SU}(n,1)$, $\mathrm{Stab}_{M(N)_0}(\Gamma)$ is a non-uniform lattices in $M(N)_0$, the identity component of the group of measure preserving automorphisms of N. However, if $G = \mathrm{Sp}(n,1)$, or the exceptional group our construction gives a uniform lattice in $M(N)_0$ (see [47] and [4]).

8.4   Fundamental Domains

In our final section of this chapter we define and then construct a fundamental domain for a discrete subgroup Γ of a connected unimodular Lie group G. Although lattices are the most interesting case, here G/Γ need not have finite volume.

Let the Lie group G act smoothly on a manifold X and Γ be a discrete subgroup of G. A *fundamental domain* for Γ with respect to this action is a closed set $\mathcal{D} \subseteq X$ satisfying the conditions listed below.

(1) If γ and γ' are distinct points of Γ then $\gamma\mathcal{D}$ and $\gamma'\mathcal{D}$ are disjoint.

(2) The union $\bigcup_{\gamma\in\Gamma} \gamma\bar{\mathcal{D}} = X$.

The idea here is to have exactly one representative from each orbit, but we will have to compromise about boundary points. If we were willing to have a measurable fundamental domain we could just take a measurable cross section (measurable axiom of choice). Of course, the measure of a fundamental domain will be unaffected by what we choose to do on the boundary since this has measure zero.

This representation $x \in X$ as γd is essentially unique. That is, except for the Γ orbit of the boundary, $\partial(\mathcal{D})$. If X has a G-invariant volume it is usually the case that $\partial(\mathcal{D})$ is lower dimensional and therefore $\mathrm{vol}(\partial(\mathcal{D})) = 0$.

Now suppose G is a unimodular Lie group and Γ is a discrete subgroup of G. Then G/Γ has an essentially unique G-invariant measure μ and Γ is a lattice if and only if $\mu(\mathcal{D}) < \infty$.

An example of a fundamental domain for the subgroup $\mathrm{SL}(2,\mathbb{Z}) = \Gamma$ in $\mathrm{SL}(2,\mathbb{R}) = G$ under the action of G on the upper half plane $H^+ = G/K$. Now since K is compact the question of compactness, or finite volume of G/Γ is the same as compactness, or finite volume of a fundamental domain in H^+. Such a fundamental domain for the modular group has been known for a long time. Since G operates by isometries in the Poincaré metric and the curvature is constant -1. By Gauss-Bonnet, or rather just Gauss, Area $= \frac{\pi}{3}$. Or one could use the invariant area $dA = \frac{dxdy}{y^2}$ coming from the invariant metric $ds^2 = \frac{dx^2+dy^2}{y^2}$ and estimate the area of a strip bounded away from zero in the vertical direction.

We let G be a Lie group and Γ be a discrete subgroup. We shall say Ω, an open subset of G, is a *fundamental domain for* Γ if Ω is a fundamental domain for the standard of the action of Γ on G by left traslation *i.e.*

(1) If for γ_1 and γ_2 different elements of Γ, $\gamma_1\Omega$ and $\gamma_2\Omega$ are disjoint.

(2) $\bigcup_{\gamma\in\Gamma} \gamma\overline{\Omega} = G$.

This definition is equivalent to the statement that for any $x \in G$ we can find $\gamma \in \Gamma$ and $\omega \in \Omega$ such that $x = \gamma\omega$. This representation is essentially unique except for points in $\overline{\Omega} \setminus \Omega$.

In order to proceed we first construct a left invariant Riemannian metric. This can be done by choosing an inner product for the tangent space of the Lie algebra and transferring by left translation to the tangent spaces of any other point in G. The metric d on G and the resulting topology determined by this left G-invariant Riemannian metric is the same as that of the Lie topology. This is because the equivalence of the two topologies is a local question and in a neighborhood of the identity is equivalent to the fact that exponential map is a local diffeomorphism. By the Hopf-Rinow theorem (see [23]) and the fact that the Lie topology is always complete (because it is locally compact) any two points of G can be joined by a minimal geodesic.

We now construct a fundamental domain for Γ in G containing 1. We let Ω be the set of all points in $g \in G$ such that

$$d(g, 1) < d(g, \Gamma \setminus \{1\}).$$

First of all Ω is open. If Ω is not open at $g \in \Omega$ then there would be a sequence $g_n \in G$ and $\gamma_n \in \Gamma$ such that

$$d(g_n, 1) \geq d(g_n, \gamma_n) \tag{8.4}$$

for all n and $g_n \to g$. Since $g_n \to g$ and by (8.4) we have $d(g_n, \gamma_n)$ is bounded and therefore $d(g, \gamma_n)$ is itself bounded. Since Γ is discrete the only way that can happen is that for that sequence γ_n is a finite set. In particular there is subsequence γ_{n_i} which is a constant γ_0. By inserting this subsequence in (8.4) we get

$$d(g_{n_i}, 1) \geq d(g_{n_i}, \gamma_0) \tag{8.5}$$

and then by taking limits we have $d(g, 1) \geq d(g, \gamma_0)$ which is a contradiction.

We now verify the conditions (i) and (ii). So for condition (i) suppose that $\gamma_0 \Omega \cap \Omega \neq \emptyset$ for some $\gamma_0 \neq 1$. Then there are ω_0 and ω_1 in Ω such that $\gamma_0 \omega_0 = \omega_1$. Since we have

$$d(\omega_0, 1) < d(\omega_0, \gamma) \text{ and } d(\omega_1, 1) < d(\omega_1, \gamma) \tag{8.6}$$

for all $\gamma \in \Gamma \setminus \{1\}$ we know that $d(\omega_1, 1) < d(\omega_1, \gamma_0)$ or $d(\gamma_0 \omega_0, 1) < d(\gamma_0 \omega_0, \gamma_0)$. Hence by left invariance we have that $d(\omega_0, \gamma_0^{-1}) < d(\omega_0, 1)$ which contradicts (8.6) for $\gamma = \gamma_0^{-1}$.

As for condition (ii), first note that $\overline{\Omega}$ is the set of all $g \in G$ such that $d(g, 1) \leq d(\gamma, g)$ for all $\gamma \in \Gamma$. Let g be an arbitrary point in G then $\gamma \mapsto d(g, \gamma)$, $\gamma \in \Gamma$, has a minimum since Γ is discrete. Therefore there is $\gamma_0 \in \Gamma$ such that $d(\gamma_0, g) \leq d(\gamma, g)$ for all γ. By left invariance we get $d(1, g) \leq d(\gamma_0^{-1} \gamma, \gamma_0^{-1} g)$ for all the γ. Hence $d(1, \gamma_0^{-1} g) \leq d(\gamma, \gamma_0^{-1} g)$ for all $\gamma \in \Gamma$. Therefore $\gamma_0^{-1} g \in \overline{\Omega}$. So $g \in \Gamma \overline{\Omega}$.

We remark that similar arguments work for a discrete group acting properly discontinuously by isometries on a complete Riemannian manifold. For example if G is a semisimple linear Lie group without compact factor, $X = G/K$ where K is a maximal compact subgroup of G and Γ is a torsion-free subgroup of G, then X/Γ is a complete Riemannian manifold on which Γ operates properly discontinuously by isometries. The case of Γ acting properly discontinuously on a compact metric space is treated in Appendix C.

Chapter 9

Density Results for Cofinite Volume Subgroups

9.1 Introduction

In this chapter we will study the situation of a connected Lie group G and a closed subgroup H where G/H has finite volume (see Section 2.3). Often, but not always, H will actually be discrete. We shall study the extent to which features of G are determined by those of H. To do so we will occasionally have to use notions of algebraic groups. So for example, in some appropriate context, we might say that H is *Zariski dense* in G as defined in Section 9.3. The earliest result along these, is the well-known Borel density theorem [8], which can be stated as follows:

Theorem 9.1.1. *Let G be a connected semisimple Lie group without compact factors, H a cofinite volume subgroup and ρ a smooth finite dimensional representation of G on V. Then every subspace of V invariant under H must be invariant under G.*

Actually, in [8] H was discrete.

Exercise 9.1.2. Show that the hypothesis that G have no compact factors is necessary.

We shall provisionally take the conclusion of Theorem 9.1.1 as our principal goal in extending this result. It should be mentioned that the conclusion of Theorem 9.1.1 need not hold when G/H is merely compact, but does not have finite volume. Thus for example if $B = AN$ is a Borel subgroup of $G = KAN$, then G/B is compact. But B does not have cofinite volume (see Theorem 2.3.5 and Exercise 2.1.8). Moreover taking a nontrivial character of B gives a B-invariant line which cannot be G-invariant since $G = [G, G]$ and so G has no nontrivial characters. G/B is a typical example of a compact homogeneous space with no finite G-invariant measure.

An interesting consequence of Borel density is the theorem of Hurwitz on finiteness of automorphism group of a compact (or finite volume) Riemann surface, S. Moreover, it also shows the only cofinite volume subgroups in a non-compact simple group are lattices. The proof of these statements is as follows:

Proof. Let G be a non-compact simple group and H a cofinite volume subgroup. A direct calculation shows that H normalizes $N_G(H)$ and therefore by continuity H normalizes $N_G(H)_0$, its identity component. Taking differentials we see that $\mathrm{Ad}(H)$ acts on $\mathfrak{n}_{\mathfrak{g}}(\mathfrak{h}) \subseteq \mathfrak{g}$. By the Borel density theorem $\mathrm{Ad}\, G$ leaves this subspace invariant. Thus $\mathfrak{n}_{\mathfrak{g}}(\mathfrak{h})$ is an ideal in $\mathfrak{g}$. Since $\mathfrak{g}$ is simple this ideal is either trivial or $\mathfrak{g}$ itself. In the latter case since $\mathfrak{g}$ normalizes $\mathfrak{h}$ hence H and therefore H_0 are normal in G. Since G is simple and H_0 is connected, H_0 must be trivial and therefore H is discrete. Hence H, as a discrete normal subgroup of G, is central by Lemma 0.3.6. Since G/H is finite volume group therefore by Corollary 2.1.3 it is compact. This means $G/Z(G)$ is also compact and therefore G is of compact type, a contradiction. Thus $\mathfrak{n}_{\mathfrak{g}}(\mathfrak{h}) = \{0\}$ and $N_G(H)$ is discrete and therefore so is H. Moreover, $N_G(H)/H$ is finite (Proposition 2.4.8). When $G = \mathrm{SL}(2, \mathbb{R})$ and H is the fundamental group of the Riemann surface, S, this quotient is the automorphism group of S. □

In the next section we will prove a generalization of Theorem 9.1.1 due to one of the present authors.

9.2   A Density Theorem for Cofinite Volume Subgroups

We now turn to a series of results ([51] and [45]) which generalize the Borel density theorem. This is based on an extension of the basic method of Furstenberg [22] together with a number of additional observations.

In what follows, V will denote a vector space over k of finite dimension, n, where $k = \mathbb{R}$, or actually any subfield of $\mathbb{C}$. $\mathrm{GL}(V)$ denotes, as usual, the general linear group of V and $P(V)$ its projective space. If r is an integer between 1 and n, then $\wedge^r V$ is the r-fold exterior product and $\mathcal{G}^r(V)$ the Grassmann space of r-dimensional subspaces of V. Of course, $P(V)$ is $\mathcal{G}^1(V)$. Each $\mathcal{G}^r(V)$ is a compact manifold (see Section 0.4) and $\mathcal{G}(V)$, the Grassmann space of V, is a disjoint union of these open submanifolds. We let $\pi : V \setminus \{0\} \to P(V)$ denote the canonical map $v \mapsto \bar{v}$. If W is a subspace of V of dimension ≥ 1, then $\bar{W}$ denotes the corresponding subvariety of $P(V)$. A finite union, $\cup \bar{W}_i$, will be called a *quasi-linear variety* (qlv). Since each $\bar{W}$ is compact, a qlv is a closed subspace of $P(V)$.

We begin with some lemmas needed to prove Proposition 9.2.6 below.

Lemma 9.2.1. *If $A \subseteq P(V)$, then there exists a unique minimal qlv S containing A.*

Proof. By considering $\pi^{-1}(A)$ it is enough to show that any subset $B \subseteq V$ is contained in a unique minimal set of the form $\cup \{W_i : i = 1, \ldots, r\}$. Now, B is contained in one such set, namely V. If we show there exists a smallest such set this will also imply uniqueness. Since each W_i is a linear subspace and hence is algebraic, a finite union of such sets is also algebraic. But an infinitely descending chain of algebraic sets in V would correspond to an infinitely ascending sequence of ideals in $k[x_1, \ldots, x_n]$ which is impossible by the Hilbert basis theorem [82]. $\square$

Now, for $g \in \mathrm{GL}(V)$ define $\bar{g} : P(V) \to P(V)$ by $\bar{g}(\bar{v}) = \overline{g(v)}$. Routine calculations prove that $\bar{g}$ is well-defined, $\bar{g}\pi = \pi\bar{g}$, $\overline{(gh)} = \bar{g}\bar{h}$ and if

$\lambda \neq 0$, then $\overline{(\lambda g)} = \bar{g}$. Now, suppose one has a linear representation, or module $G \times V \to V$ of G on V. Then this induces a compatible action of G on $P(V)$, making the diagram below commutative.

$$
\begin{array}{ccc}
G \times (V \setminus \{0\}) & \longrightarrow & V \setminus \{0\} \\
\downarrow {\scriptstyle (id,\pi)} & & \downarrow {\scriptstyle \pi} \\
G \times P(V) & \longrightarrow & P(V)
\end{array}
\tag{9.1}
$$

Lemma 9.2.2. *Let A be a G-invariant subset of $P(V)$ and $\cup\{\overline{W_i} : i = 1, \dots, r\}$ be the minimal qlv containing A. Then G permutes $\{W_i : i = 1, \dots, r\}$.*

Proof. Since $\bar{g}\pi = \pi\bar{g}$ for $g \in \mathrm{GL}(V)$, we know $\pi^{-1}(A)$ is also G-invariant, $\pi^{-1}(A) \subseteq \cup\{W_i : i = 1, \dots, r\}$ and this is the minimal linear variety containing it. But then

$$
g(\pi^{-1}(A)) = \pi^{-1}(A) \subseteq g.(\cup\{W_i : i = 1, \dots, r\}) = \cup\{gW_i : i = 1, \dots, r\}.
$$

The latter is a linear variety for each $g \in G$. By minimality

$$
\cup\{W_i : i = 1, \dots, r\} \subseteq \cup\{gW_i : i = 1, \dots, r\}.
$$

But this means the two sets are equal for each $g \in G$. The spaces involved in the unique linear variety containing a set are clearly also unique and gW_i is one of them. Therefore, $gW_i = W_j$ for some j. $\square$

Lemma 9.2.3. *Let $\{g_k\}$ be a sequence in $\mathrm{GL}(V)$ and suppose $\frac{\det g_k}{\|g_k\|^n} \to 0$, where $\|\cdot\|$ is any convenient Banach algebra norm on $\mathrm{End}(V)$. Then there exists a map $\phi : P(V) \to P(V)$ such that $\phi(P(V))$ is a proper qlv of $P(V)$ and a subsequence of $\{\bar{g}_k\}$ which converges to ϕ pointwise on $P(V)$.*

Proof. Let W be a nonzero subspace of V and consider $g_k|_W : W \to V$. Denote $\frac{1}{\|g_k|_W\|}$ by $\gamma_{k,W}$. Then $\|\gamma_{k,W} g_k|_W\| = 1$ for all k. Since $\{A : A_W : W \to V, \|A\| = 1\}$ is a compact set, there is a subsequence, which we again call $\gamma_{k,W} g_k|_W$, such that $\gamma_{k,W} g_k|_W$ converges in norm and, therefore, pointwise on W to σ_W. Here σ_W is a linear map $W \to V$. Since $\|\sigma_W\| = 1$, $\sigma_W \neq 0$. Now, since π is continuous and for $w \in W$, $\gamma_{k,W} g_k|_W(w) \to \sigma_W(w)$, we have $\overline{\gamma_{k,W} g_k|_W(w)} \to \overline{\sigma_W(w)}$. But $\overline{\gamma_{k,W} g_k|_W(w)} = \bar{g}_k(\bar{w})$ so $\bar{g}_k(\bar{w}) \to \bar{\sigma}_W(\bar{w})$ pointwise on W, and in particular for w outside of $\text{Ker}\,\sigma_W$.

In particular, if $W = V$, we have $\gamma_k g_k \to \sigma_V$ and so

$$\det(\gamma_k g_k) = (\gamma_k)^n \det g_k = \frac{\det g_k}{\|g_k\|^n} \to \det \sigma_V.$$

Since this sequence tends to 0, σ_V is singular. Now, inductively define subspaces $W_0, W_1, \ldots,$ of V by $W_0 = V$, $W_{i+1} = \text{Ker}\,\sigma_{W_i}$, $i \geq 0$. Then $\text{Ker}\,\sigma_V < V$ since $\sigma_V \neq 0$. Similarly, $W_{i+1} < W_i$ since $\sigma_{W_i} \neq 0$. Thus, the sequence $V = W_0 > W_1 > \ldots$ must terminate at $\{0\}$ after a certain number of steps; $W_{i_0} = \{0\}$ for some i_0. For each i and finer and finer subsequences, which are again called g_k, we have $g_k(\bar{w}_i) \to \overline{\sigma_{W_i}(w_i)}$ pointwise for $\bar{w}_i \in \bar{W}_i$. Define $\phi : P(V) \to P(V)$ by $\phi(\bar{v}) = \overline{\sigma_{W_i}(v)}$, if $v \in W_i$, but not in W_{i+1}, $i = 0, \ldots, i_0 - 1$. If $\bar{v} = \bar{u}$, then $v = \gamma u$, $\gamma \neq 0$. If $v \in W_i - W_{i+1}$, the same is true of u, so

$$\overline{\sigma_{W_i}(v)} = \overline{\sigma_{W_i}(\gamma u)} = \overline{\gamma \sigma_{W_i}(u)} = \overline{\sigma_{W_i}(u)},$$

since σ_{W_i} is linear. Thus $\phi(\bar{v}) = \phi(\bar{u})$, ϕ is well-defined and $\bar{g}_k$ converges to ϕ pointwise on $P(V)$. Moreover,

$$\phi(P(V)) = \cup\{\overline{\sigma_{W_i}(W_i)} : i = 0, \ldots, i_0 - 1\},$$

so the range of ϕ is a qlv. Since σ_V is singular, $\sigma_V(V) < V$. For $i > 0$, $\sigma_{W_i} : W_i \to V$ so $\dim \sigma_{W_i}(W_i) \leq \dim W_i < \dim V$ and $\sigma_{W_i}(W_i) < V$ for $i \geq 0$. Now the union of a finite (or even countable) number of subspaces each of strictly lower dimension cannot equal V. To see this it is clearly sufficient to take $k = \mathbb{R}$. Now this follows from the Baire category theorem, but it is more in the spirit of our subject to argue as follows: If $V = \cup\{V_i : i \in \mathbb{Z}\}$, then take a finite positive measure μ on V which is absolutely continuous with respect to Lebesgue measure, e.g., $d\mu = \exp(-\|x\|^2)dx$. Then, by countable subadditivity, since each $\mu(V_i) = 0$ we see that $\mu(V) = 0$, a contradiction. This means that $P(V) \neq \cup\{\overline{\sigma_{W_i}(W_i)} : i = 0, \ldots, i_0 - 1\}$ and the range of ϕ is proper. $\quad\square$

Lemma 9.2.4. *Let $G \times X \to X$ be an action of a topological group, G, on a metric space X. Suppose there exists a sequence g_k and a closed subspace Y of X such that for each $x \in X$, $g_k(x)$ converges to $y(x) \in Y$ pointwise on X. Then each finite G-invariant measure μ on X has $\operatorname{Supp} \mu \subseteq Y$.*

Proof. Let $D(x) = \operatorname{dist}(x, Y)$, where dist is an equivalent *bounded* metric on X. Then D is a bounded continuous nonnegative function on X and $D(x) = 0$ if and only if $x \in Y$. Now, for all k,

$$\int_X D(g_k x) d\mu(x) = \int_X D(x) d\mu(x).$$

Since $g_k x \to y(x)$, by continuity of D we have $D(g_k x) \to D(y(x))$ pointwise on X. Because D is bounded, there is a c such that for all $k \in \mathbb{Z}$ and $x \in X$, $|D(g_k x)| \leq c$. The finiteness of μ together with the dominated convergence theorem [73] shows $\int_X D(g_k x) d\mu(x)$ tends to 0. Therefore, $\int_X D(x) d\mu(x) = 0$, so $D \equiv 0$ on $\operatorname{Supp} \mu$. Since $D = 0$ exactly on Y, $\operatorname{Supp} \mu \subseteq Y$.   □

The following definition will play an important role in what follows.

Definition 9.2.5. Let G be a topological group and $\rho : G \to \operatorname{GL}(V)$ be a continuous representation. We shall say ρ is *admissible* if there is a family $\{H_i\}$ of subgroups of G which together generate G and each restriction has the following properties, where here again everything is done with respect to some convenient Banach algebra norm on $\operatorname{End}_k(W)$.

(1) For each i, H_i has no closed subgroup of finite index.
(2) For each i and a H_i-invariant subspace W of V, either H_i acts on W by scalars, or else there is a sequence $g_k \in \rho(H_i)$ such that

$$\frac{\det(g_k|_W)}{\|(g_k|_W)\|^{\dim W}} \to 0.$$

Furthermore we shall say ρ is *strongly admissible* if each rth exterior power $\wedge^r \rho$ acting on $\wedge^r V$ is admissible for $r = 1, \ldots, n = \dim V$.

Proposition 9.2.6. *Let ρ be an admissible representation of G on V and $G \times P(V) \to P(V)$ be the associated action on projective space. Then each finite G-invariant measure μ on $P(V)$ has $\operatorname{Supp}\mu \subseteq P(V)^G$, the G-fixed points.*

Proof. First assume that G satisfies the two conditions above. That is, G is one of the H_i. If G acts on V by scalars, then $P(V) = P(V)^G$ and we are done. Otherwise, by the second condition there exists a sequence g_k in G with the property that $\frac{\det(g_k)}{\|g_k\|^n} \to 0$. By Lemma 9.2.3, there exists $\phi : P(V) \to P(V)$ such that $\phi(P(V)) = Q$ is a proper qlv of $P(V)$ and we can assume $\bar{g}_k$ converges to ϕ pointwise on $P(V)$. Since Q is closed, $\operatorname{Supp}\mu \subseteq Q$, by Lemma 9.2.4. By Lemma 9.2.1, there exists a smallest qlv, which we shall call $S = \cup\{\overline{W_i} : i = 1, \dots, m\}$, containing $\operatorname{Supp}\mu$. Thus

$$\operatorname{Supp}\mu \subseteq S \subseteq Q < P(V).$$

Since μ is G-invariant, so is $\operatorname{Supp}\mu$. By Lemma 9.2.2, G permutes W_i. But there are only a finite number of W_i so each has a stability group of finite index. Moreover, since $G \times V \to V$ is continuous and the W_i are closed, the stability groups are also closed. By the first condition G must leave each W_i stable. Let W be any one of the W_i and consider the action of G on W. The two conditions above are clearly satisfied for this action. If we let $\mu' = \mu_{\overline{W}}$, then we get a G-invariant measure on $P(W)$ and argue as before. Unless G acts on W by scalars, we know there exists a proper qlv T of $\overline{W} = P(W)$ such that $\operatorname{Supp}\mu' \subseteq T$. This contradicts the minimality of S. Otherwise G acts on W_i by scalars for each i. But then each $\overline{W_i}$ is G-fixed. This means

$$\operatorname{Supp}\mu \subseteq S \subseteq \cup\{\overline{W_i} : i = 1, \dots, m\} \subseteq P(V)^G.$$

We have just shown that for each H_i, $\operatorname{Supp}\mu \subseteq P(V)^{H_i}$. Hence $\operatorname{Supp}\mu \subseteq \cap\{P(V)^{H_i}\} = P(V)^G$. $\qquad\qquad\square$

We now pass from projective space to the Grassmann space, $\mathcal{G}^r(V)$. There is a canonical map $\phi : \mathcal{G}^r(V) \to P(\wedge^r V)$, defined as follows: For an r-dimensional subspace W of V, choose a basis $\{w_1, \dots, w_r\}$. Then $w_1 \wedge \dots \wedge w_r$ is a nonzero element of $\wedge^r V$ and so the line through it gives a point in $P(\wedge^r V)$. This is a well-defined map because if $u_1, ..., u_r$ is another basis for W then $u_1 \wedge ... \wedge u_r$ is a multiple of $w_1 \wedge ... \wedge w_r$. Moreover this map is injective. To see this suppose that for $w_1 \wedge ... \wedge w_r = \lambda u_1 \wedge ... \wedge u_r$ where $\{w_i\}$ is a basis for W and $\{u_i\}$ is a basis for U. Then consider the subspace $T_W = \{v \in V | v \wedge w_1 \wedge \cdots \wedge w_r = 0\}$. Indeed $T_W = W$ and similarly $T_U = U$ and by assumption $T_U = T_W$ so $U = W$. Since this map is clearly smooth with respect to quotient structure (see Section 0.4) we get:

Proposition 9.2.7. *The map* $\phi : \mathcal{G}^r(V) \to P(\wedge^r V)$ *is well-defined, smooth and injective.*

We now come to our first theorem.

Theorem 9.2.8. *Let ρ be a strongly admissible representation of G on V. Then under the induced action of G on the Grassmann space, $\mathcal{G}(V)$, each finite G-invariant measure μ on $\mathcal{G}(V)$, has $\operatorname{Supp} \mu \subseteq \mathcal{G}(V)^G$, the G fixed points.*

Proof. Since $\mathcal{G}(V) = \cup \mathcal{G}^r(V)$ a disjoint union of open G-invariant sets, it clearly suffices, by restricting the measure and the action to $\mathcal{G}^r(V)$, to prove the theorem for that case. Now, since, as explained above, $\mathrm{GL}(V)$ acts transitively and continuously on $\mathcal{G}^r(V)$, the latter is a quotient space $\mathrm{GL}(V)/\operatorname{Stab}_{\mathrm{GL}(V)}(W)$ where W is some fixed r-dimensional subspace of V. If $\gamma : \mathrm{GL}(V) \to \mathcal{G}^r(V)$ denotes the corresponding projection and $\{w_1, \ldots, w_r\}$ is a basis of W, then since $\{gw_1, \ldots, gw_r\}$ are linearly independent for each $g \in G$, $g \mapsto gw_1 \wedge, \ldots, \wedge gw_r$ is a map $\psi : \mathrm{GL}(V) \to \wedge^r V \setminus \{0\}$. Clearly, ϕ factors as $\phi_1 \pi$ where $\phi_1 : \mathcal{G}^r(V) \to \wedge^r V \setminus \{0\}$ and $\pi : \wedge^r V \setminus \{0\} \to P(\wedge^r V)$ is the natural map. The diagram below, including the map ϕ is commutative, since $\phi\gamma(g) = \phi(gW) = gw_1 \wedge, \ldots, \wedge gw_r = \psi(g)$.

$$
\begin{array}{ccc}
\mathrm{GL}(V) & \xrightarrow{\ \psi\ } & \wedge^r V \setminus \{0\} \\
\downarrow{\scriptstyle \gamma} & & \downarrow{\scriptstyle \pi} \\
\mathcal{G}^r(V) & \xrightarrow[\ \phi\ ]{} & P(\wedge^r V)
\end{array}
$$

To see that ϕ is continuous, note that γ is continuous, open and surjective and that ψ and π are both continuous. Then it follows from the commutativity of the diagram above that ϕ is continuous.

For each $g \in G$ we get a commutative diagram as follows:

$$
\begin{array}{ccc}
\mathcal{G}^r(V) & \xrightarrow{\ \phi\ } & P(\wedge^r V) \\
g \downarrow & & \overline{(\wedge^r g)} \downarrow \\
\mathcal{G}^r(V) & \xrightarrow[\phi]{} & P(\wedge^r V)
\end{array}
$$

where $g : \mathcal{G}^r(V) \to \mathcal{G}^r(V)$ is the induced map $\mathcal{G}^r(V)$ by $g \in \mathrm{GL}(V)$.

For, let $W = \mathrm{l.s.}\{w_i\}$ be any point of $\mathcal{G}^r(V)$ and $g \in G$. Then $\phi(g(W)) = \overline{(gw_1 \wedge, \dots, \wedge w_r)}$. While

$$
\begin{aligned}
\overline{(g \wedge \dots \wedge g)}(\phi(W)) &= \overline{(g \wedge \dots \wedge g)}\overline{(w_1 \wedge, \dots, \wedge w_r)} \\
&= \overline{[(g \wedge \dots \wedge g)(w_1 \wedge, \dots, \wedge w_r)]} \\
&= \overline{(gw_1 \wedge, \dots, \wedge gw_r)}.
\end{aligned}
$$

Because ϕ is a G-equivariant measurable function, the measure μ can be pushed forward, by Proposition 2.3.6, and be regarded as a finite G-invariant measure on $P(\wedge^r V)$, supported on the image of $\mathcal{G}^r(V)$. Since ρ is strongly admissible, $\wedge^r \rho$ is admissible. By Proposition 9.2.6 $\mathrm{Supp}\,\mu \subseteq P(\wedge^r V)^G$. Therefore, by G-equivariance and Proposition 9.2.7, $\mathrm{Supp}\,\mu \subseteq \mathcal{G}^r(V)^G$. $\qquad\square$

We now turn cofinite volume subgroups.

Theorem 9.2.9. *Let G be a locally compact group and ρ a strongly admissible representation of G on V. If H is a closed subgroup with G/H of finite volume, then each H-invariant subspace of V is G-invariant.*

Proof. If the dimension of the subspace W is r, form $\mathcal{G}^r(V)$ and consider the action $G \times \mathcal{G}^r(V) \to \mathcal{G}^r(V)$. Here W corresponds to a point $p \in \mathcal{G}^r(V)$. Because H leaves W stable, the point p is H-fixed. So $H \subseteq \mathrm{Stab}_G(p)$. Now since G/H has a finite G-invariant measure, by pushing the measure forward the same is true of $G/\operatorname{Stab}_G(p)$. This means that $p \in \operatorname{Supp}\mu$ for an appropriate G-invariant measure μ on $\mathcal{G}^r(V)$. By Theorem 9.2.8, p is fixed under G. This means W is G-stable. $\square$

As an immediate corollary we get:

Corollary 9.2.10. *Under the assumptions of Theorem 9.2.9, if ρ is irreducible, then so is ρ_H.*

We now seek conditions for a representation to be admissible, or strongly admissible so we can apply our results. To state one of these we define *minimally almost periodic groups*. This definition is due to J. von Neumann.

Definition 9.2.11. A *minimally almost periodic group* G is a locally compact group which has no nontrivial finite dimensional continuous unitary representations.

Exercise 9.2.12. In particular, minimally almost periodic groups include semisimple Lie groups having no compact factors. (In this way Theorem 9.2.14 contains a generalization of the Borel density theorem.)

Remark 9.2.13. Also a minimally almost periodic group has no closed subgroups of finite index. This is because if had, it would also have a normal subgroup of finite index and this finite quotient would have a faithful unitary representation.

Theorem 9.2.14. *Let G be a locally compact group and $\rho : G \to \operatorname{GL}(V)$ be a continuous finite dimensional linear representation. Suppose that either*

(1) G is minimally almost periodic and ρ is arbitrary or

(2) G is a complex connected Lie group and ρ is holomorphic or

(3) G is a connected Lie group with G/R having no compact factors and the radical R acts under ρ with only real eigenvalues,

then ρ is strongly admissible. In particular, if H is a closed subgroup with G/H of finite volume, then each H-invariant subspace of V is G-invariant.

Proof. Notice that in all three cases it is sufficient to show that ρ is admissible. This is because $\wedge^r \rho$ would be continuous (respectively holomorphic) if ρ were. So in cases 1 and 2, ρ would be strongly admissible. In case 3, if R acts with only real eigenvalues under ρ, then the same is true for $\wedge^r \rho$. This is because the tensor product of operators has as its spectrum the set of products of elements from the spectra of the individual operators and, hence, is real, and since the wedge product of operators is induced by their tensor product, the spectrum here is a subset of that of the tensor product and so is also real. Thus, in case 3, as well, ρ would be strongly admissible.

Proof of case (i) (Furstenberg's case): Here we need take only one H_i, namely, G itself. By Remark 9.2.13, G has no closed subgroup of finite index. Regarding the second condition, since $g \mapsto \det g|_W$ is a homomorphism into an abelian group and such groups have finite dimensional unitary characters, it is clear that this homomorphism is trivial. We are therefore looking at $\frac{1}{\|g|_W\|^{\dim W}}$, which, of course, tends to zero if we merely select g's so $\|g|_W\|$ tends to ∞. This can be done, since $G|_W$ is not a bounded group, since G is minimally almost periodic. As we shall see, case 1 is simpler than the others because there is just one H_i and the alternative that G acts on W by scalars in condition 2 does not arise.

Case(ii): Let H_i's be all the 1-parameter subgroups $\{\exp zX : z \in \mathbb{C}\}$ where $X \in \mathfrak{g}$. Since the H_i are connected, condition 1 is automatic. Let W be an $\exp zX$-invariant subspace of V. Then W is X-invariant and we may as well assume $W = V$. We show $\frac{\det g}{\|g\|^n} \to 0$ for some sequence of the g's in $\{\rho(\exp zX)\} = \{\operatorname{Exp} z\rho'(X)\}$, which we write henceforth as $\{\operatorname{Exp} z(X)\}$. But

$$\left\| \frac{g^n}{\det(g)} \right\| \leq \frac{\|g\|^n}{|\det(g)|}$$

so it suffices to show $\|\frac{g^n}{\det(g)}\| \to \infty$. Now $g = \operatorname{Exp} zX$ so

$$\frac{g^n}{\det(g)} = \frac{\operatorname{Exp} nzX}{\det(\operatorname{Exp} zX)}.$$

This is a vector valued holomorphic function of $z \in \mathbb{C}$. By the maximum principle, it tends to ∞ as $|z|$ does, or it is constant. In the latter case, $\frac{g^n}{\det(g)} = A \in \mathrm{End}_{\mathbb{C}}(V)$. Taking $g = 1$, we see that $A = I$ and $g^n = \det(g)I$. Since each (fixed) n-th power of every element on the 1-parameter group acts as a scalar and every such element has an n-th root, the entire 1-parameter group acts as scalars.

Case (iii): We let $G = RS$ be a Levi decomposition and take for the H_i the various 1-parameter groups of R together with S itself. If $H_i = S$, then we are done, by case (i). Thus we may assume we have some 1-parameter group, which we write $\mathrm{Exp}\, tX$ as above, acting with only real eigenvalues. As before, by connectedness, condition 1 is satisfied. Now

$$X = \begin{pmatrix} \lambda_1 & 0 & \cdots & 0 \\ * & \lambda_2 & \cdots & 0 \\ * & \cdots & * & \lambda_n \end{pmatrix}$$

also has only real eigenvalues. This is because if $\lambda = a + bi$ and $e^{\lambda t} = e^{at}e^{ibt}$ is real for all t, then bt is an integer multiple of π for all t. This means $b = 0$, for otherwise t would lie in a discrete set. As above, we may assume $W = V$ and that $|t| \to \infty$. We show

$$\frac{\det(\mathrm{Exp}\, tX)}{\|\mathrm{Exp}\, t(X)\|^n} \to 0.$$

Now since

$$\mathrm{Exp}\, t(X) = \begin{pmatrix} \exp t\lambda_1 & 0 & \cdots & 0 \\ * & \exp t\lambda_2 & \cdots & 0 \\ * & & \cdots & * & \exp t\lambda_n \end{pmatrix}$$

we see that

$$\|\mathrm{Exp}\, t(X)\| \geq \|\operatorname{diag}(\exp t\lambda_1, \cdots, \exp t\lambda_n)\| = \max(|\exp t\lambda_i|),$$

while $\det \mathrm{Exp}\, t(X) = e^{t\,\mathrm{tr}\, X}$.

Let λ_i be the largest eigenvalue. Then $n\lambda_i \geq \mathrm{tr}\, X$ and $\exp t(n\lambda_i - \mathrm{tr}\, X) \to \infty$ as $t \to \infty$ unless all λ_i's are equal. Similarly, if $-t \to \infty$, choose λ_i to be the smallest eigenvalue. If all λ_i are equal, then $\mathrm{Exp}\, tX$ acts as scalars. $\qquad\square$

Remark 9.2.15. Case 3 applies, in particular, to a solvable group acting with real eigenvalues and in particular to a unipotent action.

This concludes the proof of our generalization of the Borel density theorem.

9.3   Consequences and Extensions of the Density Theorem

Various other generalizations and extensions of Theorem 9.1.1 have been proved by Mostow, S.P. Wang, Rothman, Mosak and Moskowitz, Moskowitz, and Dani some of which we will describe, but without proofs. As Theorem 9.2.14 shows these all also remove the hypothesis that G is semisimple without compact factors and consider more general groups. We will also derive a number of consequences of Theorem 9.2.14 (mostly with proofs).

We now define the algebraic hull, $G^{\#}$, of a linear group $G \subseteq \mathrm{GL}(V)$. This is the smallest algebraic subgroup of $\mathrm{GL}(V)$ containing G. Equivalently its closure of G in $\mathrm{GL}(V)$ with respect to the Zariski topology. All this with respect to the field k of definition.

The density Theorem 9.3.3 [45] below extended and unified the results we have gotten so far. To state this version of the result we need the following two definitions. If $G \subseteq \mathrm{GL}(V)$ is a linear Lie group, a representation ρ of G on W is called k-rational, if $G^{\#}$ is an algebraic k-group and ρ is the restriction of a k-rational morphism $G^{\#} \to \mathrm{GL}(W_{\mathbb{C}})$. In [45] such a linear group G is called k-minimally almost periodic if, for each k-rational representation, ρ of G, if $\rho(G)$ is bounded, then it must be trivial. In effect, this is what is being verified in the three cases of Theorem 9.3.3.

Corollary 9.3.1. *Let G be a connected subgroup of $\mathrm{GL}(V)$ which is either*

(1) minimally almost periodic, or

(2) complex connected Lie group, or

(3) a group with G/R having no compact factors and R acts on V with real eigenvalues.

Then

(1) If ρ is a k-rational representation of its algebraic hull $G^{\#}$, then $\rho|_G$ is strongly admissible.

(2) If G/H has finite volume, then $\mathrm{l.s.}_k\, G = \mathrm{l.s.}_k\, H$ and $Z_{\mathrm{End}\,V}(H) = Z_{\mathrm{End}\,V}(G)$.

(3) Any connected subgroup of G normalized by H is normal in G.

Proof. The first statement is clear from Theorem 9.2.14. As for the second, let $W = \mathrm{l.s.}_k\, H$. We first show that $G \subseteq W$. Then $\mathrm{l.s.}_k\, G \subseteq W$. Consider the k-rational representation ρ of $G^{\#}$ on $\mathrm{End}\,V$ given by $(g,T) \mapsto gT$. Since $\rho_h(\sum_i c_i h_i) = \sum c_i hh_i$, we see that W is H-invariant. By the first statement together with Theorem 9.2.14, W is G-invariant. Since $I \in W$, we see that $G \subseteq W$ and this proves the second statement. If $T \in \mathrm{End}\,V$ and $Th = hT$ for all h, then T commutes with any linear combination of h's and hence with any $g \in G$.

Finally, if L is a connected subgroup of G normalized by H and $\mathfrak{l}$ is its Lie algebra, let ρ be the adjoint representation of $G^{\#}$ on its Lie algebra. Then $\rho_G = \mathrm{Ad}\,G$ is strongly admissible and, since L is normalized by H, $\mathrm{Ad}_G(H)$ leaves $\mathfrak{l}$ stable. By Theorem 9.2.14, $\mathfrak{l}$ is also $\mathrm{Ad}\,G$ stable, so L is normal in G. $\qquad\qquad\square$

In fact, we can extend Corollary 9.3.1 to nonlinear groups. To do so requires the observation that if G acts on V with real eigenvalues, then $\mathrm{Ad}\,G$ must act on $\mathfrak{g}$ also with real eigenvalues as its eigenvalues are the exponentials of the eigenvalues of $\mathfrak{g}$ which are real by Exercise 0.5.13. We recall that the radical $\mathrm{Rad}(G)$ of a Lie group G largest connected normal solvable subgroup of G. It is the connected Lie subgroup whose Lie algebra is $\mathfrak{r}$ the radical of $\mathfrak{g}$.

Corollary 9.3.2. *Let G be a connected group and H be a closed subgroup with G/H of finite volume. Suppose that either*

(1) G is minimally almost periodic or,

(2) G is a complex connected group or,

(3) G/R has no compact factors and $\mathrm{Ad}_G(\mathrm{Rad}(G))$ *acts on* $\mathfrak{g}$ *with real eigenvalues.*

Then any connected subgroup L of G normalized by H is normal in G. In particular, if A is a closed subgroup of G containing H, then A is normal. Also, if $G/N_G(L)$ *has finite volume where L is a connected subgroup of G, then L is normal.*

Proof. L is normalized by H if and only if $\mathfrak{l}$, its lie algebra, is $\mathrm{Ad}_G(H)$-stable, that is, if and only if $\mathfrak{l}$ is stable under $\overline{\mathrm{Ad}_G(H)}$, the Euclidean closure in $\mathrm{GL}(\mathfrak{g})$. Since $\mathrm{Ad}\, G/\overline{\mathrm{Ad}(H)}$ has finite volume by pushing the measure forward (Proposition 2.3.6), the result follows from Corollary 9.3.2. $\qquad\square$

We now turn to the density theorem in the context of algebraic groups.

Theorem 9.3.3. *Let G be a Lie subgroup of* $\mathrm{GL}(V)$ *and H be a closed subgroup with G/H of finite volume. Suppose that either*

(1) G is minimally almost periodic or,

(2) G is a complex connected Lie group or,

(3) G is a real connected Lie group with G/R having no compact factors and R acts on V with real eigenvalues.

Then $H^{\#} = G^{\#}$.

Proof. Since $H^{\#}$ is an algebraic group defined over $\mathbb{C}$, we know by a theorem of Chevalley (see [8]) that there is a $\mathbb{C}$-space $W_{\mathbb{C}}$, a line $l_{\mathbb{C}}$ defined over $\mathbb{C}$ in it and a $\mathbb{C}$-morphism $\rho : G^{\#} \to \mathrm{GL}(W_{\mathbb{C}})$ such that $H^{\#} = \{g \in G^{\#} : \rho(g)l_{\mathbb{C}} = l_{\mathbb{C}}\}$. Then ρ_G is a rational morphism of G on $W = W_{\mathbb{C}}$ and the line $l_{\mathbb{C}}$ is $\rho(H)$-stable. By Theorem 9.2.9, it is also $\rho(G)$-stable so $G \subseteq H^{\#}$ and hence $G^{\#} = H^{\#}$. $\qquad\square$

Example 9.3.4. To appreciate the significance of a subgroup being merely Zariski dense we now give an example of a simply connected abelian, in fact diagonal subgroup G of $\mathrm{GL}(2, \mathbb{R})$ and a connected Lie subgroup which is therefore closed subgroup H of G. Here H is Zariski dense, but not of cofinite volume. Let $G = \{\operatorname{diag}(\lambda, \mu) : \lambda, \mu > 0\}$ and H be the 1-parameter subgroup $\{\operatorname{diag}(e^t, e^{\alpha t}) : t \in \mathbb{R}\}$ where α is an irrational number. Then G is the Euclidean identity component of the real points of an algebraic group defined over $\mathbb{Q}$. Let

$$p(\lambda, \mu) = \sum_{i,j} a_{i,j} \lambda^i \mu^j$$

be one of the polynomials defining $H^{\#}$. Since $\lambda^{\alpha} = \mu$ on H, $\sum_{i,j} a_{i,j} \lambda^{i+\alpha j} \equiv 0$. Now, the exponents $i + \alpha j$ are all distinct because α is irrational. If $\sum_k^m \beta_k \lambda^{\alpha_k} \equiv 0$ for all $\lambda > 0$ where α_k and β_k are real and $\alpha_1 < \ldots < \alpha_m$, then all β_k must be 0. For the latter equals

$$\lambda^{\alpha_1}(\beta_1 + \beta_2 \lambda^{\alpha_2 - \alpha_1} + \ldots + \beta_m \lambda^{\alpha_m - \alpha_1}).$$

Since the first factor is positive, the second must be identically 0. Letting $\lambda \to 0$, we see that $\beta_1 = 0$ and then reason by induction on m. We conclude that $a_{i,j} = 0$ for all i, j. Since $p = 0$ and p was arbitrary, $H^{\#} = G^{\#}$. On the other hand, H is a Lie subgroup of lower dimension then that of G and so is proper, and since G is simply connected and solvable, G/H is non-compact and has no finite invariant measure.

Definition 9.3.5. We shall say a subgroup H of a Lie group G is *analytically dense* in G if the only connected Lie subgroup of G containing H is G itself.

Theorem 9.3.6. *([45]) Let G be a connected linear Lie group whose radical is simply connected and whose Levi factor has no compact part. If $\operatorname{Rad}(G)$ acts with real eigenvalues, then any closed subgroup H with G/H of finite volume is analytically dense.*

Closely related to the previous theorem is the following:

Theorem 9.3.7. *([45]) Let G be a non-compact exponential Lie group (such as the adjoint group of a classical real rank 1 simple group, or an almost direct product of such things) with Lie algebra $\mathfrak{g}$ and suppose G/H has finite volume. Then every $X \in \mathfrak{g}$ is a finite linear combination of elements of $\mathfrak{g}$ that exponentiate into H.*

We shall say that a representation ρ of a group G on a vector space V is *completely reducible* if every G-invariant subspace of V has a complementary G-invariant subspace.

Theorem 9.3.8. *([54]) Let G be a locally compact group, H be a closed subgroup such that G/H is compact, or has finite volume and ρ be a continuous finite dimensional real or complex representation of G on V. If its restriction to H is completely reducible, then ρ itself is completely reducible.*

Theorem 9.3.8 can even be extended to infinite dimensional representations on a Hilbert space, but for that one needs G/H to be both compact and of finite volume.

We shall now make the following provisional definition: A subgroup A of an algebraic $\mathbb{Q}$-group G is called *arithmetic* if it is commensurable with $G_{\mathbb{Z}}$.

Theorem 9.3.8 above can be freed of these assumptions entirely if the group is an algebraic $\mathbb{Q}$-group, the representation is rational and the subgroup is arithmetic.

Theorem 9.3.9. *([54]) Let ρ be a rational representation of a linear algebraic group G defined over $\mathbb{Q}$ and A be an arithmetic subgroup of G. Then ρ is completely reducible if and only if its restriction to A is completely reducible.*

By combining Theorem 9.3.3 with a result of S. Rothman [72], we get a generalization of the density theorem of Mostow [60], where it was assumed that $G = [G, G]$, $G/\operatorname{Rad}(G)$ has no compact factors and $\operatorname{Rad}(G)$ is *abelian*.

Corollary 9.3.10. *Let G be a connected Lie subgroup of $\operatorname{GL}(n, \mathbb{R})$ and H be a closed subgroup of G with G/H of finite volume. If $G/\operatorname{Rad}(G)$ has no compact factors and $G = \overline{[G, G]}$, then $G^\# = H^\#$.*

This is because these conditions on G characterize connected Lie groups which are maps (see [72]).

We can formulate a more general form of our density theorem as Theorem 9.3.11 below. The reader who is not comfortable with this can usually just work with Theorem 9.3.3 instead.

Theorem 9.3.11. *Let G be a connected Lie subgroup of $\mathrm{GL}(V)$ which is k-minimally almost periodic and H be a closed subgroup of G of cofinite volume. Then $H^\# = G^\#$.*

We conclude with three further applications of the density theorem. This result comes from [45] and can be used to prove that under these hypotheses, for a lattice Γ in G, the orbit $\mathrm{Aut}(G) \circ \Gamma$ is locally compact in the Chabauty topology. For details see [45].

Proposition 9.3.12. *Let G be a solvable connected Lie subgroup of $\mathrm{GL}(n, \mathbb{R})$ having only real eigenvalues, H be a closed uniform subgroup of G and $\rho : G \to \mathrm{GL}(W)$ be an $\mathbb{R}$-rational representation. Then the cohomology restriction maps $H^p(G, W) \to H^p(H, W)$ are isomorphisms for all $p \geq 0$.*

Proof. Let K be a maximal compact subgroup of G. Since K is connected, it is contained in $\mathrm{SO}(n, \mathbb{R})$ (in appropriate coordinates) and each element of K lies on a 1-parameter group of K. It follows that each element of K can be put into block diagonal form with rotations (or I) in the blocks. Since the eigenvalues of the elements of K are real, $K = \{1\}$. This means that G is simply connected. The result will follow from [59], Theorem 8.1, if we can show that H is ρ-ample in G, that is, $(\rho \circ \mathrm{Ad}\, G)(H)$ is Zariski dense in $(\rho \circ \mathrm{Ad}\, G)(G)$. Since H is a uniform subgroup of G, it follows that G/H carries a finite invariant measure and so this follows from Theorem 9.3.3.   □

Our second result deals with non-amenability of lattices. It relies on the well-known Tits alternative for linear groups.

Definition 9.3.13. For our purposes we shall say an abstract group is amenable if it does not contain a non-abelian free group.

Corollary 9.3.14. *A lattice Γ in a connected minimally almost periodic Lie group G is never amenable.*

Proof. Let G be such a group. Then $B(G) = Z(G)$ by [72]. Hence (see [29]) $\mathrm{Ad}(\Gamma)$ is a lattice in $\mathrm{Ad}\, G$. As $\mathrm{Ad}(\Gamma)$ is a linear group the Tits alternative [77] tells us that either it contains a free group on 2 generators, or it has a solvable subgroup H of finite index. In the latter case by transitivity H is also a lattice in $\operatorname{Ad} G$. Since $\operatorname{Ad} G$ is also a map we see by Theorem 9.3.3 that $H^\# = \operatorname{Ad} G^\#$. But $H^\#$ is solvable since H is. It follows that $\operatorname{Ad} G$ is itself solvable and hence so is G. Hence $G > [G, G]^-$. This is impossible since G is a minimally almost periodic group. Thus $\operatorname{Ad}(\Gamma)$ contains a free group and so is not amenable and so neither is Γ. $\square$

Our final application of the density theorem requires the Borel Harish-Chandra theorem, Theorem 8.2.8. Corollary 9.3.15 applies when G is the complexification of a non-compact simple group such as $\operatorname{SL}(n, \mathbb{C})$, or $\operatorname{Sp}(n, \mathbb{C})$, but not when it is the complexification of a compact simple group such as $\operatorname{SO}(n, \mathbb{C})$. For a proof we refer the reader to [54].

Corollary 9.3.15. *([54]) Let G be a Zariski connected linear algebraic group defined over $\mathbb{Q}$. Then $G_\mathbb{Z}$ is Zariski dense in G if and only if $X_\mathbb{Q}(G)$ is trivial and $G_\mathbb{R}$ is $\mathbb{Q}$-minimally almost periodic.*

Appendix A

Vector Fields

Here we recall some basic notions in differential topology, a full account of the subject can be found in [30]. We begin with the definition of $T_p(M)$, the tangent space of a smooth manifold M at a point p. For an open $U \subset \mathbb{R}^n$ the tangent space at $x \in U$ is defined to be $T_x U = \{x\} \times \mathbb{R}^n$. Let $\{(U_i, \phi_i)\}_i$ be an atlas for M where each $\phi_i : U \to \phi(U) \subset M$ is a homeomorphism. Then $T_p M$ is defined to be the set of equivalence classes $[p, v, i]$ where $v \in T_{\phi_i^{-1}(p)} U_i$ and $[p, v, i] = [p, v', j]$ if $v_i = d_{\phi_j^{-1}(p)}(\phi_i^{-1} \circ \phi_j)(v_j)$. Then $T_p M$ can be made into a linear vector space by defining

(1) $[v, i] + [u, i] = [v + u, i]$,

(2) $k[v, i] = [kv, i]$ for $k \in \mathbb{R}$.

The tangent space TM is the union $\bigcup_{x \in M} T_x M$ and can be made into a manifold. One can give an atlas for TM by declaring $(TU_i, d\phi_i)$ a chart where

$$d\phi_i(x, v) = [\phi(x), v, i]$$

and the change of coordinates are $(\phi_i^{-1} \circ \phi_j, d(\phi_i^{-1} \circ \phi_j))$. A *vector field* is a smooth map $X : M \to TM$ such that $X(p) \in T_p M$ or, as is customary, we say that X is a smooth section of the vector bundle $\pi : TM \to M$ where $\pi([p, v, i]) = p$. Notice that the space of vector fields $\chi(M)$ on M is a module over $C^\infty(M)$ where the scalar product is defined by pointwise multiplication *i.e.*

$$(f \cdot X)(p) = f(p)X(p),$$

and using (2) above. One can consider the derivative of a smooth map $f : M \to N$ at a point p, which is a linear map

$$d_p f : T_p M \to T_{f(p)} N$$

defined using charts (U_i, ϕ_i) and (V_j, ψ_j) for a neighborhood of p and $f(p)$ respectively,

$$d_p f([p, v, i]) = [f(p), d_{\phi^{-1}(p)}(\psi_j^{-1} \circ f \circ \phi_j)(v), j].$$

The vector fields on M act on $C^\infty(M)$ as first-order differential operators by

$$(Xf)(p) = d_p f(X(p))$$

for $f \in C^\infty(M)$ and $p \in M$. It is a direct check that

(1) $X(fg) = fXg + gXf$ for all $f, g \in C^\infty(M)$,

(2) $X(f + \lambda g) = Xf + \lambda Xg$ for all $\lambda \in \mathbb{R}$,

which says that a vector field on M defines a first-order differential operator on $C^\infty(M)$. In fact one can prove that any first-order differential operator is given by a vector field. Given an atlas (U_i, ϕ_i) on M, then a vector field X on $\phi_i(U_i)$ has the local expression $X(p) = \sum_{i=1}^n \zeta_i(p) \partial/\partial x_i$ where $\{\partial/\partial x_i(p)\}_{i=1}^n$ is thought of as a basis for the tangent space $T_p M$ induced by the trivialization $TU_i = U_i \times \mathbb{R}^n$, and ζ_i are smooth functions defined on $\phi_i(U_i)$. We have

$$(Xf)(p) = \sum_{i=1}^n \zeta_i(p) \frac{\partial f}{\partial x_i}(p),$$

which gives a local expression for the operator defined by X. One can compose two such operators (vector fields) but of course the result is not a first-order differential operator. We now define the bracket of two vector fields X and Y as

$$[X,Y] = XY - YX,$$

the commutator of differential operators X, Y. Since $[\cdot,\cdot]$ is clearly skew symmetric and the Jacobi identity, $[[X,Y],Z]+[[Y,Z],X]+[[Z,X],Y] = 0$ is a formal verification, we see that the set of all vector fields is an infinite dimensional real Lie algebra if the bracket of two vector fields is also a vector field. The miracle is:

Proposition A.1. $[X,Y]$ *is a vector field.*

For purposes of comparison we give two proofs for this, one of which is classical and one modern in spirit. Typically, the classical one is more elaborate. It is full of sturm und drang. But, in recompense, it gives more insight. The modern one is quick and machine-like and has little insight. It is merely the verification of a previously established criterion.

Proof. Modern Proof. We use the isomorphism established between tangent vectors and vector fields. Clearly, $[X,Y]$ is a linear operator on functions. We therefore need only verify $[X,Y](fg) = f[X,Y](g) + [X,Y](f)g$. Indeed,

$$\begin{aligned}
[X,Y](fg) &= (XY - YX)(fg) = X(Y(fg)) - Y(X(fg)) \\
&= X(Y(f)g) + X(fY(g)) - Y(X(f)g) - Y(fX(g)) \\
&= XY(f)g + Yf(Xg) + fY(Xg) + X(fY)g \\
&\quad - YX(f)g - Xf(Yg) - fX(Yg) - Y(fX)g \\
&= XY(f)g + X(fY)g - YX(f)g - Y(fX)g \\
&= [X,Y](f)g + f[X,Y](g).
\end{aligned}$$

Classical Proof. To see this we need only see that in local coordinates so defined $[X,Y]$ is a smooth *first* order differential operator. Let f be a smooth function on M and U a neighborhood of p, $X(p) = \sum_i \eta_i(p)\partial/\partial x_i$ and $Y(p) = \sum_i \zeta_i(p)\partial/\partial x_i$. Then

$$([X,Y]f)(p) = (\sum_i \eta_i \frac{\partial}{\partial x_i} \sum_j \zeta_j \frac{\partial f}{\partial x_j} - \sum_j \zeta_j \frac{\partial}{\partial x_j} \sum_i \eta_i \frac{\partial f}{\partial x_i})(p)$$

$$= \sum_{i,j} \eta_i(p) \frac{\partial \zeta_j}{\partial x_i}(p) \frac{\partial f}{\partial x_j}(p) - \sum_{i,j} \zeta_j(p) \frac{\partial \eta_i}{\partial x_j}(p) \frac{\partial f}{\partial x_i}(p)$$

$$+ \sum_{i,j} \eta_i(p) \zeta_j(p) \frac{\partial^2 f}{\partial x_i \partial x_j}(p) - \sum_{i,j} \zeta_i(p) \eta_j(p) \frac{\partial^2 f}{\partial x_j \partial x_i}(p).$$

Since f is smooth, the mixed second partials are equal and so the second order terms cancel leaving a first-order operator,

$$\sum_{i,j} [\eta_i(p) \frac{\zeta_j}{\partial x_i}(p) \frac{\partial f}{\partial x_j}(p) - \zeta_j(p) \frac{\eta_i}{\partial x_j}(p) \frac{\partial f}{\partial x_i}(p)].$$

$\square$

A curve in M is a smooth map $x : \mathbb{R} \to M$, the tangent vector at every point on the curve is the vector

$$x'(t) = d_t x(1)$$

where 1 is thought of as a generator for $T_t \mathbb{R} \simeq \mathbb{R}$. Given a vector field and a point p, if we can find a smooth curve through p whose tangent vector at every point coincides with the vector field, we call the curve an *integral curve*. This amounts to solving a differential equation with an initial condition. If we can only find a local curve then we have a local solution to our differential equation with initial condition.

We now give the form of the fundamental theorem of ordinary differential equations which will be of use to us. A proof of this can be found in [37] or [69].

Theorem A.2. *Let $U \subseteq M$ and $V \subseteq \mathbb{R}^m$ be neighborhoods of 0 and y_0 respectively and $v(x,y)$ be a vector field in M which depends smoothly on (x,y). For each fixed $y \in V$ consider the initial value problem $f'(t) = v(f(t), y)$, $f(0) = 0$, where $f : \mathbb{R} \to M$. Then there is an $\epsilon > 0$ and a* *neighborhood V' of y such that there is a unique solution for $t \in (-\epsilon, \epsilon)$ and $y' \in V'$ to the initial value problem. It depends smoothly on t and $y \in V'$.*

Corollary A.3. *Let $v(x)$ be a smooth vector field defined in a neighborhood U of 0 in M. Consider the initial value problem $f'(t) = v(f(t))$, $f(0) = 0$. Then there is an $\epsilon > 0$ and a unique smooth solution for $t \in (-\epsilon, \epsilon)$ to the initial value problem.*

We recall that a 1-parameter group of diffeomorphisms is a map, $\phi : \mathbb{R} \times M \to M$, where we write $\phi_t(p)$ instead of $\phi(t, p)$, such that ϕ_t is a diffeomorphism for each $t \in \mathbb{R}$ and $t \mapsto \phi_t$ is a homomorphism from $\mathbb{R} \to \text{Diff}(M)$ and $\phi_0 = I$. A similar definition holds for local 1-parameter groups of diffeomorphisms. Namely, there is an interval I about 0 in $\mathbb{R}$ such that for all $p \in M$, $\phi_t(\phi_s(p)) = \phi_{t+s}(p)$ whenever s, t and $s + t \in I$. Now a local 1-parameter group of diffeomorphisms gives rise to a vector field on M as follows. For each point $p_0 \in M$ consider the smooth curve $\phi_t(p_0)$ through p_0. Taking its tangent vector at each point gives a vector field on M. Conversely, given a vector field on M and a point p_0, there is always a local 1-parameter group of local diffeomorphisms ϕ_t which is the integral curve to this vector field and for any smooth function f, $\lim_{t \to 0}(f \circ \phi_t - f) = Xf$.

Proof. Let $U, x_1, \ldots, x_n$ be local coordinates around p_0 and assume for simplicity that for $i = 1, \ldots, n$, $x_i(p_0) = 0$. Let $X = \sum_i \eta_i(x_1, \ldots, x_n) \frac{\partial}{\partial x_i}$ in U. Consider the following system of ODE, where $i = 1, \ldots, n$ and $f^1(t), \ldots, f^n(t)$ are the unknown functions,

$$\frac{df^i}{dt} = \eta_i(f^1(t), \ldots, f^n(t)).$$

By the fundamental theorem of ODE, there exists a unique set of functions $f^1(t, x_1, \ldots, x_n), \ldots, f^n(t, x_1, \ldots, x_n)$, defined for $|(x_1, \ldots, x_n)| < \delta$ and $|t| < \epsilon$ such that for all i, $f^i(0, x_1, \ldots, x_n) = x_i$. Let $x = (x_1, \ldots, x_n)$ and $\phi_t(x) = (f^1(t, x), \ldots, f^n(t, x))$. Clearly, $\phi_0 = I$ on this neighborhood. If $|x| < \delta$ and $|t|$, $|s|$ and $|t + s|$ are all less than ϵ, then x and $\phi_s(x)$, (where s is considered fixed), are both in this neighborhood. Hence the n-tuple of functions, $g^i(t,x) = f^i(t+s,x)$ are also in the neighborhood and satisfy the same ODE, but with initial conditions, $g^i(0,x) = f^i(s,x)$. By the uniqueness it follows that $g^i(t) = f^i(t, \phi_s(x))$. Hence $\phi_t \phi_s = \phi_{t+s}$ on this neighborhood. Thus we have a local 1-parameter group of local diffeomorphisms which is the integral curve to our original vector field. $\qquad\square$

Let ϕ be a diffeomorphism of M and $d\phi$ its differential. For a vector field X on M, $\phi_* X$ will denote the vector field induced by the action of $\mathrm{Diff}(M)$ on $\mathcal{X}(M)$ mentioned above. If the 1-parameter group generated by X is ϕ_t, then the smooth vector field $\phi_* X$ also generates a 1-parameter group. It is

$$\phi \circ \phi_t \circ \phi^{-1}.$$

Proof. Now $\phi \circ \phi_t \circ (\phi)^{-1}$ is clearly a 1-parameter group of diffeomorphisms, so let Y be its vector field. We must show $Y = \phi_* X$. Let $p \in M$ and $q = \phi^{-1}(p)$. Since ϕ_t induces X, the vector $X_q \in T_q$ is tangent to the curve $\phi_t(q)$ at $t = 0$. Therefore $(\phi_* X)_p = \phi_*(X_q) \in T_p$ is tangent to $\phi \circ \phi_t(q) = \phi \circ \phi_t \circ \phi^{-1}(p)$. $\qquad\square$

Corollary A.4. *Let ϕ be a diffeomorphism of M. A vector field is ϕ fixed (i.e. $\phi_* X = X$) if and only if ϕ commutes with all ϕ_t in $\mathrm{Diff}(M)$ as t varies.*

Appendix B

The Kronecker Approximation Theorem

Let n be a fixed integer $n \geq 1$ and consider n-tuples $\{\alpha_1, \ldots, \alpha_n\}$, where $\alpha_i \in \mathbb{R}$. We shall say that $\{\alpha_1, \ldots, \alpha_n\}$ is *generic* if whenever $\sum_{i=1}^{n} k_i \alpha_i \in \mathbb{Z}$ for $k_i \in \mathbb{Z}$, then all $k_i = 0$.

Here is an example of a generic set. Let θ be a transcendental real number and consider the powers, $\alpha_i = \theta^i$. Then for *any* positive integer n, $\{\theta^1, \ldots, \theta^n\}$ is generic. For if $k_1 \theta^1 + \ldots k_n \theta^n = k$, where the k_i and k are integers then since $\mathbb{Z} \subseteq \mathbb{Q}$, this is polynomial relation of degree between 1 and n which θ satisfies. This is a contradiction.

Proposition B.1. *The set $\{\alpha_1, \ldots, \alpha_n\}$ is generic if and only if $\{1, \alpha_1, \ldots, \alpha_n\}$ is linearly independent over $\mathbb{Q}$.*

Proof. Suppose $\{1, \alpha_1, \ldots, \alpha_n\}$ is linearly independent over $\mathbb{Q}$. Let $\sum_{i=1}^{n} k_i \alpha_i = k$, where $k \in \mathbb{Z}$. We may assume $k \neq 0$. For if $k = 0$ then since the subset $\{\alpha_1, \ldots, \alpha_n\}$ is linearly independent over $\mathbb{Q}$ and $k_i \in \mathbb{Q}$ for each i we get $k_i = 0$. On the other hand if $k \neq 0$ we divide and get $\sum_{i=1}^{n} \frac{k_i}{k} \alpha_i = 1$. So 1 is a $\mathbb{Q}$-linear combination of α_i's. This contradicts our hypothesis regarding linear independence.

Conversely suppose $\{\alpha_1, \ldots, \alpha_n\}$ is generic and $q + q_1 \alpha_1 + \ldots + q_n \alpha_n = 0$, where q and all the $q_i \in \mathbb{Q}$. If $q = 0$, then clearing denominators gives a relation $k_1 \alpha_1 + \ldots + k_n \alpha_n = 0$, where $k_i \in \mathbb{Z}$. Since the

403

α_i are generic and $0 \in \mathbb{Z}$ we get each $k_i = 0$. Hence also each $q_i = 0$. Thus $\{1, \alpha_1, \dots, \alpha_n\}$ is linearly independent over $\mathbb{Q}$. On the other hand if $q \neq 0$, by dividing by q we get, $1 + s_1\alpha_1 + \dots + s_n\alpha_n = 0$, where $s_i \in \mathbb{Q}$. Again clear denominators and get $k + k_1\alpha_1 + \dots + k_n\alpha_n = 0$, where k and $k_i \in \mathbb{Z}$. Since $k_1\alpha_1 + \dots + k_n\alpha_n = -k$ and the original α_i is generic, each $k_i = 0$. Therefore each s_i is also 0 and thus $1 = 0$, a contradiction.   □

Here is another way to "find" generic sets. We consider $\mathbb{R}$ to be a vector space over $\mathbb{Q}$. Let B be a basis for this vector space. Then any finite subset of this basis gives a generic set after removing 1.

Before proving *Kronecker's approximation theorem* we define the *character group* $\widehat{G}$ of a locally compact abelian group G. Here

$$\widehat{G} = \mathrm{Hom}(G, \mathbb{T})$$

consists of continous homomorphisms and is equipped with the compact-open topology and pointwise multiplication. $\widehat{G}$ is a locally compact abelian topological group.

Proposition B.2. *Let G and H be locally compact abelian groups (written additively) and $\beta : G \times H \to \mathbb{T}$ be a nondegenerate, jointly continuous bilinear function. Consider the induced map $\omega_G : G \to \widehat{H}$ given by $\omega_G(g)(h) = \beta(g, h)$. Then ω_G is a continuous injective homomorphism with dense range. Similarly, $\omega_H : H \to \widehat{G}$ given by $\omega_H(h)(g) = \beta(g, h)$ is also a continuous injective homomorphism with dense range.*

Proof. By symmetry we need only consider the case of ω_G. Clearly $\omega_G : G \to \widehat{H}$ is a continuous homomorphism. If $\omega_G(g) = 0$ then for all $h \in H$, $\beta(g, h) = 0$. Hence $g = 0$ so ω_G is injective. To prove that $\omega_G(G)$ is a dense subgroup of $\widehat{H}$ we show that its annihilator in $\widehat{\widehat{H}}$ is trivial. Identifying H with its second dual $\widehat{\widehat{H}}$, its annihilator consists of all $h \in H$ so that $\beta(g, h) = 0$ for all $g \in G$. By nondegeneracy (this time on the other side) the annihilator of $\omega_G(G)$ is trivial. Hence $\omega_G(G)$ is dense in $\widehat{H}$ (see [70]).   □

We now come to the Kronecker theorem itself. What it says is that one can simultaneously approximate $(x_1, \ldots, x_n) \bmod(1)$ by $k(\alpha_1, \ldots, \alpha_n)$. If we denote by $\pi : \mathbb{R} \to \mathbb{T}$ the canonical projection with $\operatorname{Ker} \pi = \mathbb{Z}$, the Kronecker theorem says that any point, $(\pi(x_1), \ldots, \pi(x_n))$ on the n-torus, $\mathbb{T}^n$, can be approximated to any required degree of accuracy by integer multiples of $(\pi(\alpha_1), \ldots, \pi(\alpha_n))$.

Of course a fortiori any point on the torus can be approximated to any degree of accuracy by real multiples of $(\pi(\alpha_1), \ldots, \pi(\alpha_n))$. The image under π of such a line (namely the real multiples of $(\alpha_1, \ldots, \alpha_n)$) is called the winding line on the torus. So winding lines and generic sets always exist.

Theorem B.3. *Let $\{\alpha_1, \ldots, \alpha_n\}$ be a generic set, $\{x_1, \ldots, x_n\} \in \mathbb{R}$ and $\epsilon > 0$. Then there exists a $k \in \mathbb{Z}$ and $k_i \in \mathbb{Z}$ such that $|k\alpha_i - x_i - k_i| < \epsilon$.*

Proof. Consider the bilinear form $\beta : \mathbb{Z} \times \mathbb{Z}^n \to \mathbb{T}$ given by $\beta(k, (k_1, \ldots k_n)) = \pi(k \sum_{i=1}^n k_i \alpha_i)$. Then β is additive in each variable separately and of course is jointly continuous since here the groups are discrete. The statement is equivalent to saying that image of the map $\omega_G : \mathbb{Z} \to \widehat{\mathbb{Z}^n} \simeq \mathbb{T}^n$ is dense.

We prove that β is nondegenerate. That is if $\beta(k, (k_1, \ldots, k_n)) = 0$ for all k then $(k_1, \ldots, k_n) = 0$ and if $\beta(k, (k_1, \ldots, k_n)) = 0$ for all $(k_1, \ldots, k_n)$ then $k = 0$.

If $\beta(k, (k_1, \ldots, k_n)) = 0$ for all k, then $(k_1, \ldots, k_n) = 0$. The hypothesis here means just that $\pi(k \sum_{i=1}^n k_i \alpha_i) = 0$, or $k \sum_{i=1}^n k_i \alpha_i$ is an integer. Choose any $k \neq 0$. Then $\sum_{i=1}^n k k_i \alpha_i$ is an integer, because of our hypothesis regarding the α's we conclude all $k k_i = 0$ therefore $k_i = 0$. On the other hand, suppose $\beta(k, (k_1, \ldots, k_n)) = 0$ for all $(k_1, \ldots, k_n)$, then we show $k = 0$. Hence we have $k \sum_{i=1}^n k_i \alpha_i$ is an integer for all choices of $(k_1, \ldots, k_n)$. Arguing as before suppose $k \neq 0$. Choose k_i not all zero. This gives $k k_0 = 0$ as the α_i is a generic set therefore $k = 0$.

Hence by Proposition B.2 we get an injective homomorphism $\omega_G : \mathbb{Z} \to \mathbb{Z}^n = \mathbb{T}^n$ with dense range. Thus the cyclic subgroup $\omega(\mathbb{Z})$ in dense in $\mathbb{T}^n$. $\square$

Exercise B.4.   (1) Show that in $\mathbb{R}^2$ a line is winding if and only if it has irrational slope.

(2) Find the generic sets when $n = 1$. What does this say about dense subgroups of $\mathbb{T}$?

Appendix C

Properly Discontinuous Actions

Let $\Gamma \times X \to X$ be a (continuous) group action of a locally compact group Γ on a locally compact space X. We shall say the action is *properly discontinuous* if given a compact set C of X there is a finite subset F_C of Γ so that $C \cap (\bigcup_{\gamma \in C \backslash F_C} \gamma C)$ is empty. In particular, for each point $x \in X$, the orbit, Γx, has no accumulation point. In particular, Γ must be discrete. Also clearly the isotopy group Γ_x of each point $x \in X$ is finite.

We now look at the converse in the case of an *isometric* action.

Proposition C.1. *Let (X, d) be a metric space on which Γ acts isometrically. Suppose each orbit, Γx, has no accumulation points and each isotopy group Γ_x is finite. Then Γ acts properly discontinuously.*

Proof. If not, there is some compact set $C \subseteq X$ so that $C \cap \gamma \cdot C$ is non-empty for infinitely many $\gamma \in \Gamma$. Thus there is a sequence γ_i of distinct elements of Γ with $\gamma_i(c_i) \in C$, where $c_i \in C$. By compactness there is a convergent subsequence which we relabel $\gamma_i(c_i) \to c \in C$. Again passing to a subsequence, using compactness of C and relabeling we find $c_i \to c'$, $c' \in C$. Now

$$d(\gamma_i c, c') \leq d(\gamma_i c, \gamma_i c_i) + d(\gamma_i c_i, c'),$$

since Γ acts isometrically $d(\gamma_i c, \gamma_i c_i) = d(c, c_i)$ which therefore tends to zero. Also $d(\gamma_i c_i, c')$ tends to zero. Hence $\gamma_i c \to c'$. Since Γ_c is finite, for each i there are only finitely many j with $\gamma_i c = \gamma_j c$. Hence by choosing a subsequence there is a sequence $\gamma_i c \to c'$ where the terms are distinct. This contradicts the second condition and proves the result. $\qquad\square$

However, being properly discontinuous is stronger than being discrete. For example consider the action of $\mathbb{Z}$ on $\mathbb{T}^n$ where $n \geq 2$. This action is one in which a discrete group acts by isometries on a (compact) metric space. If we have an *irrational flow*, then every orbit is dense by Kronecker's approximation theorem. Therefore this action is not properly discontinuous. Now consider a rational flow. Since it is an action on a metric space by isometries we only have to check the orbits are discrete and the isotropy groups are finite. In this case both these conditions are satisfied so the action is properly discontinuous,
We require the following lemma. Here the group, Homeo(X), the homeomorphisms of X takes the topology of uniform convergence on compacta which we call *the compact open topology*.

Lemma C.2. *Let* $\Gamma \times X \to X$ *be a continuous group action where* (X, d) *is a compact metric space and the countable discrete group,* Γ, *acts isometrically. Then the image of* $\Gamma \in \mathrm{Homeo}(X)$ *is also discrete.*

Proof. Denote the map $\gamma \mapsto \Phi(\gamma)$ by Φ, where $\Phi(\gamma)(x) = \gamma \cdot x$, $x \in X$. Then for each $\gamma \in \Gamma$, $\Phi(\gamma)$ is a homeomorphism, in fact an isometry, of X. Notice that $\Phi(\gamma)(X) = X$. For if it were smaller, then applying $\Phi(\gamma^{-1})$ would yield a contradiction. Also Φ is evidently a continuous homomorphism $\Gamma \to Homeo(X)$. To complete the proof we need to show this map is open. Since Γ is countable discrete the open mapping theorem will do this if we know the image is locally compact. Now in the compact open topology a neighborhood of I in the image is given by $N(C, \epsilon)$, together with the inverses, where C is compact and $\epsilon > 0$. However, since X is compact we can always take a smaller neighborhood $N_0 = N(X, \epsilon)$ of I. These are the homeomorphisms (actually isometries) h such that $d(h(x), x) < \epsilon$ for all $x \in X$. The condition of being an isometry automatically shows any such N, in fact all of Isom(X), is equicontinuous. Evidently N_0 is pointwise bounded. Hence by the Ascoli theorem N_0 has compact closure so $\Phi(\Gamma)$ is locally compact. The open mapping theorem says Φ is open and therefore $\Phi(\Gamma)$ is discrete. $\square$

Let G be a connected semisimple Lie group of non-compact type, $X = G/K$ the associated symmetric space. Then G is the connected component of the isometry group of X. Let Γ be a torsion free discrete cocompact subgroup of G. Then Γ, the fundamental group of $S = X/\Gamma$ acts on S and S is a smooth connected manifold locally isometric with X so S is also metric and Γ acts by isometries. The cocompactness of Γ implies S is compact.

Proposition C.3. *The action of Γ on a compact locally symmetric space S is properly discontinuous.*

Proof. If not, there is a point $s \in S$ and an infinite number of distinct γ_i so that $\Phi(\gamma_i)(s)$ converges to something in S. By Lemma C.2 $\Phi(\Gamma)$ is a discrete subgroup of $\mathrm{Homeo}(S)$. Now the set $\Phi(\Gamma_1)$ of the γ_i is equicontinuous since all of $\mathrm{Isom}(S)$ acts equicontinuously. Let $t \in S$ be fixed. Then $\Gamma_1(t) \subseteq \overline{N(\gamma_i s, d(s,t))}$ which is compact since S is. Hence Γ_1 is uniformly bounded. Since it is also equicontinuous Γ_1 has compact closure. On the other hand $\Phi(\Gamma)$ is discrete. Therefore Γ_1 is finite, a contradiction. This means Γ acts properly discontinuously. $\square$

Appendix D

The Analyticity of Smooth Lie Groups

Here we sketch the proof of the analyticity of a connected smooth Lie group G. In the complex case this is just a fact of complex analysis so here we focus on the real case, although the proof that follows works equally well in the case of complex Lie groups.

If left and right translations are analytic, to prove the claim it is sufficient to prove that multiplication and inversion are analytic in a neighborhood of 1 in G. For suppose we were at a neighborhood of (p, q). Let $x_1 = p^{-1}x$ and $y_1 = q^{-1}y$. Then $xy^{-1} = px_1y_1^{-1}q^{-1} = L_pR_{q^{-1}}x_1y_1^{-1}$. If the function $(x, y) \mapsto xy^{-1}$ is analytic at the origin and, as above, left and right translations are analytic on G, then as a composition of analytic functions $(x, y) \mapsto xy^{-1}$ is analytic at (p, q).

Now we prove the analyticity in a neighborhood of 1. Since this is a local question and any Lie group is locally isomorphic to a *linear* Lie group, as mentioned in Section 1.7, we may assume G is linear. Let U be a canonical neighborhood of 1 in G. We identify U with an open ball B about 0 in $\mathfrak{g}$ using Exp which is analytic. Since $\text{Exp}(x) \mapsto \text{Exp}(-x)$ is evidently analytic, *i.e.* $x \mapsto -x$ being linear, it is sufficient to prove multiplication is analytic on U. We can consider each $u = (u_1, \ldots, u_n) \in B$, where $n = \dim G$. Let $z = xy$, where x and $y \in U$. Then for each i, $z_i = f_i(x_1, \ldots, x_n, y_1, \ldots, y_n)$, $f_i \in C^\infty(U \times U)$. Now $\frac{\partial f_i}{\partial y_j} = \delta_{ij}$ at

411

$(x,y) = (1,1)$. However, at $y = 1$ with x varying $\frac{\partial f_i}{\partial y_j}$ is a function of x: $v_{ij}(x) = v_{ij}(x_1, \ldots, x_n)$. If $b = (b_1, \ldots, b_n) \in B$, the 1-parameter group $\mathrm{Exp}(tb)$ satisfies the system of differential equations,

$$\frac{dx_i}{dt} = \Sigma_{i=1}^n b_i v_{ij}(x_1(t), \ldots, x_n(t)), x_i(0) = 0.$$

Since $\mathrm{Exp}(tb)$ is the unique solution, this system of equations is nothing more than the matrix differential equation $\frac{dx}{dt} = b\,\mathrm{Exp}(tb)$, $x(0) = I$. Thus the matrix, $(v_{ij}(x)) = \mathrm{Exp}(x)$ and since $x \mapsto \mathrm{Exp}(x)$ is analytic so are the v_{ij}.

Now the product functions $z_i = f_i(x,y)$ satisfy a system of partial differential equations:

$$\Sigma_j v_{ij}(z) \frac{\partial z_j}{\partial x_k}(x) = v_{ik}(x), i, k = 1, \ldots, n,$$

called the *fundamental differential equations of the group*, G which determine the z's if the v's are known and certain integrability conditions are satisfied. These link the v's and their derivatives to the structure constants of $\mathfrak{g}$. Since these conditions are necessary and sufficient and G is a smooth Lie group, the v_{ij} certainly satisfy these integrability conditions. The only question remaining is whether the z_i are analytic. But since we know the v's are analytic, so are the z's. This follows from the Frobenius theorem (see [66] Theorem 211.9).

Finally we prove that left and right translations are analytic. Multiplication is analytic in a neighborhood U of 1 in G. Hence so is left translation L_g on U when $g \in U$. Therefore, because of the way we put the manifold structure on G, such L_g's are analytic on all of G. Now let $g \in G$ be arbitrary. Then $g = g_1 \ldots g_n$, where each $g_i \in U$. Hence $L_g = L_{g_1} \ldots L_{g_n}$, a composition of analytic functions and therefore each L_g is analytic. Similarly each R_g is analytic.

Bibliography

[1] Adams, J.F., *Lectures on Lie groups*, W. A. Benjamin Inc., New York-Amsterdam, 1969.

[2] Auslander L., *Lecture notes on nil-theta functions*, CBMS Reg. Conf. Series Math. **34**, Amer. Math. Soc., Providence, 1977.

[3] Ballmann W., Gromov M. and Schroeder V., *Manifolds of nonpositive curvature*, Progress in Mathematics **61**, Birkhaüser Boston Inc., Boston, MA, 1985.

[4] Barbano P., *Automorphisms and quasiconformal mappings of Heisenberg type groups*, J. of Lie Theory **8** (1998), 255–277.

[5] Borel A., *Density properties for certain subgroups of semi-simple groups without compact components*, Ann. of Math. (2) **72** (1960), 179–188.

[6] Borel A. and Harish-Chandra, *Arithmetic subgroups of algebraic groups*, Ann. of Math. (2) **75** (1962), 485–535.

[7] Borel A., *Compact Clifford-Klein forms of symmetric spaces*, Topology (2) 1963, 111–122.

[8] Borel A., *Introduction aux groupes arithmètiques*, Publications de l'Institut de Mathèmatique de l'Universitè de Strasbourg, XV, Actualitès Scientifiques et Industrielles, No. **1341**, Hermann, Paris, 1969, 125 pp.

413

[9] Borel A., *Linear algebraic groups*, Second edition Graduate Texts in Mathematics **126**, Springer-Verlag, New York, 1991.

[10] Borel A., *Semisimple Lie Groups and Riemannian Symmetric Spaces*, Hindustan Book Agency, 1998.

[11] Bröcker T. and Tom Dieck T., *Representations of compact Lie groups.* Translated from the German manuscript. Corrected reprint of the 1985 translation.Graduate Texts in Mathematics **98**, Springer-Verlag, New York, 1995.

[12] Cartan E., *Groupes simples clos et ouverts et geometrie Riemannienne*, Journal de Math. Pures et Appliqués **8** (1929), 1–33.

[13] Chabauty C., *Limites d'ensembles et géométrie des nombres*, Bull Soc. Math. de France **78** (1950), 143–151.

[14] Cheeger J. and Ebin D., *Comparison Theorems in Riemannian Geometry*, North Holland, Amsterdam, 1975.

[15] Chevalley C., *The Theory of Lie Groups*, Princeton University Press, Princeton, 1946.

[16] Corwin L. and Moskowitz M., *A note on the exponential map of a real or p-adic Lie group*, J. Pure Appl. Algebra **96** (1994), no. 2, 113–132.

[17] Djokovic D. and Thang N., *On the exponential map of almost simple real algebraic groups*, J. of Lie Theory **5** (1996), 275–291.

[18] Dubrovin B.A., Fomenko A.T. and Novikov S.P., *Modern geometry—methods and applications.* Part II. The geometry and topology of manifolds, *Graduate Texts in Mathematics*, 104, Springer-Verlag, New York, 1985.

[19] Faraut J., *Analyse harmonique sur les espaces hyperboliques*, Topics in modern harmonic analysis, Vol. I, II (Turin/Milan, 1982), 445–473, Ist. Naz. Alta Mat. Francesco Severi, Rome.

[20] Farkas H. and Kra I., *Riemann surfaces* Graduate Texts in Mathematics **71**, Springer-Verlag, New York-Berlin, 1980.

[21] Furstenberg H., *A Poisson formula for semi-simple Lie groups*, Ann. of Math. (2) **77** (1963), 335–386.

[22] Furstenberg H., *A note on Borel's density theorem*, Proc. AMS **55** (1976), 209–212.

[23] Gallot S., Hulin D. and Lafontaine J., *Riemannian geometry*, Second Edition, Springer-Verlag, Berlin, 1990.

[24] Garland H. and Goto M., *Lattices and the adjoint group of a Lie group*, Trans. Amer. Math. Soc. **124** (1966), 450–460.

[25] Gelfand, I., Graev I. and Pyatetskii-Shapiro I., *Representation theory and automorphic functions*, Translated from the Russian by K. A. Hirsch. Reprint of the 1969 edition. Generalized Functions, 6. Academic Press, Inc., Boston, MA, 1990.

[26] Glushkov V.M., *The structure of locally compact groups and Hilbert's fifth problem*, AMS Translations **15** (1960), 55–93.

[27] Grosser S. and Moskowitz M., *Representation theory of central topological groups*, Trans. Amer. Math. Soc. **129** (1967), 361–390.

[28] Grosser S. and Moskowitz M., *Harmonic analysis on central topological groups*, Trans. Amer. Math. Soc. **156** (1971), 419–454.

[29] Greenleaf F., Moskowitz M. and Rothschild L., *Compactness of certain homogeneous spaces of finite volume* Amer. J. Math. **97** no. 1 (1975), 248–259.

[30] Guillemin V. and Pollack A., *Differential topology*, Prentice-Hall Inc., Englewood Cliffs, NJ, 1974.

[31] Hausner M. and Schwarz J.T., *Lie groups; Lie algebras*, Gordon and Breach Science Publishers, New York-London-Paris, 1968.

[32] Helgason S., *Differential Geometry, Lie Groups and Symmetric Spaces*, Pure and Applied Mathematics **80**, Academic Press, New York, 1978.

[33] Hochschild G.P., *The Structure of Lie Groups*, Holden Day Inc., San Francisco-London-Amsterdam, 1965.

[34] Hochschild G.P. and Mostow G., *Representations and representative functions of Lie groups*, Ann. of Math (2) **66** (1957), 495–542.

[35] Hunt G., *A Theorem of E. Cartan*, Proc. Amer. Math. Soc. **7** (1956), 307–308.

[36] Kaplansky I., *Infinite abelian groups*, University of Michigan Press, Ann Arbor, 1954.

[37] Kolmogorov A.N. and Fomin S.V., *Elements of the Theory of Functions and Functional Analysis*, Translated from the 1st (1954) Russian ed. by Leo F. Boron, Rochester, N.Y., Graylock Press.

[38] Mahler K., *On Lattice Points in n-dimensional Star Bodies I, Existence theorems*, Proc. Roy. Soc. London Ser. A **187** (1946) 151–187.

[39] Malcev A., *On a class of homogeneous spaces*, Amer. Math. Soc. Translation **1951** (1951) no. 39.

[40] Massey W.S., *A basic course in algebraic topology* Graduate Texts in Mathematics **127**, Springer-Verlag, New York, 1991.

[41] Margulis G., *Discrete Subgroups of Semisimple Lie Groups*, Springer-Verlag, Ergebnisse der Mathematik 3, Folge Bd 17., Berlin-Heidelberg-New York, 1990.

[42] Milnor J., *Morse theory*, Based on lecture notes by M. Spivak and R. Wells, Annals of Mathematics Studies, No. 51, Princeton University Press, Princeton, NJ, 1963.

[43] Moore C.C., *Decomposition of unitary representations defined by discrete subgroups of nilpotent groups*, Ann. of Math. (2) **82** (1965), 146–182.

[44] Moore C.C., *Ergodicitiy of flows on homogeneous spaces.*, Amer. J. Math. **88** (1966) 154–178.

[45] Mosak R.D. and Moskowitz M., *Zariski density in Lie groups*, Israel J. Math. **52** (1985), no. 1-2, 1–14.

[46] Mosak R.D. and Moskowitz M., *Analytic density in Lie groups*, Israel J. Math. **58** (1987), no. 1, 1–9.

[47] Mosak R.D. and Moskowitz M., *Stabilizers of lattices in Lie groups* J. Lie Theory **4** (1994), no. 1, 1–16.

[48] Moskowitz M., *A remark on faithful representations* Atti della Accademia Nationale dei Lincei, ser. 8, **52** (1972), 829–831.

[49] Moskowitz M., *Faithful representations and a local property of Lie groups* Math. Zeitschrift **143** (1975), 193–198.

[50] Moskowitz M., *Some Remarks on Automorphisms of Bounded Displacement and Bounded Cocycles*, Monatshefte für Math. **85** (1978), 323–336.

[51] Moskowitz M., *On the density theorems of Borel and Furstenberg*, Ark. Mat. **16** (1978), no. 1, 11–27.

[52] Moskowitz M., *On the surjectivity of the exponential map in certain Lie groups*, Annali di Matematica Pura ed Applicata, Serie IV-Tomo CLXVI (1994), 129–143.

[53] Moskowitz M., Correction and addenda to: *"On the surjectivity of the exponential map for certain Lie groups"* [Ann. Mat. Pura Appl. (4) **166** (1994), 129–143, Ann. Mat. Pura Appl. (4) **173** (1997), 351–358.

[54] Moskowitz M., *Complete reducibility and Zariski density in linear Lie groups* Math. Z. **232** (1999), no. 2, 357–365.

[55] Moskowitz M., *A course in complex analysis in one variable*, World Scientific Publishing Co., Inc., River Edge, NJ, 2002.

[56] Moskowitz M. and Wüstner M., *Exponentiality of certain real solvable Lie groups*, Canad. Math. Bull. **41** (1998), no. 3, 368–373.

[57] Mostow G.D., *Self-adjoint Groups*, Ann. of Math. (2) **62** (1955), 44–55.

[58] Mostow G.D., *Equivariant embeddings in Euclidean space*, Ann. of Math. (2) **65** (1957), 432–446.

[59] Mostow G.D., *Cohomology of topological groups and solvable manifolds* Ann. of Math. **73** (1961), 20–48.

[60] Mostow, G.D., *Homogeneous spaces with finite invariant measure*, Ann. of Math. (2) **75** (1962), 17–37.

[61] Mostow G.D., *Strong rigidity of locally symmetric spaces*, Annals of Mathematics Studies, No. 78. Princeton University Press, Princeton, NJ; University of Tokyo Press, Tokyo, 1973.

[62] Mostow G.D., *Discrete subgroups of Lie groups*, Advances in Math. **16** (1975), 112–123.

[63] Mostow G.D., *Discrete subgroups of Lie groups*, The mathematical héritage of Élie Cartan (Lyon, 1984). Astérisque (1985), Numero Hors Serie, 289–309.

[64] Mostow G.D. and Tamagawa T., *On the compactness of arithmetically defined homogeneous spaces*, Ann. of Math. (2) **76** (1962), 446–463.

[65] Nachbin L., *The Haar integral*, D. Van Nostrand Co. Inc., Princeton, NJ-Toronto-London, 1965.

[66] Narasimhan R., *Analysis on Real and Complex Manifolds*, Advanced Studies in Pure Mathematics, Masson & Cie-Paris, 1973.

[67] Myers S. and Steenrod N., *The group of isometries of a riemannian manifold*, Ann. of Math. **40** (1939), 400–416.

[68] Palais R., *Imbedding of compact, differentiable transformation groups in orthogonal representations*, J. Math. Mech. **6** (1957), 673–678.

[69] Perko, L., *Differential equations and dynamical systems*, Second edition. Texts in Applied Mathematics **7**, Springer-Verlag, New York, 1996.

[70] Pontryagin L.S., *Topological groups*, Translated from the second Russian edition by Arlen Brown, Gordon and Breach Science Publishers Inc., New York-London-Paris, 1966.

[71] Raghunathan M.S., *Discrete subgroups of Lie groups*, Ergebnisse der Mathematik und ihrer Grenzgebiete, Band 68, Springer-Verlag, New York-Heidelberg, 1972.

[72] Rothman R., *The von Neumann kernel and minimally almost periodic groups*, Trans. Amer. Math. Soc. **259** (1980), no. 2, 401–421.

[73] Royden H.L., *Real analysis*, Third edition, Macmillan Publishing Company, New York, 1988.

[74] Rudin, W., *Principles of mathematical analysis*, International Series in Pure and Applied Mathematics, McGraw-Hill Book Co., New York-Auckland-Düsseldorf, 1976.

[75] Siegel C.L., *Über Gitterpunkte in convenxen Körpern und ein damit zusammenhängendes Extremalproblem*, Acta Mathematica vol **65** (1935), 307–323.

[76] Steenrod N., *The Topology of Fibre Bundles*, Princeton Mathematical Series vol. **14**, Princeton University Press, Princeton, NJ, 1951.

[77] Tits J., *Free subgroups in linear groups*, J. Algebra **20** (1972), 250–270.

[78] Varadarajan V.S., *Lie groups, Lie algebras, and their representations*, Prentice-Hall Series in Modern Analysis, Prentice-Hall, Inc., Englewood Cliffs, NJ, 1974.

[79] Wang H.C., *On the deformations of lattices in a Lie group*, Amer. J. Math. **89** (1963), 189–212.

[80] Warner F.W., *Foundations of differentiable manifolds and Lie groups*, Corrected reprint of the 1971 edition, Graduate Texts in Mathematics **94**, Springer-Verlag, New York-Berlin, 1983.

[81] Whitney H., *Elementary structure of real algebraic varieties*, Ann. of Math. **66** (1957), 545–556.

[82] Zariski O. and Samuel P., *Commutative algebra*, Vol. 1. With the cooperation of I. S. Cohen. Corrected reprinting of the 1958 edition. Graduate Texts in Mathematics, No. **28**. Springer-Verlag, New York-Heidelberg-Berlin, 1975.

Index